D0722858

Jewels are Forever

Springer
Berlin
Heidelberg
New York
Barcelona
Hong Kong
London
Milan
Paris
Singapore
Tokyo

Arto Salomaa

Juhani Karhumäki • Hermann Maurer
Gheorghe Păun • Grzegorz Rozenberg (Eds.)

Jewels are Forever

Contributions on Theoretical Computer Science in Honor of Arto Salomaa

With 20 Figures

Springer

Editors

Professor Dr. Juhani Karhumäki
Department of Mathematics
and Turku Centre for Computer Science
University of Turku
FIN-20014 Turku, Finland
karhumak@utu.fi

Dr. Gheorghe Păun
Institute of Mathematics
Romanian Academy of Sciences
PO Box 1-764
RO-70700 Bucharest, Romania
gpaun@imar.ro

Professor Dr. Hermann Maurer
Institutes for Information Processing
and Computer Supported Media (IICM)
Technical University of Graz
Schiessstattgasse 4a
A-8010 Graz, Austria
hmaurer@iicm.tu-graz.ac.at

Professor Dr. Grzegorz Rozenberg
Department of Computer Science
Leiden University
PO Box 9512
2300 RA Leiden, The Netherlands
rozenber@wi.leidenuniv.nl

Frontispiece (p. II): Digital photographic image
by Henrik Duncker and Yrjö Tuunanen, 1994,
©Media Lab, University of Art and Design Helsinki

Library of Congress Cataloging-in-Publication Data applied for

Die Deutsche Bibliothek – CIP-Einheitsaufnahme

Jewels are forever: contributions on theoretical computer science in
honor of Arto Salomaa/Juhani Karhumäki... (ed.). - Berlin;
Heidelberg; New York; Barcelona; Hong Kong; London; Milan;
Paris; Singapore; Tokyo: Springer, 1999
ISBN 3-540-65984-6

ISBN 3-540-65984-6 Springer-Verlag Berlin Heidelberg New York

Cover Design: Künkel + Lopka, Werbeagentur, Heidelberg
Typesetting: Camera ready by the editors
SPIN 10727442   45/3142 – 5 4 3 2 1 0 – Printed on acid-free paper

Preface

This volume is dedicated to Professor Arto Salomaa, a towering figure of theoretical computer science, on the occasion of his 65th birthday. His scientific contributions and influence are enormous – we refer the reader to the following Laudation for an account of his scientific achievements.

The book is a tribute to Arto by the theoretical computer science community. It includes the contributions of many outstanding scientists, and it covers most of Arto's many research areas. Due to its representative selection of classic topics and cutting edge trends in theoretical computer science it constitutes a comprehensive state-of-the-art survey.

We feel very honored that we have been given the opportunity to edit this book – we also feel privileged to be his friends. We are indebted to all contributors for paying tribute to Arto through this book. We thank Dr. H. Wössner and Mrs. I. Mayer from Springer-Verlag for a perfect cooperation in producing this volume. Our special thanks go to Mr. M. Hirvensalo and Mr. A. Lepistö, the "2nd generation Ph.D. students" of Arto, for their work on this book.

Turku, April 1999

Juhani Karhumäki
Hermann Maurer
Gheorghe Păun
Grzegorz Rozenberg

Contents

Laudation for Arto Salomaa

Arto Kustaa Salomaa was born in Turku, Finland, on June 6, 1934. He published his first paper in 1959, and since then he has published about 300 papers and 11 books, and has edited 15 books – see the following Bibliography. He has had enormous influence on theoretical computer science and is certainly one of the founding fathers of formal language and automata theory.

His influence on theoretical computer science extends far beyond his writings. He has supervised 24 Ph.D. students in Finland, Denmark and Canada, and many of them have become top scientists in various areas of computer science.

He is on the editorial board of some fifteen international journals and book series and has been a member of program committees for almost all important conferences in theoretical computer science in Europe.

Arto is a very stimulating colleague, and he has cooperated with many scientists all over the world – his list of co-authors includes some 53 names.

His contributions to theoretical computer science have been recognized in many ways. He must be one of the most decorated computer scientists – he has been awarded six Honorary Doctorates (in four countries) and has received many scientific awards. He is a member of Academia Europaea, the Academy of Sciences of Finland, and the Hungarian Academy of Sciences. He was the President of the European Association for Theoretical Computer Science (EATCS) from 1979 until 1986, and had an enormous influence on this organisation through many other functions. Some of the honors he has received are quite unusual. On one of his trips a Maori group north of Auckland put on a Powhiri, a ceremony usually conducted as a way of honoring persons of the importance of the Queen of England. Arto may well be the only computer scientist to whom such a tribute has ever been paid.

From all the facts listed above it must be clear that in celebrating Arto's 65th birthday we pay tribute to a towering figure of theoretical computer science. There is no doubt that theoretical computer science owes a lot to Arto Salomaa.

Throughout his scientific career Arto has had many offers from top universities all over the world. However, he has decided that Turku is his home town and that this is where he wants to have his base. He received his Master of Science from the University of Turku in 1956, and although he did his graduate studies in Berkeley, he defended his Ph.D. at the University of Turku in 1960. He became a full professor at Turku University in 1966. Since then he has made Turku a real place of excellence in theoretical computer science, and certainly the world capital in formal language and automata theory.

Although Turku was always his home base, he has traveled extensively. He spent long periods of time at the universities of Western Ontario and Waterloo in Canada and the University of Aarhus in Denmark. He has also had quite lengthy research stays, among others, at the universities of Auckland, Graz, Karlsruhe, Leiden, Szeged, and Vienna.

He has lectured at more than 150 universities and research centers all over the world – he is a very inspiring lecturer. His lectures and his books demonstrate a unique ability: he is always able to pick out from the sea of results the real jewels, and then present them in the way that jewels deserve (quite often the jewels are also created by him). As a matter of fact, one of his most frequently cited books is "Jewels of Formal Language Theory". That's why we thought that the title "Jewels are Forever" fits this book very well.

Turku, April 1999

Juhani Karhumäki
Hermann Maurer
Gheorghe Păun
Grzegorz Rozenberg

Bibliography of Arto Salomaa

A. Books

1. *Theory of Automata.* International Series of Monographs in Pure and Applied Mathematics, vol. 100, Pergamon Press, Oxford, 1969, 276 pp. (Japanese translation in 1974)
2. *Formal Languages.* Academic Press, New York, 1973, 335 pp. (German translation by Springer-Verlag in 1979)
3. (with M. Soittola) *Automata-Theoretic Aspects of Formal Power Series.* Springer-Verlag, 1978, 181 pp.
4. (with G. Rozenberg) *The Mathematical Theory of L Systems.* Academic Press, New York, 1980, XVI+352 pp.
5. *Jewels of Formal Language Theory.* Computer Science Press, Potomac, Maryland, 1981, X+144 pp. (Russian translation in 1986)
6. *Computation and Automata.* Encyclopedia of Mathematics and Its Applications, vol. 25. Cambridge University Press, Cambridge and New York, 1985, XIII+282 pp. (Japanese translation in 1988, French translation in 1990, and Vietnamese translation in 1992)
7. (with W. Kuich) *Semirings, Automata, Languages.* EATCS Monographs on Theoretical Computer Science, vol. 5, Springer-Verlag, 1986, VI+374 pp.
8. *Public-Key Cryptography.* Springer-Verlag, 1990, X+245 pp.; second, enlarged edition, 1996, X+271 pp. (Japanese translation in 1992, Romanian translation in 1993, Russian translation in 1996, and Chinese translation in 1998)
9. (with G. Rozenberg) *Cornerstones of Undecidability.* Prentice Hall, New York, London, Toronto, Sydney, Tokyo, Singapore, 1994, XVI+197 pp.
10. (with C. Ding and D. Pei) *Chinese Remainder Theorem. Applications in Computing, Coding, Cryptography.* World Scientific, Singapore, 1996, VIII+213 pp.
11. (with Gh. Păun and G. Rozenberg) *DNA Computing. New Computing Paradigms.* Springer-Verlag, 1998, X+402 pp.

B. Edited Books

1. (with G. Rozenberg) *L Systems*. Springer-Verlag, *Lecture Notes in Computer Science*, 15 (1974), VI+338 pp.
2. (with M. Steinby) *Automata, Languages and Programming*. Proc. of ICALP-77, Springer-Verlag, *Lecture Notes in Computer Science*, 52, (1977), X+569 pp.
3. (with G. Rozenberg) *The Book of L*. Springer-Verlag, 1985, XV+471 pp.
4. (with J. Demetrovics and G. Katona) *Algebra, Combinatorics and Logic in Computer Science*, I-II. North-Holland, Amsterdam, New York, 1986, 887 pp.
5. (with T. Lepistö) *Automata, Languages and Programming*. Proc. of ICALP-88, Springer-Verlag, *Lecture Notes in Computer Science*, 317 (1988), XI+741 pp.
6. (with G. Rozenberg) *Lindenmayer Systems*. Springer-Verlag, 1992, X + 514 pp.
7. (with G. Rozenberg) *Current Trends in Theoretical Computer Science*. World Scientific, Singapore, 1993, IX+628 pp.
8. (with G. Rozenberg) *Developments in Language Theory*. World Scientific, Singapore, New Jersey, London, Hong Kong, 1994, XII+492 pp.
9. (with J. Dassow and G. Rozenberg) *Developments in Language Theory II. At the Crossroads of Mathematics, Computer Science and Biology*. World Scientific, Singapore, 1996, X+491 pp.
10. (with G. Rozenberg) *Handbook of Formal Languages*, Vol. 1: *Word, Language, Grammar*. Springer-Verlag, 1997, XXIV+873 pp.
11. (with G. Rozenberg) *Handbook of Formal Languages*, Vol. 2: *Linear Modeling: Background and Application*. Springer-Verlag, 1997, XXII + 528 pp.
12. (with G. Rozenberg) *Handbook of Formal Languages*, Vol. 3: *Beyond Words*. Springer-Verlag, 1997, XX+625 pp.
13. (with Gh. Păun) *New Trends in Formal Languages. Control, Communication, and Combinatorics*. Springer-Verlag, *Lecture Notes in Computer Science*, 1218 (1997), X+466 pp.
14. (with J. Mycielski and G. Rozenberg) *Structures in Logic and Computer Science*. Springer-Verlag, *Lecture Notes in Computer Science*, 1261 (1997), X+370 pp.
15. (with Gh. Păun) *Grammatical Models of Multi-Agent Systems*. Gordon and Breach, Amsterdam, 1999, VIII+356 pp.

C. Papers

1. On many-valued systems of logic. *Ajatus*, 22 (1959), 115–159
2. On the composition of functions of several variables ranging over a finite set. *Annales Universitatis Turkuensis*, Ser. A I, 41 (1960), 48 pp.
3. A theorem concerning the composition of functions of several variables ranging over a finite set. *Journal of Symbolic Logic*, 25 (1960), 203–208
4. On the number of simple bases of the set of functions over a finite domain. *Annales Universitatis Turkuensis*, Ser. A I, 52 (1962), 4 pp.
5. Some completeness criteria for sets of functions over a finite domain, I. *Annales Universitatis Turkuensis*, Ser. A I, 53 (1962), 10 pp.
6. Some analogues of Sheffer functions in infinite-valued logics. *Proc. Colloq. Modal and Many-valued Logics*, Helsinki, 1962, 227–235
7. Some completeness criteria for sets of functions over a finite domain, II. *Annales Universitatis Turkuensis*, Ser. A I, 63 (1963), 19 pp. (Russian translations of the previous two papers in *Kibernetitseskii Sbornik*, 8 (1964), 8–32.)
8. On sequences of functions over an arbitrary domain. *Annales Universitatis Turkuensis*, Ser. A I, 62 (1963), 5 pp.
9. On basic groups for the set of functions over a finite domain. *Annales Academiae Scientiarum Fennicae*, Ser. A I, 338 (1963), 15 pp.
10. On essential variables of functions, especially in the algebra of logic. *Annales Academiae Scientiarum Fennicae*, Ser. A I, 339 (1963), 11 pp.
11. Theorems on the representation of events in Moore automata. *Annales Universitatis Turkuensis*, Ser. A I, 69 (1964), 14 pp.
12. On infinitely generated sets of operations in finite algebras. *Annales Universitatis Turkuensis*, Ser. A I, 74 (1964), 13 pp.
13. Axiom systems for regular expressions of finite automata. *Annales Universitatis Turkuensis*, Ser. A I, 75 (1964), 29 pp.
14. On the reducibility of events represented in automata. *Annales Academiae Scientiarum Fennicae*, Ser. A I, 353 (1964), 16 pp.
15. On the heights of closed sets of operations in finite algebras. *Annales Academiae Scientiarum Fennicae*, Ser. A I, 363 (1965), 12 pp.
16. On some algebraic notions in the theory of truth functions. *Acta Philosophiae Fennica*, 18 (1965), 193–202
17. On probabilistic automata with one input letter. *Annales Universitatis Turkuensis*, Ser. A I, 85 (1965), 16 pp.
18. Automaattien teoriasta. *Arkhimedes*, 1965, 7–20
19. Two complete axiom systems for the algebra of regular events. *Journal of the Association for Computing Machinery*, 13 (1966), 158–169
20. Aksiomatizatsija algebry sobytii, realizuemyh logitseskimi setjami. *Problemy Kibernetiki*, 17 (1966), 237–246
21. On m-adic probabilistic automata. *Information and Control*, 10 (1967), 215–219
22. On events represented by probabilistic automata of different types. *Canadian Journal of Mathematics*, 20 (1968), 242–251

23. On languages accepted by probabilistic and time-variant automata. *Proc. II Princeton Conf. on Information Sciences and Systems*, 1968, 184–188
24. (with V. Tixier) Two complete axiom systems for the extended language of regular expressions. *IEEE Computer Trans*, C-17 (1968), 700–701
25. On finite automata with a time-variant structure. *Information and Control*, 13 (1968), 85–98
26. On finite time-variant automata with monitors of different types. *Annales Universitatis Turkuensis*, Ser. A I, 118 (1968), 12 pp.
27. On regular expressions and regular canonical systems. *Mathematical Systems Theory*, 2 (1968), 341–355
28. Matematiikka ja tietokone. *Arkhimedes*, 1968, 5–10
29. On the index of a context-free grammar and language. *Information and Control*, 14 (1969), 474–477
30. Probabilistic and time-variant grammars and languages. *Avh. Första Nordiska Logikersymposiet*, 1969, 115–133
31. On grammars with restricted use of productions. *Annales Academiae Scientiarum Fennicae*, Ser. A I, 454 (1969), 32 pp.
32. On some families of formal languages obtained by regulated derivations. *Annales Academiae Scientiarum Fennicae*, Ser. A I, 479 (1970), 18 pp.
33. Probabilistic and weighted grammars. *Information and Control*, 15 (1970), 529–544
34. Periodically time-variant context-free grammars. *Information and Control*, 17 (1970), 294–311
35. The generative capacity of transformational grammars of Ginsburg and Partee. *Information and Control*, 18 (1971), 227–232
36. Theories of abstract automata (review). *Information and Control*, 19 (1971), 476–478
37. Matrix grammars with a leftmost restriction. *Information and Control*, 20 (1972), 143–149
38. On a homomorphic characterization of recursively enumerable languages. *Annales Academiae Scientiarum Fennicae*, Ser. A I, 525 (1972), 10 pp.
39. On exponential growth in Lindenmayer systems. *Indagationes Mathematicae*, 35 (1973), 23–30
40. On sentential forms of context-free grammars. *Acta Informatica*, 2 (1973), 40–49
41. (with A. Paz) Integral sequential word functions and growth equivalence of Lindenmayer systems. *Information and Control*, 23 (1973), 313–343
42. Growth functions associated with some new types of grammars. *Proc. Conf. on Algebraic Theory of Automata*, Szeged, 1973, 27–31
43. On some recent problems concerning developmental languages. *Proc. First Fachtagung über Automatentheorie und formale Sprachen*, Springer-Verlag, *Lecture Notes in Computer Science*, 2 (1973), 23–34
44. L-systems: a device in biologically motivated automata theory. *Proc. Conf. on Mathematical Foundations of Computer Science*, Slovak Academy of Sciences, 1973, 147–151

45. Developmental languages: a new type of formal languages. *Annales Universitatis Turkuensis*, Ser. B, 126 (1973), 183–189
46. Solution of a decision problem concerning unary Lindenmayer systems. *Discrete Mathematics*, 9 (1974), 71–77
47. Some remarks concerning many-valued propositional logics. In: S. Stenlund (ed.), *Logical Theory and Semantical Analysis*, D. Reidel Publ. Co., 1974, 15–21
48. (with M. Nielsen, G. Rozenberg, and S. Skyum) Nonterminals, homomorphisms and codings in different variations of OL-systems I, II. *Acta Informatica*, 3 (1974), 357–364, and 4 (1974), 87–106
49. (with G. Rozenberg) The mathematical theory of L systems. *Aarhus University DAIMI Publications*, 33 (1974), 67 pp.; an extended version appears also in: J. Tou (ed.), *Advances in Information Systems Science*, vol. 6, Plenum Press, 1976, 161–206
50. Recent results on L-systems. *Proc. Conf. on Biologically Motivated Automata Theory*, IEEE Publications no. 74 CH0 889-6 C (1974), 38–45
51. Parallelism in rewriting systems. *Proc. ICALP-74*, Springer-Verlag, *Lecture Notes in Computer Science*, 14 (1974), 523–533
52. Iteration grammars and Lindenmayer AFL's. In: G. Rozenberg and A. Salomaa (eds.), *L Systems*, Springer-Verlag, *Lecture Notes in Computer Science*, 15 (1974), 250–253.
53. Comparative decision problems between sequential and parallel rewriting. *Proc. Intern. Symp. Uniformly Structured Automata and Logic*, IEEE Publications 75 CH1 052-0 C (1975), 62–66
54. On some decidability problems concerning developmental languages. *Proc. 3rd Scandinavian Logic Symposium 73*, North-Holland Publ. Co., 1975, pp. 144–153
55. Formal power series and growth functions of Lindenmayer systems. Springer-Verlag, *Lecture Notes in Computer Science*, 32 (1975), 101–113
56. Tietokoneiden tulo. In: *Luonnontieteellisen tutkimuksen historia*, WSOY, Porvoo, Finland, 1975, 245–256
57. Growth functions of Lindenmayer systems: some new approaches. In: *Automata, Languages and Development*, North-Holland, 1976, 271–282
58. (with G. Rozenberg) Context-free grammars with graph-controlled tables. *Journal of Computer and System Sciences*, 13 (1976), 90–99
59. (with G. Rozenberg and K. Ruohonen) Developmental systems with fragmentation. *International Journal of Computer Mathematics*, 5 (1976), 177–191
60. L systems: A parallel way of looking at formal languages. New ideas and recent developments. *Mathematical Centre Tracts*, Amsterdam, 82 (1976), 65–107
61. Sequential and parallel rewriting. In: R. Aguilar (ed.), *Formal Languages and Programming*, North-Holland, 1976, 111–129

62. Undecidable problems concerning growth in informationless Lindenmayer systems. *Elektronische Informationsverarbeitung und Kybernetik*, 12 (1976), 331–335
63. Recent results on L systems. Springer-Verlag, *Lecture Notes in Computer Science*, 45 (1976), 115–123
64. (with G. Rozenberg) New squeezing mechanisms for L systems. *Information Sciences*, 12 (1977), 187–201
65. Formal power series and language theory. *Nanyang University Publications*, 1977, 23 pp.
66. (with H. Maurer and D. Wood) E0L forms. *Acta Informatica*, 8 (1977), 75–96.
67. (with H. Maurer and Th. Ottman) On the form equivalence of L forms. *Theoretical Computer Science*, 4 (1977), 199–225
68. (with M. Penttonen and G. Rozenberg) Bibliography of L systems. *Theoretical Computer Science*, 5 (1977), 339–354
69. (with H. Maurer and D. Wood) On good E0L forms. *SIAM Journal of Computing*, 7 (1978), 158–166
70. (with H. Maurer and D. Wood) Uniform interpretations of L forms. *Information and Control*, 36 (1978), 157–173
71. (with H. Maurer and D. Wood) ET0L forms. *Journal of Computer and System Sciences*, 16 (1978), 345–361
72. (with K. Culik, H. Maurer, Th. Ottman, and K. Ruohonen) Isomorphism, form equivalence and sequence equivalence of PD0L forms. *Theoretical Computer Science*, 6 (1978), 143–173
73. (with H. Maurer and D. Wood) Relative goodness of E0L forms. *RAIRO, Theoretical Computer Science*, 12 (1978), 291–304
74. D0L equivalence: The problem of iterated morphisms. *EATCS Bulletin*, 4 (1978), 5–12
75. L systems and L forms. *Journal of the Computer Society of India*, 8 (1978), 23–30
76. Equality sets for homomorphisms of free monoids. *Acta Cybernetica*, 4 (1978), 127–139
77. (with K. Culik) On the decidability of homomorphism equivalence for languages. *Journal of Computer and System Sciences*, 17 (1978), 163–175
78. (with H. Maurer, M. Penttonen, and D. Wood) On non context-free grammar forms. *Mathematical Systems Theory*, 12 (1979), 297–324
79. D0L language equivalence. *EATCS Bulletin*, 8 (1979), 4–12
80. Power from power series. Springer-Verlag, *Lecture Notes in Computer Science*, 74 (1979), 170–181
81. Language theory based on parallelism: old and new results about L systems. *Proc. of the Fourth IBM Symposium on Mathematical Foundations of Computer Science*, Oiso, Japan, 1979, 1–20
82. (with H. Maurer, G. Rozenberg, and D. Wood) Pure interpretations of E0L forms. *RAIRO, Theoretical Informatics*, 13 (1979), 347–362

83. (with H. Maurer and D. Wood) Context-dependent L forms. *Information and Control*, 42 (1979), 97–118
84. Sata vuotta matemaattista logiikkaa: päättelysäännöistä tietokoneohjelmointiin. In: *Muuttuvat ajat*, WSOY, Porvoo, Finland, 1979, 116–130
85. Morphisms on free monoids and language theory. In: R. Book (ed.), *Formal Language Theory*, Academic Press, 1980, 141–166
86. (with H. Maurer and D. Wood) Synchronized E0L forms. *Theoretical Computer Science*, 12 (1980), 135–159
87. (with H. Maurer and D. Wood) Pure grammars. *Information and Control*, 44 (1980), 47–72
88. (with H. Maurer and D. Wood) On generators and generative capacity of E0L forms. *Acta Informatica*, 13 (1980), 87–107
89. (with K. Culik) Test sets and checking words for homomorphism equivalence. *Journal of Computer and System Sciences*, 20 (1980), 379–395
90. (with H. Maurer and D. Wood) Context-free grammar forms with strict interpretations. *Journal of Computer and System Sciences*, 21 (1980), 110–135
91. Grammatical families. Springer-Verlag, *Lecture Notes in Computer Science*, 85 (1980), 543–554
92. (with H. Maurer and D. Wood) MSW spaces. *Information and Control*, 46 (1980), 187–199
93. (with H. Maurer and D. Wood) Derivation languages of grammar forms, *Journal of Computer Mathematics*, 9 (1981), 117–130
94. (with H. Maurer and D. Wood) Colorings and interpretations: a connection between graphs and grammar forms, *Discrete Applied Mathematics*, 3 (1981), 119–135
95. (with H. Maurer and D. Wood) Decidability and density in two-symbol grammar forms, *Discrete Applied Mathematics*, 3 (1981), 289–299
96. (with H. Maurer and D. Wood) Uniform interpretations of grammar forms, *SIAM Journal of Computing*, 10 (1981), 483–502
97. (with Th. Ottman and D. Wood) Sub-regular grammar forms, *Information Processing Letters*, 12 (1981), 184–187
98. Salakirjoitus ja tietosuoja – näkymiä kryptografian tutkimuksesta. *Arkhimedes*, 33 (1981), 129–135
99. What computer scientists should know about sauna? *EATCS Bulletin*, 15 (1981), 8–21
100. (with H. Maurer and D. Wood) Synchronized E0L forms under uniform interpretation. *RAIRO, Theoretical Informatics*, 15 (1981), 337–353
101. (with H. Maurer and D. Wood) Completeness of context-free grammar forms, *Journal of Computer and System Sciences*, 23 (1981), 1–10
102. (with G. Rozenberg) Table systems with unconditional transfer. *Discrete Applied Mathematics*, 3 (1981), 319–322
103. Formal power series in noncommuting variables. *Proc. of the 18th Scandinavian Congress for Mathematicians*, Birkhäuser, 1981, 104–124
104. On color-families of graphs. *Annales Academiae Scientiarum Fennicae*, Ser. A I, 6 (1981), 135–148

105. (with H. Maurer and D. Wood) Dense hierarchies of grammatical families. *Journal of the Association for Computing Machinery*, 29 (1982), 118–126
106. (with K. Culik and F.E. Fich) A homomorphic characterization of regular languages. *Discrete Applied Mathematics*, 4 (1982), 149–152
107. (with K. Culik and J. Gruska) On non-regular context-free languages and pumping. *EATCS Bulletin*, 16 (1982), 22–24
108. (with H. Maurer and D. Wood) On predecessors of finite languages. *Information and Control*, 50 (1982), 259–275
109. (with K. Culik) On infinite words obtained by iterating morphisms. *Theoretical Computer Science*, 19 (1982), 29–38
110. (with H. Maurer and D. Wood) Finitary and infinitary interpretations of languages. *Mathematical Systems Theory*, 15 (1982), 251–265
111. (with K. Culik and J. Gruska) Systolic automata for VLSI on balanced trees. *Acta Informatica*, 18 (1983), 335–344
112. (with H. Maurer and D. Wood) L codes and number systems. *Theoretical Computer Science*, 22 (1983), 331–346
113. (with H. Maurer and D. Wood) A supernormal-form theorem for context-free grammars. *Journal of the Association for Computing Machinery*, 30 (1983), 95–102
114. (with J. Honkala) How do you define the complement of a language. *EATCS Bulletin*, 20 (1983), 68–69
115. (with H. Maurer and D. Wood) On finite grammar forms. *International Journal of Computer Mathematics*, 12 (1983), 227–240
116. (with K. Culik and J. Gruska) On a family of L languages resulting from systolic tree automata. *Theoretical Computer Science*, 23 (1983), 231–242
117. (with K. Culik) Ambiguity and decision problems concerning number systems. Springer-Verlag, *Lecture Notes in Computer Science*, 154 (1983), 137–146
118. (with K. Culik and J. Gruska) Systolic trellis automata, I and II. *International Journal of Computer Mathematics*, 15 (1984), 195–212, and 16 (1984), 3–22
119. Trapdoors and protocols: recent trends in cryptography. In: H. Maurer (ed.) *Überblicke Informationsverarbeitung 1984*, Bibliographisches Institut Mannheim-Wien-Zürich, 1984, 275–320
120. (with K. Culik and D. Wood) Systolic tree acceptors. *RAIRO, Theoretical Informatics*, 18 (1984), 53–69
121. (with K. Culik) Ambiguity and decision problems concerning number systems. *Information and Control*, 56 (1984), 139–153
122. (with H. Jürgensen) Syntactic monoids in the construction of systolic tree automata. *International Journal of Computer and Information Sciences*, 14 (1985), 35–49
123. On a public-key cryptosystem based on parallel rewriting. *Parcella-84, Proc. of the International Conference on Parallel Processing*, Berlin, 1985, 209–214

124. Cryptography from Caesar to DES and RSA. *EATCS Bulletin*, 26 (1985), 101–119
125. The Ehrenfeucht conjecture: a proof for language theorists. *EATCS Bulletin*, 27 (1985), 71–82
126. Generalized number systems: decidability, ambiguity, codes. *Proc. of the 19th Nordic Congress of Mathematicians*, Reykjavik, 1985, 213–214
127. Tietosuojauksen kehittäminen. *Matemaattisten aineiden aikakauskirja*, 49 (1985), 283–291
128. On meta-normal forms for algebraic power series in noncommuting variables. *Annales Academiae Scientiarum Fennicae*, Ser. A I, 10 (1985), 501–510
129. (with G. Rozenberg) When L was young. In: G. Rozenberg and A. Salomaa (eds.), *The Book of L*, Springer-Verlag, 1985, 383–392
130. Systolic tree and trellis automata. In: J. Demetrovics, G. Katona, and A. Salomaa (eds.), *Algebra, Combinatorics and Logic in Computer Science*, North-Holland, 1986, 695–710
131. (with E. Kinber and S. Yu) On the equivalence of grammars inferred from derivations. *EATCS Bulletin*, 29 (1986), 39–46
132. (with K. Culik and J. Gruska) Systolic trellis automata: stability, decidability and complexity. *Information and Control*, 71 (1986), 218–230
133. (with H. Maurer, E. Welzl, and D. Wood) Denseness, maximality and decidability of grammatical families. *Annales Academiae Scientiarum Fennicae*, Ser. A I, 11 (1986), 167–178
134. (with S. Yu) On a public-key cryptosystem based on iterated morphisms and substitutions. *Theoretical Computer Science*, 48 (1986), 283–296
135. Markov algorithms as language-defining devices. In: *The Very Knowledge of Coding*, Univ. of Turku, 1987, 120–127
136. (with S. Horvath, E. Kinber, and S. Yu) Decision problems resulting from grammatical inference. *Annales Academiae Scientiarum Fennicae*, A I, 12 (1987), 287–298
137. Two-way Thue. *EATCS Bulletin*, 32 (1987), 82–86
138. Playfair. *EATCS Bulletin*, 33 (1987), 42–53
139. On a public-key cryptosystem based on language theory. *Computers and Security*, 7 (1988), 83–87
140. L codes: variations on a theme of MSW. In: *IIG Report 260, Ten years of IIG*, 1988, 218
141. Cryptography and natural languages. *EATCS Bulletin*, 35 (1988), 92–96
142. Cryptographic transductions. *EATCS Bulletin*, 36 (1988), 85–95
143. Knapsacks and superdogs. *EATCS Bulletin*, 38 (1989), 107–123
144. Tutorial: Cryptography and data security. Springer-Verlag, *Lecture Notes in Computer Science*, 381 (1989), 220–244
145. Public-key cryptosystems and language theory. *A Perspective in Theoretical Computer Science. Commemorative Volume for Gift Siromoney*, World Scientific, Singapore, 1989, 257–266
146. (with G. Rozenberg) Complexity theory. In: *Encyclopaedia of Mathematics*, vol. 2, Kluwer Academic Publishers, 1989, 280–283

147. (with G. Rozenberg) Cryptography. In: *Encyclopaedia of Mathematics*, vol. 2, Kluwer Academic Publishers, 1989, 466–468
148. (with G. Rozenberg) Formal languages and automata. In: *Encyclopaedia of Mathematics*, vol. 4, Kluwer Academic Publishers, 1989, 53–57
149. (with G. Rozenberg) L-systems. In: *Encyclopaedia of Mathematics*, vol. 5, Kluwer Academic Publishers, 1990, 325–327
150. Formal languages and power series. In: J. van Leeuwen (ed.), *Handbook of Theoretical Computer Science*, vol. 2, Elsevier Science Publishers, 1990, 103–132
151. Decidability in finite automata. *EATCS Bulletin*, 41 (1990), 175–183
152. Decision problems arising from knapsack transformations. *Acta Cybernetica*, 9 (1990), 419–440
153. Interaction. *Japan Computer Science Association Reports*, 15 (1990), 4–8
154. Formal power series: a powerful tool for theoretical informatics. *Proc. of the 300-Year Festival Congress of the Hamburg Mathematical Association*, 1990, 1033–1048
155. (with L. Sântean) Secret selling of secrets with many buyers. *EATCS Bulletin*, 42 (1990), 178–186
156. (with G. Rozenberg) Mathematical theory of computation. *Encyclopaedia of Mathematics*, vol. 6, Kluwer Academic Publishers, 1990, 146–148
157. From number theory to cryptography: RSA. *Arkhimedes*, 42 (1990), 526–535
158. (with G. Rozenberg) Post correspondence problem. *Encyclopaedia of Mathematics*, vol. 7, Kluwer Academic Publishers, 1991, 252–253
159. (with H. Maurer and D. Wood) Bounded delay L codes. *Theoretical Computer Science*, 84 (1991), 265–279
160. A deterministic algorithm for modular knapsack problems. *Theoretical Computer Science*, 88 (1991), 127–138
161. (with H. Nurmi) A cryptographic approach to the secret ballot. *Behavioral Science*, 36 (1991), 34–40
162. Many aspects of formal languages. *Information Sciences*, 57–58 (1991) (Special issue "Information Sciences: Past, Present, Future"), 119–129
163. (with H. Nurmi) Salaiset vaalit ja matemaattinen kryptografia. *Politiikka*, 1 (1991), 11–18
164. L codes and L systems with immigration. *EATCS Bulletin*, 43 (1991), 124–130
165. (with K. Salomaa and S. Yu) Primality types of instances of the Post correspondence problem. *EATCS Bulletin*, 44 (1991), 226–241
166. (with J. Honkala) L morphisms: bounded delay and regularity of ambiguity. Springer-Verlag, *Lecture Notes in Computer Science*, 510 (1991), 566–574
167. (with H. Nurmi and L. Sântean) Secret ballot elections in computer networks. *Computers and Security*, 10 (1991), 553–560
168. Verifying and recasting secret ballots in computer networks. Springer-Verlag, *Lecture Notes in Computer Science*, 555 (1991), 283–289

169. (with M. Andraşiu, A. Atanasiu, and Gh. Păun) A new cryptosystem based on formal language theory. *Bulletin Mathématique de la Societé des Sciences Mathématiques de Roumanie*, 36 (84) (1992), 3–16
170. (with Gh. Păun and S. Vicolov) On the generative capacity of parallel communicating grammar systems. *International Journal of Computer Mathematics*, 45 (1992), 45–59
171. (with J. Honkala) Characterization results about L codes. *RAIRO, Theoretical Informatics*, 26 (1992), 287–301
172. (with L. Kari, S. Marcus, and Gh. Păun) In the prehistory of formal language theory: Gauss languages. *EATCS Bulletin*, 46 (1992), 124–139 and in: G. Rozenberg and A. Salomaa (eds.), *Current Trends in Theoretical Computer Science*, World Scientific, 1993, 551–562
173. (with Gh. Păun) Decision problems concerning the thinness of DOL languages. *EATCS Bulletin*, 46 (1992), 171–181
174. Nhung huong phat trien moi trong tin hoc ly thuyet. In the Vietnamese translation of *Computation and Automata*, 1992, 394–404
175. (with L. Kari and Gh. Păun) Semi-commutativity sets for morphisms on free monoids. *Bulletin Mathématique de la Societé des Sciences Mathématiques de Roumanie*, 36 (84) (1992), 293–307
176. (with H. Nurmi) Secret ballot elections and public-key cryptosystems. *European Journal of Political Economy*, 8 (1992), 295–303
177. (with H. Nurmi) Tietokonevaalit ja Tengvallin credo. *Politiikka*, XXXIV (1992), 199–201
178. (with Gh. Păun) Semi-commutativity sets – a cryptographically grounded topic. *Bulletin Mathématique de la Societé des Sciences Mathématiques de Roumanie*, 35 (1992), 255–270
179. Recent trends in the theory of formal languages. *Proc. Conf. "Salodays in Theoretical Computer Science"*, Bucharest, May 1992, 3 pp.
180. Different aspects of the Post correspondence problem. *EATCS Bulletin*, 47 (1992), 154–165
181. Simple reductions between D0L language and sequence equivalence problems. *Discrete Applied Mathematics*, 41 (1993), 271–274
182. (with A. Mateescu) PCP-prime words and primality types. *RAIRO, Theoretical Informatics*, 27 (1993), 57–70
183. (with M. Andraşiu, J. Dassow, and Gh. Păun) Language-theoretic problems arising from Richelieu cryptosystems. *Theoretical Computer Science*, 116 (1993), 339–357
184. (with G. Rozenberg) Undecidability. In: *Encyclopaedia of Mathematics*, vol. 9, Kluwer Academic Publishers, 1993, 310–311
185. What Emil said about the Post Correspondence Problem. In: G. Rozenberg and A. Salomaa (eds.), *Current Trends in Theoretical Computer Science*, World Scientific, 1993, 563–571
186. Decidability in finite automata. In: G. Rozenberg and A. Salomaa (eds.), *Current Trends in Theoretical Computer Science*, World Scientific, 1993, 572–578

187. (with A. Mateescu) On simplest possible solutions for Post correspondence problems. *Acta Informatica*, 30 (1993), 441–457
188. (with J. Dassow and Gh. Păun) On thinness and slenderness of L Languages. *EATCS Bulletin*, 49 (1993), 152–158
189. (with J. Dassow and Gh. Păun) Grammars based on patterns. *International Journal of Foundations of Computer Science*, 4 (1993), 1–14
190. (with J. Dassow, A. Mateescu, and Gh. Păun) Regularizing context-free languages by AFL operations: concatenation and Kleene closure. *Acta Cybernetica*, 10 (1993), 243–253
191. (with L. Kari) 50 EATCS Bulletins. *EATCS Bulletin*, 50 (1993), 5–12
192. (with Gh. Păun) Remarks concerning self-reading sequences. *EATCS Bulletin*, 50 (1993), 229–233
193. (with Gh. Păun) Closure properties of slender languages. *Theoretical Computer Science*, 120 (1993), 293–301
194. (with H. Nurmi) Cryptographic protocols for Vickrey auctions. *Annales Universitatis Turkuensis*, Series B, 200, 9–22 (1993), and *Group Decision and Negotiation*, 2 (1993), 263–273
195. (with H. Nurmi) Cancellation and reassignment of votes in secret ballot elections. *European Journal of Political Economy*, 9 (1993), 427–435
196. (with T. Jiang, K. Salomaa, and S. Yu) Inclusion is undecidable for pattern languages. Springer-Verlag, *Lecture Notes in Computer Science*, 700 (1993), 301–312
197. Pattern languages: problems of decidability and generation. Springer-Verlag, *Lecture Notes in Computer Science*, 710 (1993), 121–132
198. (with A. Mateescu) Post correspondence problem: primitivity and interrelations with complexity classes. Springer-Verlag, *Lecture Notes in Computer Science*, 711 (1993), 174–184
199. (with J. Dassow and Gh. Păun) On the union of OL languages. *Information Processing Letters*, 47 (1993), 59–63
200. (with L. Kari, A. Mateescu, and Gh. Păun) Deletion sets. *Fundamenta Informaticae*, 19 (1993), 355–370
201. (with L. Kari, A. Mateescu, and Gh. Păun) Grammars with oracles. *Annals of Iaşi University, Informatics*, 2 (1993), 3–12.
202. (with C. Calude) Algorithmically coding the universe. In: G. Rozenberg and A. Salomaa (eds.), *Developments in Language Theory*, World Scientific, 1994, 472–492
203. (with S. Marcus, A. Mateescu, and Gh. Păun) On symmetry in strings, sequences and languages. *International Journal of Computer Mathematics*, 54 (1994), 1–13
204. (with A. Mateescu and V. Mitrana) Dynamical teams of cooperating grammar systems. *Annals of Bucharest University, Mathematics-Informatics Series*, 43 (1994), 3–14
205. (with A. Mateescu) Nondeterminism in patterns. Springer-Verlag, *Lecture Notes in Computer Science*, 775 (1994), 661–668

206. (with Gh. Păun and G. Rozenberg) Contextual grammars: erasing, determinism, one-side contexts. In: G. Rozenberg and A. Salomaa (eds.), *Developments in Language Theory*, World Scientific, 1994, 370–389
207. (with T. Jiang, E. Kinber, K. Salomaa, and S. Yu) Pattern languages with and without erasing. *International Journal of Computer Mathematics*, 50 (1994), 147–163
208. (with A. Mateescu) Finite degrees of ambiguity in pattern languages. *RAIRO, Theoretical Informatics*, 28 (1994), 233–253
209. (with H. Nurmi) Conducting secret ballot elections in computer networks: problems and solutions. *Annals of Operations Research*, 5 (1994), 185–194
210. (with H. Nurmi) The nearly perfect auctioner: cryptographic protocols for auctions and bidding. In: S. Rios (ed.), *Decision Theory and Decision Analysis: Trends and Challenges*, Kluwer Academic Publishers, 1994
211. Patterns. *EATCS Bulletin*, 54 (1994), 194–206
212. Patterns and pattern languages. In: C. Calude, M. Lennon, and H. Maurer (eds.), *Proc. of "Salodays in Auckland"*, Auckland Univ., 1994, 8–12
213. Machine-oriented Post Correspondence Problem. In: C. Calude, M. Lennon, and H. Maurer (eds.) *Proc. of "Salodays in Auckland"*, Auckland Univ., 1994, 13–14
214. (with L. Kari, A. Mateescu, and Gh. Păun) Teams in cooperating grammar systems. *Journal of Experimental and Theoretical AI*, 7 (1995), 347–359
215. (with L. Kari, A. Mateescu, and Gh. Păun) Multi-pattern languages. *Theoretical Computer Science*, 141 (1995), 253–268
216. (with Gh. Păun and G. Rozenberg) Contextual grammars: modularity and leftmost derivation. In: Gh. Păun (ed.), *Mathematical Aspects of Natural and Formal Languages*, World Scientific, 1995, 375–392
217. (with T. Jiang, K. Salomaa, and S. Yu) Decision problems concerning patterns. *Journal of Computer and System Sciences*, 50 (1995), 53–63
218. (with L. Kari, A. Mateescu, and Gh. Păun) On parallel deletions applied to a word. *RAIRO, Theoretical Informatics*, 29 (1995), 129–144
219. (with Gh. Păun and G. Rozenberg) Grammars based on the shuffle operation. *Journal of Universal Computer Science*, 1 (1995), 67–82
220. (with E. Csuhaj-Varjú and Gh. Păun) Conditional tabled eco-grammar systems versus (E)TOL systems. *Journal of Universal Computer Science*, 1 (1995), 252–268
221. (with A. Mateescu, K. Salomaa, and S. Yu) Lexical analysis with a simple finite-fuzzy-automaton model. *Journal of Universal Computer Science*, 1 (1995), 292–311
222. Developmental models for artificial life: basics of L systems. In: Gh. Păun (ed.), *Artificial Life: Grammatical Models*, Black Sea University Press, Bucharest, 1995, 22–32
223. Stagnation and malignancy: growth patterns in artificial life. In: Gh. Păun (ed.), *Artificial Life: Grammatical Models*, Black Sea University Press, Bucharest, 1995, 104–115

224. Return to patterns. *EATCS Bulletin*, 55 (1995), 144–157
225. (with L. Kari and G. Rozenberg) Generalized DOL trees. *Acta Cybernetica*, 12 (1995), 1–10
226. (with A. Mateescu, Gh. Păun, and G. Rozenberg) Parikh prime words and GO-like territories. *Journal of Universal Computer Science*, 1 (1995), 790–810
227. (with A. Mateescu, K. Salomaa, and S. Yu) P, NP and Post correspondence problem. *Information and Computation*, 121 (1995), 135–142
228. (with Gh. Păun) Thin and slender languages. *Discrete Applied Mathematics*, 61 (1995), 257–270
229. Julkiset salat – tietosuojauksen matematiikkaa. In: J. Rydman (ed.), *Tutkimuksen etulinjassa*, WSOY (1995), 301–315
230. From Parikh vectors to GO territories. *EATCS Bulletin*, 56 (1995), 89–95
231. (with A. Ehrenfeucht, L. Ilie, Gh. Păun, and G. Rozenberg) On the generative capacity of certain classes of contextual grammars. In: Gh. Păun (ed.), *Mathematical Linguistics and Related Topics*, The Publishing House of the Romanian Academy, 1995, 105–118
232. (with T. Nishida) On slender OL languages. *Theoretical Computer Science*, 158 (1996), 161–176
233. (with V. Mitrana, Gh. Păun, and G. Rozenberg) Pattern systems. *Theoretical Computer Science*, 154 (1996), 183–201
234. (with A. Mateescu) Views on linguistics. *EATCS Bulletin*, 58 (1996), 148–154
235. (with Gh. Păun) Self-reading sequences. *American Mathematical Monthly*, 103 (1996), 166–168
236. Slenderness and immigration: new aspects of L systems. *Publicationes Mathematicae Debrecen*, 48 (1996), 411–420
237. (with Gh. Păun and G. Rozenberg) Contextual grammars: parallelism and blocking of derivation. *Fundamenta Informaticae*, 25 (1996), 381–397
238. (with C. Ding) Cooperatively distributed ciphering and hashing. *Computers and Artificial Intelligence*, 15 (1996), 233–245
239. (with S. Dumitrescu and Gh. Păun) Languages associated to finite and infinite sets of patterns. *Revue Roumaine de Mathématiques Pures et Appliquées*, 41 (1996), 607–625
240. (with L. Kari and Gh. Păun) The power of restricted splicing with rules from a regular language. *Journal of Universal Computer Science*, 2 (1996), 224–240
241. (with Gh. Păun and G. Rozenberg) Contextual grammars: deterministic derivations and growth functions. *Revue Roumaine de Mathématiques Pures et Appliquées*, 41 (1996), 83–108
242. (with Gh. Păun and G. Rozenberg) Computing by splicing. *Theoretical Computer Science*, 168 (1996), 321–336
243. (with Gh. Păun and G. Rozenberg) Restricted use of the splicing operation. *International Journal of Computer Mathematics*, 60 (1996), 17–32

244. (with Gh. Păun and G. Rozenberg) Pattern grammars. *Journal of Automata, Languages, and Combinatorics*, 1 (1996), 219–235
245. (with Gh. Păun) Formal languages. Chapter 16.1 in *Handbook of Discrete and Combinatorial Mathematics*, to appear.
246. Conjugate words, cuts of the deck and cryptographic protocols. *EATCS Bulletin*, 59 (1996), 137–149
247. (with Gh. Păun) DNA computing based on the splicing operation. *Mathematica Japonica*, 43 (1996), 607–632
248. (with C. Martin-Vide, A. Mateescu, and J. Miquel-Verges) Quasi shuffle Marcus grammars. *Actas del XII Congr. Lenguajes Naturales y Lenguajes Formales*, Barcelona, 1996, 495–500
249. (with M. Lipponen) Simple words in equality sets. *EATCS Bulletin*, 60 (1996), 123–143
250. (with G. Rozenberg) Watson-Crick complementarity, universal computations and genetic engineering. *Technical Report*, Leiden University, Department of Computer Science, 96–28, 1996
251. (with L. Ilie) On regular characterizations of languages using grammar systems. *Acta Cybernetica*, 12 (1996), 411–425
252. (with V. Mihalache) Growth functions and length sets of replicating systems. *Acta Cybernetica*, 12 (3) (1996), 235–247
253. (with V. Mihalache) Mathematical properties of a particular type of DNA recombination. *Proc. of the 8th International Conf. on Automata and Formal Languages*, Salgotarjan, Hungary, 1996, to appear.
254. (with A. Ehrenfeucht, A. Mateescu, Gh. Păun, and G. Rozenberg) On representing RE languages by one-sided internal contextual languages. *Acta Cybernetica*, 12 (1996), 217–233
255. (with A. Mateescu) Parallel composition of words with re-entrant symbols. *Annals of Bucharest University. Mathematics-Informatics Series*, 1 (1996), 71–80
256. (with T. Nishida) A note on slender 0L languages. *Theoretical Computer Science*, to appear.
257. (with A. Mateescu and G. Rozenberg) Geometric transformations on language families: The power of symmetry. *International Journal of Foundations of Computer Science*, 8 (1997), 1–14
258. (with J. Dassow and Gh. Păun) Grammars with controlled derivations. In: G. Rozenberg and A. Salomaa (eds.), *Handbook of Formal Languages*, vol. 2, Springer-Verlag, 1997, 101–154
259. (with L. Kari and G. Rozenberg) L systems. In: G. Rozenberg and A. Salomaa (eds.), *Handbook of Formal Languages*, vol. 1, Springer-Verlag, 1997, 253–328
260. (with A. Mateescu) Formal languages: an introduction and a synopsis. In: G. Rozenberg and A. Salomaa (eds.), *Handbook of Formal Languages*, vol. 1, Springer-Verlag, 1997, 1–39
261. (with A. Mateescu) Aspects of classical language theory. In: G. Rozenberg and A. Salomaa (eds.), *Handbook of Formal Languages*, vol. 1, Springer-Verlag, 1997, 175–251

262. (with Gh. Păun) Families generated by grammars and L systems. In: G. Rozenberg and A. Salomaa (eds.), *Handbook of Formal Languages*, vol. 1, Springer-Verlag, 1997, 811–861
263. (with E. Csuhaj-Varjú) Networks of parallel language processors. In: Gh. Păun and A. Salomaa (eds.), *New Trends in Formal Languages. Control, Cooperation, and Combinatorics*, Springer-Verlag, *Lecture Notes in Computer Science*, 1218 (1997), 299–318
264. (with A. Mateescu, G.D. Mateescu, and G. Rozenberg) Shuffle-like operations on ω-words. In: Gh. Păun and A. Salomaa (eds.), *New Trends in Formal Languages. Control, Cooperation, and Combinatorics*, Springer-Verlag, *Lecture Notes in Computer Science*, 1218 (1997), 395–411.
265. Computability paradigms based on DNA complementarity. In: V. Keränen (ed.), *Innovation in Mathematics, Proc. 2nd Intern. Mathematical Symposium, Computational Mechanics Publications*, Southhampton and Boston, 1997, 15–28
266. (with C. Ding, V. Niemi, and A. Renvall) Twoprime: A fast stream ciphering algorithm. In: E. Biham (ed.), *Fast Software Encryption*, Springer-Verlag, *Lecture Notes in Computer Science*, 1267 (1997), 88–102
267. (with S. Dumitrescu and Gh. Păun) Pattern languages versus parallel communicating grammar systems. *International Journal of Foundations of Computer Science*, 8 (1997), 67–80
268. (with V. Mihalache) Lindenmayer and DNA: Watson-Crick DOL systems. *EATCS Bulletin*, 62 (1997), 160–175
269. (with J. Dassow and V. Mitrana) Context-free evolutionary grammars and structural language of nucleic acids. *BioSystems*, 43 (1997), 169–177
270. (with R. Freund, Gh. Păun, and G. Rozenberg) Bidirectional sticker systems. In: R. B. Altman, A. K. Dunker, L. Hunter, and T. E. Klein (eds.), *Proc. of Third Annual Pacific Conference on Biocomputing*, Hawaii, World Scientific, 1998, 535–546
271. (with R. Freund, Gh. Păun, and G. Rozenberg) Watson-Crick finite automata. *Proc. of Third DIMACS DNA Based Computers Meeting*, Philadelphia, 1997, 305–317
272. (with Gh. Păun and G. Rozenberg) Computing by splicing. Programmed and evolving splicing rules. *Proc. of IEEE International Conference on Evolutionary Computing*, Indianapolis, 1997, 273–277
273. (with Gh. Păun) From DNA recombination to DNA computing via formal languages. In: R. Hofestädt, T. Lengauer, M. Löffler, and D. Schomburn (eds.), *Bioinformatics*, Springer-Verlag, *Lecture Notes in Computer Science*, 1278 (1997), 210–220
274. (with A. Mateescu and G. Rozenberg) Syntactic and semantic aspects of parallelism. In: C. Freksa, M. Jantzen, and R. Valk (eds.), *Foundations of Computer Science; Potential - Theory - Cognition*, Springer-Verlag, *Lecture Notes in Computer Science*, 1337 (1997), 79–105

275. (with Gh. Păun) Characterizations of recursively enumerable languages by using copy languages. *Revue Roumaine de Mathématiques Pures et Appliquées*, to appear.
276. (with C. Martin-Vide, J. Miquel-Verges, and Gh. Păun) Attempting to define the ambiguity of internal contextual languages. In: C. Martin-Vide (ed.), *Mathematical and Computational Analysis of Natural Language*, John Benjamins, Amsterdam, 1998, 59–81.
277. (with C. Martin-Vide, Gh. Păun, and G. Rozenberg) Universality results for finite H systems and Watson-Crick finite automata. In: Gh. Păun (ed.), *Computing with Bio-Molecules. Theory and Experiments*, Springer-Verlag, 1998, 200–220
278. Events and languages. In: C. Calude (ed.), *Theoretical Computer Science. People and Ideas*, Springer-Verlag, 1998.
279. (with A. Mateescu) Abstract family of languages. In: M. Hezinwinkel (ed.), *Encyclopaedia of Mathematics*, suppl. vol. 1, Kluwer Academic Publishers, 1998, 12–13
280. (with A. Mateescu) Grammar form. In: M. Hezinwinkel (ed.), *Encyclopaedia of Mathematics*, suppl. vol. 1, Kluwer Academic Publishers, 1998, 272–273
281. (with A. Mateescu, Gh. Păun, and G. Rozenberg) Characterizations of recursively enumerable languages starting from internal contextual languages. *International Journal of Computer Mathematics*, 66 (1998), 179–197
282. (with A. Mateescu and G. Rozenberg) Shuffle on trajectories: Syntactic constraints. *Theoretical Computer Science*, 197 (1998), 1–56 (Fundamental study)
283. (with A. Mateescu, Gh. Păun, and G. Rozenberg) Simple splicing systems. *Discrete Applied Mathematics*, 84 (1998), 145–163
284. (with V. Mihalache) Language-theoretic aspects of string replication. *International Journal for Computer Mathematics*, 66 (1998), 163–177
285. (with L. Ilie) On well quasi orders of free monoids. *Theoretical Computer Science*, 204 (1998), 131–152
286. (with L. Ilie) 2-Testability and relabelings produce everything. *Journal of Computer and System Sciences*, 56 (1998), 253–262
287. (with L. Kari, Gh. Păun, G. Rozenberg, and S. Yu) DNA computing, sticker systems, and universality. *Acta Informatica*, 35 (1998), 401–420
288. (with E. Csuhaj-Varjú) Networks of language processors: parallel communicating systems. *EATCS Bulletin*, 66 (1998), 122–138
289. Turing, Watson-Crick and Lindenmayer. Aspects of DNA complementarity. In: C. S. Calude, J. Casti, and M. J. Dinneen (eds.), *Unconventional Models of Computation*, Springer-Verlag, 1998, 94–107
290. (with T. Harju and A. Mateescu) Shuffle on trajectories: The Schützenberger product and related operations. In: L. Brim, J. Gruska, and J. Zlatuska (eds.), *Proc. of MFCS'98*, Springer-Verlag, *Lecture Notes in Computer Science*, 1450 (1998), 503–511

291. (with V. Mihalache) Language-theoretic aspects of DNA complementarity. *Theoretical Computer Science*, to appear.
292. Watson-Crick walks and roads on DOL graphs. *Acta Cybernetica*, to appear
293. (with C. Martin-Vide and Gh. Păun) Characterizations of recursively enumerable languages by means of insertion grammars, *Theoretical Computer Science*, 205 (1998), 195–205
294. (with Gh. Păun and G. Rozenberg) Complementarity versus universality: Keynotes of DNA computing. *Complexity*, 4 (September–October 1998), 14–19

Part I

Automata I: Finite State Machines

Semilattices of Fault Semiautomata

Janusz A. Brzozowski and Helmut Jürgensen

Summary. We study defects affecting state transitions in sequential circuits. The fault-free circuit is modeled by a semiautomaton M, and 'simple' defects, called *single faults*, by a set $S = \{M^1, ..., M^k\}$ of 'faulty' semiautomata. To define *multiple* faults from S, we need a binary composition operation, say $\odot$, on semiautomata, which is idempotent, commutative, and associative. Thus, one has the free semilattice $S^{\odot}$ generated by S. In general, however, the single faults are not independent; a finite set E of equations of the form $M^{i_1} \odot \ldots \odot M^{i_h} = M^{j_1} \odot \ldots \odot M^{j_k}$ describes the relations among them. The pair (S, E) is a finite presentation of the quotient semilattice $S^{\odot}/\eta$, where η is the smallest semilattice congruence containing E. In this paper, we first characterize such abstract quotient semilattices. We then survey the known results about random-access memories (RAMs) for the Thatte-Abraham fault model consisting of stuck-at, transition, and coupling faults. We present these results in a simplified semiautomaton model and give new characterizations of two fault semilattices.

1 Single and Multiple Faults

In this paper we study the following general problem. We have a *fault-free object*, which is completely abstract at this stage. When the object is manufactured, zero or more 'simple defects' may be introduced. Each of these defects produces a faulty version of the object, called a *single fault*. Let $D = \{d_1, \ldots, d_k\}$ be the set of simple defects and let $S = \{s_1, \ldots, s_k\}$, $k \geq 0$, be the corresponding set of single faults.

If several simple defects are present in the faulty object at the same time, we have a *multiple fault*. Our objective is to characterize such multiple faults in terms of the single faults.

In the simplest case, the single faults are not related in any way. Here, a multiple fault is simply a set of single faults. For example, suppose there are two single faults. The fault-free object corresponds to the empty set of single faults; the objects containing one defect, to $\{s_1\}$ and $\{s_2\}$; and the double fault, to the set $\{s_1, s_2\}$. The composition operation on single faults that yields multiple faults is set union. Our multiple fault model is the algebra $(2^S, \cup, \emptyset)$. Note that this algebra is a finite semilattice, with identity $\emptyset$ and zero S, generated by the finite set S under the union operation.

In practice a set A of defects may result in a faulty object that is indistinguishable from the faulty object corresponding to defects from a different set B. We will write $A \equiv B$ to indicate that A and B result in the same object. Thus, in general, we have a finite set S of single faults, the set 2^S of potential multiple faults, and a finite set $\mathcal{E} = \{(A \equiv B) \mid A, B \in 2^S\}$ of equations. Our model for multiple faults is the quotient semilattice $2^S/\eta_{\mathcal{E}}$, where $\eta_{\mathcal{E}}$ is the smallest semilattice congruence containing $\mathcal{E}$.

2 Finitely Presented Semilattices

Abstractly we have the following setting for our problem. We are given a finite set S of generators and a finite set E of equations of the form $a \equiv b$ where a and b are words over the alphabet S. Let S^* be the free monoid with identity 1 generated by S, and let η be the smallest semilattice congruence on S^* containing E. Then the quotient semilattice $S^\bullet = S^*/\eta$ is the semilattice defined by the presentation (S, E). For any word u over S, let $[u]_\eta$ be the η-class of u. The multiplication $\bullet$ on $S^\bullet$ is given by $[u]_\eta \bullet [v]_\eta = [uv]_\eta$.

For a word u over S, let $\alpha(u)$ be the set of elements of S appearing in u. If u and v are words over S such that $\alpha(u) = \alpha(v)$, then u and v represent the same element of the free semilattice generated by S, hence also of $S^\bullet$. Thus, via α, the free semilattice with identity generated by S is isomorphic with the semilattice $(2^S, \cup, \emptyset)$. For an equation $a \equiv b$ in E, let $\alpha(a \equiv b)$ be the equation $\alpha(a) \equiv \alpha(b)$; moreover, let $\alpha(E)$ be the set of equations $\alpha(a \equiv b)$ where $a \equiv b$ is in E. Let θ be the smallest congruence relation on 2^S containing $\alpha(E)$. Then the semilattice $S^\bullet$ is isomorphic with the quotient semilattice $2^S/\theta$. Thus to compute the product of two elements in $S^\bullet$, it is equivalent to compute the product of the corresponding sets in $2^S/\theta$.

In general, there may be several different subsets of S representing the same element of $2^S/\theta$. However, for every $T \subseteq S$ one can choose a canonical representative as follows.

Proposition 1. *Let $T \subseteq S$. Then there is a unique set rep(T) in the θ-class $[T]_\theta$ of T such that $T' \subseteq$ rep(T) for all $T' \in [T]_\theta$.*

Proof. The class $[T]_\theta$ is closed under union. Hence, the union of all sets $T' \in [T]_\theta$ is the desired canonical representative.

We now give an algorithm for constructing the canonical representative *rep*(T) of any set $T \subseteq S$. Given that α is a semilattice isomorphism, instead of words and equations on words, we consider only subsets of S and equations on sets. An equation $e_i = A_i \equiv B_i$ is *applicable* to T if $A_i \subseteq T$ or $B_i \subseteq T$. The set of equations will be represented by an array $\mathcal{E} = (e_1, \ldots, e_m)$. Variable $\mathcal{E}'$ represents equations that have not yet been eliminated. Variable T' represents the set obtained from T by the equations used so far. The size of the array $\mathcal{E}'$ is denoted by $|\mathcal{E}'|$. An *application* of an applicable equation e_i to T' consists of replacing T' by $T' \cup A_i \cup B_i$ and of deleting e_i from $\mathcal{E}'$. With these assumptions we have the algorithm shown in Fig. 1.

```
function rep(ℰ:array[1..m] of equations, T: subset of S): subset of S;
var ℰ': array[1..m] of equations, T': subset of S;
ℰ' ← ℰ; T' ← T;
while ℰ' ≠ ∅ do
        i ← 1;
        while e_i not applicable do
                i ← i + 1;
                if i > |ℰ'| then
                        rep ← T';
                        exit {rep};
        ℰ' ← ℰ' with e_i deleted; T' ← T' ∪ A_i ∪ B_i;
rep ← T';
```

Fig. 1. Function *rep*. Scoping is indicated by indentation.

Let $S^{\vee}$ be the set of distinct representatives obtained by *rep*. Let $\vee$ be defined as follows: For any sets $A, B \in S^{\vee}$, $A \vee B = rep(A \cup B)$.

Theorem 1. *The algebra* $(S^{\vee}, \vee, \emptyset)$ *is a semilattice with identity. The three semilattices* $S^{\bullet}$, $2^S/\theta$ *and* $S^{\vee}$ *are isomorphic.*

Proof. The mapping *rep* from 2^S to $S^{\vee}$ induces an isomorphism μ from $2^S/\theta$ onto $2^{\vee}$ by $\mu([T]_\theta) = rep(T)$ for $T \subseteq S$.

Note that Theorem 1 provides an algorithm for solving the word problem in finitely presented semilattices. The solvability of the word problem itself follows already from some very general theorems about finitely presented algebras [4]. Surprisingly, however, we failed to find our simple solution to this problem in the literature.

As in any finite semilattice, one can derive a second semilattice operation $\wedge$ on $S^{\vee}$ such that $(S^{\vee}, \vee, \wedge, \emptyset, S)$ is a lattice, where

$$T \wedge T' = \bigvee_{U \in S^{\vee}, U \subseteq T \cap T'} U.$$

It turns out that $T \wedge T' = T \cap T'$, that is, the set of canonical representatives is closed under intersection.

Example 1. Let $S = \{s_0, s_1, t_0, t_1\}$, and

$$E = \{s_i t_j \equiv s_i \mid i, j \in \{0,1\}\} \cup \{s_0 s_1 \equiv s_0 s_1 t_0 t_1\}.$$

One verifies that $S^{\bullet} = S^*/\eta$ has the 7 elements $\emptyset$, t_0, t_1, t_0t_1, $s_0t_0t_1$, $s_1t_0t_1$, and $s_0s_1t_0t_1$. The Hasse diagram of $S^{\bullet}$ is shown in Fig. 2.

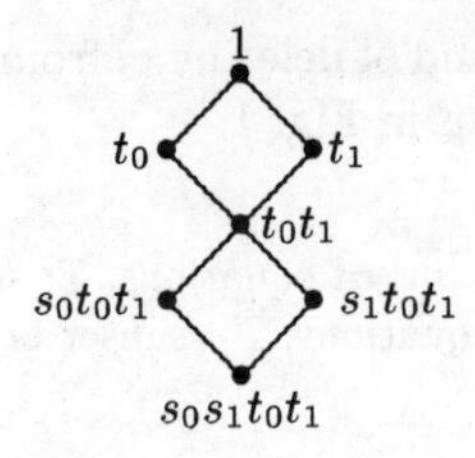

Fig. 2. Semilattice of Example 1

3 Faults in Sequential Circuits

We now make our problem more concrete by considering sequential circuits. Such circuits can be modeled by Mealy automata defined as follows. A *finite deterministic Mealy automaton* is a quintuple $M = (Q, X, Y, \delta, \lambda)$, where Q is a finite nonempty set called the *state set*, X is a finite nonempty set called the *input alphabet*, δ is a function, called the *transition function*, from $Q \times X$ into Q, set Y is a finite nonempty set called the *output alphabet*, and λ is a function, called the *output function*, from $Q \times X$ into Y. When a defect is present in a sequential circuit, we assume that the faulty circuit is still represented by an automaton. A faulty automaton may differ from the fault-free one in the state set, transition function, and output function.

In this paper we consider only defects affecting the state set and the transition function; hence, we can use semiautomata. A *finite deterministic semiautomaton* is denoted by $M = (Q, X, \delta)$, where Q, X, and δ are as above. The *empty semiautomaton* is $\mathbf{0} = (\emptyset, X, \emptyset)$.

Example 2. We give an example involving partial semiautomata, where the transition function is defined only on a subset of $Q \times X$. Consider an arbitrary partial semiautomaton $M = (Q, X, \delta)$ and the set S of faulty partial semiautomata of the following kind. For each transition $\delta(q, x)$ of M there is a faulty semiautomaton $M_{q,x} = (Q, X, \delta_{q,x})$, such that $\delta_{q,x}$ coincides with δ, except that $\delta_{q,x}(q, x)$ is undefined. The composition of two partial semiautomata $M^1 = (Q, X, \delta^1)$ and $M^2 = (Q, X, \delta^2)$ with $\delta^1, \delta^2 \subseteq \delta$, is $M^{1\odot 2} = M^1 \odot M^2 = (Q, X, \delta^{1\odot 2})$, where $\delta^{1\odot 2} = \delta^1 \cap \delta^2$. Clearly, M is the identity element, and the semiautomaton $(Q, X, \emptyset)$ with no transitions is the zero element. The algebra $(S^{\odot}, \odot, M)$ is isomorphic with the intersection semilattice of the subsets of δ. Note that this is the free semilattice with identity generated by S.

There has been extensive research on products of automata with output [5]. However, very little is known about composition operations on *semiautomata* in general. Hence, in the rest of the paper we turn to semiautomata with more structure, where several results about multiple faults are known.

4 Single Faults in RAMs

A ***component semiautomaton*** is a finite deterministic semiautomaton $M = (Q, X, \delta)$, where Q has the following property: There exists an integer $n > 0$ and n finite nonempty sets $Q_1, \ldots, Q_n$ such that $Q \subseteq Q_1 \times \ldots \times Q_n$. Thus each state $q \in Q$ is an n-tuple $q = (q_1, \ldots, q_n)$, where $q_i \in Q_i$ for $i = 1, \ldots, n$. In the sequel, Q_i is assumed to be minimal for each i, that is, for every $a \in Q_i$, there is $q \in Q$ such that $q_i = a$.

A component semiautomaton is ***binary*** if $Q_1, \ldots, Q_n$ are subsets of $\{0, 1\}$. In this paper we consider only binary semiautomata.

We use semiautomata to model random access memories (RAMs). We assume that a RAM consists of n one-bit cells. The *fault-free* n-cell RAM is a semiautomaton $M = (Q, X, \delta)$, where $Q = \{0, 1\}^n$, $X = \bigcup_{i=1}^n X_i$ with $X_i = \{w_0^i, w_1^i\}$, and δ is defined by $\delta((q_1, \ldots, q_i, \ldots, q_n), w_a^i) = (q_1, \ldots, a, \ldots, q_n)$, for $a \in \{0, 1\}$. Here $q = (q_1, \ldots, q_n)$ is the present state of the RAM, and $\delta(q, x)$ is the next state of the RAM after input $x \in X$ has been applied. Inputs w_0^i and w_1^i are the 'write 0 to cell i' and the 'write 1 to cell i' operations, respectively[1].

In the sequel, the number n of cells in a RAM is a fixed integer. Also, since all the semiautomata considered here are over the alphabet X given above, this alphabet will not be explicitly mentioned, and a semiautomaton will be denoted simply by $M = (Q, \delta)$. One-cell and two-cell RAMs are shown in Figs. 3 and 4, respectively; they are labelled M.

When a RAM is manufactured, a defect resulting in a faulty RAM may occur. Many defects that occur in practice can still be represented by normal semiautomata [3]. If that is not the case, however, we represent the defective RAM by the special empty semiautomaton $\mathbf{0}$.

To model defective RAMs, we associate with the fault-free n-cell RAM $M = (Q, \delta)$, a family Δ of semiautomata $M' = (Q', \delta')$, called *delta faults* or simply *faults* of M, which may differ from M in the state set and the transition function. The state set Q' is a subset of Q and the transition function δ' is arbitrary. Both M and $\mathbf{0}$ satisfy the definition of a fault.

We now give some examples of common faults: the transition, stuck-at, and coupling faults. These faults constitute the set $\mathcal{T}$ of *Thatte-Abraham faults* [7], and are the main object of our investigation here.

The complement of $a \in \{0, 1\}$ is denoted by $\overline{a}$. A RAM with a *transition* fault in cell i is a fault $M^{i\neg a} = (Q, \delta^{i\neg a})$, where $a \in \{0, 1\}$ and $\delta^{i\neg a}$ behaves like δ *except* that, when the input w_a^i is applied to any state in which cell i is in state $\overline{a}$, the write operation is unsuccessful, that is,

$$\delta^{i\neg a}((q_1, \ldots, q_{i-1}, \overline{a}, q_{i+1}, \ldots, q_n), w_a^i) = (q_1, \ldots, q_{i-1}, \overline{a}, q_{i+1}, \ldots, q_n).$$

Figure 3 shows the transition fault $M^{i\neg 1}$. All operations to other cells are correct and not shown. An n-cell RAM has $2n$ transition faults.

[1] A RAM also has read operations, an output alphabet, and an output function, but these concepts are irrelevant to this paper.

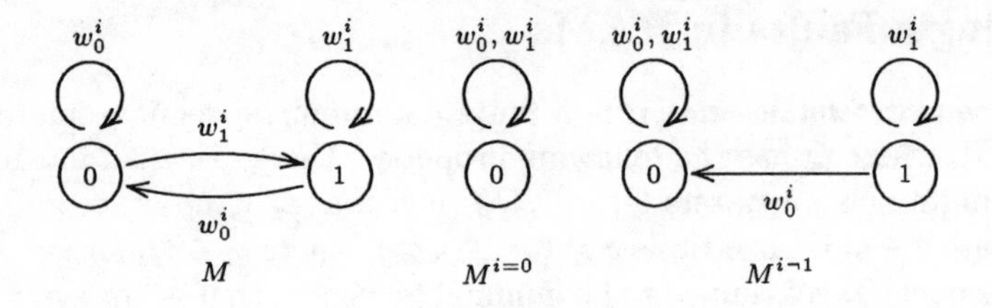

Fig. 3. Transition and stuck-at faults of a one-cell RAM M.

For $a \in \{0,1\}$, a *stuck-at-a* fault in cell i is a fault $M^{i=a} = (Q^{i=a}, \delta^{i=a})$, where $Q^{i=a} = \{q \mid q = (q_1, \ldots, q_{i-1}, a, q_{i+1}, \ldots, q_n)\}$, and $\delta^{i=a}$ is the restriction of δ to $Q^{i=a} \times X$, *except* for the transitions under input $w_{\overline{a}}^i$. This input fails to write $\overline{a}$ in cell i, that is,

$$\delta^{i=a}((q_1, \ldots, q_{i-1}, a, q_{i+1}, \ldots, q_n), w_{\overline{a}}^i) = (q_1, \ldots, q_{i-1}, a, q_{i+1}, \ldots, q_n).$$

Figure 3 illustrates the stuck-at fault $M^{i=0}$. All operations to other cells are correct and not shown. An n-cell RAM has $2n$ stuck-at faults.

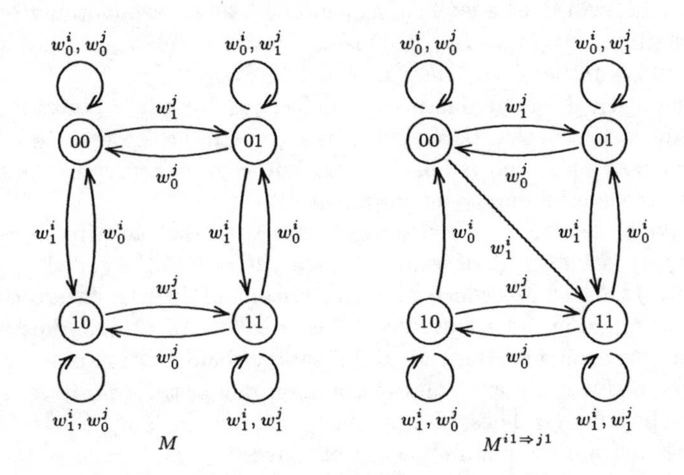

Fig. 4. Coupling fault $M^{i1 \Rightarrow j1}$ in a two-cell RAM M. The first state component refers to cell i, the second one to cell j.

A *coupling* fault involves two cells, the *coupling* cell i and the *coupled* cell j. When the present states of cells i and j are $\overline{a}$ and $\overline{b}$, respectively, where $a, b \in \{0,1\}$, and a is written into cell i, then cell j, as well as cell i, changes state. The fault is denoted $M^{ia \Rightarrow jb}$. Figure 4 illustrates the fault $M^{i1 \Rightarrow j1}$; only the states of cells i and j are shown. All operations to other cells are correct and not shown. An n-cell RAM has $4n(n-1)$ coupling faults.

5 Multiple Faults in RAMs

A 'simple' defect represented by a semiautomaton is considered to be a *single* fault. For example, each transition, stuck-at, and coupling fault is considered to be a single fault in the literature on circuit testing. Suppose now that two defects D^1 and D^2 are present in a RAM, and are represented by semiautomata M^1 and M^2. It should be possible to represent the simultaneous presence of both defects by another semiautomaton. What is needed is an *composition operation on semiautomata,* say $\odot$, such that $M^1 \odot M^2$ would correctly represent the double fault. Although there are cases where it is quite obvious how to construct the semiautomaton for the double fault composed of M^1 and M^2, in general this problem is very difficult.

We make a number of assumptions about the nature of faults. First, whenever the presence of two defects, represented by semiautomata M^1 and M^2, leads to a circuit outside the normal semiautomaton model, we represent the resulting circuit by $\mathbf{0}$; thus $M^1 \odot M^2 = \mathbf{0}$. Second, we assume that the composition of any fault M' with a fault outside the semiautomaton model is outside the model, that is, $\mathbf{0} \odot M' = M' \odot \mathbf{0} = \mathbf{0}$. Third, if defects D^1 and D^2 are represented by the same semiautomaton M', the composite semiautomaton should be exactly the same as each of the two components. Thus, the composition should be *idempotent*. Fourth, the order in which we consider two defects when they are both present in the same circuit should be immaterial. Thus, the composition should be *commutative*. Fifth, the order in which we compose three defects should not matter; the final result should be a circuit with all three defects. Thus, the composition should be *associative*. Fault composition was first considered in [1]. Major results related to this paper were presented in [2,3]. For further justification of the assumptions above see [3].

Altogether, for any given set of single faults, we have the *generic* hypothesis that the composition operation should be a semilattice operation, and that the semilattice of multiple faults may contain a zero element $\mathbf{0}$, and, by convention, contains the identity element M.

We now attack the composition problem for the Thatte-Abraham faults. A set $Q' \subseteq Q_1 \times \ldots \times Q_n$ is called a *subproduct* if $Q' = Q'_1 \times \cdots \times Q'_n$, for some sets $Q'_1 \subseteq Q_1, \ldots, Q'_n \subseteq Q_n$. A semiautomaton is called a *subproduct* semiautomaton if its state set is a subproduct. Note that all the faults in $\mathcal{T}$, including M and $\mathbf{0}$ are (binary) subproduct semiautomata. Let $\mathcal{P}$ be the set of all subproduct faults; from now on, we consider only faults from $\mathcal{P}$. An example of a fault that is not in $\mathcal{P}$ is the stuck-equal fault of Fig. 5.

5.1 The Semilattice $\mathcal{T}^\circ$

In addition to the generic properties of fault composition, there are also conditions derived from the physical nature of defects. We now state a set of such *physical* conditions for the Thatte-Abraham fault model. Condition ρ_1

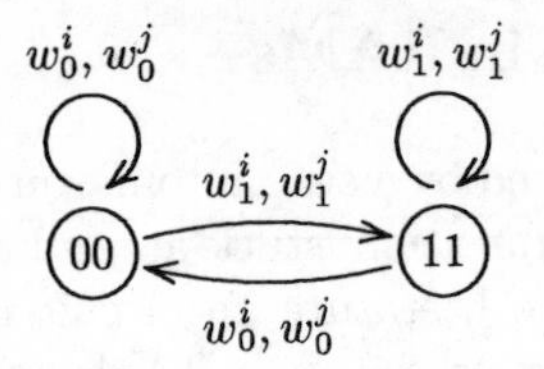

Fig. 5. Stuck-equal fault $M^{i=j}$.

states that a cell that is both stuck-at-0 and stuck-at-1 results in a fault that is outside the semiautomaton model. Condition ρ_2 affirms that a stuck-at fault in a cell is stronger than any transition fault in that cell. Condition ρ_3 declares that a stuck-at fault in a cell is stronger than any coupling fault in which that cell is the coupled cell. For a justification of these conditions see [3,6].

(ρ_1) $M^{i=a} \odot M^{i=\overline{a}} = \mathbf{0}$,
(ρ_2) $M^{i=a} \odot M^{i\neg b} = M^{i=a}$,
(ρ_3) $M^{i=a} \odot M^{jb \Rightarrow ic} = M^{i=a}$.

Our main problem now is to find a binary composition operation on semiautomata that satisfies both the generic conditions and the physical conditions above. Let $M = (Q, \delta)$ be the fault-free n-cell RAM, and let $M^1 = (Q^1, \delta^1)$ and $M^2 = (Q^2, \delta^2)$ be two faults of M. We define the composition $M^1 \circ M^2$ to be the semiautomaton $M^{1\circ 2} = (Q^{1\circ 2}, \delta^{1\circ 2})$, where $Q^{1\circ 2} = Q^1 \cap Q^2$, and $\delta^{1\circ 2}$ is defined as follows. Consider a state $q = (q_1, \ldots, q_n) \in Q^{1\circ 2}$ and an input $x \in X_i$. Let $\delta(q,x) = (p_1, \ldots, p_n)$, $\delta^1(q,x) = (p_1^1, \ldots, p_n^1)$, $\delta^2(q,x) = (p_1^2, \ldots, p_n^2)$, and $\delta^{1\circ 2}(q,x) = (p_1^{1\circ 2}, \ldots, p_n^{1\circ 2})$. To determine $\delta^{1\circ 2}$ one applies the rule:

Circle Rule. For all j, (a) $p_j^{1\circ 2} = a$, if $Q_j^{1\circ 2} = \{a\}$, and (b) $p_j^{1\circ 2} = p_j \oplus [(p_j \oplus p_j^1) + (p_j \oplus p_j^2)]$, if $Q_j^{1\circ 2} = \{0,1\}$, where $+$ and $\oplus$ denote the inclusive and exclusive OR operations, respectively.

In the following theorem, $(\mathcal{T}^\circ, \circ)$ denotes the semilattice generated by $\mathcal{T}$ under the operation $\circ$.

Theorem 2. *The algebra* $(\mathcal{P}, \circ)$ *is a semilattice with* $\mathbf{0}$ *as zero element and* M *as the identity, and* $\circ$ *satisfies* ρ_1–ρ_3. *The algebra* $(\mathcal{T}^\circ, \circ)$ *is a subsemilattice of* $(\mathcal{P}, \circ)$, *and* ρ_1–ρ_3 *form a minimal set of defining relations for* $\mathcal{T}^\circ$. *Furthermore,* $\mathcal{T}^\circ$ *is a proper subsemilattice of* $\mathcal{P}$.

Proof. The first two claims of the theorem were proved in [2]. In that paper, delta faults were defined more narrowly, since they were required to have state sets which were subproducts of Q. Thus the family of delta faults in [2] is precisely $\mathcal{P}$. Semiautomaton M^a of Fig. 6 is not in $\mathcal{T}^\circ$. This follows from the characterization of $\mathcal{T}^\circ$ given in Theorem 4 in Section 6.

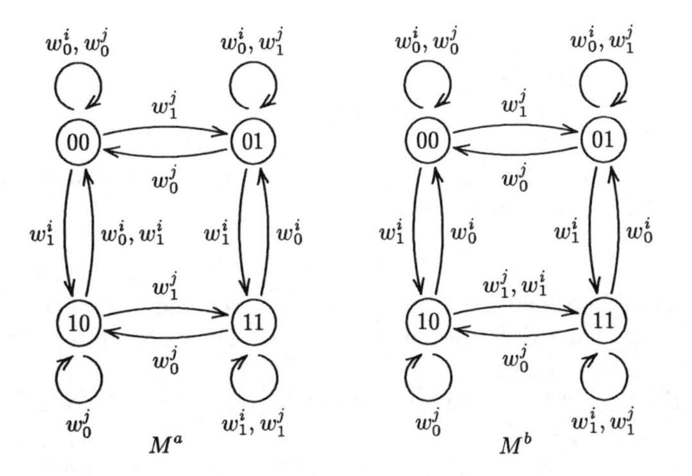

Fig. 6. Faults M^a and M^b.

5.2 The Semilattice $\mathcal{T}^\diamond$

A different circuit implementation of a RAM may lead to a different set of physical conditions. In fact, in another useful fault model the composition must satisfy, in addition to ρ_1–ρ_3, also ρ_4 and ρ_5, where

(ρ_4) $M^{i\neg a} \odot M^{ia\Rightarrow jb} = M^{i\neg a}$,
(ρ_5) $M^{i=a} \odot M^{ib\Rightarrow jc} = M^{i=a}$.

The fourth condition requires that a transition fault $M^{i\neg a}$ be stronger than any coupling fault $M^{ia\Rightarrow jb}$. The fifth condition states that the stuck-at fault $M^{i=a}$ is stronger than any coupling fault $M^{ib\Rightarrow jc}$. We again refer the reader to [3,6] for the justification of these conditions. One can verify that ρ_5 is a consequence of ρ_1–ρ_4.

We now define a second composition operation $\diamond$ as follows. Let $M = (Q, \delta)$ be the fault-free n-cell RAM, and let $M^1 = (Q^1, \delta^1)$ and $M^2 = (Q^2, \delta^2)$ be two faults of M. We define the composition $M^1 \diamond M^2$ to be the semiautomaton $M^{1\diamond 2} = (Q^{1\diamond 2}, \delta^{1\diamond 2})$, where $Q^{1\diamond 2} = Q^1 \cap Q^2$ and $\delta^{1\diamond 2}$ is defined as follows. Consider a state $q = (q_1, \ldots, q_n) \in Q^{1\diamond 2}$ and an input $x \in X_i$. Use the same notation as in the Circle Rule, and let $\delta^{1\diamond 2}(q, x) = (p_1^{1\diamond 2}, \ldots, p_n^{1\diamond 2})$. To determine $\delta^{1\diamond 2}$ one applies the rule:

Diamond Rule. If $p_i^1 = p_i^2 = p_i$, then (a) for all j including i, $p_j^{1\diamond 2} = a$, if $Q_j^{1\diamond 2} = \{a\}$, and (b) $p_j^{1\diamond 2} = p_j \oplus [(p_j \oplus p_j^1) + (p_j \oplus p_j^2)]$, if $Q_j^{1\diamond 2} = \{0, 1\}$. Otherwise, $\delta^{1\diamond 2}(q, x) = q$.

The algebra $(\mathcal{P}, \diamond)$ is an idempotent and commutative grupoid with $\mathbf{0}$ as zero element and M as the identity, but it is not associative. The verification of idempotency and commutativity is routine. Figure 6 shows two faults M^a

and M^b. Letting $M^1 = M^a$, and $M^2 = M^3 = M^b$, we find that the transition from state 10 under input w_1^i results in $p_j^{(1\diamond 2)\diamond 3} = 1$ and $p_j^{1\diamond(2\diamond 3)} = 0$.

Since our main interest is in the Thatte-Abraham faults, we add a restriction to obtain a class of faults in which $\diamond$ is associative.

A RAM delta fault $M' = (Q', \delta')$ is *change-attempt-activated* if for all $q \in Q'$, $i \in \{1, \ldots, n\}$, and $x \in X_i$,

$$\delta_i(q,x) = q_i \Longrightarrow \delta'(q,x) = q.$$

In such a fault, no change of state is possible, unless the state of the addressed cell is being changed in the fault-free RAM. Let $\mathcal{A}$ be the set of change-attempt-activated faults.

Let $\mathcal{C}$ be the set of delta faults satisfying, for all $q \in Q'$, $i \in \{1, \ldots, n\}$, and $x \in X_i$,

$$\delta_i(q,x) \neq q_i = \delta'_i(q,x) \Longrightarrow \delta'(q,x) = q.$$

Let $\mathcal{S} = \mathcal{A} \cap \mathcal{C}$ be the set of *change-success-activated* faults. One verifies that all the faults in $\mathcal{T}$ are in $\mathcal{S}$.

In the theorem below, $(\mathcal{T}^\diamond, \diamond)$ denotes the semilattice generated by $\mathcal{T}$ under the operation $\circ$.

Theorem 3. *The algebra $(\mathcal{S} \cap \mathcal{P}, \diamond)$ is a semilattice with* $\mathbf{0}$ *as zero element and M as the identity, and $\diamond$ satisfies ρ_1–ρ_5. The algebra $(\mathcal{T}^\diamond, \diamond)$ is a proper subsemilattice of $(\mathcal{S} \cap \mathcal{P}, \diamond)$. Furthermore, equations ρ_1–ρ_4 form a minimal set of defining relations for $\mathcal{T}^\diamond$.*

Proof. The diamond operation was introduced in [3], as an composition operation on delta faults. However, we later discovered that this operation was ill-defined for some delta faults; consequently, several results in [3] are incorrect as stated. The subproduct condition should be added, as it is done in the present paper, to obtain correct statements. With this added condition, the proofs of [3] apply for the first two claims of the theorem. The proof of the third claim is similar to the proof of the corresponding statement in Theorem 2.

6 Characterization of $\mathcal{T}^\circ$ and $\mathcal{T}^\diamond$

Roughly speaking, a RAM delta fault involving a set C of cells is pattern-sensitive if it occurs only for some values of the cells that are not in C. More precisely, A delta fault $M' = (Q', \delta')$ is *pattern-sensitive* if one of the following conditions holds:
(1) There exist $i \in \{1, \ldots, n\}$, $x \in X_i$, and $q, q' \in Q'$ such that $q'_i = q_i$ and $\delta'_i(q,x) \neq \delta'_i(q',x)$.
(2) There exist $i, j \in \{1, \ldots, n\}$, $x \in X_i$, and $q, q' \in Q'$ such that $q'_i = q_i$, $q'_j = q_j$, and $\delta'_j(q,x) \neq \delta'_j(q',x)$.
Otherwise, it is *pattern-insensitive*. Let $\mathcal{I}$ be the set of pattern-insensitive faults.

Lemma 1. *The sets $\mathcal{A}$, $\mathcal{S}$ and $\mathcal{I}$ are closed under $\circ$ and $\diamond$.*

Lemma 2. *Let $M^1, M^2 \in \Delta$ with $M^1 = (Q^1, \delta^1)$ and $M^2 = (Q^2, \delta^2)$, and let $M = (Q, \delta)$ be the corresponding fault-free automaton.*

1. *For all $q \in Q^{1\circ 2}$ and all $x \in X$, if $\delta^1(q,x) = \delta(q,x)$ and $\delta^2(q,x) \in Q^{1\circ 2}$ then $\delta^{1\circ 2}(q,x) = \delta^2(q,x)$.*
2. *If $M^2 \in \mathcal{C}$ then, for all $q \in Q^{1\diamond 2}$ and all $x \in X$, if $\delta^1(q,x) = \delta(q,x)$ and $\delta^2(q,x) \in Q^{1\diamond 2}$ then $\delta^{1\diamond 2}(q,x) = \delta^2(q,x)$.*

Theorem 4. *(a) A delta fault is change-attempt-activated, pattern-insensitive, and a subproduct fault if and only if it is equivalent to a $\circ$-composition of Thatte-Abraham faults, that is, $\mathcal{T}^\circ = \mathcal{A} \cap \mathcal{I} \cap \mathcal{P}$. (b) A delta fault is change-success-activated, pattern-insensitive, and a subproduct fault if and only if it is equivalent to a $\diamond$-composition of Thatte-Abraham faults, that is, $\mathcal{T}^\diamond = \mathcal{C} \cap \mathcal{T}^\circ$.*

Proof. By the definition of $\mathcal{T}$ one has $\mathcal{T} \subseteq \mathcal{A} \cap \mathcal{I} \cap \mathcal{P} \cap \mathcal{C}$. By Lemma 1 and Theorems 2 and 3, $\mathcal{A} \cap \mathcal{I} \cap \mathcal{P}$ is closed under $\circ$ and $\mathcal{A} \cap \mathcal{I} \cap \mathcal{P} \cap \mathcal{C}$ is closed under $\diamond$. Therefore, $\mathcal{T}^\circ \subseteq \mathcal{A} \cap \mathcal{I} \cap \mathcal{P}$ and $\mathcal{T}^\diamond \subseteq \mathcal{A} \cap \mathcal{I} \cap \mathcal{P} \cap \mathcal{C}$.

We first show that $\mathcal{T}^\circ \supseteq \mathcal{A} \cap \mathcal{I} \cap \mathcal{P}$. Consider $M' = (Q', \delta') \in \mathcal{A} \cap \mathcal{I} \cap \mathcal{P}$. As $M' \in \mathcal{A}$, errors can arise only from writing a new value. Let $\sigma(M')$ be the number of state set components Q'_j which are singleton sets; let $\tau(M')$ be the number of pairs (i,a) with $i \in \{1, \ldots, n\}$ and $a \in \{0,1\}$ such that, for some $q \in Q'$ with $q_i = \overline{a}$, one has $\delta'_i(q, w^i_a) \neq a$; let $\gamma(M')$ be the number of quadruples (i,a,j,b) with $i,j \in \{1, \ldots, n\}$, $i \neq j$, and $a,b \in \{0,1\}$ such that, for some $q \in Q'$ with $q_i = \overline{a}$ and $q_j = \overline{b}$, one has $\delta'_j(q, w^i_a) = b$.

We show that M' can be represented as $M' = S \circ T \circ C \circ M$ where S is a $\circ$-composition of stuck-at faults, T is a $\circ$-composition of transition faults, and C is a $\circ$-composition of coupling faults. In all three cases, the $\circ$-composition of zero faults is M.

We first determine the component S. By splitting S off, we may then assume that $\sigma(M') = 0$. We then determine the component T. By splitting T off, we may assume that also $\tau(M') = 0$. Third, we split C off, leaving a semiautomaton M' with $\sigma(M') = \tau(M') = \gamma(M') = 0$. One verifies that this implies $M' = M$, given that $M' \in \mathcal{A} \cap \mathcal{I} \cap \mathcal{P}$. We proceed by induction, first on $\sigma(M')$, then on $\tau(M')$, and then on $\gamma(M')$. If $M' \in \mathcal{T}$ or $M' = \mathbf{0}$, nothing needs to be proved. Assume, therefore, that $M' \notin \mathcal{T}$ and $M' \neq \mathbf{0}$.

First suppose that $\sigma(M') > 0$. Then $Q'_j = \{a\}$ for some j and some a. Let $M^1 = M^{j=a}$ and let $M^2 = (Q^2, \delta^2)$ be defined as follows. Let

$$Q^2 = Q'_1 \times \cdots \times Q'_{j-1} \times \{0,1\} \times Q'_{j+1} \times \cdots \times Q'_n.$$

For $q \in Q^2$, let $q' = (q_1, \ldots, q_{j-1}, a, q_{j+1}, \ldots, q_n)$. For all inputs x and for all $i \neq j$, $\delta^2_i(q,x) = \delta'_i(q',x)$. For component j, $\delta^2_j(q,x)$ is $\overline{q}_j$, if $x = w^j_{\overline{q}_j}$, and q_j otherwise.

We verify that $M^1 \circ M^2 = M'$. Clearly, $Q' = Q^1 \cap Q^2$. Consider $q \in Q'$ and $x \in X$. If $x \notin X_j$ then $\delta^1(q,x) = \delta(q,x)$ and $\delta^2(q,x) \in Q'$; hence $\delta^{1\circ 2}(q,x) = \delta^2(q,x)$ by Lemma 2 and $\delta^2(q,x) = \delta'(q,x)$. Now consider $x \in X_j$. If $x = w^j_a$ then $\delta'(q,x) = q$ as $M' \in \mathcal{A}$ and $\delta^{1\circ 2}(q,x) = q = \delta'(q,x)$. Now let $x = w^j_{\overline{a}}$. Then, by the definition of $\circ$, $\delta^{1\circ 2}(q,x) = \delta'(q,x)$.

By construction, $M^2 \in \mathcal{A} \cap \mathcal{I} \cap \mathcal{P}$ and $\sigma(M^2) < \sigma(M')$. Hence, by induction, M^2 can be written in the form $M^2 = M'' \circ M'''$ where M'' is a $\circ$-composition of stuck-at faults and $M''' \in \mathcal{A} \cap \mathcal{I} \cap \mathcal{P}$ with $\sigma(M''') = 0$. Then $(M^1 \circ M'') \circ M'''$ is the required representation of M' with $S = M^1 \circ M''$.

Now we consider the case of $\sigma(M') = 0$. Then $Q' = Q = \{0,1\}^n$. Suppose that $\tau(M') > 0$. Then there is a cell i such w^i_a fails, that is, there is a state q with $q_i = \overline{a}$ such that $\delta'_i(q, w^i_a) = \overline{a}$. Let $M^1 = M^{i\neg a}$ and let $M^2 = (Q, \delta^2)$ be defined as

$$\delta^2(q,x) = \begin{cases} \delta'(q,x), & \text{if } x \neq w^i_a, \\ (\delta'_1(q,x), \ldots, \delta'_{i-1}(q,x), a, \delta'_{i+1}(q,x), \ldots, \delta'_n(q,x)), & \text{if } x = w^i_a, \end{cases}$$

for $q \in Q$ and $x \in X$. Using the fact that $M' \in \mathcal{A} \cap \mathcal{I}$ one verifies that $M' = M^1 \circ M^2$. By construction, $M^2 \in \mathcal{A} \cap \mathcal{I} \cap \mathcal{P}$, $\sigma(M^2) = 0$, and $\tau(M^2) < \tau(M')$ as $M' \in \mathcal{I}$. Hence, by induction, M^2 can be written in the form $M^2 = M'' \circ M'''$ where M'' is a $\circ$-composition of transition faults and $M''' \in \mathcal{A} \cap \mathcal{I} \cap \mathcal{P}$ with $\sigma(M''') = 0 = \tau(M''')$. Then $(M^1 \circ M'') \circ M'''$ is the required representation of M' with $T = M^1 \circ M''$.

Finally, assume that $\sigma(M') = \tau(M') = 0$ and $\gamma(M') > 0$. Then there are distinct cells i and j and values $a, b \in \{0,1\}$ such that for some state $q \in Q$ with $q_i = \overline{a}$ and $q_j = \overline{b}$ one has $\delta'_j(q, w^i_a) = b$. Let $M^1 = M^{ia \Rightarrow jb}$ and let $M^2 = (Q, \delta^2)$ be given by

$$\delta^2_k(q,x) = \begin{cases} b, & \text{if } x = w^i_a,\ q_i = \overline{a},\ q_j = \overline{b}, \text{ and } k = j, \\ \delta'_k(q,x), & \text{otherwise}, \end{cases}$$

for $q \in Q$, $k = 1, \ldots, n$, and $x \in X$. We prove that $M' = M^1 \circ M^2$.

Clearly, $Q' = Q^{1\circ 2} = Q$. Consider $q \in Q$ and $x \in X$. If $x \neq w^i_a$ or $q_i \neq \overline{a}$ or $q_j \neq \overline{b}$ then $\delta^1(q,x) = \delta(q,x)$; hence $\delta^{1\circ 2}(q,x) = \delta^2(q,x) = \delta'(q,x)$.

Now suppose that $x = w^i_a$, $q_i = \overline{a}$ and $q_j = \overline{b}$. Then

$$\delta^{1\circ 2}_k(q,x) = \begin{cases} \delta'_k(q,x), & \text{if } k \neq j, \\ b, & \text{if } k = j, \end{cases}$$

for $k = 1, \ldots, n$. As $M' \in \mathcal{I}$, this is equal to $\delta'_k(q,x)$.

By construction, $M^2 \in \mathcal{A} \cap \mathcal{I} \cap \mathcal{P}$, $\sigma(M^2) = 0 = \tau(M')$, and $\gamma(M^2) < \gamma(M')$. Hence, by induction, M^2 can be written in the form of a $\circ$-composition of coupling faults and, therefore, also M' can be represented in this way. This completes the proof of statement (a).

The proof of the inclusion $\mathcal{T}^\diamond \supseteq \mathcal{A} \cap \mathcal{I} \cap \mathcal{P} \cap \mathcal{C}$ is analogous to the one for $\circ$ except that the faults to be considered have to be also in $\mathcal{C}$ at every induction step.

Proposition 2. *There is a homomorphism of $\mathcal{T}^{\circ}$ onto $\mathcal{T}^{\diamond}$ such that its restriction to $\mathcal{T}^{\diamond}$ is the identity mapping. For $M^1, M^2 \in \mathcal{T}^{\diamond}$, if $M^1 \circ M^2 \in \mathcal{C}$, then $M^1 \circ M^2 = M^1 \diamond M^2$.*

Proof. The two semilattices have equations ρ_1–ρ_3 in common.

Acknowledgments

This work was supported by the Natural Sciences and Engineering Research Council of Canada, under Grants OGP0000871 and OGP0000243. The authors thank Jacques Sakarovitch for his suggestions regarding the construction of representatives in Section 2, and Gerard Lallement for pointing out reference [4].

References

1. J. A. Brzozowski and B. F. Cockburn: Detection of Coupling Faults in RAMs. *J. of Electronic Testing: Theory and Applications* **1** (1990), 151–162.
2. J. A. Brzozowski, H. Jürgensen: Component Automata and RAM Faults. *Proc. 2nd International Colloquium on Words, Languages, and Combinatorics, Kyoto, Japan, 1992*, edited by M. Ito and H. Jürgensen. World Scientific, Singapore, 1994, 49–67.
3. J. A. Brzozowski, H. Jürgensen: An Algebra of Multiple Faults in RAMs. *J. of Electronic Testing: Theory and Applications* **8** (1996), 129–142.
4. T. Evans: Some Connections between Residual Finiteness, Finite Embeddability and the Word Problem. *J. London Math. Soc. (2)* **1** (1969), 399–403.
5. F. Gécseg: *Products of Automata*, EACTS Monographs on Theoretical Computer Science, Vol. 7, Springer-Verlag, Berlin, 1986.
6. A. J. van de Goor: *Testing Semiconductor Memories, Theory and Practice*, Wiley, Chichester, England, 1991.
7. S. M. Thatte, J. A. Abraham: Testing of Semiconductor Random-Access Memories. *Digest of Papers, 7th Int. Conf. on Fault-Tolerant Computing*, Los Angeles, June 28–30, 1977, 81–87.

Thompson Languages

Dora Giammarresi, Jean-Luc Ponty, and Derick Wood

Summary. A finite-state machine is called a Thompson machine if it can be constructed from a regular expression using Thompson's construction. We assign distinct bracket pairs to the source and target states of all machine units that make up a Thompson machine. Each path from the start state to the final state of such a Thompson machine spells out a Dyck string. The collection of all such Dyck strings is the Thompson language of the given machine and the collection of such strings that are spelled out by simple paths is the simple Thompson language of the given machine. We characterize simple Thompson languages and Thompson languages, and we investigate their relationship.

1 Introduction

In 1968, Thompson [6] introduced an elegant and simple inductive construction of a finite-state machine from a regular expression. The resulting finite-state machine has size linear in the size of the original expression and each state has no more than two ingoing transitions and two outgoing transitions. A resurge of interest in the implementation of machines has resulted in some new discoveries about the Thompson construction [2,4].

Recently [5] we characterized the underlying digraphs of the machines resulting from the Thompson construction, **Thompson digraphs** and **Thompson machines,** respectively. The characterization is different from the one of Caron and Ziadi [3] for Glushkov digraphs although it is necessarily related to theirs indirectly.

The development of the characterization of Thompson digraphs [5] led us to the idea of associating labeled left and right brackets with the source and target states, respectively, of the unit machines that are used in the inductive Thompson construction. (We use the designation source state and target state when the corresponding states are not a start state and not a final state which occurs when we discuss a unit that is part of a larger machine.) We can view the bracketing as a means of linearizing the induction tree given by the Thompson construction. Once we have labeled the states in this way each path from the start state to the final state spells out a Dyck string. The set of all Dyck strings spelled out by such paths is a subset of the Dyck language for the appropriate alphabet and we call it a **Thompson language.**

Although Thompson languages appear to be nonregular, they are regular languages as they are specified by the duals of finite-state machines. We establish a Kleene-like theorem for them that is analogous to the standard Kleene theorem for regular languages; see Section 4.

In Section 5, we characterize **simple Thompson languages.** The set of all strings spelled out by simple paths from the start state to the final state of a Thompson machine is a simple Thompson language. We then characterize Thompson languages in terms of simple Thompson languages—our main result.

2 Notation and terminology

We recall the basics of finite-state machines and regular expressions and introduce the notation that we use.

A **finite-state machine**[1] consists of a finite set Q of states, an input alphabet Σ, a start state $s \in Q$, a final state $f \in Q$ and a transition relation $\delta \subseteq Q \times \Sigma_\lambda \times Q$, where λ denotes the null string and $\Sigma_\lambda = \Sigma \cup \{\lambda\}$. Clearly, we can depict the transition relation of such a machine as an edge-labeled digraph (the labels are symbols from Σ_λ); it is usually called the state or transition digraph of the machine.

Let Σ be an alphabet. Then, we define a regular expression E over Σ inductively as follows:

$E = \emptyset$, where $\emptyset$ is the empty-set symbol;
$E = \lambda$, where λ is the null-string symbol;
$E = a$, where a is in Σ;
$E = (F + G)$, where F and G are regular expressions;
$E = (F \cdot G)$, where F and G are regular expressions;
$E = (F^*)$, where F is a regular expression.

We define the language of E inductively, in the usual way [1,7].

We say that a regular expression E is **empty free** if E does not contain any appearance of the empty-set symbol.

3 Thompson machines and languages

Thompson designed a novel method [6] to compile regular expressions into a form that is suitable for pattern matching in text files. The construction is defined inductively (see Fig. 1). We treat only empty-free regular expressions; therefore, we never have to consider the finite-state machine given by Fig. 1(a). The Thompson construction gives a finite-state machine that has a number of pleasing properties:

[1] Normally, a finite-state machine is allowed to have more than one final state and, sometimes, more than one start state. The formulation we have chosen is appropriate for the study of Thompson machines.

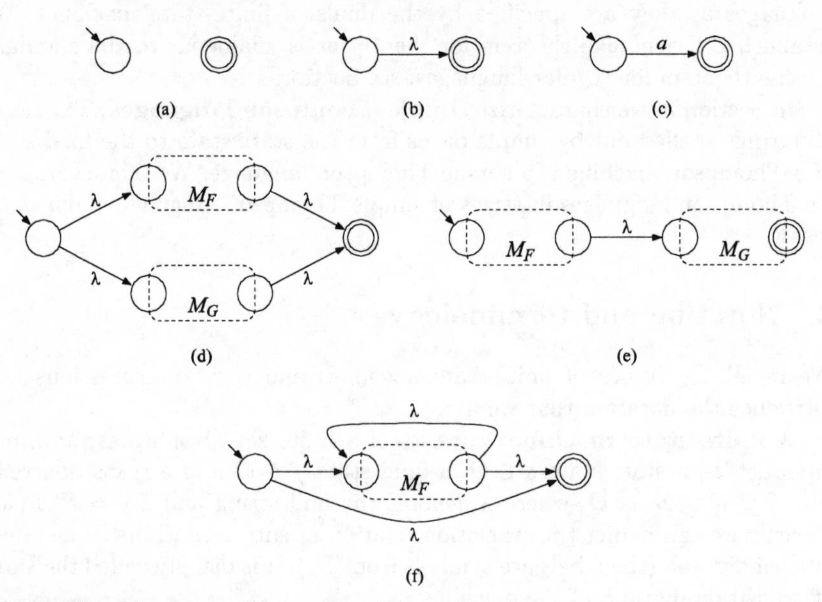

Fig. 1. The Thompson construction. The order of the figures corresponds to the order of the cases in the definition of regular expressions. The finite-state machines correspond to the regular expressions: a. $E = \emptyset$; b. $E = \lambda$; c. $E = a$, $a \in \Sigma$; d. $E = (F + G)$; e. $E = (F \cdot G)$; and f. $E = (F^*)$.

1. It has an even number of states.
2. It has one start state that is **only exiting;** that is, there are no transitions that enter it.
3. It has one final state that is **only entering;** that is, there are no transitions that leave it.
4. Every state is on a simple path from the start state to the final state.
5. Every state has at most two ingoing transitions and at most two outgoing transitions.
6. Its size is at most three times the size of the given regular expression.

In Fig. 2, we give the result of the Thompson construction on the regular expression $(((a + b)^*) \cdot ((b + \lambda) \cdot a))$.

We define a **Thompson machine** to be a finite-state machine that is obtained by the Thompson construction from an empty-free expression. We name the units that the Thompson construction uses to assemble a Thompson machine as follows: the **base units** (Figs. 1(b) and (c)); the **plus unit** (Fig. 1(d)); the **dot unit** (Fig. 1(e)); and the **star unit** (Fig. 1(f)). In Fig. 1, the source and target states for each unit correspond to their start and final states, respectively. Observe that the uniqueness of the start and final states implies that each Thompson machine has an even number of states.

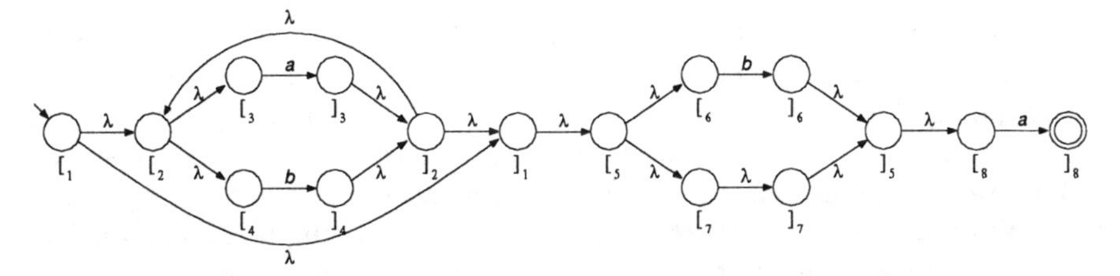

Fig. 2. The result of the Thompson construction on the running example expression $(((a+b)^*)\cdot((b+\lambda)\cdot a))$.

We now define Thompson strings and languages. Define a countably infinite alphabet Γ of bracket pairs that we enumerate as $[_1,]_1, \ldots, [_i,]_i, \ldots$. For each $n > 0$, we define $\Gamma_n = \{[_i,]_i : 1 \le i \le n\}$. Given a language L over Γ_n, the image of L under an injection (or renaming) $h : \Gamma_n \longrightarrow \Gamma$, namely $h(L)$, clearly has the same structure as L. (Indeed, this observation holds for any injection or bijection of L into an appropriately sized alphabet.) Thus, when we write L, we really mean the family $\mathcal{L}$ of languages $\{h(L) : h \text{ is an injection } \Gamma_n \longrightarrow \Gamma\}$; all languages in this family are structurally equivalent.

Given a Thompson machine M that has $2n$ states, we label the states arbitrarily but uniquely with brackets from Γ_n in the following way: For each unit submachine of M, we label its source state with left bracket $[_i$ and its target state with right bracket $]_i$, for some i, $1 \le i \le n$.

Having labeled the states of M with the $2n$ brackets in Γ_n, we define a **Thompson string** as any string that is spelled out by some path in M from its start state to its final state. For example, if we label the source and target states of the machine of Fig. 2 with $[_1, \ldots,]_8$ as indicated, the strings

$$[_1\, [_2\, [_3\,]_3\,]_2\, [_2\, [_4\,]_4\,]_2\,]_1\, [_5\, [_6\,]_6\,]_5\, [_8\,]_8$$

and

$$[_1\, [_2\, [_4\,]_4\,]_2\,]_1\, [_5\, [_6\,]_6\,]_5\, [_8\,]_8$$

are Thompson strings of the given machine. Clearly, Thompson strings are Dyck strings over Γ_n; however, not all Dyck strings over Γ_n are Thompson strings. Two simple examples are the strings $x = [_1\,]_1\, [_1\,]_1$ and $y = [_1\, [_1\,]_1\,]_1$; we leave the proofs of their non-Thompson-ness to you. Similarly, we define a **simple Thompson string** as any string that is spelled out by some simple path in M from its start state to its final state (recall that a simple path visits each state at most once). Thus, each simple Thompson string contains at most one pair of each of the n bracket pairs. Of the two preceding examples of Thompson strings, the first string is not a simple Thompson string, but the second string is.

Given a Thompson machine M, its **Thompson language** is the set of all Thompson strings given by M. Similarly, M's **simple Thompson language** is the set of all simple Thompson strings given by M. By definition, every simple Thompson language is finite and some Thompson languages are infinite. Clearly, each Thompson language $L \subseteq {\Gamma_n}^*$ is a subset of the corresponding Dyck language $D(\Gamma_n)$. That each Thompson language does not contain all Dyck strings follows from the previous examples of Dyck strings that are not Thompson. But, Thompson languages are even more restrictive.

We investigate the structure of Thompson and simple Thompson languages and characterize them in terms of their structural properties.

4 A Kleene Theorem

Let $\mathcal{T}$ and $\mathcal{ST}$ denote the families of Thompson and simple Thompson languages, respectively. We introduce variants of the usual regular operations that lead to a Kleene theorem for the two families. Given two Thompson languages $K \subseteq D(\Gamma_m)$ and $L \subseteq D(\Gamma_n)$, we define bracketed versions of the regular operations as follows:

Bracketed union $\oplus$:

$$K \oplus L = \{[_{m+n+1}\}(K \cup h(L))\{]_{m+n+1}\},$$

where $h : \Gamma_n \longrightarrow \{[_i,]_i : m+1 \leq i \leq m+n\}$ is a renaming.

Bracketed product $\odot$:

$$K \odot L = Kh(L),$$

where $h : \Gamma_n \longrightarrow \{[_i,]_i : m+1 \leq i \leq m+n\}$ is a renaming.

Bracketed star $\otimes$:

$$K^{\otimes} = \{[_{m+1}\}K^*\{]_{m+1}\}.$$

Similarly, given two simple Thompson languages $K \subseteq D(\Gamma_m)$ and $L \subseteq D(\Gamma_n)$, we use the bracketed versions of union and product but we replace bracketed star with bracketed prestar defined as follows:

Bracketed prestar $\oslash$:

$$K^{\oslash} = \{[_{m+1}\}(K \cup \{\lambda\})\{]_{m+1}\}.$$

The proof of the following theorem is straightforward.

Theorem 1. *The family $\mathcal{T}$ of Thompson languages is closed under bracketed union, bracketed product and bracketed star.*

The family $\mathcal{ST}$ of simple Thompson languages is closed under bracketed union, bracketed product and bracketed prestar.

As a result of the preceding theorem we can provide an inductive definition of Thompson languages that parallels the inductive definition of Thompson machines. Given an alphabet Γ_n, we can inductively define a Thompson language as follows:

1. The language $\{[_1\,]_1\} \subseteq D(\Gamma_i)$ is a Thompson language.
2. If $K \subseteq D(\Gamma_m)$ and $L \subseteq D(\Gamma_n)$ are Thompson languages, then $K \oplus L$ is a Thompson language.
3. If $K \subseteq D(\Gamma_m)$ and $L \subseteq D(\Gamma_n)$ are Thompson languages, then $K \odot L$ is a Thompson language.
4. If $K \subseteq D(\Gamma_m)$ is a Thompson language, then $K^{\otimes}$ is a Thompson language.

If we replace $\otimes$ with $\oslash$, then we obtain an inductive definition of simple Thompson languages.

The basis for these definitions is the following Kleene theorem.

Theorem 2. *The family $\mathcal{T}$ of Thompson languages is the smallest family of languages that contains $\{[_1\,]_1\}$ and is closed under bracketed union, bracketed product and bracketed star.*

The family $\mathcal{ST}$ of simple Thompson languages is the smallest family of languages that contains $\{[_1\,]_1\}$ and is closed under bracketed union, bracketed product and bracketed prestar.

5 The characterization theorem

We develop a characterization of Thompson languages by first considering simple Thompson languages. We present some necessary conditions for a language $L \subseteq D(\Gamma_n)$ to be a simple Thompson language or a Thompson language.

Property 1: Bracket uniqueness. Each pair $[_i,]_i$ appears at most once in each string in a simple Thompson language L.

The following four properties apply to both simple Thompson and Thompson languages.

Property 2: Starting and ending uniqueness. All strings in L begin with the same left bracket $[_i$, say, and end with the same right bracket $]_j$, say.

Given a Thompson machine M of $2n$ states that are labeled with the brackets in Γ_n, let $L \subseteq D(\Gamma_n)$ be its simple Thompson or its Thompson language. Given a Thompson language $L \subseteq D(\Gamma_n)$, we define $L(i)$ to be the set of all strings that appear between the brackets $[_i,]_i$ in strings of L. (Note that the brackets may appear more than once in Thompson strings, but each appearance can contain only the same substrings.) More formally,

$$\begin{aligned} L(i) = \{x : & \text{ there are strings } w \text{ and } y \text{ such that} \\ & w[_i x]_i y \in L \text{ and } x \text{ does not contain } [_i \text{ or }]_i\}. \end{aligned}$$

Similarly, we define $L[i]$ to be the set of all substrings of strings in L that begin with $[_i$ and end with $]_i$; that is,

$$\begin{aligned} L[i] = \{[_i x]_i : & \text{ there are strings } w \text{ and } y \text{ such that} \\ & w[_i x]_i y \in L \text{ and } x \text{ does not contain } [_i \text{ or }]_i\}. \end{aligned}$$

Property 3: Unit conditions.

1. If a pair $[_i,]_i$ labels a base unit, then $L(i) = \{\lambda\}$.
2. If a pair $[_i,]_i$ labels a star unit, then $L(i) \supset \{\lambda\}$.
3. If a pair $[_i,]_i$ labels a plus unit, then $L(i) \neq \emptyset$ and $\{\lambda\} \notin L(i)$.

We consider the restrictions on substrings that occur with plus and star units:

Property 4: Multiplicity conditions.

1. If a pair $[_i,]_i$ labels a star unit, then all strings in $L(i) - \{\lambda\}$ begin with some bracket $[_j$, $j \neq i$, and end with some bracket $]_k$, $k \neq i$.
2. If a pair $[_i,]_i$ labels a plus unit, then all strings in $L(i)$ begin either with $[_j$ or with $[_m$, where $i \neq j$, $i \neq m$ and $j \neq m$, and end either with $]_k$ or with $]_n$, respectively, where $i \neq k$, $i \neq n$ and $k \neq n$.

The next property captures the sequential nature of the product of two machines:

Property 5: Product sequentiality. Let the pairs $[_i,]_i$ and $[_j,]_j$ label two units in M that have been combined by product, then, for all strings u, v, w and x, neither $u[_iv]_i[_nw]_nx$ nor $u[_mv]_m[_jw]_jx$ are in L, where $i \neq m$ and $j \neq n$.

The following result follows immediately from the definitions of the properties.

Lemma 1. *Let L be a simple Thompson language. Then, L satisfies Properties 1, 2, 3, 4 and 5.*

Let L be a Thompson language. Then, L satisfies Properties 2, 3, 4 and 5.

We are now in a position to characterize simple Thompson languages.

Theorem 3. *A language $L \subseteq D(\Gamma_n)$ is a simple Thompson language if and only if L satisfies Properties 1, 2, 3, 4 and 5.*

Proof. We already observed, by Lemma 1 that, if L is a simple Thompson language, then it satisfies Properties 1, 2, 3, 4 and 5. We now prove the converse. Let $L \subseteq D(\Gamma_n)$ be a language that satisfies Properties 1, 2, 3, 4 and 5. To demonstrate that L is a Thompson language, we sketch the construction of a Thompson machine M such that the Thompson language of M is isomorphic to L. The proof is by induction on the size n of the alphabet Γ_n of L.

Basis: In this case $n = 1$, and $L = \{[_1\,]_1\}$ by Properties 1 and 2.

Induction hypothesis: We assume the result holds for all languages $L \subseteq D(\Gamma_m)$ that satisfy Properties 1, 2, 3, 4 and 5, where $1 \leq m \leq n$, for some $n \geq 1$.

Induction step: Consider a language $L \subseteq D(\Gamma_{n+1})$ that satisfies Properties 1, 2, 3, 4 and 5. Since $n \geq 1$, the strings in L must contain at least two different brackets. Consider a string x in L. Then, either x begins with a left bracket $[_i$ and ends with the corresponding right bracket $]_i$ or it does not. In the first case, all strings in L must begin and end the same way by Property 2; therefore, by Property 3, the strings correspond either to a bracketed union or to a bracketed prestar (they cannot correspond to a base unit because that would imply that $n + 1 = 1$, which contradicts the assumption that $n + 1 \geq 2$). By Property 3, we can uniquely identify which alternative the strings correspond to; therefore, we can replace L either with two languages each with fewer brackets than L or with one language that has fewer brackets than L. A corresponding Thompson machine (with state labels but without edge labels) can be constructed inductively in this case.

In the second case, x starts with a left bracket $[_i$ but $]_i$ does not end x; therefore, we can partition x into uv, where u begins with $[_i$ and ends with $]_i$, by the structure of these strings. Since $v \neq \lambda$, we have discovered a product decomposition; namely, by Property 5, there is a left bracket $[_i$ such that, for all strings x in L, x can be decomposed into strings u and v such that $x = uv$, u begins with left bracket $[_i$ and ends with $]_i$, and $v \neq \lambda$. This decomposition of strings induces a decomposition of L into two languages J and K such that $J = L[i]$ and $K = \{v : uv \in L \text{ and } u \in L[i]\}$. Both J and K have fewer brackets than L and both satisfy Properties 1, 2, 3, 4 and 5; therefore, we can inductively construct a Thompson machine that corresponds to L.

The proof of the preceding theorem provides an inductive construction of a Thompson machine M (with state labels but without edge labels) from a simple Thompson language L such that the simple Thompson language of M is L. Based on this construction, we now characterize Thompson languages.

First, observe that, for each Thompson language L, the simple Thompson subset S of L consists of all the strings in L that satisfy Property 1. Second, observe that if the bracket pair $[_i,]_i$ corresponds to a bracketed prestar, then, conceptually, we can replace every string from $S[i]$ in strings of S with $[_i S(i)^*]_i$. We can then repeat this expansion for each bracketed prestar in turn. Unfortunately, after the first expansion, we obtain an infinite language. Therefore, we construct the Thompson machine M_S for the language S, where each bracketed prestar gives a star unit. The Thompson language of M_S is the language L if and only if L is a Thompson language. This proof sketch provides our last result, the sought characterization of Thompson languages.

Theorem 4. *Given a language $L \subseteq D(\Gamma_n)$ and its simple Thompson subset S, L is a Thompson language if and only if L satisfies Properties 2, 3, 4 and 5, and the Thompson language of M_S is L.*

Observe that the characterization is effective if we have an effective specification of L.

References

1. A.V. Aho and J.D. Ullman. *The Theory of Parsing, Translation, and Compiling, Vol. I: Parsing.* Prentice-Hall, Inc., Englewood Cliffs, NJ, 1972.
2. A. Brüggemann-Klein and D. Wood. The validation of SGML content models. *Mathematical and Computer Modelling*, 25:73–84, 1997.
3. P. Caron and D. Ziadi. Characterization of Glushkov automata. *Theoretical Computer Science*, 1998. To appear.
4. D. Giammarresi, J.-L Ponty, and D. Wood. The Glushkov and Thompson constructions: A synthesis. Submitted for publication, October 1998.
5. D. Giammarresi, J.-L Ponty, and D. Wood. A characterization of Thompson digraphs. Submitted for publication, January 1999.
6. K. Thompson. Regular expression search algorithm. *Communications of the ACM*, 11:419–422, 1968.
7. D. Wood. *Theory of Computation.* John Wiley & Sons, Inc., New York, NY, second edition, 1998. In preparation.

On Some Special Classes of Regular Languages*

Balázs Imreh and Masami Ito

Summary. In [5], the directable nondeterministic automata are studied. For nondeterministic automata, the directability can be defined in several ways. In each case considered, the directing words constitute a regular language and one can study the families of such regular languages. In this paper, six classes of regular languages are defined in accordance with the different definitions of directability given in [5], and we investigate the properties of the classes considered.

1 Introduction

An input word of an automaton is called *directing* or *synchronizing word* if it brings the automaton from every state into the same state. An automaton is *directable* if it has a directing word. The directable automata and directing words have been studied from different points of view (see [2], [3], [4], [7], [9], [10], for example). For nondeterministic (n.d.) automata, the directability can be defined in several ways. We study here the following three notions of directability which are introduced in [5]. Namely, an input word w of an n.d. automaton $\mathcal{A}$ is

(1) D1-directing if the set of states aw in which $\mathcal{A}$ may be after reading w consists of the same single state c whatever the initial state a is;

(2) D2-directing if the set aw is independent of the initial state a;

(3) D3-directing if there exists a state c which appears in all sets aw.

It is worth noting that the D1-directability of complete n.d. automata was already studied by Burkhard [1], where he gave an exact exponential bound for the length of minimum-length D1-directing words of complete n.d. automata. Here, we study the classes of languages consisting of D1-, D2- and D-3 directing words of n.d. automata and complete n.d. automata. The next section contains a general preliminaries and the formal definitions of these languages, and Section 3 presents some properties of the languages considered.

* This work has been supported by the Hungarian National Foundation for Scientific Research, Grant T 014888, the Ministry of Culture and Education of Hungary, Grant FKFP 0704/1997, the Japanese Ministry of Education, Mombusho International Scientific Research Program, Joint Research 10044098

2 Preliminaries

Let X be a finite nonempty alphabet. The set of all (finite) words over X is denoted by X^* and the empty word by ε. The length of a word w is denoted by $|w|$.

By an *automaton* we mean a system $\mathcal{A} = (A, X, \delta)$, where A is a finite nonempty set of *states*, X is the *input alphabet*, and $\delta : A \times X \to A$ is the *transition function*. The transition function can be extended to $A \times X^*$ in the usual way. By a *recognizer* we mean a system $\mathbf{A} = (A, X, \delta, a_0, F)$, where (A, X, δ) is an automaton, $a_0 (\in A)$ is the *initial state*, and $F (\subseteq A)$ is the set of *final states*. The *language recognized* by $\mathbf{A}$ is the set

$$L(\mathbf{A}) = \{w \in X^* : \delta(a_0, w) \in F\}.$$

A language is called *recognizable*, or *regular*, if it is recognized by some recognizer. The set of all recognizable languages over the alphabet X is denoted by $\mathrm{Rec}(X)$.

An automaton $\mathcal{A} = (A, X, \delta)$ can also be defined as an algebra $\mathcal{A} = (A, X)$ in which each input letter x is realized as the unary operation $x^{\mathcal{A}} : A \to A$, $a \mapsto \delta(a, x)$. Now, nondeterministic automata can be introduced as generalized automata in which the unary operations are replaced by binary relations. Thus, by a *nondeterministic (n.d.) automaton* we mean a system $\mathcal{A} = (A, X)$ where A is a finite nonempty set of *states*, X is the *input alphabet*, and each letter $x (\in X)$ is realized as a binary relation $x^{\mathcal{A}} (\subseteq A \times A)$ on A. For any $a \in A$ and $x \in X$, $ax^{\mathcal{A}} = \{b \in A : (a, b) \in x^{\mathcal{A}}\}$ is the set of states into which $\mathcal{A}$ may enter from state a by reading the input letter x. For any $C \subseteq A$ and $x \in X$, we set $Cx^{\mathcal{A}} = \bigcup\{ax^{\mathcal{A}} : a \in C\}$. This transition can be extended for arbitrary $w \in X^*$ and $C \subseteq A$. If $w \in X^*$ and $a \in A$, let $aw^{\mathcal{A}} = \{a\}w^{\mathcal{A}}$. This means that if $w = x_1 x_2 \ldots x_k$, then $w^{\mathcal{A}} = x_1^{\mathcal{A}} x_2^{\mathcal{A}} \ldots x_k^{\mathcal{A}} (\subseteq A \times A)$. When $\mathcal{A}$ is known from the context, we usually write simply aw and Cw for $aw^{\mathcal{A}}$ and $Cw^{\mathcal{A}}$, respectively.

An n.d. automaton $\mathcal{A} = (A, X)$ is *complete* if $ax^{\mathcal{A}} \neq \emptyset$, for all $a \in A$ and $x \in X$. Complete n.d. automata are called c.n.d. automata for short. In what follows, we denote a deterministic automaton by $\mathcal{A} = (A, X, \delta)$ and a nondeterministic automaton by $\mathcal{A} = (A, X)$.

The idea of directability of deterministic automata can be extended to n.d. automata in several nonequivalent ways. In [5], three definitions are introduced as follows. Let $\mathcal{A} = (A, X)$ be an n.d. automaton. For any word $w \in X^*$ we consider the following three conditions:

(D1) $(\exists c \in A)(\forall a \in A)(aw = \{c\})$;

(D2) $(\forall a, b \in A)(aw = bw)$;

(D3) $(\exists c \in A)(\forall a \in A)(c \in aw)$.

If w satisfies (Di), then w is a Di-*directing word* of $\mathcal{A}$ $(i = 1, 2, 3)$. For each $i = 1, 2, 3$, the set of Di-directing words of $\mathcal{A}$ is denoted by $\mathrm{D}_i(\mathcal{A})$, and

$\mathcal{A}$ is called Di-*directable* if $\mathrm{D}_i(\mathcal{A}) \neq \emptyset$. It is proved (see [5]) that $\mathrm{D}_i(\mathcal{A})$ is recognizable, for every n.d. automaton $\mathcal{A}$ and i, $i = 1, 2, 3$. The classes of Di-directable n.d. automata and c.n.d. automata are denoted by $\mathbf{Dir}(i)$ and $\mathbf{CDir}(i)$, respectively.

Now, we can define the following classes of languages: For $i = 1, 2, 3$, let

$$\mathcal{L}_{\mathrm{ND(i)}} = \{\mathrm{D}_i(\mathcal{A}) : \mathcal{A} \in \mathbf{Dir}(i)\},$$

$$\mathcal{L}_{\mathrm{CND(i)}} = \{\mathrm{D}_i(\mathcal{A}) : \mathcal{A} \in \mathbf{CDir}(i)\}.$$

Since all of the languages occuring in the definitions above are recognizable, the defined classes are subclasses of $\mathrm{Rec}(X)$.

Finally, let us denote by $\mathbf{D}$ the class of directable deterministic automata, and for any $\mathcal{A} \in \mathbf{D}$, let $\mathrm{D}(\mathcal{A})$ be the set of the directing words of $\mathcal{A}$. Furthermore, let

$$\mathcal{L}_{\mathrm{D}} = \{\mathrm{D}(\mathcal{A}) : \mathcal{A} \in \mathbf{D}\}.$$

3 Some properties of the languages of directing words of n.d. automata

In [5], the following observation is presented, which is used below.

Lemma 1. *For any n.d. automaton $\mathcal{A} = (A, X)$, $\mathrm{D}_2(\mathcal{A})X^* = \mathrm{D}_2(\mathcal{A})$. If $\mathcal{A}$ is complete, then $X^*\mathrm{D}_1(\mathcal{A}) = \mathrm{D}_1(\mathcal{A})$, $X^*\mathrm{D}_2(\mathcal{A})X^* = \mathrm{D}_2(\mathcal{A})$, and $X^*\mathrm{D}_3(\mathcal{A})X^* = \mathrm{D}_3(\mathcal{A})$.*

The following assertion gives a characterization of $\mathcal{L}_{\mathrm{D}}$.

Proposition 1. *For a language $L \subseteq X^*$, $L \in \mathcal{L}_{\mathrm{D}}$ if and only if $L \neq \emptyset$, L is regular, and $X^*LX^* = L$.*

Proof. To prove the necessity, let us suppose that $L \in \mathcal{L}_{\mathrm{D}}$. It is known, (*cf.* [4]) that then $L \neq \emptyset$ and L is regular. On the other hand, it is easy to see that if w is a directing word of an automaton $\mathcal{A}$, then pwq is also a directing word of $\mathcal{A}$ for every $p, q \in X^*$. These observations yield the necessity.

In order to prove the sufficiency, let us suppose that $L \neq \emptyset$, L is regular, and $X^*LX^* = L$. Since L is regular, there exists a recognizer $\mathbf{A} = (A, X, \delta, a_0, F)$ such that $L = L(\mathbf{A})$. Without loss of generality, we may assume that the automaton $\mathcal{A} = (A, X, \delta)$ is generated by one element, namely by a_0. Since $L \neq \emptyset$, we have that $F \neq \emptyset$. On the other hand, by our assumption, $X^*LX^* = L$, which implies that $\delta(a, x) \in F$, for every $a \in F$ and $x \in X$, *i.e.*, $(F, X, \bar{\delta})$ is a subautomaton of $\mathcal{A}$ where $\bar{\delta}$ denotes the restriction of δ to $F \times X$. Let us denote by $\mathcal{A}'$ the Rees factor automaton defined by

$\mathcal{A}' = (A/F, X, \delta')$ where $A/F = (A \setminus F) \cup \{F\}$ and for every $a \in A \setminus F$ and $x \in X$,

$$\delta'(a,x) = \begin{cases} \delta(a,x) & \text{if } \delta(a,x) \in A \setminus F, \\ F & \text{if } \delta(a,x) \in F. \end{cases}$$

and

$$\delta'(F,x) = F.$$

Then, it is easy to see that $L = L(\mathbf{A}')$ where $\mathbf{A}' = (A/F, X, \delta', a_0, \{F\})$.

Now, we show that $\mathcal{A}' = (A/F, X, \delta')$ is a directable automaton and $\mathrm{D}(\mathcal{A}') = L$. Indeed, let $w \in L$ and $a \in A/F$ be arbitrary elements. Since $\mathcal{A}$ is generated by a_0, $\mathcal{A}'$ is also generated by a_0. This implies that there is a word $p \in X^*$ such that $\delta'(a_0, p) = a$. Since $X^*LX^* = L$, $pw \in L$, but then, $\delta'(a_0, pw) = F$, and thus $\delta'(a, w) = F$. Therefore, w is a directing word of $\mathcal{A}'$, and thus $L \subseteq \mathrm{D}(\mathcal{A}')$. Finally, we have to prove that if q is a directing word of $\mathcal{A}'$, then $q \in L$. If q is a directing word of $\mathcal{A}'$, then $\delta'(b,q) = \tilde{b}$ is valid for every $b \in A/F$ where $\tilde{b}$ is a fixed element of A/F. Then, $\delta'(F,q) = \tilde{b}$, and thus, $\tilde{b} = F$ since $\delta'(F,q) = F$. This yields that $q \in L$, and hence, $\mathrm{D}(\mathcal{A}') \subseteq L$. Consequently, $L = D(\mathcal{A}')$, and thus, $L \in \mathcal{L}_{\mathrm{D}}$.

Corollary 1. *$\mathcal{L}_{\mathrm{D}}$ is closed under union.*

Now, by Lemma 1 and Proposition 1, we can prove the following statement.

Proposition 2. $\mathcal{L}_{\mathrm{CND}(2)} = \mathcal{L}_{\mathrm{D}}$, $\mathcal{L}_{\mathrm{CND}(3)} = \mathcal{L}_{\mathrm{D}}$, $\mathcal{L}_{\mathrm{CND}(1)} \cap \mathcal{L}_{\mathrm{ND}(2)} = \mathcal{L}_{\mathrm{D}}$, *and* $\mathcal{L}_{\mathrm{CND}(1)} \cap \mathcal{L}_{\mathrm{ND}(3)} = \mathcal{L}_{\mathrm{D}}$.

Proof. Obviously, $\mathcal{L}_{\mathrm{D}} \subseteq \mathcal{L}_{\mathrm{CND}(2)}$. Now, let $L \in \mathcal{L}_{\mathrm{CND}(2)}$ be arbitrary. Then, $L \neq \emptyset$ and L is regular, furthermore, by Lemma 1, $X^*LX^* = L$, and hence, by Proposition 1, $L \in \mathcal{L}_{\mathrm{D}}$. This implies the validity of the first equality. The second equality can be proved in a similar way. Regarding the third equality, it is obvious again that $\mathcal{L}_{\mathrm{D}} \subseteq \mathcal{L}_{\mathrm{CND}(1)} \cap \mathcal{L}_{\mathrm{ND}(2)}$. Now, let $L \in \mathcal{L}_{\mathrm{CND}(1)} \cap \mathcal{L}_{\mathrm{ND}(2)}$ be arbitrary. By Lemma 1, $X^*L = L$ and $LX^* = L$, and hence $X^*LX^* = L$. On the other hand, $L \neq \emptyset$ and L is regular, and thus, by Proposition 1, $L \in \mathcal{L}_{\mathrm{D}}$, which implies the required equality. Finally, we show the validity of the last equality. Obviously, $\mathcal{L}_{\mathrm{D}} \subseteq \mathcal{L}_{\mathrm{CND}(1)} \cap \mathcal{L}_{\mathrm{ND}(3)}$. Now, let $L \in \mathcal{L}_{\mathrm{CND}(1)} \cap \mathcal{L}_{\mathrm{ND}(3)}$. Since $L \in \mathcal{L}_{\mathrm{CND}(1)}$, by Lemma 1, $X^*L = L$. Since $L \in \mathcal{L}_{\mathrm{ND}(3)}$, there is an n.d. automaton $\mathcal{A}$ such that $L = \mathrm{D}_3(\mathcal{A})$. Now, we show that $X^*L = L$ implies that $\mathcal{A}$ is a complete n.d. automaton. Contrary, let us suppose that $a'x^{\mathcal{A}} = \emptyset$ for some state a' and input sign x of $\mathcal{A}$. Furthermore, let w be an arbitrary D3-directing word of $\mathcal{A}$. Then, by $X^*L = L$, xw must be also a D3-directing word which is a contradiction since $a'xw^{\mathcal{A}} = \emptyset$. Consequently, $\mathcal{A}$ is a c.n.d. automaton, and thus, $L \in \mathcal{L}_{\mathrm{CND}(3)} = \mathcal{L}_{\mathrm{D}}$.

For the remaining 5 classes we have the proper inclusions presented in Figure 1.

To show that all of the inclusions above are proper, we give examples.

1. $\mathcal{L}_{\mathrm{D}} \subset \mathcal{L}_{\mathrm{CND}(1)}$. Let us consider the n.d. automaton $\mathcal{A}_1 = (\{1,2\},\{x,y\})$ where $x^{\mathcal{A}_1} = \{(1,1),(1,2),(2,1),(2,2)\}$ and $y^{\mathcal{A}_1} = \{(1,2),(2,2)\}$. Then, $\mathcal{A}_1$ is a D1-directable complete n.d. automaton, and y is a D1-directing word of $\mathcal{A}_1$. Now, let us suppose that $\mathrm{D}_1(\mathcal{A}_1) \in \mathcal{L}_{\mathrm{D}}$. Then, $yp \in \mathrm{D}_1(\mathcal{A}_1)$ must hold, for every $p \in X^*$. On the other hand, obviously $yx \notin \mathrm{D}_1(\mathcal{A}_1)$ which is a contradiction. Therefore, $\mathrm{D}_1(\mathcal{A}_1) \notin \mathcal{L}_{\mathrm{D}}$.

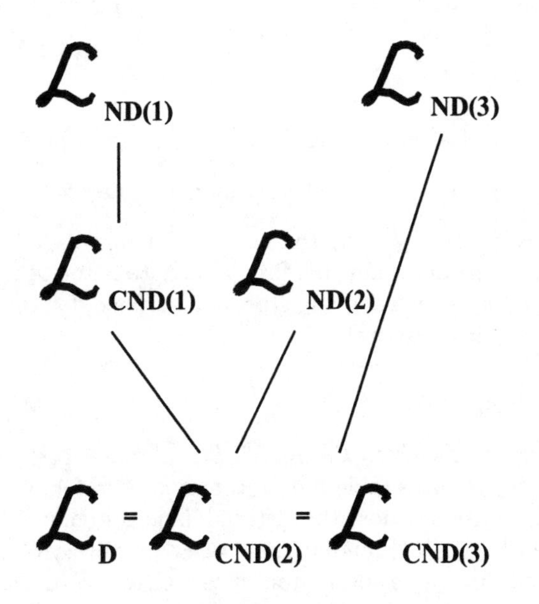

Figure 1.

2. $\mathcal{L}_{\mathrm{CND}(1)} \subset \mathcal{L}_{\mathrm{ND}(1)}$. Let the n.d. automaton $\mathcal{A}_2 = (\{1,2\},\{x,y\})$ be defined by $x^{\mathcal{A}_2} = \{(1,2),(2,2)\}$ and $y^{\mathcal{A}_2} = \{(2,1)\}$. Then, x is a D1-directing word of $\mathcal{A}_2$. Assume that $\mathrm{D}_1(\mathcal{A}_2) \in \mathcal{L}_{\mathrm{CND}(1)}$. In this case, Lemma 1 implies that $yx \in \mathrm{D}_1(\mathcal{A}_2)$ which is a contradiction. Thus, $\mathrm{D}_1(\mathcal{A}_2) \notin \mathcal{L}_{\mathrm{CND}(1)}$.

3. $\mathcal{L}_{\mathrm{D}} \subset \mathcal{L}_{\mathrm{ND}(2)}$. Let us define the n.d. automaton $\mathcal{A}_3 = (\{1,2\},\{x,y\})$ by $x^{\mathcal{A}_3} = \{(1,1),(1,2),(2,1),(2,2)\}$ and $y^{\mathcal{A}_3} = \{(1,1)\}$. Now, x is a D2-directing word. Suppose that $\mathrm{D}_2(\mathcal{A}_3) \in \mathcal{L}_{\mathrm{D}}$. Then, $px \in \mathrm{D}_2(\mathcal{A}_3)$ must hold for every $p \in X^*$ which is a contradiction since $yx \notin \mathrm{D}_2(\mathcal{A}_3)$. Consequently, $\mathrm{D}_2(\mathcal{A}_3) \notin \mathcal{L}_{\mathrm{D}}$.

4. $\mathcal{L}_{\mathrm{D}} \subset \mathcal{L}_{\mathrm{ND}(3)}$. Let us consider the n.d. automaton $\mathcal{A}_4 = (\{1,2\},\{x,y\})$ where $x^{\mathcal{A}_4} = \{(1,2),(2,1),(2,2)\}$ and $y^{\mathcal{A}_4} = \{(1,1)\}$. Obviously, x is a D3-directing word of $\mathcal{A}_4$, while yx is not a D3-directing word of $\mathcal{A}_4$. Using these

two observations and Lemma 1, we obtain that the inclusion considered is proper.

In the following statement, we give a description of $\mathcal{L}_{\mathrm{CND}(1)}$ based on the deterministic directable automata. Namely, if we consider some sublanguages of the languages belonging to $\mathcal{L}_{\mathrm{D}}$, then we can get all of the languages from $\mathcal{L}_{\mathrm{CND}(1)}$. To present this description, we need some preparation. Let $\mathcal{A} = (A, X, \delta)$ be a deterministic directable automaton. Then, a state a' of A is called a *receiving state* of $\mathcal{A}$ if $\delta(A, w) = \{a'\}$ for some directing word w of $\mathcal{A}$ where $\delta(A, w) = \{aw : a \in A\}$. Let us denote by $R_{\mathcal{A}}$ the set of all receiving states of $\mathcal{A}$. Obviously, $R_{\mathcal{A}} \neq \emptyset$. Then, the following fact is obvious.

Fact. Let $\emptyset \neq Q \subseteq R_{\mathcal{A}}$. Then,

$$\mathrm{D}_Q(\mathcal{A}) = \{w : w \in X^* \ \&\ \delta(A, w) = \{a'\} \text{ for some } a' \in Q\}$$

is regular, $\mathrm{D}_Q(\mathcal{A}) \subseteq \mathrm{D}(\mathcal{A})$, and in particular, $\mathrm{D}_{R_{\mathcal{A}}}(\mathcal{A}) = \mathrm{D}(\mathcal{A})$.

Now, let us denote by $\mathcal{L}_{\mathrm{GD}}$ the class of languages $L \subseteq X^*$ such that $L = \mathrm{D}_Q(\mathcal{A})$ for some deterministic directable automaton $\mathcal{A}$ and a nonempty subset Q of the set of the receiving states of $\mathcal{A}$. Then, for the class $\mathcal{L}_{\mathrm{CND}(1)}$, we have the following description.

Proposition 3. $\mathcal{L}_{\mathrm{CND}(1)} = \mathcal{L}_{\mathrm{GD}}$.

Proof. To prove the inclusion $\mathcal{L}_{\mathrm{CND}(1)} \subseteq \mathcal{L}_{\mathrm{GD}}$, let $L \in \mathcal{L}_{\mathrm{CND}(1)}$. Then, there exists a complete D1-directable n.d. automaton $\mathcal{A} = (A, X)$ such that $L = \mathrm{D}_1(\mathcal{A})$. Let us construct now the deterministic automaton $\mathcal{B} = (B, X, \delta)$ where $B = \{Ap^{\mathcal{A}} : p \in X^*\}$, and $\delta(C, x) = Cx^{\mathcal{A}}$, for every $C \in B$ and $x \in X$. Obviously, $\mathcal{B}$ is a directable automaton since $\delta(C, w) = \delta(C', w) = \delta(A, w)$ for every $C, C' \in B$ and $w \in L$. Let $Q = \{\delta(A, w) : w \in L\}$. Then, $L = \mathrm{D}_Q(\mathcal{B})$, and therefore, $L \in \mathcal{L}_{\mathrm{GD}}$.

For proving the converse inclusion, let $L \in \mathcal{L}_{\mathrm{GD}}$. Then, there exist a deterministic directable automaton $\mathcal{A} = (A, X, \delta)$ and a nonempty subset Q of its receiving states such that $L = \mathrm{D}_Q(\mathcal{A})$. Let us construct the $\mathcal{B} = (B, X)$ c.n.d. automaton as follows:

(1) $B = A \cup \{a^\dagger : a \in R_{\mathcal{A}} \setminus Q\}$ where $a^\dagger \notin A$ for every $a \in R_{\mathcal{A}} \setminus Q$ and $a^\dagger \neq b^\dagger$ if $a \neq b$.
(2) For every $b \in B$ and $x \in X$, let

$$bx^{\mathcal{B}} = \begin{cases} \{\delta(b, x)\} & \text{if } b \in A \ \&\ \delta(b, x) \notin R_{\mathcal{A}} \setminus Q, \\ \{u, u^\dagger\} & \text{if } b \in A \ \&\ \delta(b, x) = u \in R_{\mathcal{A}} \setminus Q, \\ \{\delta(a, x)\} & \text{if } b = a^\dagger \ \&\ \delta(a, x) \notin R_{\mathcal{A}} \setminus Q, \\ \{u, u^\dagger\} & \text{if } b = a^\dagger \ \&\ \delta(a, x) = u \in R_{\mathcal{A}} \setminus Q. \end{cases}$$

Now, it is easy to prove by induction on the length of words that, for every $b \in B$ and $p \in X^+$,

$$bp^{\mathcal{B}} = \begin{cases} \{\delta(b,p)\} & \text{if } b \in A \ \& \ \delta(b,p) \notin R_{\mathcal{A}} \setminus Q, \\ \{u, u^\dagger\} & \text{if } b \in A \ \& \ \delta(b,p) = u \in R_{\mathcal{A}} \setminus Q, \\ \{\delta(a,p)\} & \text{if } b = a^\dagger \ \& \ \delta(a,p) \notin R \setminus Q, \\ \{u, u^\dagger\} & \text{if } b = a^\dagger \ \& \ \delta(b,p) = u \in R_{\mathcal{A}} \setminus Q. \end{cases}$$

Now, we show that $L = \mathrm{D}_1(\mathcal{B})$. First, let us suppose that $w \in L$. Since $L = \mathrm{D}_Q(\mathcal{A})$, there exists a state $u \in Q$ such that $\delta(a,w) = u$, for all $a \in A$. Then, by the above observation, $aw^{\mathcal{B}} = \{u\}$, for all $a \in A$. On the other hand, by the above observation again, $a^\dagger w^{\mathcal{B}} = \{\delta(a,w)\} = \{u\}$. Consequently, $bw^{\mathcal{B}} = \{u\}$, for all $b \in B$, and thus $w \in \mathrm{D}_1(\mathcal{B})$.

Conversely, let $w \in \mathrm{D}_1(\mathcal{B})$. Then, there is a state $u \in B$ such that $bw^{\mathcal{B}} = \{u\}$ is valid, for all $b \in B$. By the observation above, if $u \in R_{\mathcal{A}} \setminus Q$ or $u = v^\dagger$ for some $v \in R_{\mathcal{A}} \setminus Q$, then $|uw^{\mathcal{B}}| = 2$. Consequently, $u \in (A \setminus R_{\mathcal{A}}) \cup Q$. On the other hand, $bw^{\mathcal{B}} = \{u\}$, for all $b \in B$, implies that $\delta(a,w) = u$, for all $a \in A$, which means that u is a receiving state of $\mathcal{A}$, and hence, $u \in R_{\mathcal{A}}$. From the two relations concerning u, it follows that $u \in Q$. Therefore, we have that $u \in Q$ and $\delta(A,w) = \{u\}$, which means that $w \in \mathrm{D}_Q(\mathcal{A}) = L$. Consequently, $L = \mathrm{D}_1(\mathcal{B})$.

It is known (see [6] or [8]) that the regular languages satisfying $X^*L = L$ are the regular ultimate definite languages. The next observation shows that $\mathcal{L}_{\mathrm{CND}(1)}$ is a proper subclass of the class of the regular ultimate definite languages.

Proposition 4. *There exists a nonempty regular language L satisfying $X^*L = L$ such that $L \notin \mathcal{L}_{\mathrm{CND}(1)}$.*

Proof. Let $X = \{x,y\}$ and $L = X^*x \cup X^*yxy$. Obviously, $L \neq \emptyset$, moreover, L is regular and $X^*L = L$. Now, let us suppose that $L \in \mathcal{L}_{\mathrm{CND}(1)}$. Then, there is a c.n.d. automaton $\mathcal{A} = (A,X)$ with $L = \mathrm{D}_1(\mathcal{A})$. Since $x, yxy \in L$, $Ax^{\mathcal{A}} = \{c\}$ and $A(yxy)^{\mathcal{A}} = \{d\}$ for some states $c, d \in A$. Let us observe that $\{c\}x^{\mathcal{A}} = \{c\}$ and $\{c\}y^{\mathcal{A}} = \{d\}$. Now, consider the word $yxxy$. $A(yxxy)^{\mathcal{A}} = A((yx)^{\mathcal{A}}x^{\mathcal{A}})y^{\mathcal{A}} = \{c\}y^{\mathcal{A}} = \{d\}$. Consequently, $yxxy \in \mathrm{D}_1(\mathcal{A}) = L$, which is a contradiction.

Lemma 2. *Let $L \in \mathcal{L}_{\mathrm{CND}(1)}$ and $y \in X$. Then, $Ly \in \mathcal{L}_{\mathrm{CND}(1)}$.*

Proof. By Proposition 3, from $L \in \mathcal{L}_{\mathrm{CND}(1)}$ it follows that $L \in \mathcal{L}_{\mathrm{GD}}$. Hence, there exists a deterministic directable automata $\mathcal{A} = (A, X, \delta)$ and a nonempty set $Q \subseteq R_{\mathcal{A}}$ such that $L = \mathrm{D}_Q(\mathcal{A})$. Now, let us define the automaton $\mathcal{B} = (B, X, \delta')$ as follows.

(1) $B = A \cup \{a_y : a \in Q\}$ where $a_y \notin A$ and $a_y \neq a'_y$ if $a \neq a'$.

(2) $\delta'(a,y) = a_y$, for all $a \in Q$,

$\delta'(a,x) = \delta(a,x)$, for all $a \in Q$ and $x \in X \setminus \{y\}$,

$\delta'(a_y,x) = \delta(a,yx)$, for every $a \in Q$ and $x \in X$,

$\delta'(a,x) = \delta(a,x)$, for every $a \in A \setminus Q$ and $x \in X$.

Obviously, $\mathcal{B}$ is a deterministic directable automaton. Let $P = \{a_y : a \in Q\}$. Then, $P \subseteq R_{\mathcal{B}}$ and $Ly = \mathrm{D}_P(\mathcal{B})$, and thus, by Proposition 3, $Ly \in \mathcal{L}_{\mathrm{CND}(1)}$.

Proposition 5. *Let $w \in X^*$ and $L = X^*w$. Then, $L \in \mathcal{L}_{\mathrm{CND}(1)}$.*

Proof. Let $\mathcal{A} = (\{1\}, X)$ be the c.n.d. automaton with $x^{\mathcal{A}} = \{(1,1)\}$, for all $x \in X$. Obviously, $X^* = \mathrm{D}_1(\mathcal{A}) \in \mathcal{L}_{\mathrm{CND}(1)}$. By a consecutive application of Lemma 2, one can obtain that $X^*w \in \mathcal{L}_{\mathrm{CND}(1)}$.

Corollary 2. *$\mathcal{L}_{\mathrm{CND}(1)}$ is not closed under union.*

Proof. By Proposition 5, $X^*x, X^*yxy \in \mathcal{L}_{\mathrm{CND}(1)}$. On the other hand, by the proof of Proposition 4, we have that the union of these two languages is not included in $\mathcal{L}_{\mathrm{CND}(1)}$.

Unlike the case of $\mathcal{L}_{\mathrm{CND}(1)}$, $wX^* \in \mathcal{L}_{\mathrm{ND}(2)}$ does not hold in general. To show this, for every $x \in X$, let us denote by x^+ the language $\{x^t : t = 1,2,\ldots\}$. Then, the following assertion is valid.

Proposition 6. *Let $x \in X$ and $w \in X^+ \setminus (\cup\{x^+ : x \in X\})$, furthermore, let n be an arbitrarily fixed positive integer. Then, $x^nX^* \in \mathcal{L}_{\mathrm{ND}(2)}$ and $wX^* \notin \mathcal{L}_{\mathrm{ND}(2)}$.*

Proof. Let us consider the automaton $\mathcal{A} = (A, X)$ defined by

$$A = \{1, 2, \ldots, n, n+1\},$$

$$x^{\mathcal{A}} = \{(1,2), (2,3), \ldots, (n, n+1), (n+1, n+1)\},$$

$$y^{\mathcal{A}} = \{(n+1, n+1)\}, \text{ for all } y \in X \setminus \{x\}.$$

It is easy to see that $x^nX^* = \mathrm{D}_2(\mathcal{A})$, *i.e.*, $x^nX^* \in \mathcal{L}_{\mathrm{ND}(2)}$.

Let $w \in X^+ \setminus (\cup\{x^+ : x \in X\})$. Then, $w = yw' = w_1zw_2$ for some $y \neq z \in X$ and a nonempty word $w_1 \in X^*$. Now, we show that $wX^* \notin \mathcal{L}_{\mathrm{ND}(2)}$. Contrary, let us suppose that $wX^* \in \mathcal{L}_{\mathrm{ND}(2)}$. Then, $wX^* = \mathrm{D}_2(\mathcal{B})$ for some n.d. automaton $\mathcal{B} = (B, X)$. First, we prove that $bw_1^{\mathcal{B}} \neq \emptyset$ is valid, for all $b \in B$. In the opposite case, $b'w_1^{\mathcal{B}} = \emptyset$ for some $b' \in B$. Then, $bw^{\mathcal{B}} = \emptyset$, for all $b \in B$, since $w \in \mathrm{D}_2(\mathcal{B})$. This implies that $b(zw)^{\mathcal{B}} = \emptyset$, for every $b \in B$, and therefore, $zw \in \mathrm{D}_2(\mathcal{B})$ which is a contradiction since the first letter of any words from wX^* is y and $y \neq z$. Consequently, $bw_1^{\mathcal{B}} \neq \emptyset$, for all $b \in B$. Now, let $b \in B$ be arbitrary and let us consider $b(w_1w)^{\mathcal{B}}$. Since

$bw_1^{\mathcal{B}} \neq \emptyset$ and $w \in \mathrm{D}_2(\mathcal{B})$, $b(w_1w)^{\mathcal{B}} = bw^{\mathcal{B}}$. Since b is arbitrary, this means that $w_1w \in \mathrm{D}_2(\mathcal{B})$, and hence, $w_1w \in wX^*$. On the other hand, $w_1w = w_1yw'$ and $w = w_1zw_2$, where $y \neq z$. This results in $w_1w \notin wX^*$ which is a contradiction. Consequently, $wX^* \notin \mathcal{L}_{\mathrm{ND}(2)}$.

It is known (see [6]) that the regular languages satisfying $LX^* = L$ are the regular reverse ultimate definite languages. The next observation shows that $\mathcal{L}_{\mathrm{ND}(2)}$ is a proper subclass of the class of the regular reverse ultimate definite languages.

Corollary 3. *There exists a nonempty regular language L satisfying $LX^* = L$ such that $L \notin \mathcal{L}_{\mathrm{ND}(2)}$.*

Proof. Let $w \in X^+ \setminus (\cup\{x^+ : x \in X\})$. Then, by Proposition 6, $wX^* \notin \mathcal{L}_{\mathrm{ND}(2)}$ though wX^* is a nonempty regular language satisfying $(wX^*)X^* = wX^*$.

Corollary 4. *$\mathcal{L}_{\mathrm{ND}(2)}$ is not closed under union.*

Proof. Let $x, y \in X$ with $x \neq y$. Then, by Proposition 6, x^2X^*, $y^2X^* \in \mathcal{L}_{\mathrm{ND}(2)}$. Now, we show that the language $L = x^2X^* \cup y^2X^*$ is not included in $\mathcal{L}_{\mathrm{ND}(2)}$. Contrary, let us suppose that $L \in \mathcal{L}_{\mathrm{ND}(2)}$. Then, there exists an n.d. automaton $\mathcal{A} = (A, X)$ such that $L = \mathrm{D}_2(\mathcal{A})$. Since $x^2 \in L$, $a(xx)^{\mathcal{A}} = a'(xx)^{\mathcal{A}}$, for every pair $a, a' \in A$. Therefore, if $a(xx)^{\mathcal{A}} = \emptyset$ for some $a \in A$, then $a(xx)^{\mathcal{A}} = \emptyset$, for all $a \in A$. But in this case, $a(yxx)^{\mathcal{A}} = \emptyset$, for all $a \in A$. This yields $yxx \in L$ which is a contradiction. Therefore, $a(xx)^{\mathcal{A}} \neq \emptyset$, for all $a \in A$, which implies that $ax^{\mathcal{A}} \neq \emptyset$, for all $a \in A$. From the latter observation it follows that $a(xyy)^{\mathcal{A}} = a(yy)^{\mathcal{A}} = a'(yy)^{\mathcal{A}} = a'(xyy)^{\mathcal{A}}$ is valid, for every pair $a, a' \in A$. This means that $xyy \in L$ which is a contradiction. Therefore, $L \notin \mathcal{L}_{\mathrm{ND}(2)}$.

Regarding the intersection, the classes of languages considered are closed under nonempty intersection as the following assertion shows.

Proposition 7. *Each of the classes of languages considerd is closed under nonempty intersection.*

Proof. The validity of the statement for $\mathcal{L}_{\mathrm{D}}$ follows from Proposition 1. Now, let us consider $\mathcal{L}_{\mathrm{ND}(2)}$. Let $L_1, L_2 \in \mathcal{L}_{\mathrm{ND}(2)}$, and let us suppose that $L_1 \cap L_2 \neq \emptyset$. Then, there are n.d. automata $\mathcal{A}$ and $\mathcal{B}$ such that $L_1 = \mathrm{D}_2(\mathcal{A})$ and $L_2 = \mathrm{D}_2(\mathcal{B})$. Let us construct the direct product $\mathcal{A} \times \mathcal{B}$. This direct product is also an n.d. automaton. Now, it is easy to check that $L_1 \cap L_2 = \mathrm{D}_2(\mathcal{A} \times \mathcal{B})$ which yields that $\mathcal{L}_{\mathrm{ND}(2)}$ is closed under the nonempty intersection. As far as the remaining classes are concerned, one can prove the validity of the statement in the same way as above.

References

1. H.V. Burkhard, Zum Längenproblem homogener experimente an determinierten und nicht-deterministischen automaten, *Elektronische Informationsverarbeitung und Kybernetik, EIK* **12** (1976), 301-306.
2. J. Černý, Poznámka k homogénnym experimentom s konečnými automatmi. *Matematicko-fysikalny Časopis SAV* **14** (1964), 208-215.
3. J. Černý, A. Piricka & B. Rosenauerová, On directable automata, *Kybernetika (Praha)* **7** (1971), 289-297.
4. B. Imreh, M. Steinby, Some remarks on directable automata, *Acta Cybernetica* **12** (1995), 23-35.
5. B. Imreh, M. Steinby, Directable nondeterministic automata, *Acta Cybernetica*, submitted for publication.
6. A. Paz, B. Peleg, Ultimate-definite and symmetric definite events and automata, *J. ACM* **12** (1965), 399-410.
7. J.-E. Pin, Sur les mots synchronisants dans un automata fini, *Elekronische Informationsverarbeitung und Kybernetik, EIK* **14** (1978), 297-303.
8. R.G. Reynolds, W.F. Cutlip, Synchronization and general repetitive machines, with applications to ultimate definite automata, *J. ACM* **16** (1969), 226-234.
9. I. Rystsov, Reset words for commutative and solvable automata, *Theoretical Computer Science* **172** (1997), 273-279.
10. P.H. Starke, *Abstrakte Automaten*, VEB Deutscher Verlag der Wissenschaften, Berlin 1969.

Synchronized Shuffle and Regular Languages

Michel Latteux and Yves Roos

Summary. New representation results for three families of regular languages are stated, using a special kind of shuffle operation, namely the *synchronized shuffle*. First, it is proved that the family of regular star languages is the smallest family containing the language $(a+bc)^*$ and closed under synchronized shuffle and length preserving morphism. The second representation result states that the family of ε-free regular languages is the smallest family containing the language $(a+bc)^*d$ and closed under synchronized shuffle, union and length preserving morphism. At last, it is proved that Reg is the smallest family containing the two languages $(a+bb)^*$ and $a+(ab)^+$, closed under synchronized shuffle, union and length preserving morphism.

1 Introduction

Finite automata are very popular objects in Computer Science and are now used in several other scientific domains. The family of regular languages, called Reg, is a basic class in Chomsky hierarchy. The significance of this family is strengthened by a great number of its nice characterizations. Some of these characterizations state that Reg is equal to the closure of a *small* family under a given set of operations. Besides the famous Kleene's theorem, one can recall the morphic compositional representation of regular languages establishing that Reg is the closure of the family reduced to the single language a^*b under morphism and inverse morphism. This nice result was proved in 1982 by Culik, Fich and Salomaa [1]. After the existence of such a morphic representation had been proved there followed a sequence of papers improving and generalizing this theorem (see [4], [6] and [9]).

We shall prove here similar characterizations of Reg involving a special kind of shuffle. The shuffle operation appears as a fundamental operation in the theory of concurrency and admits several variations such as literal shuffle, insertion, etc.. Recently, Mateescu, Rozenberg and Salomaa introduced the notion of shuffle on trajectories [7] that provides a way to find again the most of these variations. Synchronized shuffle was introduced in [2] by De Simone under the name of *produit de mixage*(see also [5], where a closely related operation was introduced). This special shuffle can bee seen as a shuffle with *rendez-vous* and is very useful in modelling the behaviours of parallel processes (see [3] and [8]). This powerful operation is quite amazing. For example, contrary to what was stated in [2], it is not associative.

The study of synchronized shuffle in connection with union and length preserving morphism provides us to obtain new representation results for three families of regular languages.

First, we prove that the family of regular star languages is the smallest family containing the language $(a+bc)^*$ and closed under synchronized shuffle and length preserving morphism. Our second representation result states that the family of ε-free regular languages is the smallest family containing the language $(a+bc)^*d$ and closed under synchronized shuffle, union and length preserving morphism. At last, we prove that Reg is the smallest family containing the two languages $(a+bb)^*$ and $a+(ab)^+$, closed under synchronized shuffle, union and length preserving morphism.

2 Preliminaries

2.1 Notations

Let X be an alphabet. For any word $w \in X^*$, we shall denote by $|w|$ the length of the word w and for any letter $a \in X$, we shall denote by $|w|_a$ the number of occurrences of the letter a that appear in the word w.

For any language $L \subseteq X^*$, we denote by $\alpha(L)$, the *alphabet of* L that is :

$$\alpha(L) = \{x \in X \mid \exists u \in L : |u|_x > 0\}$$

For example we get : $\alpha(a+ab) = \alpha(a^2+b^3) = \{a,b\}$, $\alpha(\emptyset) = \alpha(\varepsilon) = \emptyset$.

We denote by $\Pi_{X,Y}$ the *projection* over the *sub-alphabet* Y, i.e. the homomorphism from X^* to Y^* defined by:

$$\forall x \in X, \text{ if } x \in Y \text{ then } \Pi_{X,Y}(x) = x \text{ , else } \Pi_{X,Y}(x) = \varepsilon$$

When there is no ambiguity, we shall use notation Π_Y instead of $\Pi_{X,Y}$.

$u \sqcup\!\sqcup v$ is the *shuffle* of the two words u and v that is

$$u \sqcup\!\sqcup v = \{u_1v_1u_2v_2...u_nv_n \mid u_i \in \Sigma^*, v_i \in \Sigma^*, u = u_1u_2...u_n, v = v_1v_2...v_n\}$$

This definition is extended over languages by :

$$\forall L_1, L_2 \subseteq X^*, L_1 \sqcup\!\sqcup L_2 = \bigcup_{\substack{w_1 \in L_1 \\ w_2 \in L_2}} w_1 \sqcup\!\sqcup w_2$$

When there is no ambiguity, for a language reduced to a single word, we shall write this language u rather than $\{u\}$.

2.2 Synchronized shuffle

Definition 1. [2] Let X be an alphabet and L_1 , L_2 be two languages included in X^*. The ***synchronized shuffle*** of L_1 and L_2, denoted by $L_1 \sqcap\!\!\sqcap L_2$ is defined by :

$$L_1 \sqcap\!\!\sqcap L_2 = \{w \in (\alpha(L_1) \cup \alpha(L2))^* \mid \Pi_{\alpha(L_i)}(w) \in L_i, i \in \{1,2\}\}$$

Examples :

1. $bab \sqcap\!\!\sqcap cac = \{bcabc, bcacb, cbabc, cbacb\}$
2. $bac \sqcap\!\!\sqcap cab = \emptyset$
3. $\{a, abb\} \sqcap\!\!\sqcap ab = \emptyset$ and $a \sqcap\!\!\sqcap ab = ab$
4. $(a \sqcap\!\!\sqcap \{a,b\}) \sqcap\!\!\sqcap ab = a \sqcap\!\!\sqcap ab = ab$
5. $a \sqcap\!\!\sqcap (\{a,b\} \sqcap\!\!\sqcap ab) = a \sqcap\!\!\sqcap \emptyset = \emptyset$

Example 3 shows that it generally does not hold that

$$L_1 \sqcap\!\!\sqcap L_2 = \bigcup_{\substack{w_1 \in L_1 \\ w_2 \in L_2}} w_1 \sqcap\!\!\sqcap w_2$$

Examples 4 and 5 show that synchronized shuffle is not an associative operation.

The following properties and proposition easily come from definition :

- the synchronized shuffle is commutative :$L_1 \sqcap\!\!\sqcap L_2 = L_2 \sqcap\!\!\sqcap L_1$
- $L \sqcap\!\!\sqcap \varepsilon = L$
- $L \sqcap\!\!\sqcap \emptyset = \emptyset$
- if $\alpha(L_1) = \alpha(L_2)$ then $L_1 \sqcap\!\!\sqcap L_2 = L_1 \cap L_2$
- if $\alpha(L_1) \cap \alpha(L_2) = \emptyset$ then $L_1 \sqcap\!\!\sqcap L_2 = L_1 \sqcup\!\!\sqcup L_2$
- $(L_1^* \sqcap\!\!\sqcap L_2^*) = (L_1^* \sqcap\!\!\sqcap L_2^*)^*$
- $(\varepsilon \in L_1 \sqcap\!\!\sqcap L_2) \Longleftrightarrow (\varepsilon \in L_1 \cap L_2)$
- if L_1 and L_2 are ε-free languages then for any letter a :$(a \in L_1 \sqcap\!\!\sqcap L_2) \Longleftrightarrow (a \in L_1 \cap L_2)$

Proposition 1. *Let L_1 and L_2 be languages with $X_i = \alpha(L_i)$ for $i \in \{1,2\}$. Then the following equality holds :*

$$L_1 \sqcap\!\!\sqcap L_2 = (L_1 \sqcup\!\!\sqcup (X_2 \setminus X_1)^*) \bigcap (L_2 \sqcup\!\!\sqcup (X_1 \setminus X_2)^*)$$

2.3 synchronized shuffle and associativity

As we have seen in examples, the synchronized shuffle is not associative. We shall give now a sufficient condition for the synchronized shuffle to be associative over a ($\sqcap\!\!\sqcap$-closed) family of language. For every family of language $\mathcal{L}$, we denote by $(\mathcal{L})_{\sqcap\!\sqcap}$ the least family of languages containing $\mathcal{L}$ and closed under $\sqcap\!\!\sqcap$. First we can state :

Proposition 2. *Let $\mathcal{L}$ be a family of languages. If for every languages L_1, $L_2 \in (\mathcal{L})_{\sqcap\!\sqcap}$, the equality $\alpha(L_1 \mathbin{\sqcap\!\sqcap} L_2) = \alpha(L_1) \cup \alpha(L_2)$ holds then $\sqcap\!\sqcap$ is associative over $(\mathcal{L})_{\sqcap\!\sqcap}$.*

Proof. Let $L_1, L_2, L_3 \in (\mathcal{L})_{\sqcap\!\sqcap}$ then, from definition, we have

$$u \in (L_1 \mathbin{\sqcap\!\sqcap} L_2) \mathbin{\sqcap\!\sqcap} L_3 \Longleftrightarrow (\Pi_{\alpha(L_1 \sqcap\!\sqcap L_2)}(u) \in L_1 \mathbin{\sqcap\!\sqcap} L_2) \wedge (\Pi_{\alpha(L_3)}(u) \in L_3)$$

Since $\alpha(L_1 \mathbin{\sqcap\!\sqcap} L_2) = \alpha(L_1) \cup \alpha(L_2)$, we get :

$$u \in (L_1 \mathbin{\sqcap\!\sqcap} L_2) \mathbin{\sqcap\!\sqcap} L_3 \Longleftrightarrow (\Pi_{\alpha(L_1 \cup L_2)}(u) \in L_1 \mathbin{\sqcap\!\sqcap} L_2) \wedge (\Pi_{\alpha(L_3)}(u) \in L_3)$$

Moreover $\Pi_{\alpha(L_i)}(u) = \Pi_{\alpha(L_i)}(\Pi_{\alpha(L_1) \cup \alpha(L_2)}(u))$ for $i \in \{1, 2\}$, hence we have :

$$u \in (L_1 \mathbin{\sqcap\!\sqcap} L_2) \mathbin{\sqcap\!\sqcap} L_3$$
$$\Updownarrow$$
$$(\Pi_{\alpha(L_1)}(u) \in L_1) \wedge (\Pi_{\alpha(L_2)}(u) \in L_2) \wedge (\Pi_{\alpha(L_3)}(u) \in L_3)$$

and $\sqcap\!\sqcap$ is associative over $(\mathcal{L})_{\sqcap\!\sqcap}$.

Definition 2. Let L be a language. A language R is said to be L-compatible if the two following conditions are satisfied :

1. $\Pi_{\alpha(R)}(L) \subseteq R$
2. $\alpha(R) \subseteq \alpha(L)$

A family of languages $\mathcal{L}$ is said to be L-compatible if every language R in $\mathcal{L}$ is L-compatible.

Lemma 1. *Let L be a language and R_1, R_2 be two languages which are L-compatible. Then :*

1. *$\alpha(R_1 \mathbin{\sqcap\!\sqcap} R_2) = \alpha(R_1) \cup \alpha(R_2)$.*
2. *$R_1 \mathbin{\sqcap\!\sqcap} R_2$ is L-compatible.*

Proof. From 1 of definition 2, $\Pi_{\alpha(R_1) \cup \alpha(R_2)}(L) \subseteq R_1 \mathbin{\sqcap\!\sqcap} R_2$ and from 2 of definition 2, we also have

$$\alpha(\Pi_{\alpha(R_1) \cup \alpha(R_2)}(L)) = \alpha(R_1) \cup \alpha(R2)$$

thus $\alpha(R_1) \cup \alpha(R_2) \subseteq \alpha(R_1 \mathbin{\sqcap\!\sqcap} R_2)$. From definition of synchronized shuffle, $\alpha(R_1 \mathbin{\sqcap\!\sqcap} R_2) \subseteq \alpha(R_1) \cup \alpha(R_2)$ then $\alpha(R_1 \mathbin{\sqcap\!\sqcap} R_2) = \alpha(R_1) \cup \alpha(R_2) \subseteq \alpha(L)$. It follows that

$$\Pi_{\alpha(R_1 \sqcap\!\sqcap R_2)}(L) = \Pi_{\alpha(R_1) \cup \alpha(R_2)}(L) \subseteq R_1 \mathbin{\sqcap\!\sqcap} R_2$$

hence $R_1 \mathbin{\sqcap\!\sqcap} R_2$ is L-compatible.

Then, by induction and from proposition 2 it directly follows :

Proposition 3. *Let $\mathcal{L}$ be a family of languages. If there exists a language L such that $\mathcal{L}$ is L-compatible, then $(\mathcal{L})_{\sqcap\!\sqcap}$ is L-compatible and $\sqcap\!\sqcap$ is associative over $(\mathcal{L})_{\sqcap\!\sqcap}$*

In the following, we shall omit use of parenthesis in expressions involving synchronized shuffle over such families of languages.

3 A new characterization of Reg

Here, we shall prove characterization results involving synchronized shuffle. These results concern several families of regular languages. First, let us consider Fin_ε, the family of ε-free finite languages.

Proposition 4. *The family* Fin_ε *is the smallest family containing the language* $a + ab$ *and closed under union, length preserving morphisms and synchronized shuffle.*

Proof. Let $\mathcal{L}$ be the smallest family containing the language $a+ab$ and closed under the three above operations. Clearly, $\mathcal{L} \subseteq \text{Fin}_\varepsilon$. Conversely, since union is allowed it is sufficient to prove that :

1. $\emptyset$ is in $\mathcal{L}$
2. for every alphabet $X = \{x_1, x_2, \ldots, x_n\}, n > 0$, the language reduced to the single word $x_1 x_2 \ldots x_n$ is in $\mathcal{L}$.

For 1, we get $\emptyset = (a + ab) \sqcap\!\!\sqcap (b + ba)$. For 2, we make an induction on n. If $n < 3$, we get $a = [(a{+}ab){+}(b{+}bc)] \sqcap\!\!\sqcap [(a{+}ac){+}(c{+}cb)]$ and $ab = (a{+}ab) \sqcap\!\!\sqcap b$. If $n \geq 3$, it is easily seen that $x_1 x_2 \ldots x_n = x_1 x_2 \ldots x_{n-1} \sqcap\!\!\sqcap x_2 \ldots x_n$.

We shall prove a similar result for Reg_*, the family of regular star languages. The starting language is $(a + bc)^*$ and we use only length preserving morphisms and synchronized shuffle. First, we consider monoids generated by special finite languages.

Definition 3. A *marked* language F is a finite language such that :

$$\forall u, v \in F, \forall x \in \alpha(F), (|u|_x > 0 \wedge |v|_x > 0) \Longrightarrow (u = v \wedge |u|_x = 1)$$

Next, we show that if F is a marked language, then F^* can be obtained from languages of the type $(a + bc)^*$ using synchronized shuffle.

Definition 4. Two languages L_1 and L_2 are said equivalent if there is a length preserving morphism which maps bijectively L_1 onto L_2.

Lemma 2. *For every marked language* F, F^* *can be obtained from languages equivalent to* $(a + bc)^*$ *using synchronized shuffle.*

Proof. Let $\mathcal{L}$ be the family of languages which can be obtained by synchronized shuffle from languages equivalent to $(a + bc)^*$. The languages $a^* = (a + bc)^* \sqcap\!\!\sqcap (a + cb)^*$ and $(ab + cd)^* = (a + cd)^* \sqcap\!\!\sqcap (c + ab)^*$ belong to $\mathcal{L}$. Hence, $(a+b)^* = a^* \sqcap\!\!\sqcap b^*$, $(ab)^* = (ab+cd)^* \sqcap\!\!\sqcap (ab+dc)^*$, $\varepsilon = (ab)^* \sqcap\!\!\sqcap (ba)^*$ are in $\mathcal{L}$.

Let us consider now a marked language $F = u+v$ with $|v| \geq |u| \geq 0$. From the above equalities, if $|v| \leq 2$ then $F^* \in \mathcal{L}$. Assume that $v = x_1 x_2 \ldots x_k$ with $k \geq 3$. The languages $L_1 = (u + x_1 \ldots x_{k-1})^*$, $L_2 = (u + x_2 \ldots x_k)^*$ and $L_3 =$

$(u + x_1x_k)^*$ are F^*-compatible. From the equality $(u + v)^* = L_1 \sqcap\!\!\sqcap L_2 \sqcap\!\!\sqcap L_3$, we get, by induction, that $(u + v)^* \in \mathcal{L}$. At last, if $F = u_1 + u_2 + \ldots + u_n$ is a marked language with $n \geq 3$, the languages $R_1 = (u_1 + \ldots + u_{n-1})^*$, $R_2 = (u_2 + \ldots + u_n)^*$ and $R_3 = (u_1 + u_n)^*$ are F^*-compatible. Once again, the equality $F^* = R_1 \sqcap\!\!\sqcap R_2 \sqcap\!\!\sqcap R_3$ implies by induction that $F^* \in \mathcal{L}$.

We can now get easily our characterization result for the family Reg_*.

Proposition 5. *The family of regular star languages is the smallest family containing the language $(a + bc)^*$ and closed under length preserving morphisms and synchronized shuffle.*

Proof. Clearly, Reg_* is closed under the above operations. For the reverse inclusion, let R^* be a regular language. It is known (see [4], [6]) that there exist two finite languages F_1 and F_2 and a length preserving morphism such that $R^* = g(F_1^* \cap F_2^*)$. One can assume that $\alpha(F_1) = \alpha(F_2)$. Hence, $R^* = g(F_1^* \sqcap\!\!\sqcap F_2^*)$. Moreover, F_1 and F_2 are the image by a length preserving morphism of marked languages. Thus, lemma 2 implies the result.

We are now able to state our first characterization of Reg.

Proposition 6. *The family of regular languages Reg is the smallest family containing the languages $(a+bc)^*$ and a, closed under union, length preserving morphisms and synchronized shuffle.*

Proof. Let $\mathcal{L}$ be the smallest family containing the languages $(a + bc)^*$ and a, closed under the three above operations. Clearly $\mathcal{L}$ is included in Reg. For the reverse inclusion, let us consider a language $R \in \mathrm{Reg}$. From proposition 5, we know that ε is in $\mathcal{L}$, so we may suppose that $\varepsilon \notin R$ since union is allowed. Hence one may assume, without loss of generality, that $R \subseteq A^*A'$ where A and A' are disjoint alphabets. Then $R = R^* \sqcap\!\!\sqcap A'$. From proposition 5, R^* can be obtained from $(a + bc)^*$ using length preserving morphisms and synchronized shuffle. On the other hand, A' can be obtained from a using union and length preserving morphisms.

We can observe that it is not possible to enunciate a similar result with a single *generator*. Indeed, since $\varepsilon \in L_1 \sqcap\!\!\sqcap L_2$ if and only if $\varepsilon \in L_1 \cap L_2$, we need to start from a family containing at least two languages L_1 and L_2 with $\varepsilon \in L_1$ and $\varepsilon \notin L_2$. The following result concerns the family of ε-free regular languages, that is regular languages which do not contain the empty word. For this family, it is possible to start from a single generator.

Proposition 7. *The family of regular ε-free languages Reg_ε is the smallest family containing the language $(a + bc)^*d$ and closed under union, length preserving morphisms and synchronized shuffle.*

Proof. Let $\mathcal{L}$ be the smallest family containing the language $(a+bc)^*d$ and closed under the three above operations. Clearly $\mathcal{L}$ is included in Reg_ε. For the reverse inclusion, let us consider a language $R \in \mathrm{Reg}_\varepsilon$. Without loss of generality, one may assume that R is a finite union of languages in the form Kd with $K \in \mathrm{Reg}$ and d a letter not in $\alpha(K)$. It remains to prove that such a language Kd is in $\mathcal{L}$. This can be done by induction over the construction of K with respect to proposition 6.

If $K = (a+bc)^*$ then Kd is in $\mathcal{L}$. If $K = a$, observe first that $d = (a+ab)^*d \mathbin{\sqcap\!\!\sqcap} (b+ba)^*d$ is in $\mathcal{L}$. Moreover a^*d, which can be obtained from $(a+bc)^*d$ using a length preserving morphism, is in $\mathcal{L}$. Then we get $ad = a \mathbin{\sqcap\!\!\sqcap} a^*d$ belongs to $\mathcal{L}$.

Now,

- if $K = h(K')$ for some $K' \in \mathcal{L}$ and some length preserving morphism h, we may suppose that $d \notin \alpha(K')$ then $K'd \in \mathcal{L}$ which implies $Kd \in \mathcal{L}$.
- if $K = K_1 + K_2$ with $K_i \in \mathcal{L}$ for $i = 1, 2$ then $Kd = K_1d + K_2d \in \mathcal{L}$.
- if $K = K_1 \mathbin{\sqcap\!\!\sqcap} K_2$ with $K_i \in \mathcal{L}$ for $i = 1, 2$. We may suppose that $d \notin \alpha(K_i)$ for $i = 1, 2$ then $Kd = K_1d \mathbin{\sqcap\!\!\sqcap} K_2d \in \mathcal{L}$.

4 Binary generators

The single generator for the family of ε-free regular languages Reg_ε used in proposition 7 is built over a four letter alphabet. The following proposition shows that it is possible to start from a language defined over a three letter alphabet :

Proposition 8. *The family of regular ε-free languages Reg_ε is the smallest family containing the language $(a+bc)^*b$ and closed under union, length preserving morphisms and synchronized shuffle.*

Proof. Let $\mathcal{L}$ be the smallest family containing the language $(a+bc)^*b$ and closed under union, length preserving morphisms and synchronized shuffle. From proposition 7, we have to prove that $(a+bc)^*d$ is in $\mathcal{L}$. Observe first that $b = (a+bc)^*b \mathbin{\sqcap\!\!\sqcap} (c+ba)^*b$ is in $\mathcal{L}$. Then $a^*b = (a+bc)^*b \mathbin{\sqcap\!\!\sqcap} b$ and $(a+bc)^*bd = (a+bc)^*b \mathbin{\sqcap\!\!\sqcap} b^*d$ are in $\mathcal{L}$. It follows that $(a+bc)^*bc$ is also in $\mathcal{L}$. Let us consider now the length preserving morphism h defined by : $h(a) = a, h(b) = h(d) = b$ and $h(c) = c$. We get

$$(a+bc)^*bc \mathbin{\sqcap\!\!\sqcap} h((a+cb)^*c \mathbin{\sqcap\!\!\sqcap} d) = a^*(bc)^+ \in \mathcal{L}$$

then $a^*(bb)^+ \in \mathcal{L}$ and $a^*b^+ = a^*b + a^*(bb)^+ + h(a^*(bb)^+ \mathbin{\sqcap\!\!\sqcap} a^*d)$ is in $\mathcal{L}$. Moreover $a^+b^* = a^+ + (a^*b^+ \mathbin{\sqcap\!\!\sqcap} a^+)$ is also in $\mathcal{L}$ since, clearly, $a^+ \in \mathcal{L}$. Now, with the length preserving morphism g defined by : $g(a) = g(e) = a, g(b) = b, g(c) = c$ and $g(d) = d$, we obtain

$$(a+bc)^*bca^*d = g((a+bc)^*bc \mathbin{\sqcap\!\!\sqcap} c^+e^* \mathbin{\sqcap\!\!\sqcap} e^*d \mathbin{\sqcap\!\!\sqcap} c^*d) \in \mathcal{L}$$

At last, we get $(a+bc)^*d = a^*d + (a+bc)^*bca^*d \in \mathcal{L}$.

We have seen that the family of (ε-free) regular languages can be obtained from languages whose cardinality alphabet is less or equal to three. A natural question is whether it is possible to start from binary languages which are languages built over two letter alphabets. In a first time, we shall establish a negative answer for the family of star regular languages.

Definition 5. A star language L satisfies the (P) property if there exist three distinct letters a, b, c such that the words a and bc are in L, but the word bac is not in L.

Lemma 3. *Let S be a set of star languages and $\mathcal{L}$ be the smallest family of languages, containing S and closed under length preserving morphism and synchronized shuffle. If $\mathcal{L}$ contains a language which satisfies the (P) property then so do S.*

Proof. First, it is clear that $\mathcal{L}$ contains only star languages. We shall prove this lemma by induction. Let L be a language in $\mathcal{L}$ with $a, bc \in L$ and $bac \notin L$.

If there exist some language $L' \in \mathcal{L}$ and some length preserving morphism h such that $L = h(L')$ then it is obvious that L' satisfies the (P) property.

Let us suppose now that $L = L_1 ⧢ L_2$ such that neither L_1 nor L_2 satisfies (P). We shall show that it will lead to the following contradiction $bac \in L$. For $i = 1, 2$, we denote $\alpha_i = \alpha(L_i) \cap \{a, b, c\}$. Observe that $bac \in L$ if and only if $\Pi_{\alpha_i}(bac) \in L_i$ for $i = 1, 2$. Moreover, since L is a star language, the words abc and bca are in L. Now, for $i = 1, 2$:

- if $\alpha_i \subsetneq \{a, b, c\}$, we get $\Pi_{\alpha_i}(bac) \in \Pi_{\alpha_i}(abc + bca) \subseteq L_i$
- if $\alpha_i = \{a, b, c\}$, then $a \in L_i$ and $bc \in L_i$. Since L_i does not satisfy (P), it follows that $\Pi_{\alpha_i}(bac) = bac \in L_i$.

The contradiction $bac \in L$ implies that L_1 satisfies (P) or L_2 satisfies (P).

Since the language $(a + bc)^*$ satisfies the (P) property, we can state, as a corollary of the above lemma :

Proposition 9. *The family of regular star languages can not be obtained as the closure of a set of binary languages under length preserving morphism and synchronized shuffle.*

For the family of (ε-free) regular languages, the answer is positive. In order to establish this result, we shall first prove the following lemma :

Lemma 4. *Let $\mathcal{L}$ be a family of languages containing the language a^*b^* and closed under length preserving morphism and synchronized shuffle, then $\mathcal{L}$ is closed under product.*

Proof. Let L_1 and L_2 be two languages of $\mathcal{L}$. We may suppose that $\alpha_1 = \alpha(L_1)$ and $\alpha_2 = \alpha(L_2)$ are disjoint. The family $\{x^*y^* \mid x \in \alpha_1, y \in \alpha_2\}$

is clearly L_1L_2-compatible, then synchronized shuffle is associative over this family and

$$\coprod_{\substack{x \in \alpha_1 \\ y \in \alpha_2}} x^*y^* = \alpha_1^*\alpha_2^* \in \mathcal{L}$$

Then $L_1L_2 = (L_1 \sqcap\!\sqcap L_2) \sqcap\!\sqcap \alpha_1^*\alpha_2^*$ is in $\mathcal{L}$.

We are now able to enunciate our last proposition which states that Reg, the family of regular languages can be obtained from binary generators.

Proposition 10. *The family of regular languages* Reg *is the smallest family containing the languages* $a + (ab)^+$ *and* $(a + bb)^*$*, and closed under union, length preserving morphism and synchronized shuffle.*

Proof. Let $\mathcal{L}$ be the smallest family containing the languages $a + (ab)^+$ and $(a + bb)^*$, closed under the three above operations. Clearly $\mathcal{L}$ is included in Reg.

For the reverse inclusion, observe first that

$$(a + bb)^* \sqcap\!\sqcap (a + (ab)^+) = a \in \mathcal{L}$$

It remains to prove that $(a + bc)^* \in \mathcal{L}$. Let h be the length preserving morphism defined by $h(a) = h(b) = a$, we get $h((a + bb)^*) = a^* \in \mathcal{L}$ and $h(a^* \sqcap\!\sqcap b) = a^+ \in \mathcal{L}$. It follows that $(a + (ab)^+) \sqcap\!\sqcap b = ab \in \mathcal{L}$ and $(a + (ab)^+) \sqcap\!\sqcap b^+ = (ab)^+ \in \mathcal{L}$.

Now, if g is the length preserving morphism defined from by $g(a) = g(d) = g(e) = a$, $g(b) = b$ and $g(c) = c$, we get

$$g((ac)^+ \sqcap\!\sqcap bd) \sqcap\!\sqcap g((ca)^+ \sqcap\!\sqcap be) = b(ac)^+a \in \mathcal{L}$$

Then $b(aa)^+a$ in $\mathcal{L}$, $g(b(aa)^+a \sqcap\!\sqcap bd) = b(aa)^+aa \in \mathcal{L}$ and $g(ba \sqcap\!\sqcap ad) = baa \in \mathcal{L}$ so :

$$ba^* = b + ba + baa + b(aa)^+a + b(aa)^+aa \in \mathcal{L}$$

In a same way we can prove $a^*b \in \mathcal{L}$. It follows that $a^*b \sqcap\!\sqcap bc^* = a^*bc^*$ is in $\mathcal{L}$ then a^*b^+ is in $\mathcal{L}$. Now, since $a^*b^+ + a^* = a^*b^* \in \mathcal{L}$ and from lemma 4, we get that $\mathcal{L}$ is closed under product.

We are now able to obtain $(a + bc)^*$:

$$(((a + bb)^* \sqcap\!\sqcap (a + cc)^*) \sqcap\!\sqcap (bc)^+) + a^* = (a + bcbc)^* \in \mathcal{L}$$

The family $\mathcal{L}$ is closed under product then we also have $bc(a + bcbc)^*bc \in \mathcal{L}$. Then

$$L_1 = (a + bcbc)^* \sqcap\!\sqcap bc(d + bcbc)^*bc = a^*bc(d^*bca^*bc)^*d^*bca^* \in \mathcal{L}$$

and, since product is allowed $L_2 = a^*bcL_1 \in \mathcal{L}$. We finally get that

$$g(L_1) + g(L_2) + a^*bca^* + a^* = (a + bc)^* \in \mathcal{L}$$

References

1. Culik II K., Fich F.E. and Salomaa A. (1982) A homomorphic characterization of regular languages. Discrete Applied Mathematics **4**, 149–152
2. De Simone R. (1984) Langages infinitaires et produit de mixage. Theoretical Computer Science **31**, 83–100
3. Duboc C. (1986) Commutation dans les monoïdes libres : Un cadre théorique pour l'étude du parallélisme. Thèse de doctorat. Université de Rouen.
4. Karhumaki J., Linna M. (1983) A note on morphic characterization of languages. Discrete Applied Mathematics **5**, 243–246
5. Kimura T. (1976) An algebraic system for process structuring and interprocess communication. 8th ACM SIGACTS Symposium on Theory of Computing. 92–100
6. Latteux M., Leguy J. (1983) On the composition of morphisms and inverse morphisms. Lecture Notes in Computer Science, **154**, Springer-Verlag, pp. 420–432
7. Mateescu A., Rozenberg G. and Salomaa A. (1998) Shuffle on Trajectories : Syntactic Constraints. Theoretical Computer Science, TCS, Fundamental Study, **197, 1-2**, 1–56
8. Ryl I. (1998) Langages de synchronisation. Thèse de doctorat. Université de Lille 1.
9. Turakainen P. (1982) A homomorphic characterization of principal semi-AFLs without using intersection with regular sets. Inform. Sci. **27**, 141–149

Synchronization Expressions: Characterization Results and Implementation

Kai Salomaa and Sheng Yu

Summary. Synchronization expressions are defined as restricted regular expressions that specify synchronization constraints between parallel processes and their semantics is defined using the synchronization languages. In this paper we survey results on synchronization languages, in particular, various approaches to obtain a characterization of this language family using closure under a set of rewriting rules. Also, we discuss the use and implementation of synchronization expressions in a programming language designed for a parallel or distributed computing environment.

1 Introduction

Synchronization expressions (SE) were originally introduced in the parallel programming language ParC [8] as high-level constructs for specifying synchronization constraints between parallel processes. Synchronization requests are specified as expressions of tags for statements that are to be constrained by the synchronization requirements.

The semantics of SEs is defined using synchronization languages. The corresponding synchronization language (SL) describes intuitively how an SE is implemented in the parallel programming language ParC. Using this formalism, relations such as equivalence and inclusion between SEs can be easily understood and tested. Thus the synchronization languages provide a systematic approach for the implementation and simplification of SEs in parallel or distributed programming languages.

A synchronization language can be seen as the set of correct executions (as controlled by the SE) of a distributed application where each action is split into two atomic parts, its start and termination. It can be argued [9,10] that synchronization languages are a more suitable semantic model for SEs than traditional models such as traces, Petri nets, or process algebra [11,19]. An important feature of the semantics defined by synchronization languages is that it splits each action into two parts, the start and the termination. For instance, in trace semantics [7] parallel execution is interpreted as $a \parallel b = ab + ba$, but in order to control the execution of parallel processes it is necessary to distinguish sequences like ab and ba from real parallel execution. By having the start and termination of an action as separate instantaneous parts such distinctions can be made. Furthermore, synchronization languages can specify that certain occurrences of given processes are parallel whereas other occurrences need to be executed in a specified sequential order. The relationship of and the differences between synchronization languages and the other formalisms used in concurrency will be discussed further in [3].

It was shown in [9,10] that synchronization languages are closed under certain naturally defined rewriting rules and there it was also conjectured that synchronization languages would consist of the subset of regular languages closed under these rules and satisfying the so called start-termination property. The conjecture was proved for languages expressing synchronization between two distinct actions (i.e, for a two-letter alphabet) by Clerbout, Roos and Ryl [5,14] and they showed that the conjecture is false in general. Furthermore, [15] establishes a more general negative result showing that these languages cannot be characterized by any set of rewriting rules.

By generalizing the syntactic definition (and modifying the semantics) of SEs [18] considered a new definition of synchronization languages. The new definition has the advantage that it allows us to eliminate the intuitively less well motivated transformations (rewriting rules) describing closure properties of the synchronization languages and, furthermore, the new set of rules always preserves regularity. This gives us hope to avoid the negative results that were obtained for the original definition of synchronization languages. The approach allows us to obtain an exact characterization for the finite synchronization languages in terms of simple semi-commutation rules [18] and a characterization of the images of synchronization languages under st-morphisms [16]. However, it is not known whether the family of synchronization languages is closed under st-morphisms and, thus, it remains an open question whether the rewriting rules can be used to characterize synchronization languages in general.

Besides giving rise to interesting questions in formal language theory, synchronization languages provide us with a systematic method for implementing SEs in a parallel programming language such as ParC. This gives one more example on the usefulness of the interaction between formal language theory and practical computer science. In the final section we discuss the practical use and implementation of SEs and SLs. A description of the implementation has not been presented previously in published form. Roughly speaking the idea is as follows. For an SE, we construct at compile time a finite automaton that accepts the prefix language of the corresponding SL. The constructed automaton is then used to impose at runtime the synchronization constraints specified by the SE.

2 Synchronization Expressions and Languages

We assume that the reader is familiar with the basic notions related to regular expressions and rewriting systems [1,17,20]. The set of finite words over a finite alphabet Σ is denoted Σ^*, λ is the empty word, and $\Sigma^+ = \Sigma^* - \{\lambda\}$. The catenation of languages $L_1, L_2 \subseteq \Sigma^*$ is $L_1 \cdot L_2 = \{w \in \Sigma^* \mid (\exists v_i \in L_i, i = 1,2)\ w = v_1v_2\}$ and the catenation of n copies of $L \subseteq \Sigma^*$ is L^n $(n \geq 0)$. Note that $L^0 = \{\lambda\}$. The Kleene star of a language L is $L^* = \cup_{i=0}^{\infty} L^i$.

The *shuffle* of words $u, v \in \Sigma^*$ is the language $\omega(u,v) \subseteq \Sigma^*$ consisting of all words that can be written in the form $u_1v_1 \cdots u_nv_n$, $n \geq 0$, where $u = u_1 \cdots u_n$, $v = v_1 \cdots v_n$, $u_i, v_i \in \Sigma^*$, $i = 1, \ldots, n$. The shuffle operation is extended for languages in the natural way: $\omega(L_1, L_2) = \bigcup_{w_1 \in L_1, w_2 \in L_2} \omega(w_1, w_2)$, where $L_1, L_2 \subseteq \Sigma^*$.

A *string-rewriting system* (or Thue system) over Σ is a finite set R of rules $u \rightarrow v$, $u, v \in \Sigma^*$. The rules of R define the (single step) reduction relation $\rightarrow_R \subseteq \Sigma^* \times \Sigma^*$ as follows. For $w_1, w_2 \in \Sigma^*$, $w_1 \rightarrow_R w_2$ if and only if there exists a rule $u_1 \rightarrow u_2 \in R$ and $r, s \in \Sigma^*$ such that $w_i = ru_is$, $i = 1, 2$. The *reduction relation* of R is the reflexive and transitive closure of $\rightarrow_R$ and it is denoted $\rightarrow_R^*$.

We define synchronization expressions using the extended definition from [18]. The operators $\rightarrow$, $\|$, $|$, & and $*$ are called, respectively, the *sequencing, join, selection, intersection,* and *repetition* operators. Later we give also the original more restricted definition that differs only in the use of the join operator.

The intuitive meaning of the operations will be apparent from the semantic interpretation given below. For easier readability we omit the outermost parentheses of an expression. All the four binary operations will be associative, so usually also other parentheses may be omitted.

Let Σ be an alphabet such that each symbol of Σ denotes a distinct process. In order to define the meaning of synchronization expressions, for each $a \in \Sigma$ we associate symbols a_s and a_t to denote, respectively, the *start* and the *termination* of the process. Also, we denote

$$\Sigma_s = \{a_s \mid a \in \Sigma\}, \quad \Sigma_t = \{a_t \mid a \in \Sigma\}.$$

The synchronization languages satisfy the intuitively natural condition that the start of an occurrence of a process always precedes the termination of the same occurrence. The condition is defined formally as follows. For $a \in \Sigma$, let $p_a : (\Sigma_s \cup \Sigma_t)^* \longrightarrow \{a_s, a_t\}^*$ be the morphism determined by the conditions

$$p_a(x) = \begin{cases} x, & \text{if } x \in \{a_s, a_t\}, \\ \lambda, & \text{if } x \notin \{a_s, a_t\}. \end{cases}$$

A word $w \in (\Sigma_s \cup \Sigma_t)^*$ is said to satisfy the start-termination condition (or st-condition for short) if for all $a \in \Sigma$, $p_a(w) \in (a_sa_t)^*$. If w satisfies the st-condition, then w belongs to the shuffle of some languages $((a_1)_s(a_1)_t)^*$, $\ldots$, $((a_k)_s(a_k)_t)^*$ where $a_i \neq a_j$, when $i \neq j$, $i, j = 1, \ldots k$.

The set of all words over $\Sigma_s \cup \Sigma_t$ satisfying the st-condition is denoted W_Σ^{st} and subsets of W_Σ^{st} are called *st-languages.*

The synchronization expressions over an alphabet Σ and the languages denoted by the expressions are defined inductively as follows.

Definition 1. The set of *synchronization expressions* over an alphabet Σ, $\mathrm{SE}(\Sigma)$, is the smallest subset of $(\Sigma \cup \{\phi, \to, \&, |, \|, *, (,)\})^*$ defined inductively by the following rules. The synchronization language denoted by $\alpha \in \mathrm{SE}(\Sigma)$ is $L(\alpha)$.

(i) $\Sigma \cup \{\phi\} \subseteq \mathrm{SE}(\Sigma)$. $L(\phi) = \emptyset$ and $L(a) = \{a_s a_t\}$ when $a \in \Sigma$.
(ii) If $\alpha_1, \alpha_2 \in \mathrm{SE}(\Sigma)$ then $(\alpha_1 \to \alpha_2) \in \mathrm{SE}(\Sigma)$ and $L(\alpha_1 \to \alpha_2) = L(\alpha_1) \cdot L(\alpha_2)$.
(iii) If $\alpha_1, \alpha_2 \in \mathrm{SE}(\Sigma)$ then $(\alpha_1 \| \alpha_2) \in \mathrm{SE}(\Sigma)$ and $L(\alpha_1 \| \alpha_2) = \omega(L(\alpha_1), L(\alpha_2)) \cap W_\Sigma^{\mathrm{st}}$.
(iv) If $\alpha_1, \alpha_2 \in \mathrm{SE}(\Sigma)$ then $(\alpha_1 \mid \alpha_2) \in \mathrm{SE}(\Sigma)$ and $L(\alpha_1 \mid \alpha_2) = L(\alpha_1) \cup L(\alpha_2)$.
(v) If $\alpha_1, \alpha_2 \in \mathrm{SE}(\Sigma)$ then $(\alpha_1 \& \alpha_2) \in \mathrm{SE}(\Sigma)$ and $L(\alpha_1 \& \alpha_2) = L(\alpha_1) \cap L(\alpha_2)$.
(vi) If $\alpha \in \mathrm{SE}(\Sigma)$ then $\alpha^* \in \mathrm{SE}(\Sigma)$ and $L(\alpha^*) = L(\alpha)^*$.

For concrete examples of synchronization expressions see the last section or [9,10]. When defining the interpretation of the parallelization (join) operator in (iii) we use intersection with the set of st-words. If we would allow arbitrary shuffles, this would produce words like $a_s a_s a_t a_t$ that do not have any meaningful interpretation. The definition implies that $L(a \| a) = L(a \to a)$.

The original definition of synchronization expressions required that the expressions appearing as arguments of the parallelization operator must be over disjoint alphabets. The set of *restricted synchronization expressions*, $\mathrm{SE}_r(\Sigma)$, is defined as in Definition 1 by substituting $\mathrm{SE}_r(\Sigma)$ everywhere for $\mathrm{SE}(\Sigma)$ and modifying the condition (iii) as follows. We denote by $\mathrm{alph}(\alpha)$ the set of symbols of Σ appearing in the expression α. ($\mathrm{alph}(\alpha)$ can naturally be defined inductively following Definition 1.)

(iii)' If $\alpha_1, \alpha_2 \in \mathrm{SE}_r(\Sigma)$ and $\mathrm{alph}(\alpha_1) \cap \mathrm{alph}(\alpha_2) = \emptyset$ then $(\alpha_1 \| \alpha_2) \in \mathrm{SE}_r(\Sigma)$ and $L(\alpha_1 \| \alpha_2) = \omega(L(\alpha_1), L(\alpha_2))$.

In restricted synchronization expressions, always when $\alpha_1 \| \alpha_2$ is defined each word of $\omega(L(\alpha_1), L(\alpha_2))$ satisfies the st-condition so we do not need the intersection with the set W_Σ^{st}.

The family of *synchronization languages* $\mathcal{L}(\mathrm{SE})$ (respectively, *restricted synchoronization languages* $\mathcal{L}(\mathrm{SE}_r)$) consists of all languages $L(\alpha)$ where $\alpha \in \mathrm{SE}(\Sigma)$ (respectively, $\alpha \in \mathrm{SE}_r(\Sigma)$) for some alphabet Σ. Directly from the definition it follows that $\mathcal{L}(\mathrm{SE}_r) \subseteq \mathcal{L}(\mathrm{SE})$. Furthermore, the inclusion is strict since, for instance, the expression $(b \to a^*) \| (a^* \to c)$ denotes a language that is not in $\mathcal{L}(\mathrm{SE}_r)$ [14,18].

3 Partial Characterization Results

Synchronization languages are regular languages satisfying the st-condition and furthermore they are closed under certain types of semi-commutation rules and their extensions. It would be very useful if we could exactly characterize the synchronization languages in terms of closure under some simple rewriting rules. A characterization could yield more efficient algorithms for testing the equivalence of SEs since in many cases it would be easy to test whether a given finite automaton accepts a language satisfying the given semi-commutation properties.

Definition 2. Let Σ be an alphabet. We define the rewriting system $R_\Sigma \subseteq (\Sigma_s \cup \Sigma_t)^* \times (\Sigma_s \cup \Sigma_t)^*$ to consist of the following rules, where $a, b \in \Sigma$, $a \neq b$:

(i) $a_s b_t \to b_t a_s$,
(ii) $a_s b_s \to b_s a_s$,
(iii) $a_t b_t \to b_t a_t$.

The set S_Σ is defined to consist of all rules

$$(a_1)_t \cdots (a_i)_t (a_1)_s \cdots (a_i)_s (b_1)_t \cdots (b_j)_t (b_1)_s \cdots (b_j)_s \qquad (1)$$
$$\to (b_1)_t \cdots (b_j)_t (b_1)_s \cdots (b_j)_s (a_1)_t \cdots (a_i)_t (a_1)_s \cdots (a_i)_s,$$

where $a_1, \ldots, a_i, b_1, \ldots, b_j$ are pairwise distinct elements of Σ, $i, j \geq 1$. Then we denote

$$R'_\Sigma = R_\Sigma \cup S_\Sigma.$$

Note that the rules R_Σ define a *semi-commutation relation* on Σ [4]. By symmetry the rules (ii) and (iii) could be written also as two-directional rules.

If R is a rewriting system over Σ and $L \subseteq \Sigma^*$ we denote $\Delta_R^*(L) = \{u \in \Sigma^* \mid (\exists v \in L) v \to_R^* u\}$. We say that L is closed under R if $L = \Delta_R^*(L)$. The following results hold.

Theorem 1. *Let $L \subseteq \Sigma^*$.*

(a) [18] *If L is a synchronization language, then L is closed under R_Σ.*
(b) [9,10] *If L is a restricted synchronization language, then L is closed under R'_Σ.*

The intuitive explanation why restricted synchronization languages are closed under rules (1) is that if we have a subword of the form $\ldots a_t a_s b_t b_s \ldots$ (or the given more general form), then all the following pairs of the given occurrences of the symbols are parallel: (a_t, b_t), (a_s, b_s) and (a_s, b_t), i.e., they correspond to symbols occuring in different arguments of the parallelization operator. On the other hand, since the different arguments may not contain occurrences of the same symbol of Σ, the only possibility is that the symbols a_t, a_s correspond to occurrences in one of the arguments, and b_t, b_s in the other argument. The proof showing closure under the rules of R_Σ is similar but simpler.

It was conjectured [10] that closure under the rewriting system R'_Σ together with the st-condition exactly characterizes the restricted synchronization languages. The conjecture was proved in [5] for alphabets corresponding to two actions.

Theorem 2. [5] *A language* $L \subseteq \{a_s, a_t, b_s, b_t\}^*$ *is a restricted synchronization language if and only if L is regular, satisfies the st-condition and is closed under* R'_{Σ}.

At the same time [5,14] established that the conjecture does not hold in general. The counter-example given there is the language

$$L_1 = \Delta^*_{R'_{\Sigma}}(b_s(a_s a_t)^* c_s b_t (a_s a_t)^* c_t)$$

which is a regular st-language and it can be shown that L_1 is not a restricted synchronization language. The intuitive idea is that restricted synchronization expressions cannot differentiate in L_1 between instances of the action a that occur during b and during c, respectively. Formally the fact that $L_1 \notin \mathcal{L}(\mathrm{SE}_r)$ is proved by establishing a so called switching property for restricted synchronization languages [5,14].

One can show that $\mathcal{L}(\mathrm{SE}_r)$ is closed under certain extensions of R'_{Σ} satisfying nice projection properties onto subalphabets [14,15]. However, the hope of obtaining a rewriting characterization for this class was destroyed by the following result.

Theorem 3. [15,6] *There does not exist any rewriting system Q such that each* $L \in \mathcal{L}(\mathrm{SE}_r)$ *is closed under Q and each regular st-language closed under Q is in* $\mathcal{L}(\mathrm{SE}_r)$.

It may be noted that another undesirable property of the restricted synchronization languages is that the corresponding rewriting rules (1) do not, in general, preserve regularity of st-languages. Consider the language $L_2 = a_s b_s (a_t a_s b_t b_s)^* b_t a_t$. Using the rule $a_t a_s b_t b_s \to b_t b_s a_t a_s$ (and the rules (i)–(iii) of Definition 2) every word of L_2 can be rewritten to a word $(b_s b_t)^m (a_s a_t)^m$, $m \geq 1$, and this observation easily implies that $\Delta^*_{R'_{\Sigma}}(L_2)$ is not regular.

Due to the negative result of Theorem 3 it is natural to consider the more general definition of SEs given in Definition 1. This allows us to drop the rules (1) from the rewriting system describing closure properties of the synchronization languages. Making use of general properties of semi-commutations [4] we can then prove:

Proposition 1. [18] *If* $L \subseteq (\Sigma_s \cup \Sigma_t)^*$ *is a regular st-language, then* $\Delta^*_{R_{\Sigma}}(L)$ *is also regular.*

Using the more general definition, in the case of finite languages we can strengthen the result of Theorem 1(a) into an "if and only if" condition.

Theorem 4. [18] *Assume that L is a finite language over* $\Sigma_s \cup \Sigma_t$. *Then* $L \in \mathcal{L}(\mathrm{SE})$ *if and only if L is an st-language closed under* R_{Σ}.

Interestingly it turns out that closure under R_{Σ} exactly characterizes the images of synchronization languages under functions called st-morphisms [15,16].

Definition 3. Let Σ and Ω be alphabets and $\varphi : \Sigma^* \longrightarrow \Omega^*$ a strictly alphabetical morphism (that is, $\varphi(a) \in \Omega$ for each $a \in \Sigma$). The morphism φ is extended in the natural way to a morphism $\bar{\varphi} : (\Sigma_s \cup \Sigma_t)^* \longrightarrow (\Omega_s \cup \Omega_t)^*$ by setting $\bar{\varphi}(a_s) = \varphi(a)_s$ and $\bar{\varphi}(a_t) = \varphi(a)_t$ for each $a \in \Sigma$.

With each strictly alphabetical morphism $\varphi : \Sigma^* \longrightarrow \Omega^*$, we associate a partial function $\hat{\varphi}$ called an *st-morphism:*

$$\hat{\varphi} = \{(u, \bar{\varphi}(u)) \mid u \in W_\Sigma^{\mathrm{st}} \text{ and } \bar{\varphi}(u) \in W_\Omega^{\mathrm{st}}\}.$$

The family of st-morphisms is denoted Φ_{st}.

Note that $\hat{\varphi}$ is equal to the relation $(\cap W_\Omega^{\mathrm{st}}) \circ \bar{\varphi} \circ (\cap W_\Sigma^{\mathrm{st}})$ and an st-morphism is not, strictly speaking, a morphism. For instance, if $\varphi(a) = \varphi(b)$ then $\hat{\varphi}(a_s b_s a_t b_t)$ is not defined.

Theorem 5. [16] *$\Phi_{\mathrm{st}}(\mathcal{L}(\mathrm{SE}))$ equals to the family of regular st-languages $L \subseteq \Sigma^*$ that are closed under the rewriting system R_Σ.*

We conjecture that in general a regular st-language is a synchronization language if and only if it is closed under R_Σ, i.e., that we could drop the finiteness condition from Theorem 4. However, we do not have a proof for this conjecture. According to Theorem 5, the conjecture is equivalent to showing that

$$\Phi_{\mathrm{st}}(\mathcal{L}(\mathrm{SE})) = \mathcal{L}(\mathrm{SE}),$$

that is, that the family of synchronization languages is closed under st-morphisms. In [15] it is shown that $\Phi_{\mathrm{st}}(\mathcal{L}(\mathrm{SE}_r)) = \Phi_{\mathrm{st}}(\mathcal{L}(\mathrm{SE}))$ which implies that the family $\mathcal{L}(\mathrm{SE}_r)$ is not closed under st-morphisms since it is strictly included in $\mathcal{L}(\mathrm{SE})$.

4 Implementation of SEs in Concurrent Programming Languages

In this section, we explain how the ideas of SEs and SLs can be used and implemented in a programming language that is designed for a parallel or/and distributed computing environment. The implementation follows strictly our semantic interpretation of SEs using SLs. More specifically, for an SE, we construct at compile time a finite automaton that accepts the prefix language of the SL denoted by the SE. Then the constructed automaton is used to impose the synchronization constraints specified by the SE at runtime.

In the following, we will use examples written in the *ParC* concurrent programming language [8]. The examples will be explained in detail assuming that the reader knows the C programming language but not *ParC*. In *ParC*, (simple or compound) statements are considered as the basic elements of synchronization. Statements that are involved in synchronization are represented in SEs by ***statement tags***. In other words, the alphabet of an SE is a set of statement tags. This is a design decision specific to *ParC* and need not be true in general. Especially in an object-oriented concurrent programming language, it may be more appropriate that object operations rather than statements should be the basic elements of synchronization specified by SEs. However, the implementation principle described below would still be applicable as long as a symbol in an SE represents an entire execution of a process.

We provide in the following a simple example to show how a ParC program with synchronization expressions is translated into a Sequent C program with system calls [13]. We first give the ParC program with a detailed explanation. Then we give the generated code in Sequent C with system calls.

```
main(){
    shared int w;
    int r;
    int f();
    tag a, b;
    restrict (a->b)*;

    pexec{
        {/* ...... */
         b:: r = w;
         /* ...... */
        }
        {/* ...... */
         a:: w = f();
         /* ...... */
        }
    }
}
```

We first look at the `pexec` statement, i.e., the *parallel execution* statement. This language construct denotes that all statements that are directly under the scope can be executed in parallel. In the above program, there are two compound statements directly under the `pexec` statement, which can be considered as two parallel processes. In the first process, the statement "r = w;" is tagged with `b`, and in the second process, "w = f();" is tagged with `a`. We assume that each of the two statements is in a loop. The `restrict` statement above the `pexec` statement specifies a synchronization expression `(a->b)*`, meaning that the `a` and `b` statements can be executed only in the following sequence: `a, b, a, b,` ..., i.e., the ith instance of `b` cannot start its execution before the ith instance of `a` finishes and the $(i+1)$th instance of `a` cannot start its execution before the ith instance of `b` finishes, $i \geq 1$. Also in the declarations, `w` is declared as a shared integer variable, which means that the two appearances of `w` in the two parallel processes, respectively, are the same variable. The integer variable `r` is not shared. So, it has different copies in different parallel processes in the scope. And the statement `tag a, b;` declares that both `a` and `b` are statement tags.

A deterministic finite automaton (DFA) is built according to the SE `a->b`, which accepts the prefix language of the SL defined by the SE. The DFA is shown in Figure 1.

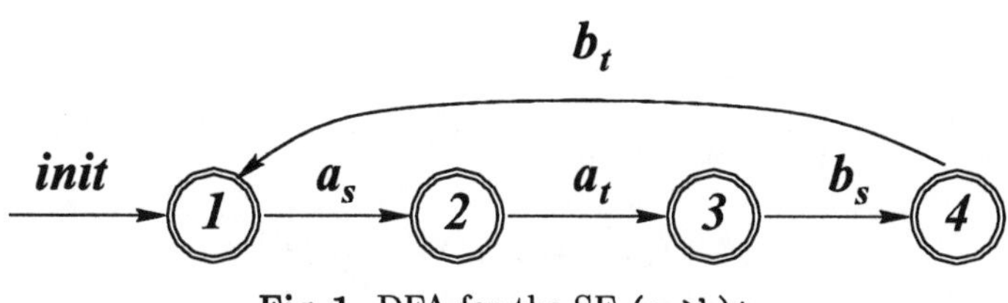

Fig. 1. DFA for the SE `(a->b)*`

To ensure that the state update can only be performed exclusively by one process, we build a macro called

`wait_and_set(int state; int` i_1, j_1`; ...; int` i_n, j_n`).`

for `n > 0`. The macro waits until the condition "`state ==` i_t", for some t, $1 \le t \le n$, becomes true and then sets "`state =` j_t". Instead of using a variable-length list `int` i_1, j_1`; ...; int` i_n, j_n, the actual implementation of **`wait_and_set`** may use a pointer to a table (an array) for the transitions. Here we use the list in order to make the discussion easier to follow. The macro is an indivisible (exclusive) operation and can be implemented using primitive synchronization mechanisms of the system.

Then the state transitions of the DFA correspond to the following Sequent C statements or macros:

- Initialization
  `_state = 1;`
- The a_s transition
  `wait_and_set(_state; 1, 2);`
- The a_t transition
  `wait_and_set(_state; 2, 3);`
- The b_s transition
  `wait_and_set(_state; 3, 4);`
- The b_t transition
  `wait_and_set(_state; 4, 1);`

The following is the code generated from the ParC program:

```
#include<parallel/microtask.h>
#include<parallel/parallel.h>
#include<sys/wait.h>
```

```
#include<sys/types.h>
#include<stdio.h>

main() {
shared int w;
int k, r;
shared int _state=1;
int f();

    {
    int _pw, _pid;

        if ((_pid=fork()) == 0){
            /* ...... */
            wait_and_set(_state; 3, 4);
                r = w;
            wait_and_set(_state; 4, 1);
            /* ...... */
            exit(0);
        }
        else if (_pid < 0){
            printf("fork error\n");
            exit(1);
        }
        /* ...... */
        wait_and_set(_state; 1, 2);
        w = f(k);
        wait_and_set(_state; 2, 3);
        /* ...... */
    }
}
```

For a slightly more complicated SE "`(a || b)->c`", a corresponding DFA is shown in Figure 2.

Then the code generated for the `a` statement would be

```
wait_and_set(_state; 1, 2; 3, 4; 6, 8);
............ /* the a statement itself */
wait_and_set(_state; 2, 5; 4, 7; 8, 9);
```

Similarly, the code generated for `b` statement would be

```
wait_and_set(_state; 1, 3; 2, 4; 5, 7);
............ /* the b statement itself */
wait_and_set(_state; 3, 6; 4, 8; 7, 9);
```

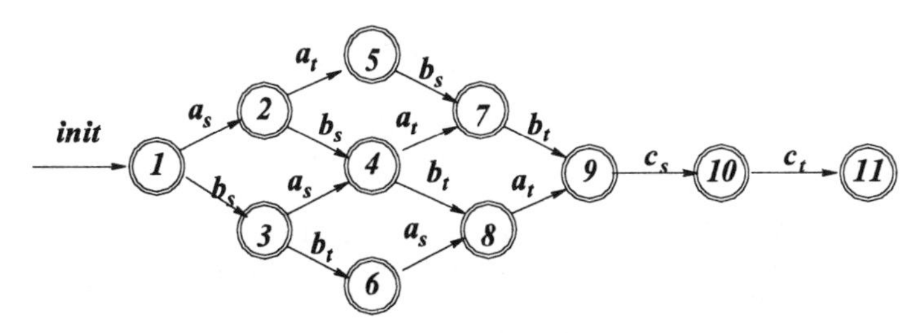

Fig. 2. DFA for the SE `(a || b)->c`

And the following is the code for `c`:

```
wait_and_set(_state; 9, 10);
............ /* the c statement itself */
wait_and_set(_state; 10, 11);
```

With the above two simple examples, the basic idea of implementing SEs using SLs should be clear. There are many other issues concerning implementation. We mention several of them briefly in the following.

The first is the *size-explosion problem* of DFA. For many practical synchronization problems, the sizes of the DFA can be too large to implement. We suggest the use of alternating finite automata (AFA) [2,12] instead of DFA in the implementation of SEs. The use of AFA can significantly increase the space efficiency.

The checking of states may become the bottleneck of synchronization since it is done sequentially. This is not necessarily true. Note that SEs within different concurrent blocks, respectively, can be implemented by automata that actually run concurrently.

Conflicting synchronization constraints and deadlock conditions caused by the definitions of SEs in the same scope can be examined by checking the intersection of the SLs defined by the SEs and the execution sequences defined by the program flow. The precise semantic definition of SEs makes such checking conceptually clear and practically possible to implement.

References

1. Book, R. V., Otto, F. (1993) String-Rewriting Systems. Texts and Monographs in Computer Science. Springer-Verlag
2. Cheung, H., Salomaa, K., Wu, X., Yu, S. (1997) An efficient implementation of regular languages using r-AFA. In: Proceedings of the Second International Workshop on Implementing Automata, WIA'97, 33–42
3. Ciobanu, G., Salomaa, K., Yu, S. Synchronization languages and ST semantics, manuscript in preparation
4. Clerbout, M. Latteux, M., Roos, Y. (1995) Semi-commutations. In: The Book of Traces. (Diekert, V., Rozenberg, G. eds.) World Scientific, Singapore, 487–552

5. Clerbout, M., Roos, Y., Ryl, I. Synchronization languages. Theoret. Comput. Sci., to appear
6. Clerbout, M., Roos, Y., Ryl, I. (1998) Langages de synchronization et systèmes de réécriture. Tech. Rep. IT-98-311, Université des Sciences et Technologies de Lille
7. Diekert, V., Métivier, Y. (1997) Partial commutation and traces. In: Handbook of Formal Languages, Vol. III. (Rozenberg, G., Salomaa, A. eds.) Springer-Verlag, 457–533
8. Govindarajan, R., Guo, L., Yu, S., Wang, P. (1991) ParC Project: Practical constructs for parallel programming languages. In: Proc. of the 15th Annual IEEE International Computer Software & Applications Conference, 183–189
9. Guo, L., Salomaa, K., Yu, S. (1994) Synchronization expressions and languages. In: Proc. of the 6th IEEE Symposium on Parallel and Distributed Processing, (Dallas, Texas). IEEE Computer Society Press, 257–264
10. Guo, L., Salomaa, K., Yu, S. (1996) On synchronization languages. Fundamenta Inform. **25**, 423–436
11. Hennessy, M. (1989) Algebraic Theory of Processes. The MIT Press, Cambridge, Mass.
12. Huerter, S., Salomaa, K., Wu, X., Yu, S. (1998) Implementing reversed alternating finite automaton (r-AFA) operations. In: Proc. of the Third International Workshop on Implementing Automata, WIA'98, 54-64
13. Osterhaug, A. (ed.) (1989) Guide to Parallel Programming — On Sequent Computer Systems. Prentice Hall, Englewood Cliffs, New Jersey
14. Ryl, I., Roos, Y., Clerbout, M. (1997) Partial characterization of synchronization languages. In: Proc. of 22nd International Symposium on Mathematical Foundations of Computer Science, MFCS'97. Lect. Notes Comput. Sci. **1295**, 209–218
15. Ryl, I., Roos, Y., Clerbout, M. (1998) About synchronization languages. In: Proc. of 23rd International Symposium on Mathematical Foundations of Computer Science, MFCS'98. Lect. Notes Comput. Sci. **1450**, 533–542
16. Ryl, I., Roos, Y., Clerbout, M. (1998) Generalized synchronization languages and commutations. Tech. Rep. IT-315, L.I.F.L., Univ. Lille 1
17. Salomaa, A. (1973) Formal Languages. Academic Press, New York
18. Salomaa, K., Yu, S. (1998) Synchronization expressions with extended join operation. Theoret. Comput. Sci. **207**, 73–88
19. Sassone, V., Nielsen, M., Winskel, G. (1996) Models of concurrency: Towards a classification. Theoret. Comput. Sci. **170**, 297–348
20. Yu, S. (1997) Regular languages. In: Handbook of Formal Languages, Vol. I. (Rozenberg, G., Salomaa, A. eds.) Springer-Verlag, 41–110

Part II

Automata II: More General Devices

Uniformization of Rational Relations

Christian Choffrut and Serge Grigorieff

Summary. Uniformizing a relation belonging to some family, consists of finding a function whose graph belongs to that family and whose domain coincides with that of the given relation. Here we focus on relations on finite or infinite strings that can be recognized by some version of finite automata.

Eilenberg showed that rational relations of finite strings have the uniformization property. We examine how this result can or cannot be extended. We show that given a length preserving n-ary rational relation and one of its component it can be uniformized, when viewed as a binary relation froma this component to the remaining components, and that binary deterministic relations also have the uniformization property. Finally, we shall show that rational relations on infinite strings can be uniformized.

1 Introduction

Uniformizing a relation belonging to some family, consists of finding a function whose graph belongs to the family and whose domain coincides with that of the given relation. Here we are particularly concerned with the relations on finite or infinite strings that can be recognized in the traditional sense by some type of finite automaton.

Eilenberg proved in 1974 a uniformization result for rational relations on finite strings. Siefkes established in 1975 that the synchronous relations on infinite strings enjoy the uniformization property as well. Actually these results can be refined to subfamilies of rational relations on finite or infinite strings or a mixture of those: to name but the two most important, say the deterministic and the synchronous relations. Our purpose is to give a survey of all the known results of this type and to show how far they can or cannot be extended. In order to more accurately evaluate how lucky we are with dealing with strings, suffice it to say that uniformization results on trees no longer hold, see [14]. Theoretical computer science and logic have studied the subject with different tools. We think it is time to present these results in a unifying framework by bringing the two approaches together. We hope the reader will be convinced that using both the methods of theoretical computer science and those of logic helps greatly simplifying and clarifying some proofs. A good illustration is the investigation of the synchronous relations, whether on finite or infinite strings, where the language of logic spares some tedious (but of course equivalent) set constructions. This is no wonder since it allows us to use "for free" the powerful theory developed by Büchi.

Historically, the uniformization result on rational relations on finite strings can be traced back to [7], i. e., over 30 years ago. It was not stated as such, rather it was given a more precise form (technically the function by which one can uniformize a given relation is obtained as a composition of a left followed by a right "sequential" function). Since then it has been reproved with different methods, [1], [16], see also [15] for an account on the subject. Eilenberg proved it as a corollary of his "cross-section" theorem, stating intuitively that it is possible to "rationally" select a representative for each equivalence class that intersects a rational subset. This result carries over to infinite strings as well.

As previously said, most of the material here can be considered as "folklore" by some (actually non so many) researchers. There is one exception however: the proof of the uniformization property for rational relations on infinite strings seems to be new, [13]. Using a different approach, a similar result was independently obtained by D. Beauquier, J. Devolder, M. Latteux and E. Timmerman, but has not been published.

2 Preliminaries

2.1 Basics on uniformization

Let us recall a few elementary definitions in order to fix the notations. Given two sets X and Y and a (partially defined) function $f : X \to Y$, the ***graph*** of f is the subset $\sharp f = \{(u,v) \mid v = f(u)\}$. The *domain* $\mathrm{dom}(R)$ of a relation $R \subseteq X \times Y$ is the set of elements $x \in X$ for which there exists an element $y \in Y$ with $(x,y) \in R$. The *composition* of relations is the operation that associates with the relations $R \subseteq X \times Y$ and $S \subseteq Y \times Z$ the relation

$$R \circ S = \{(x,z) \in X \times Z \mid \text{there exists } y \in Y, \text{ with} (x,y) \in R \text{ and } (y,z) \in S\} \tag{1}$$

We compose the functions from left to right.

Now we come to the main definition of this work. Given a family $\mathcal{F}$ of relations in $X \times Y$ and $R \in \mathcal{F}$, *uniformizing* R in $\mathcal{F}$ means finding a function $f_R : X \to Y$ such that 1) $\mathrm{dom}(f_R) = \mathrm{dom}(R)$ 2) $\sharp(f_R) \subseteq R$ and 3) $\sharp(f_R) \in \mathcal{F}$. When no particular mention is given, saying that a relation belonging to a family $\mathcal{F}$ is *uniformizable*, implicitely means that it can be uniformized in $\mathcal{F}$.

2.2 Finite and infinite strings

Given a finite *alphabet* A whose elements are *symbols* or *letters*, we denote by A^* (resp. A^ω) the set of finite (resp. infinite) strings over A. As usual we denote by $A^\infty = A^* \cup A^\omega$ the set consisting of the finite and infinite strings.

The empty string is denoted by ϵ regardless of which alphabet in relates to as no confusion usually arises and $|u|$ denotes the length of a string with the convention $|u| = \infty$ whenever $u \in A^\omega$.

This paper is concerned with direct products of sets of the form A^* and A^ω. One of the central tools that is used in this theory is that of *hierarchical ordering.* We recall that given a linear ordering $<$ on an alphabet A, we extend it to the free monoid A^* by posing $u <_{\text{hier}} v$ if $|u| < |v|$ or if $|u| = |v|$ and there exist $w, u_1, v_1 \in A^*$ and $a, b \in A^*$, such that $u = wau_1$, $v = wbv_1$ and $a < b$ holds.

The notion of *lexicographical ordering* is more general. Consider a collection of sets E_i, where i ranges over an initial segment of the integers or over $\mathbb{N}$. Assume there exists a (possibly partial) ordering $<_i$ on each set E_i. We endow the direct product $\prod_{i \in I} E_i$ with the *lexicographical ordering* $<_{\text{lex}}$ defined by $\prod_{i \in I} x_i <_{\text{lex}} \prod_{i \in I} y_i$ if there exists $i \in I$ such that $x_j = y_j$ for all $j < i$ and $x_i <_i y_i$. This construction applies in particular to A^ω. Indeed, this set can be viewed as a collection of copies of A indexed by $\mathbb{N}$. Any linear ordering on A extends to a lexicographical ordering on A^ω.

As usual we will assume the set A^ω is endowed with the product topology of the discrete topology on A where the family of subsets of the form uA^ω with $u \in A^*$ form a basis of the open sets. It is a standard result that given an arbitrary lexicographical ordering on A^ω every topologically closed subset of A^ω contains its lexicographically minimal element.

3 Relations on finite strings

3.1 Rational relations on finite strings

Given an arbitrary monoid M, the least family $\mathcal{F}$ of subsets containing all finite sets and closed under set union, concatenation (i. e., X and Y are in $\mathcal{F}$ then so is $\{xy \mid x \in X, y \in Y\}$) and Kleene closure (i. e., if X is in $\mathcal{F}$ then so is $\{\epsilon\} \cup X \cup \ldots \cup X^i \ldots$) is the family of *rational* subsets and is denoted by $\mathrm{Rat}(M)$. As a particular case, given two monoids M and N, a function of M into N is *rational* if its graph is a rational subset of the product monoid $M \times N$. We refer the reader to the two handbooks [5] and [2] for basic results in this theory.

It is well-known that in the case of a direct product of free monoids $A_1^* \times \ldots \times A_n^*$, the family of rational subsets, also called *rational relations*, is precisely the family of relations recognized by finite automata. Indeed, the notion of finite automaton designed to recognize single strings, was extended in the late fifties in such a way as to operate on n-tuples of strings. The idea is to provide an automaton with as many tapes as there are components in the tuple.

More precisely assume without loss of generality that the n alphabets A_i are disjoint and set $A = \bigcup_{1 \leq i \leq n} A_i$. We denote by π_i the projection of A^*

onto A_i^* for all $i = 1, \ldots, n$ and by π the projection onto the direct product $A_1^* \times \ldots \times A_n^*$: $\pi(w) = (\pi_1(w), \ldots, \pi_n(w))$. A finite *n-tape automaton* (we shall say also more simply an *automaton*) is a quadruple $\mathcal{A} = (Q, I, F, T)$ where Q is the finite set of *states*, $I \subseteq Q$ is the set of *initial* states, $F \subseteq Q$ is the set of *final* states and $T \subseteq Q \times A \times Q$ the set of *transitions*. The subset of $A_1^* \times \ldots \times A_n^*$ recognized by Aconsists of those n-tuples of strings $(\pi_1(w), \ldots, \pi_n(w))$ where w is the label of a *successful* path, i. e., a path starting in an initial state and ending in a final state (see [2, section III.6] for background on n-tape automata where they are called finite transducers).

The following is well-known, see [5, Thm IX. 2. 2.] or [2, Thm 4.1].

Theorem 1. *A relation $R \subseteq A_1^* \times \ldots \times A_n^*$ is rational if and only if there exist a rational subset $K \subseteq A^*$ such that*

$$R = \{\pi(w) \mid w \in K\}$$

There exists a deterministic version of such automata but contrarily to the free monoids, they are expressively less powerful than the non deterministic ones. Intuitively, there exists a decomposition $Q = \bigcup_{1 \leq i \leq n} Q_i$ where Q_i corresponds to the subalphabet A_i. In a state $q \in Q_i$, only transitions of letters of the subalphabet A_i are allowed and moreover a given letter defines at most one transition. Furthermore the ability to recognize the end of a component is required. More formally, we assume the alphabets A_i contain an extra symbol $\sharp_i$ (the "end-marker" of the i-th tape). We modify the π_i's by considering them as mappings of A^* into $(A_i - \sharp_i)^*$ satisfying $\pi_i(a) = a$ if $a \in A_i - \sharp_i$ and $\pi_i(a) = \epsilon$ otherwise.

An automaton is *deterministic* whenever the transitions satisfy the three conditions

$$\left.\begin{array}{l}
\text{for all } (q,a,p),(q,b,r) \in T \text{ if } a \in A_i \text{ and } b \in A_j \text{ then } j = i \\
\text{for all } (q,a,p),(q,b,r) \in T \text{ if } a = b \text{ then } p = r \\
\text{for all } (q,\sharp_i,r) \in T \text{ if } w \text{ is the label of a path leaving } r \\
\qquad\qquad\qquad \text{then } w \in (A - A_i)^*
\end{array}\right\} \quad (2)$$

A relation is *deterministic rational* if there exists a deterministic automaton in the above sense that recognizes it. Then the following is a paraphrase of the definition.

Proposition 1. *A relation $R \subseteq A_1^* \times \ldots \times A_n^*$ is recognized by a deterministic n-tape automaton if and only if the rational subset $K \subseteq A^*$ of Theorem 1 can be assumed to satisfy the two conditions*

$$\left.\begin{array}{l}
\text{for all } u,v,w \in A^*, \text{ if } uv, uw \in K, v \in A_iA^*, w \in A_jA^* \text{ then } i = j \\
\text{for all } u \in A^*, \text{ if } u\sharp_i v \in K \text{ then } v \in (A - A_i)^*
\end{array}\right\} \quad (3)$$

3.2 Synchronous relations on finite strings

Synchronous relations form an important subfamily the rational relations which enjoys nice closure properties. In particular it forms a Boolean algebra and some of its properties are decidable, whereas almost all properties of the general rational relations are undecidable (Post Correspondence Problem can be interpreted as a question on two rational relations).

Consider a fresh symbol $\sharp$ not belonging to the A_i's. With each n-tuple $(u_1, \ldots, u_n) \in \prod_{1 \le i \le n} A_i^*$ associate the n-tuple of strings of the same length defined as

$$(u_1, \ldots, u_n)^\sharp = (u_1 \sharp^{\ell - |u_1|}, \ldots, u_n \sharp^{\ell - |u_n|}) \text{ with } \ell = \max_i |u_i| \tag{4}$$

Extending the notation to subsets $R \subseteq A_1^* \times \ldots \times A_n^*$ in the natural way, we identify $R^\sharp$ with a subset of strings over the alphabet $\Sigma = \prod_{1 \le i \le n} (A_i \cup \{\sharp\})$. Then the relation R is *synchronous* if the subset $R^\sharp$ is recognized by a finite automaton over the alphabet Σ. It is not difficult to verify that the synchronous relations form a subfamily of the rational relations that is closed under the Boolean operations, composition of relations, direct products and projections (e. g., [7] where these relations were called FAD-relations or [8]). Finally a function $f : A_1^* \times \ldots \times A_n^* \to B_1^* \times \ldots \times B_m^*$ is *synchronous* if its graph $\sharp f$ is a synchronous relation of $A_1^* \times \ldots \times A_n^* \times B_1^* \times \ldots \times B_m^*$.

The set of synchronous relations has been logically characterized in [6]. For the reader's convenience we recall the logical language that defines it. The signature contains two symbols $<$ and E of binary predicates and a symbol T_a of unary predicate for each letter $a \in A = \bigcup_{1 \le i \le n} A_i$. The first order language in question is defined on this signature. The individual variables belong to the disjoint union of denumerable sets X_i, for $1 \le i \le n$. All formulae are interpreted as follows. The universe is the union of the A_i^*'s, for $1 \le i \le n$, and an individual variable $x \in X_i$ is interpreted as a string in A_i^*. Now $u < v$ is true if and only if u and v belong to the same free monoid A_i^* for some $1 \le i \le n$ and u is a prefix of v. Furthermore, uEv is true if and only if u and v have the same length and finally $T_a(u)$ for some $a \in A$ is true if and only if the last letter of u is a. To each formula $\phi(x_1, \ldots, x_n)$ with set of free variables $x_1 \in X_{k_1}, \ldots, x_n \in X_{k_n}$ is assigned the set R of all n-tuples $(u_1, \ldots, u_n) \in A_{k_1}^* \times \ldots \times A_{k_n}^*$ such that ϕ is true when each u_i is substituted for x_i in ϕ. It is said that ϕ *defines* R or that R satisfies ϕ.

Theorem 2. *A subset $R \subseteq A_1^* \times \ldots A_n^*$ is synchronous if and only if it is defined by some formula ϕ of the above language.*

As an immediate result we get

Corollary 1. *Let $0 \leq k \leq n$ be some integer. Each synchronous relation $R \subseteq A_1^* \times \ldots \times A_n^*$ can be uniformized by some synchronous function $f : A_1^* \times \ldots \times A_k^* \to A_{k+1}^* \times \ldots \times A_n^*$.*

Proof. Observe first that the hierarchical ordering on the free monoid can be easily expressed in the logic. Also, we can express the fact that a n-tuple of strings $x_1, \ldots, x_n$ is lexicographically less than or equal to another $y_1, \ldots, y_n$. Now, let $\phi(x_1, \ldots, x_n)$ be a formula defining R. It suffices to associate with each k-tuple of $A_1^* \times \ldots \times A_k^*$ belonging to the domain of R, the (in the lexicographical ordering) least $(n-k)$-tuple of $A_{k+1}^* \times \ldots \times A_n^*$ which is associated to it. We leave it to the reader to work out the details.

3.3 Uniformization on finite strings

Intuitively, Eilenberg's cross-section theorem, [5, Thm. IX, 7. 1.] asserts that given an morphism $f : A^* \to B^*$ and a rational subset $K \subseteq A^*$, it is possible to "rationally" select a representative among all the elements of K that map onto the same element of B^*.

Theorem 3. *Let $f : A^* \to B^*$ be a morphism and let K be a rational subset of A^*. Then there exists a rational subset $L \subseteq K$ such that f maps L bijectively onto Kf.*

This result and its appoach have been widely commented, used and reproven. Traditionally, it has two major consequences: 1) each rational function f of a free monoid into another can be recognized by some "unambiguous" 2-tape automaton (each pair of strings (u, v) with $v = f(u)$ defines at most one successful path in the automaton) and 2) all rational relations of a free monoid into another are uniformizable which is precisely the result that this paper wants to extend to infinite strings, [5, Prop. IX, 8. 2].

Proposition 2. *Each rational (resp. deterministic rational) relation can be uniformized*

Proof. For rational relations this follows from Theorem 1 where the subset L of the previous theorem is substituted for K. For deterministic rational relations it suffices to observe that condition (2) still holds for all subsets of K.

Observe that the previous result cannot be extended to two or more components. Indeed, consider the following rational relation on the direct product $\{a, b\}^* \times \{a\}^* \times \{a\}^*$

$$\{(a^n b^m, a^n, a) \mid n, m \geq 0\} \cup \{(a^n b^m, a^m, \epsilon) \mid n, m \geq 0\}$$

and assume there exists a rational function $f : \{a, b\}^* \times \{a\}^* \to \{a\}^*$ that uniformizes it. Let $X_0, X_1 \subseteq \{a, b\}^* \times \{a\}^*$ be the pre-images of a and ϵ

respectively and let $\mathcal{A}_0$ and $\mathcal{A}_1$ be finite automata recognizing X_0 and X_1. Denote by m the maximal number of states in these automata and by $\mu > m$ an integer which is a multiple of the number of occurrences of a (resp. b) in any simple cycle of $\mathcal{A}_0$ and $\mathcal{A}_1$ (a simple cycle is a path where initial and final states coincide and where no other state is visited more than once). Set

$$p_0 = \max\{n \in \mathbb{N} \mid (a^n b^n, a^n)f = a\} \text{ and } p_1 = \max\{n \in \mathbb{N} \mid (a^n b^n, a^n)f = \epsilon\}$$

Assume first $p_0 < \infty$ and let $N > \max\{m, p_0\}$. Consider $(a^N b^{N+\mu}, a^N) \in X_0$. By the pigeon-hole principle applied to the m first states visited by a path labelled by $(a^N b^{N+\mu}, a^N)$ there exists a cycle labelled by a pair (a^p, a^p) with $p < m$. Since μ is a multiple of p, the pair $(a^{N+\mu} b^{N+\mu}, a^{N+\mu})$ belongs to X_0, a contradiction. So we must assume that $p_0 = \infty$. A similar argument shows that $p_1 = \infty$. For some integer M greater than $m + \mu$ we have $(a^M b^M, a^M) \in X_1$. Then $(a^{M-\mu} b^M, a^{M-\mu}) \in X_0$. By the same pigeon-hole principle applied again to the m first states visited by a path labelled by $(a^{M-\mu} b^M, a^{M-\mu})$, there exists a cycle labelled by a pair (a^p, a^p) with $p < m$. Thus $(a^M b^M, a^M)$ belongs to X_0, a contradiction.

Using similar techniques we would prove that $X \cup Y \subseteq \{a\}^* \times \{a\}^* \times \{a, b\}^*$ with

$$X = \{(a^{2m}, a^p, a^m b^p) \mid m, p \geq 0\} \text{ and } Y = \{(a^{m+p}, a^m, b^m a^p) \mid m, p \geq 0\}$$

is a rational relation which cannot be uniformized by any rational function of $\{a\}^* \times \{a\}^*$ into $\{a, b\}^*$. However, when all alphabets are unary, rational relations can be uniformized for any subset of components. This follows trivially from the fact that such rational relations are defined by the logic of Presburger arithmetic, [9].

Proposition 3. *Let $1 \leq k \leq n$ be some integer and let A_i be unary alphabets for $i = 1, \ldots, n$. Each rational relation $R \subseteq A_1^* \times \ldots \times A_n^*$ can be uniformized by some rational function $f : A_1^* \times \ldots \times A_k^* \to A_{k+1}^* \times \ldots \times A_n^*$.*

4 Relations on infinite strings

Büchi generalized in [3] the notion of finite automaton in order to have it operate on infinite strings. The family of subsets of A^ω *recognized* by some Büchi automaton in this manner is denoted by Rat A^ω and is called the family of *rational subsets of infinite strings* (this is justified by the fact that this family is closed under extended "rational" operations, [5, Thm. XIV. 4. 1.]). In the same way traditional finite automata can be used to recognize relations on finite strings, Büchi automata can be used to recognize relations on infinite strings. We refer the interested reader to [11] for a thorough study of these relations. Here, we will only recall what is necessary for our purpose.

4.1 Rational relations on infinite strings

A finite n-tape automaton $\mathcal{A} = (Q, I, F, T)$ can be transformed into a *Büchi* automaton and used to recognize n-tuples of possibly infinite strings by interpreting F as a set of *repeated* states. The definitions of paragraph 3.1 carry over here naturally. An infinite path is *successful* if it starts in an initial state and visits infinitely often a repeated state. The subset of $A_1^\infty \times \ldots \times A_n^\infty$ recognized by the automaton is the set of n-tuples $\pi(w) = (\pi_1(w), \ldots, \pi_n(w))$ where $w \in A^\omega$ is the label of a successful path in the automaton (for all $i = 1, \ldots, n$ the projections π_i extend trivially from A^ω to A_i^∞).

As seen in section 3.3, the key argument for Eilenberg's uniformization result is the cross-section theorem along with what he calls the first factorization theorem (Theorem 1). For infinite strings we obtain the same result via an extension of his "second factorization theorem" which shows that all rational relations is the somposition of a synchronous relation followed by some rational substitution [5, Thm IX. 5.1.].

We recall that a morphism of B^* into A^* is *alphabetic* if it associates with every letter of B a letter of A. A *substitution* of B^* into a monoid M is a morphism of B^* into the power set of all subsets of M. In the case where M is a direct product of free monoids $A_1^* \times \ldots \times A_n^*$, the substitution can be extended from B^ω to $A_1^\infty \times \ldots \times A_n^\infty$. Hereafter we deal with substitutions into $A_1^* \times \ldots \times A_n^*$ that are *rational*, i. e., that map B^* into $\mathrm{Rat}(A_1^* \times \ldots \times A_n^*)$

Theorem 4. *[10, Prop. 2. 1.] Given a relation $R \subseteq A_1^\infty \times \ldots \times A_n^\infty$ the following conditions are equivalent.*

1) R is recognized by some Büchi automaton.

2) there exist a finite alphabet B, a rational subset $K \subseteq B^\omega$, an alphabetic morphism $\varphi : B^* \to A_1^*$ and a rational substitution $\psi : B^* \to \mathrm{Rat}(A_2^* \times \ldots \times A_n^*)$ such that $R = \varphi^{-1} \circ \cap K \circ \psi$ holds, where $\cap K$ is the retriction of the identity to the subset K.

The subsets of $A_1^\infty \times \ldots \times A_n^\infty$ recognized in this manner are called *rational relations*. It can be readily verified that for all rational relations $R \subseteq A_1^\infty \times \ldots \times A_n^\infty$ the relation $R \cap A_1^\omega \times \ldots \times A_n^\omega$ is also rational. More generally, the following holds (e. g., [11]).

Proposition 4. *For $i = 1, \ldots, n$ let $S_i = A_i^*$ or $S_i = A_i^\omega$. Let R be the relation recognized by a Büchi automaton A. Then $R \cap S_1 \times \ldots \times S_n$ is rational.*

From now on we deal with "purely infinite relations" only, i. e., with relations in $A_1^\omega \times \ldots \times A_n^\omega$. As for finite strings, the notion of deterministic automaton exists. A *deterministic* automaton satisfies the following conditions

$$\left.\begin{array}{l} \text{1) for all } (q,a,p),(q,b,r) \in T \text{ if } a \in A_i \text{ and } b \in A_j \text{ then } j = i \\ \text{2) for all } (q,a,p),(q,b,r) \in T \text{ if } a = b \text{ then } p = r \end{array}\right\} \quad (5)$$

A relation $R \subseteq A_1^\omega \times \ldots \times A_n^\omega$ is *deterministic* if there exists a deterministic automaton that recognizes it. It is *synchronous* (resp. *deterministic synchronous*) if viewed as a subset of $(A_1 \times \ldots\ldots \times A_n)^\omega$, it is recognizable by some Büchi (resp. deterministic Büchi) automaton on the alphabet $A_1 \times \ldots \times A_n$.

4.2 Uniformization on infinite strings

The main result of this paper (Theorem 5) is based on the following property which shows that synchronous relations can be uniformized.

Proposition 5. *Let $0 < k < n$ be some integer. Each synchronous relation $R \subseteq A_1^\omega \times \ldots \times A_n^\omega$ can be uniformized by some synchronous function $f : A_1^\omega \times \ldots \times A_k^\omega \to A_{k+1}^\omega \times \ldots \times A_n^\omega$.*

Proof. We first verify that it suffices to consider the case $n = 2, k = 1$. Indeed, consider two bijections $\alpha : A_1 \times \ldots \times A_k \to A$ and $\beta : A_{k+1} \times \ldots \times A_n \to B$ where A and B are new subsets. By identifying $A_1^\omega \times \ldots \times A_k^\omega$ with $(A_1 \times \ldots \times A_k)^\omega$, we may extend α to an isomorphism of $A_1^\omega \times \ldots \times A_k^\omega$ onto A^ω. Similarly we extend β to an isomorphism of $A_{k+1}^\omega \times \ldots \times A_n^\omega$ onto B^ω. The relation $\alpha^{-1} \circ R \circ \beta \subseteq A^\omega \times B^\omega$ is synchronous and can be uniformized by some function $f : A^\omega \to B^\omega$. Then the function $\alpha \circ f \circ \beta^{-1} : A_1^\omega \times \ldots \times A_k^\omega \to A_{k+1}^\omega \times \ldots \times A_n^\omega$ uniformizes R as it can be readily verified.

From now on we deal with a synchronous relation $R \subseteq A^\omega \times B^\omega$ recognized by some Büchi automaton $\mathcal{A} = (Q, I, F, T)$. It is convenient, given a pair $(u, v) \in A^\omega \times B^\omega$, to say u is the *input* and v is the *output* component. A *run* is a finite or infinite sequence of states $(q_i)_{i<n}$, $n \leq \infty$, visited in a path of the automaton, i. e., for which there exist $(u_i, v_i) \in A \times B$ such that $(q_i, u_i, v_i, q_{i+1}) \in T$ holds for all $i < n$. Without loss of generality we may enforce the following additional condition which guarantees that the output is uniquely defined by an input and a run

$$\left.\begin{array}{l}\text{for all } q, p \in Q, a \in A \text{ and } b_1, b_2 \in B \\ \text{if } (q, a, b_1, p), (q, a, b_2, p) \in T \text{ then } b_1 = b_2\end{array}\right\} \quad (6)$$

Our proof follows the usual pattern. It consists of selecting for each input string $u \in A^\omega$ a specific image $v \in B^\omega$ satifying $(u, v) \in R$ in such a way that the selection can be performed by a finite automaton. The initial idea of Eilenberg of choosing v as the minimal string in some prescribed lexicographical ordering does not carry over to infinite strings since whatever the ordering chosen, there might not exist a minimal element associated with an input (e. g., the relation consisting of the pairs $(a^\omega, a^n b^\omega)$ and $(b^\omega, b^n a^\omega)$ for all $n \geq 0$).

In the present situation we show that to any arbitrary string $u \in A^\omega$ in the domain of the relation, we can assign a unique string $v \in B^\omega$ with $(u, v) \in R$

through a second order monadic formula. However, contrarily to Eilenberg's approach, instead of selecting the image through some of its properties we choose it via a run of the automaton. Among the runs determined by the input string, we choose that which visits repeated states earliest (hence the term "greedy ordering" see below), and whenever this does not suffice to single out one run, we will choose the minimal in the lexicographical ordering. Thus, if $\phi(u,v)$ is a monadic second order formula defining the relation R (i. e., $R = \{(u,v) \in A^\omega \times B^\omega \mid \phi(u,v) = \mathtt{true}\}$), then the uniformization is expressed by the following monadic second order formula

for all $u \in A^\omega$, $v \in B^\omega$ the three conditions hold
1) $\phi(u,v)$ is true
2) there exists a run ξ with label (u,v)
3) for all runs η with label (u,w) for some $w \neq v$, inequality $\xi < \eta$ holds

More precisely, we consider a linear ordering $<$ on Q under which the set F is an initial segment ($q \in F$ and $p < q$ implies $p \in F$) and we denote by $<_{\text{lex}}$ the lexicographical extension of $<$ to Q^ω. Let $\top$ be a new symbol and consider the ordering on the set $F \cup \{\top\}$ which is the trace of $<$ on F and for which $\top$ is the greatest element. Extend this ordering to a lexicographical ordering on the infinite sequences on the alphabet $F \cup \{\top\}$ and denote this new ordering by $<_F$. Let finally $\pi_F : Q^\omega \to (F \cup \{\top\})^\omega$ be the substitution defined by:

$$q\pi_F = \begin{cases} q & \text{if } q \in F \\ \top & \text{otherwise} \end{cases}$$

The *greedy ordering* $<_{\text{greedy}}$ on $(Q^*F)^\omega$ is defined by setting $\eta <_{\text{greedy}} \xi$ if and only if

$$\eta\pi_F <_F \xi\pi_F \text{ or } (\ \eta\pi_F = \xi\pi_F \text{ and } \eta <_{\text{lex}} \xi\ ) \tag{7}$$

We leave it to the reader to verify that $<_{\text{greedy}}$ is indeed an ordering.

Consider an input u and let Accept_u be the set of successful runs associated with it. We assume Accept_u is non empty and we shall "construct" its $<_{\text{greedy}}$-minimal element. We start with defining the ordering $\prec$ on the set $(Q-F)^*F$ by posing $x \prec y$ if one of the following conditions holds: 1) $|x| < |y|$ or 2) $|x| = |y|$ and their last occurrences $x', y' \in F$ satisfy $x' < y'$ or 3) $|x| = |y|$ and $x' = y'$ and $x <_{\text{lex}} y$.

Let $x_1 \in (Q-F)^*F$ be the $\prec$-smallest element in $(Q-F)^*F$ which is the prefix of some run in Accept_u and let $S_1 \subseteq \text{Accept}_u$ be the non empty set of successful runs starting with x_1. Now let $x_2 \in (Q-F)^*F$ be the $\prec$-smallest string such that x_1x_2 is the prefix of some successful run in S_1 and let $S_2 \subseteq S_1$ be the non empty set of successful runs starting with x_1x_2. The process continues and defines an infinite string $x_1x_2\dots$ which is the $<_{\text{greedy}}$-minimal element of Accept_u.

Because of condition (6) we have defined in this way a mapping $f : A^\omega \to B^\omega$ by setting $f(u) = v$ where v is the output associated with the run η.

Since the greedy and the lexicographical orderings are monadic second order definable, so is the function f which completes the proof.

We are now ready to prove the result for arbitrary rational relations.

Theorem 5. *Let* $0 < i < n$. *Every rational relation on* $A_1^\omega \times \ldots \times A_n^\omega$ *can be uniformized by some rational function* : $f : A_i^\omega \rightarrow \prod_{j \neq i} A_j^\omega$.

Proof. Indeed, by Theorem 4 every relation R can be factorized into $R = \varphi^{-1} \circ \cap K \circ \psi$ where B is a finite alphabet, $K \in \mathrm{Rat}B^*$, $\varphi : B^* \rightarrow A_1^*$ is an alphabetic morphism and $\psi : B^* \rightarrow \mathrm{Rat}(A_2^* \times \ldots \times A_n^*)$ a rational substitution. Choose for all $b \in B$ an arbitrary element in $b\psi$ and let $\psi' : B^* \rightarrow A_2^* \times \ldots \times A_n^*$ be the resulting morphism. It suffices to show that $\varphi^{-1} \circ \cap K \circ \psi'$ can be uniformized. However the relation $\varphi^{-1} \circ \cap K$ is synchronous. By the previous theorem, it can be uniformized by a function $f : A_1^\omega \rightarrow B^\omega$. The function $f_R = f \circ \psi'$ uniformizes the relation R.

The same counter-examples of paragraph 3.3 can be adapted to infinite strings (by completing every finite string with a special symbol infinitely repeated) to show that uniformization in more than one component does not hold in general. Another interesting consequence is the fact that the cross-section property holds for infinite strings.

Corollary 2. *Let* $f : A^\omega \rightarrow B^\omega$ *be a morphism and let* K *be a rational subset of* A^ω. *Then there exists a rational subset* $L \subseteq K$ *such that* f *maps* L *bijectively onto* Kf.

Proof. Indeed, the relation $\cap Kf \circ f^{-1} \circ \cap K \subseteq B^\omega \times A^\omega$ is rational. There exists a rational function $g : B^\omega \rightarrow A^\omega$ that uniformizes it, i. e., that selects for each element $x \in Kf$ a unique element in $y \in K$ with $x = yf$. Then $L = Kfg \in \mathrm{Rat}A^\omega$ and f maps bijectively L onto Kf.

Also, since all rational subsets of infinite strings are unambiguous, [4], we have

Corollary 3. *Every rational function* $f : A^\omega \rightarrow B_1^\omega \times \ldots \times B_n^\omega$ *is unambiguous.*

We observe, as a easy negative result, that the family of topologically closed rational relations cannot be uniformized. Indeed, consider the closed subset on the alphabets $A = B = \{a, b\}$

$$(a^\omega \times A^\omega) \cup \{(a^n bs, bs) \mid n \geq 0, s \in A^\omega\}$$

If there would exist a function that would uniformize the relation, it would be continuous, but this is clearly impossible.

4.3 A topological interpretation

The proof of uniformization for synchronous relations can also be seen as a proof of the following result.

Proposition 6. *On every rational rational subset X of A^ω there exists a rational linear ordering $\leq$ such that every non empty $Y \subseteq X$ which is relatively closed in X has a smallest element.*

Proof. Let $\mathcal{A} = (Q, I, F, T)$ be a Büchi automaton recognizing $X \subseteq A^\omega$. We assume without loss of generality that the existence of two transitions of the form $(q, a, p), (q, b, p) \in T$ implies $a = b$. A *run* is a finite or infinite sequence of states $(q_i)_{i<n}$, $n \leq \infty$, visited in by path of the automaton, i. e., for which there exist $u_i \in A$ such that $(q_i, u_i, q_{i+1}) \in T$ holds for all $i < n$. Because of the previous condition, there exists at most one string in A^ω associated with a given run. A run is *successful* if it visits infinitely often some repeated state. For every non empty subset $Y \subseteq X$ we denote by Accept_Y the set of all successful runs associated with some element in Y.

As in Theorem 5 we can construct an infinite string $x_1 x_2 \ldots \in Q^\omega$ which is the $<_{\text{greedy}}$-minimal element of Accept_Y. This string is a successful run on some input $u \in X$. Assume Y is of the form $Y = X \cap Z$ for some subset $Z \subseteq A^\omega$. Then u belongs to the adherence of Z. If Y is relatively closed, i. e., if we may assume furthermore that Z is closed, then u belongs to Y.

Now define a linear ordering $\prec$ on X as follows : $v \prec w$ if and only if the $<_{\text{greedy}}$-minimal element of Accept_v is smaller than the $<_{\text{greedy}}$-minimal element of Accept_w. If $Y \subseteq X$ is relatively closed in X then the string $u \in Y$ obtained as explained above is clearly the $\prec$-smallest element of Y.

This result can be viewed as the automaton version of a general topological result which states that on every Borel subset X of A^ω (in fact, also on every analytical subset) there exists a linear ordering such that every non empty relatively closed subset $Y \subseteq X$ has a smallest element. The proof of this result also uses an auxiliary greater topological set. A classical result (see [12, Thm 37. 1.]) states that X is a continuous image of the Baire space ω^ω hence the projection of some closed $R \subseteq A_1^\omega \times \omega^\omega$. The lexicographical ordering on $A_1^\omega \times \omega^\omega$ is a linear ordering such that every non empty closed set has a smallest element. It induces on X the wanted ordering defined as follows : $x \prec y$ iff the smallest element of $R \cap (\{x\} \times \omega^\omega)$ is smaller than the smallest element of $R \cap (\{y\} \times \omega^\omega)$.

Observe that the Baire space ω^ω cannot be replaced by any space B^ω with B finite since continuous images of compact spaces are compact.

In fact, in the proof of the above Proposition, the Baire space does occur implicitely in the definition of the greedy ordering. This can be seen as follows. There is a natural injection from the space of successful runs into the product $\omega^\omega \times F^\omega \times Q^\omega$ which maps a successful run onto the triple consisting of the sequence of positions of states in F, the sequence of successive states in F and the sequence of successive states. The greedy ordering on successful runs then corresponds to the lexicographic product of lexicographic orderings on the components.

References

1. A. Arnold and M. Latteux. A new proof of two theorems about rational transductions. *Theoretical Computer Science*, 8:261–263, 1979.
2. J. Berstel. *Transductions and context-free languages*. B. G. Teubner, 1979.
3. J.R. Büchi. Weak Second-Order Logic and Finite Automata. *Z. Math. Logik Grundlagen Math.*, 6:66–92, 1960.
4. O. Carton. Unambiguous Büchi automata, unpublished.
5. S. Eilenberg. *Automata, Languages and Machines*, volume A. Academic Press, 1974.
6. S. Eilenberg, C.C. Elgot, and J.C. Shepherdson. Sets recognized by n-tape automata. *Journal of Algebra*, 3:447–464, 1969.
7. C. C. Elgot and J. E. Mezei. On Relations Defined by Finite Automata. *IBM Journal*, 10:47–68, 1965.
8. C. Frougny and J. Sakarovitch. Synchronized rational relations of finite and infinite words. *Theoretical Computer Science*, 108:45–82, 1993.
9. S. Ginsburg and E. Spanier. Presburger formulas and languages. *Pacific Journal of Mathematics*, 16:285–296, 1966.
10. F. Gire. Two decidable problems for infinite words. *Information Processing Letters*, 22:135–140, 1986.
11. F. Gire and M. Nivat. Relations rationnelles infinitaires. *Calcolo*, XXI:91–125, 1984.
12. K. Kuratowski. *Topology*, volume 1. Academic Press, 1966.
13. S. Grigorieff. Uniformisation de relations rationnelles, unpublished.
14. S. Lifsches and S. Shelah. Uniformization and skolem functions in the class of trees. *Journal of Symbolic Logic*, 63:103–127, 1998.
15. J. Sakarovitch. A construction on finite automata that has remained hidden. *Theoretical Computer Science*, 204:205–231, 1998.
16. M. P. Schützenberger. Sur les relations rationnelles entre monoïdes libres. *Theoretical Computer Science*, 3:243–259, 1976.

Tree-Walking Pebble Automata

Joost Engelfriet and Hendrik Jan Hoogeboom

Summary. The tree languages accepted by (finite state) tree-walking automata are known to form a subclass of the regular tree languages which is not known to be proper. They include all locally first-order definable tree languages. We allow the tree-walking automaton to use a finite number of pebbles, which have to be dropped and lifted in a nested fashion. The class of tree languages accepted by these tree-walking pebble automata contains all first-order definable tree languages and is still included in the class of regular tree languages. It also contains all deterministic top-down recognizable tree languages.

1 Introduction

One of the questions in tree language theory is whether a natural sequential automaton model exists for recognizing the regular tree languages. Of course, by definition, they are recognized by the bottom-up finite tree automaton. But that automaton is essentially parallel rather than sequential: the control of the automaton is at several nodes of the input tree simultaneously, rather than at just one. The top-down finite tree automaton is also parallel and, moreover, its deterministic version does not recognize all regular tree languages, which seems unnatural when compared to the case of strings. For strings, the (1-way, on-line) finite automaton was generalized to the 2-way (off-line) finite automaton, which also recognizes exactly the regular languages, see [16,15]. Here the point of view has changed: the input is not fed into the automaton (like money into a coffee machine), but the automaton walks on the input string (like a mouse in a maze). Clearly, this can be generalized to a sequential finite automaton on trees: the tree-walking automaton, introduced in [1]. The finite control of the tree-walking automaton is always at one node of the input tree. Based on the label of that node and on its child number (which is i if it is the i-th child of its parent, with $i = 0$ for the root), the automaton changes state and steps to one of the neighbouring nodes (parent or child). The tree-walking automaton of [1] was equipped with an underlying context-free grammar and with an output tape, to model syntax-directed translation from strings to strings. It did not need the test on the child number because that information could be coded into the nonterminals of the context-free grammar. As shown in [14], without the child number test the tree-walking automaton is not able to search the input tree in a systematic way, such as by a pre-order traversal. For this and other reasons the child number test is an essential feature of the tree-walking automaton.

Unfortunately, it is an open problem whether the tree-walking automaton recognizes all regular tree languages, and we conjecture that it does not, cf. [9]. Assuming the conjecture, there are two natural questions. How does the class TWA of tree languages accepted by tree-walking automata compare to other well-known subclasses of the regular tree languages? Which natural features can be added to the tree-walking automaton to obtain an automaton that *does* recognize the regular tree languages? To start with the second question, the main trouble with the tree-walking automaton seems to be that it gets lost rather easily: in a binary tree of which all internal nodes have the same label, all nodes look pretty much the same. One way of solving this is to extend the tree-walking automaton with a synchronized pushdown, for which pushing and popping is synchronized with moving down and up in the tree, respectively. It is shown (implicitly) in [10] and (explicitly) in [14] that this automaton recognizes exactly the regular tree languages. Another, classical remedy against getting lost is to use pebbles. For instance, arbitrary mazes can be searched by "maze-walking" finite automata with two pebbles, see [5]. The main aim of this paper is to investigate the power of tree-walking automata with pebbles. Obviously, the unrestricted use of pebbles leads to a class of tree languages much larger than the regular tree languages, in fact to all tree languages in NSPACE($\log n$). Thus, we restrict the automaton to the recursive use of pebbles, in the sense that the life times of pebbles, i.e., the times between dropping a pebble and lifting it again, are properly nested. A similar, but stronger, nesting requirement is studied in [13] for 2-way automata on strings. We prove in Section 5 that our restriction indeed guarantees that all tree languages recognized by the tree-walking pebble automaton are regular, but we conjecture that the automaton is not powerful enough to recognize *all* regular tree languages. In Section 6 we generalize the notion of pebble to that of a "set-pebble", in such a way that the tree-walking set-pebble automaton recognizes exactly the regular tree languages.

To answer the first question, we compare the TWA languages with the tree languages that can be defined in first-order logic, see [19] for a survey. One of the reasons that the regular tree languages are the natural generalization of the regular string languages to trees, is that they are precisely the tree languages definable in monadic second-order logic. This was, in fact, the main motivation for the introduction of finite tree automata in [7,18]. Thus, one way of investigating the power of several types of tree-walking automata is to compare them to several types of logics on trees. We show in Section 3 that TWA contains all tree languages that are definable in locally first-order logic, where 'locally' means that the logic can talk about the parent-child relation between nodes of the tree, but not about the ancestor relation. We conjecture that it does not contain all first-order definable tree languages, but show in Section 5 that they can be recognized by the tree-walking pebble automaton. Note that TWA contains tree languages that are not first-order definable, such as the set of all trees with an even number of nodes (cf. [19]). As another answer to the first question, we conjecture in Section 4 that TWA does not contain all deterministic top-down recognizable tree languages, but show that they can be recognized by the tree-walking pebble automaton, with one pebble only.

2 Preliminaries

In this section we recall some well-known concepts and results concerning trees, logic for trees, and tree-walking automata. We use $\mathbb{N} = \{0, 1, 2, \ldots\}$, and for $m, n \in \mathbb{N}$, $[m, n] = \{i \mid m \leq i \leq n\}$.

Trees. For a ranked alphabet Σ, i.e., an alphabet Σ together with a function rank : $\Sigma \to \mathbb{N}$, the set of all trees over Σ is denoted T_Σ. A subset of T_Σ is called a tree language. As usual, a tree t over Σ is viewed as a finite, directed graph of which the nodes are labelled with symbols from Σ. By V_t we denote the set of nodes of t. Each node $v \in V_t$ has k children where k is the rank of the label of v. There is a linear order on these children which allows us to speak about the i-th child of v, and there is an edge with label i from v to its i-th child (for $1 \leq i \leq k$). The *child number* of a node v is i if it is the i-th child of its parent, and 0 if v is the root of t. For nodes u and v of t, $u \leq v$ denotes that u is an ancestor of v, i.e., that there is a directed (possibly empty) path from u to v. The yield of t is the string obtained by concatenating the labels of its leaves, from left to right. Finally, trees are denoted by terms in the usual way: $\sigma(t_1, \ldots, t_k)$ is the tree of which the root has label σ (of rank k), with direct subtrees $t_1, \ldots, t_k$.

Logic for Trees. We consider the same types of logic as in [19], viz. monadic second-order (MSO) logic, first-order (FO) logic, and locally first-order (LFO) logic. In [19] the latter two are called FO[<] logic and FO[S] logic, respectively.

For a ranked alphabet Σ, we consider monadic second-order formulas over Σ that describe properties of trees over Σ. This logical language has node variables $x, y, \ldots$, and node-set variables $X, Y, \ldots$. For a given tree t over Σ, node variables range over the elements of V_t, and node-set variables range over the subsets of V_t. There are five types of atomic formulas over Σ: $\mathrm{lab}_\sigma(x)$, for every $\sigma \in \Sigma$, meaning that x has label σ; $\mathrm{edg}_i(x, y)$, for every $i \leq$ the rank of a symbol in Σ, meaning that the i-th child of x is y; $x \leq y$, meaning that x is an ancestor of y; $x = y$, and $x \in X$, with obvious meaning. The formulas are built from the atomic formulas using the connectives $\neg$, $\wedge$, $\vee$, $\rightarrow$, and $\leftrightarrow$, as usual. Both node variables and node-set variables can be quantified with $\exists$ and $\forall$. We will use $\mathrm{edg}(x, y)$ for the disjunction of all

$\mathrm{edg}_i(x,y)$ (meaning that x is the parent of y), $\mathrm{root}(x)$ for $\neg\exists y(\mathrm{edg}(y,x))$, and $\mathrm{leaf}(x)$ for $\neg\exists y(\mathrm{edg}(x,y))$.

A *first-order* formula is an MSO formula that does not contain node-set variables, and a *locally first-order* formula is an FO formula that does not contain atomic formulas $x \leq y$.

For a closed formula ϕ over Σ and a tree $t \in T_\Sigma$, $t \models \phi$ denotes that t satisfies ϕ. If a formula ϕ has free variables, say, x, y, X, we also write $\phi(x,y,X)$. Moreover, $t \models \phi(u,v,U)$ denotes that t satisfies ϕ when x,y,X have value u,v,U, respectively (with $u,v \in V_t$ and $U \subseteq V_t$). The tree language defined by a closed formula ϕ over Σ is $L(\phi) = \{t \in T_\Sigma \mid t \models \phi\}$. $L(\phi)$ is called an MSO *definable tree language*. If ϕ is first-order, or locally first-order, then $L(\phi)$ is called an FO definable, or LFO definable tree language, respectively. According to the classical result of [7,18] (first shown for strings in [6,8]) the MSO definable tree languages are precisely the *regular tree languages*, i.e., the tree languages accepted by (bottom-up or top-down) finite tree automata. For more information on regular tree languages, see, e.g., [11,12].

We will also consider "trips" on trees (cf. [2,3,9]). A *trip*, or node relation, over Σ is a set of triples (t,u,v) where t is a tree over Σ, and $u, v \in V_t$. The trip defined by an MSO formula $\phi(x,y)$ with two free node variables x and y, is $T(\phi) = \{(t,u,v) \mid t \models \phi(u,v)\}$. $T(\phi)$ is called an MSO *definable trip*. In [9] the MSO definable trips are called the regular trips.

Tree-Walking Automata. A *tree-walking automaton*, or tw automaton, for short, is a (nondeterministic) finite state device that walks on a tree from node to node, following the edges of the tree (in either direction). At each moment of time the tw automaton is in a certain state, at a certain node of the input tree (over some ranked alphabet Σ). In one step, it can test the label and the child number of the current node, and move to the parent or to one of the children of the node, changing state. A child can be specified by its child number. The *language* $L(A)$ *accepted* by a tree-walking automaton A consists of all trees (over Σ) on which A has a computation that starts at the root of the input tree in its initial state, and ends in a final state. $L(A)$ will be called a *twa language*.

Since the tw automaton can test the child number of the current node (and hence, in particular, can test whether or not it is at the root), one of its basic capabilities is to make a pre-order traversal of the input tree (deterministically), starting and ending at the root: when entering a node v for the first time (from above), it moves up if v is a leaf, and otherwise moves down to v's first child; when entering v from below, it knows the number of the child u it just left (by testing u's child number and storing it in its finite control) and thus can move down to v's next child, or move up if u was the last child of v.

In [2,3] the tree-walking automaton is extended with logical capabilities: a *tree-walking automaton with* MSO *tests* is a finite state device as described above, with the additional possibility of testing MSO properties of the current node. An MSO property is an MSO formula $\phi(x)$ with one free node variable; for the current node u of the input tree t, the automaton can test whether $t \models \phi(u)$. A given automaton can of course use only finitely many of these MSO tests. The language $L(A)$ accepted by such an automaton A is defined as above, but the main concern in [2,3] is with the *trip* $T(A)$ *computed* by A: the set of all (t, u, v) such that A can walk from u to v on t, starting in its initial state and ending in a final state. The main result of [2,3] is the following.

Proposition. *A trip is* MSO *definable if and only if it can be computed by a tree-walking automaton with* MSO *tests.*

It is an immediate consequence of (the if-direction of) this result that all tree languages accepted by tree-walking automata with MSO tests are MSO definable: if $T(A) = T(\phi(x,y))$, then $L(A) = L(\psi)$ where ψ is the formula $\exists x, y : \mathrm{root}(x) \wedge \phi(x,y)$. In particular, all twa languages are MSO definable and hence regular. However, as mentioned in the Introduction, it is an open problem whether every regular tree language is a twa language. Note that every regular tree language is a projection of a deterministic twa language. This is because every regular tree language is a projection of the set of derivation trees of a context-free grammar (cf. [12], Section 8), and every such derivation tree language can obviously be accepted by a deterministic tw automaton: the automaton traverses the input tree and checks for each node whether the labels of the node and its children form a production of the context-free grammar.

3 Locally First-Order Logic

As mentioned above, it is not known whether the tree-walking automaton can accept all MSO definable tree languages. Here we show that it can accept all locally first-order definable tree languages. The hard work for this has already been done: the proof is based on the characterization of the LFO definable tree languages as the so-called locally threshold testable tree languages ([19], Section 4.2). Intuitively, a tree language is locally threshold testable if membership of a tree in the language can be determined by looking at local spheres around all nodes of the tree, and count how many times these spheres occur, up to some threshold. To explain this formally we need some terminology.

Let Σ be a ranked alphabet. Consider a "radius" $r \in \mathbb{N}$. For a node u of a tree t over Σ, we denote by $\mathrm{sph}_r(t, u)$ the subgraph of t induced by all nodes of distance at most r to u (where distance is measured along undirected paths), with u as a designated node. Define $S_r = \{\mathrm{sph}_r(t,u) \mid t \in T_\Sigma, u \in V_t\}$. This is a finite set because we do not distinguish between isomorphic graphs. Now consider a "threshold" $q \in \mathbb{N}$. Define $F_{r,q}$ to be the (finite) set of all partial functions $f : S_r \to [0, q]$. For a tree t, define $f^t_{r,q} \in F_{r,q}$ such that for every sphere $s \in S_r$, $f^t_{r,q}(s) = \#\{u \in V_t \mid \mathrm{sph}_r(t,u) = s\}$ provided this number is in $[0,q]$. Now, a tree language L over Σ is *locally threshold testable* if there exist $r, q \in \mathbb{N}$ and $F \subseteq F_{r,q}$ such that $L = \{t \in T_\Sigma \mid f^t_{r,q} \in F\}$.

It should be clear that every locally threshold testable tree language can be accepted by a deterministic tw automaton. The automaton A traverses the input tree t in pre-order and for each node u of t it computes the sphere $\text{sph}_r(t,u)$ with centre u, in its finite control, by just searching systematically the neighbourhood of u within distance r. Thus, counting the number of occurrences of these spheres up to the threshold q, A can compute $f^t_{r,q}$ and check whether it is in F.

This shows that the deterministic tree-walking automaton can accept all LFO definable tree languages.

4 One Pebble

We conjecture that the twa languages form a proper subclass of the regular tree languages (cf. [9]) and propose the following tree languages L_1^{sp} and L_2^{sp} as counter-examples. Let $\Sigma = \{\sigma, a, b\}$ where σ has rank 2 and a, b have rank 0. Let L_1 be the set of all trees $t \in T_\Sigma$ such that all root-to-leaf paths of t have even length, and let L_2 be the set of all trees $t \in T_\Sigma$ such that the yield of t is in a^*ba^*. For $i = 1, 2$, the language L_i^{sp} consists of all trees $\sigma(t_1, \sigma(t_2, \ldots \sigma(t_k, \tau) \ldots))$ with $k \geq 0$, $\tau \in \{a, b\}$, and $t_1, \ldots, t_k \in L_i$. Thus, a tree t in L_i^{sp} consists of a *spine*, i.e., the path from the root of t to its right-most leaf, such that the first child of each node on the spine is the root of a subtree which belongs to L_i. It should be clear that L_1^{sp} and L_2^{sp} are regular tree languages. We think that they cannot be accepted by a tw automaton.

Intuitively, the reason that no tree-walking automaton A can accept L_i^{sp} is the following. Let us say that a subtree of the input tree t is a *spine subtree* if its root is the first child of a node on the spine of t. Thus, A has to check that every spine subtree of t is in L_i. One way of doing this would be to move down node by node from the root along the spine, and for each node u on the spine check the spine subtree s of u. To do this, A has to visit all the leaves of s. But then A gets lost because it does not know when it has returned to the spine, i.e., to u. The only way for A to find out whether a node v is on the spine seems to be to move up as long as the child number is 2 and then test whether it is at the root, but then of course A has also got lost because it does not know how to return to v.

Note that it is not difficult to see that the *complement* of L_1^{sp} *can* be accepted by a tree-walking automaton: walk down nondeterministically to the root of one of the spine subtrees and then walk down nondeterministically to one of its leaves, checking that the path has odd length. Thus, if L_1^{sp} cannot be accepted by a tree-walking automaton, then the class of twa languages is not closed under complement. Moreover, it would imply that the deterministic tw automaton is less powerful than the nondeterministic one, because the class of deterministic twa languages *is* closed under complement. The latter follows from the, maybe surprising, fact that every deterministic tw automaton can be simulated by one that is always halting (see [17]).

As mentioned in the Introduction, one way of not getting lost is to use pebbles. It should be clear that $L_i^{\rm sp}$ can be accepted by a (deterministic) tree-walking automaton with one pebble: the pebble is put on a spine node while the automaton is checking the corresponding spine subtree. A *one-pebble tree-walking automaton* (shortly, 1p-tw automaton) is a tree-walking automaton that additionally carries a pebble. It can drop the pebble on the current node, it can test whether or not the pebble is at the current node, and it can lift the pebble from the current node (when it lies there, of course). At the beginning and end of a computation the pebble should not be on the input tree. A 1p-tw automaton A accepts a tree language $L(A)$ in the usual way, and $L(A)$ is called a *1p-twa language*. It is well known that 2-way finite automata (on strings) with one pebble recognize the regular languages, see [4,13]. Similarly, it is shown in (Theorem 10 of) [9] that all 1p-twa languages are regular tree languages, see also the next section.

We now note that the language $L_1^{\rm sp}$ is, in fact, a deterministic top-down recognizable (dtr, for short) tree language, i.e., can be recognized by a deterministic top-down finite tree automaton (see, e.g., [11,12]). In fact, we will show that *every* dtr tree language can be accepted by a (deterministic) one-pebble tree-walking automaton. This is based on a well-known characterization of the dtr tree languages (see, e.g., [11], Theorem II.11.6) which we now recall. For a ranked alphabet Σ, we define the (non-ranked) *path alphabet* Π_Σ to consist of all symbols σ_i with $\sigma \in \Sigma$ and $1 \leq i \leq \text{rank}(\sigma)$, plus all symbols of Σ of rank 0. Thus, for the ranked alphabet Σ of $L_{\rm sp}$, $\Pi_\Sigma = \{\sigma_1, \sigma_2, a, b\}$. For a tree t, its *path language* $\text{path}(t) \subseteq \Pi_\Sigma^*$ is defined recursively as follows: for τ of rank 0, $\text{path}(\tau) = \{\tau\}$, and for σ of rank $k > 0$, $\text{path}(\sigma(t_1, \ldots, t_k)) = \bigcup_{i=1}^k \sigma_i \cdot \text{path}(t_i)$.

The characterization is: a tree language L over Σ is a dtr tree language if and only if there is a regular (string) language R over Π_Σ such that $L = \{t \in T_\Sigma \mid \text{path}(t) \subseteq R\}$. As an example, the regular language R for $L_{\rm sp}$ is $\sigma_2^*(a \cup b \cup \sigma_1 R')$ where R' is the set of all strings of odd length.

Using this characterization, it is easy to see that all dtr tree languages are deterministic 1p-twa languages. Let R be a regular language over Π_Σ and let M be an ordinary deterministic finite automaton that accepts the mirror image of R. We describe a one-pebble tw automaton A that checks whether all paths of the input tree t are in R. The automaton A traverses t in pre-order and for each leaf u of t it executes the following subroutine. It drops its pebble on u and walks up to the root, simulating M on the corresponding string in $\text{path}(t)$. More precisely, it starts with the initial state of M and feeds M the label of u as input. For each node v on the path from u to the root, A determines the child number i of v, moves to the parent v' of v, and feeds M the symbol σ_i as input, where σ is the label of v'. At the root, A checks that M has arrived in a final state. Finally, A returns to u by searching t for the pebble (traversing t deterministically), and lifts the pebble from u.

This shows that all deterministic top-down recognizable tree languages are accepted by deterministic tree-walking automata with one pebble.

We conjecture that the 1p-twa languages are still a proper subclass of the regular tree languages. The proposed counter-example is the language L_2^{nsp} which is obtained from $L = L_2^{\mathrm{sp}}$ by substituting $\rho(L)$ for each a and $\rho(L^c)$ for each b, where ρ is a symbol of rank 1 and L^c is the complement of L. Intuitively, to recognize a tree from L_2^{nsp} a 1p-tw automaton would need its pebble to recognize the outer occurrence of a tree from L and thus would have to behave like a tw automaton to recognize the nested occurrences of trees from L and L^c. Note that it is easy to accept L_2^{nsp} with a tree-walking automaton that uses *two* pebbles.

5 Nested Pebbles

Next we note that the language L_2^{sp} of Section 4 is first-order definable (but not locally). To define it, we use $x \leq_2 y$ to mean that $x \leq y$ and that all edges on the path from x to y have label 2. The latter property is expressible by the formula $\forall x', y' : (x \leq x' \wedge \mathrm{edg}(x', y') \wedge y' \leq y) \rightarrow \mathrm{edg}_2(x', y')$. Now $L_2^{\mathrm{sp}} = L(\phi)$ where ϕ is the formula $\forall x, y, z : (\mathrm{root}(x) \wedge x \leq_2 y \wedge \mathrm{edg}_1(y, z)) \rightarrow \psi(z)$, and $\psi(z)$ is $\exists! u : z \leq u \wedge \mathrm{leaf}(u) \wedge \mathrm{lab}_b(u)$. Similarly, it is easy to show that the language L_2^{nsp}, discussed at the end of Section 4, is first-order definable too.

As observed in Section 4, L_2^{sp} can be accepted by a tw automaton with a pebble, and L_2^{nsp} by one that uses two pebbles. Since we know from Section 3 that tw automata can recognize all locally first-order definable languages, this suggests that tw automata with pebbles can recognize all first-order definable languages. The key idea here is that a quantified variable corresponds to a pebble. However, as mentioned in the Introduction, the unrestricted use of pebbles leads out of the class of regular tree languages. Since quantifiers in a formula are properly nested, it seems natural to require that the life times of the pebbles are nested, where a life time of a pebble is the time between dropping it on a node and lifting it again (note that a pebble, like a cat, usually has more than one life). The obvious automaton for L_2^{nsp} indeed has nested pebble life times.

A *tree-walking automaton with nested pebbles* (shortly, p-tw automaton) is a tree-walking automaton that carries a finite number of pebbles, each with a unique name. In one step, it can determine which pebbles are lying on the current node, lift some of them, and drop some others. Initially and finally there should be no pebbles on the input tree. Moreover, the life times of the pebbles should be nested. This can be formalized by keeping a pushdown of pebble names in the configuration of the automaton, pushing a pebble when it is dropped and popping it when it is lifted. Note that since each pebble name occurs at most once in the pushdown, the automaton may keep track of the pushdown in its finite control and thus "know" when it is allowed to drop or lift a pebble. A p-tw automaton A accepts a tree language $L(A)$ as usual, and $L(A)$ is called a *p-twa language*. We will also need the trip $T(A)$ computed by A (cf. Section 2): the set of all (t, u, v) such that A can walk from u to v on t, starting in its initial state and ending in a final state (with the above restrictions on pebbles).

In the remainder of this section we show that all first-order definable tree languages are deterministic p-twa languages, and that all p-twa languages are regular. The first proof is by induction on the structure of the formula. For each first-order formula ϕ we construct a deterministic always-halting p-tw automaton A that uses the (free or bound) variables of ϕ as pebble names. Without loss of generality we may assume that no variable occurs both free and bound in ϕ. For a given input tree t, the automaton A finds out whether or not ϕ is true for t. More precisely, let the free variables of ϕ be in the set $\{x_1, \ldots, x_k\}$, i.e., $\phi(x_1, \ldots, x_k)$. For nodes $u_1, \ldots, u_k$ of t, A receives as input the tree t with the pebbles $x_1, \ldots, x_k$ lying on the nodes $u_1, \ldots, u_k$, respectively, and starts its computation in its initial state at the root of t. During its computation it is not allowed to lift or drop the pebbles x_i (but they can of course be tested); the life times of the other pebbles should be nested. The computation of A should halt at the root of t with the pebbles $x_1, \ldots, x_k$ still lying on the nodes $u_1, \ldots, u_k$, and no other pebbles lying on t. Finally, A halts in a final state if and only if $t \models \phi(u_1, \ldots, u_k)$. The construction of A is easy for the atomic formulas. As an example, the automaton for $x \leq y$ searches the input tree for the node u with pebble x, traverses the subtree with root u to see whether it contains the pebble y, and returns to the root. In the induction step it suffices to consider negation, conjunction, and universal quantification. For negation, just interchange final and non-final states (note that the automaton is deterministic and always-halting). For conjunction, just simulate the two given automata, one after the other. Now consider a formula $\phi = \forall x : \phi'(x)$ and let A' be the automaton constructed for ϕ'. The automaton A for ϕ traverses the input tree in pre-order and executes the following subroutine for every node u: it drops pebble x on u, walks to the root, simulates A', returns to pebble x, and lifts pebble x from u.

This proves that all FO definable tree languages can be accepted by a deterministic tree-walking automaton with nested pebbles. Note that the automaton can be constructed in such a way that it uses k pebbles where k is the quantifier nesting depth of the formula.

Next we prove that all p-twa languages are regular. Let A be a p-tw automaton over the ranked alphabet Σ. We will show, by induction on the number of pebbles, that the trip $T(A)$ can be computed by a tree-walking automaton M with MSO tests, i.e., $T(M) = T(A)$. By the Proposition in Section 2 and the argument following it, this implies that all p-twa languages are MSO definable and hence regular. The basic idea is that a subcomputation of A that corresponds to the life time of a pebble on node u, can be replaced by an MSO test on u. Without loss of generality we assume that the pebble names of A are $1, \dots, n$, and that at each moment of time pebbles $1, \dots, i$ are lying on the input tree, for some $0 \leq i \leq n$. The induction is on n. For $n = 0$, A is an ordinary tw automaton (and so we take $M = A$). Assume now that the result holds for $n - 1$. Consider a subcomputation of A that corresponds to a life time of pebble 1. Obviously, during this life time, A behaves like a p-tw automaton with $n - 1$ pebbles on an input tree of which one node is "marked". Let Σ' be the ranked alphabet $\Sigma \cup \{\sigma' \mid \sigma \in \Sigma\}$, where each σ' is a new ("marked") symbol, of the same rank as σ. Define A' to be the tw automaton with pebbles $\{2, \dots, n\}$ that behaves in the same way as A, interpreting a node with label σ' as a node with label σ that contains pebble 1; A' has initial state p and final state q, where p and q are the states of A at the start and end of the considered life time, respectively. By the induction hypothesis and the Proposition in Section 2, $T(A')$ is MSO definable, i.e., there is a formula $\phi'(x_1, x_2)$ over Σ' such that $T(A') = T(\phi')$. Let $\phi(x_1, x_2, y)$ be the formula that is obtained from $\phi'(x_1, x_2)$ by changing every atomic subformula $\text{lab}_{\sigma'}(z)$ into the formula $\text{lab}_\sigma(z) \wedge z = y$ and every atomic subformula $\text{lab}_\sigma(z)$ into $\text{lab}_\sigma(z) \wedge z \neq y$. Then, for all nodes u_1, u_2, v of $t \in T_\Sigma$, $t \models \phi(u_1, u_2, v)$ iff A can walk from u_1 in state p to u_2 in state q with pebble 1 on v (without lifting pebble 1). Consequently, the considered subcomputation can be replaced by the MSO test $\phi(x, x, x)$ on the current node. Thus, A is turned into a tw automaton with MSO tests that computes the same trip. For more details see Chapter 2 of [20] where this result is shown for the special case of 2-way automata on strings.

This proves that all tree languages accepted by tree-walking automata with nested pebbles are regular. We conjecture that they form a proper subclass of the regular tree languages, and that the number of pebbles determines a proper hierarchy (by iterating the construction at the end of Section 4).

6 Set-Pebbles

It was shown in the previous section that the FO definable tree languages can be recognized by tree-walking automata that use their pebbles to implement the quantification of node variables. Thus, one naturally gets the idea that all MSO definable tree languages (i.e., all regular tree languages) can be recognized by an automaton that uses a generalized type of pebble by which the quantification of node-set variables can be implemented. We will call such a generalized pebble a "set-pebble". Let us define a (nondeterministic) ***tree-walking automaton with set-pebbles*** to be a tree-walking automaton that carries a finite number of set-pebbles (with distinct colours) that it can drop, test, and lift. Dropping a set-pebble of a certain colour means that, nondeterministically, a pebble of that colour is dropped on any number of nodes of the input tree. Lifting a set-pebble of a certain colour means that all pebbles of that colour are lifted from the tree. The automaton can test the colours of the pebbles that lie on the current node. Note that dropping and lifting are independent of the current node. Finally, as for pebbles, we require the life times of the set-pebbles to be nested.

Let us now argue that the tree-walking automaton with set-pebbles accepts exactly the regular tree languages. In one direction, it is well known (see [19]) that every regular tree language is MSO definable by a formula of the form $\phi = \exists X_1, \ldots, X_k : \phi'(X_1, \ldots, X_k)$ where ϕ' is locally first-order (in the sense that it does not contain $\exists X$, $\forall X$, and $x \leq y$). Since, by Section 3, LFO formulas can be accepted by tw automata, it should be clear that, to implement ϕ, it suffices to use set-pebbles with colours $X_1, \ldots, X_k$, dropping them at the start of the computation, then simulating the tw automaton that implements ϕ', and lifting them at the end of the computation. In the other direction we follow the proof in Section 5 that all p-twa languages are regular. As in that proof, consider a subcomputation of A that corresponds to a life time of the set-pebble with colour 1. Suppose that A drops the set-pebble in state p when it is at node u, and lifts it again in state q when it is at node v. This means that there is a set of nodes U (the nodes that are covered by the set-pebble) such that A walks from u to v behaving like a tw automaton with $n-1$ set-pebbles on an input tree t of which the nodes in U are "marked". Thus, by the induction hypothesis and the Proposition in Section 2, there is a formula $\phi(x_1, x_2, X)$ that models this behaviour (for $x_1 = u$, $x_2 = v$, and $X = U$). Hence t satisfies $\psi(u, v)$, where ψ is the formula $\exists X : \phi(x_1, x_2, X)$. In other words, (t, u, v) is in the trip $T(\psi)$. By the Proposition in Section 2 there is a tw automaton M with MSO tests such that $T(M) = T(\psi)$. Consequently, the considered subcomputation of A can be replaced by a simulation of M. This turns A into a tw automaton with MSO tests.

We finally note that is not clear how one could define a deterministic version of the tree-walking automaton with set-pebbles.

Acknowledgment. We are grateful to Jan-Pascal van Best for helping us with the proof of the second result of Section 5.

References

1. A.V. Aho and J.D. Ullman; Translations on a context free grammar, Inform. and Control 19 (1971), 439–475
2. R. Bloem, J. Engelfriet; Monadic second order logic and node relations on graphs and trees, in J. Mycielski, G. Rozenberg, and A. Salomaa, editors, *Structures in Logic and Computer Science*, Lecture Notes in Computer Science 1261, Springer-Verlag, 1997, pp.144-161
3. R. Bloem, J. Engelfriet; Characterization of properties and relations defined in monadic second order logic on the nodes of trees, Tech. Report 97–03, Leiden University, August 1997
http://www.wi.LeidenUniv.nl/TechRep/1997/tr97-03.html

4. M. Blum, C. Hewitt; Automata on a 2-dimensional tape, in Proc. 8th IEEE Symp. on Switching and Automata Theory, pp.155–160, 1967
5. M. Blum, D. Kozen; On the power of the compass (or, why mazes are easier to search than graphs), Proc. 19th FOCS (Annual Symposium on Foundations of Computer Science), 1978, pp.132–142
6. J. Büchi; Weak second-order arithmetic and finite automata, Z. Math. Logik Grundlag. Math. 6 (1960), 66–92
7. J. Doner; Tree acceptors and some of their applications, J. of Comp. Syst. Sci. 4 (1970), 406–451
8. C.C. Elgot; Decision problems of finite automata and related arithmetics, Trans. Amer. Math. Soc. 98 (1961), 21–51
9. J. Engelfriet, H.J. Hoogeboom, J.P. van Best; Trips on trees, Manuscript, 1999, to appear in Acta Cybernetica
10. J. Engelfriet, G. Rozenberg, G. Slutzki; Tree transducers, L systems, and two-way machines, J. of Comp. Syst. Sci. 20 (1980), 150-202
11. F. Gécseg, M. Steinby; *Tree Automata*, Akadémiai Kiadó, Budapest, 1984
12. F. Gécseg, M. Steinby; Tree Languages, in G. Rozenberg and A. Salomaa, editors, *Handbook of Formal Languages, Volume 3: Beyond Words*, Chapter 1, Springer-Verlag, 1997
13. N. Globerman, D. Harel; Complexity results for two-way and multi-pebble automata and their logics, Theor. Comput. Sci. 169 (1996), 161–184
14. T. Kamimura, G. Slutzki; Parallel and two-way automata on directed ordered acyclic graphs, Inf. and Control 49 (1981), 10–51
15. M.O. Rabin, D. Scott; Finite automata and their decision problems, IBM J. Res. Devel. 3 (1959), 115–125
16. J.C. Shepherdson; The reduction of two-way automata to one-way automata, IBM J. Res. Devel. 3 (1959), 198–200
17. M. Sipser; Halting space-bounded computations, Proc. 19th FOCS (Annual Symposium on Foundations of Computer Science), 1978, pp.73–74
18. J.W. Thatcher, J.B. Wright; Generalized finite automata theory with an application to a decision problem of second-order logic, Math. Systems Theory 2 (1968), 57–81
19. W. Thomas; Languages, Automata, and Logic, in G. Rozenberg and A. Salomaa, editors, *Handbook of Formal Languages, Volume 3: Beyond Words*, Chapter 7, Springer-Verlag, 1997
20. J.P. van Best; *Tree-Walking Automata and Monadic Second Order Logic*, Master's Thesis, Leiden University, July 1998
http://www.wi.LeidenUniv.nl/MScThesis/IR98-06.html

Counter Machines: Decision Problems and Applications*

Oscar H. Ibarra and Jianwen Su

Summary. We give a brief survey of the (un)decidable properties of multicounter machines. In particular, we present some of the strongest decidable results (concerning emptiness, containment, and equivalence) known to date regarding these machines. We also discuss some applications.

1 Introduction

A fundamental decision question concerning any class C of language recognizers is whether there exists an algorithm to decide the following question: given an arbitrary machine M in C, is the language accepted by M empty? This is known as the emptiness problem (for C). Decidability (existence of an algorithm) of emptiness can lead to the decidability of other questions such as containment and equivalence (given arbitrary machines M_1 and M_2 in C, is the language accepted by M_1 contained (respectively, equal) to the language accepted by M_2?) if the languages defined by C are effectively closed under union and complementation.

The simplest recognizers are the finite automata. It is well-known that all the different varieties of finite automata (one-way, two-way, etc.) are effectively equivalent, and the class has decidable emptiness, containment, and equivalence (ECE, for short) problems.

When the two-way finite automaton is augmented with a storage device, such as a counter, a pushdown stack or a Turing machine tape, the ECE problems become undecidable (no algorithms exist). In fact, it follows from a result in [12] that the emptiness problem is undecidable for two-way counter machines even over a unary input alphabet. In this paper, we look at various generalizations and restrictions of the counter machine model and give a brief survey of the (un)decidable properties of these machines. In particular, we present some of the strongest decidable results known to date concerning these machines. We also give some applications.

---

* Supported in part by NSF grant IRI-9700370.

2 One-Counter Machines

We begin with machines with only one counter. In what follows, 2DCM (1DCM) denotes a two-way (one-way) deterministic counter machine.

We already know that the emptiness problem for 2DCM is undecidable. Suppose we restrict the operation of the 2DCM so that its input head makes only a fixed number of turns on the input tape (independent of the input). Even for the case when the machine makes only one input turn, we have:

Theorem 1. *The emptiness problem for 1-input turn* 2DCM's *is undecidable.*

Proof. We sketch the proof in [8]. We reduce the emptiness problem for Turing machines to the emptiness for 1-turn 2DCM's. Given an arbitrary Turing machine M, we can find two 1DCM M_1 and M_2 and a homomorphism g_1 such that $L(M) = g(L(M_1) \cap L(M_2))$. Furthermore, we can find a 1NCM M_3 such that $L(M_3) = L(M_2)^R$. By expanding the input alphabet with markers to dictate the nondeterministic choices of M_3, we can obtain a 1DCM M_4 and a homomorphism g_2 (which maps the markers to the null string and leaves the other symbols the same) such that $L(M_3) = g_2(L(M_4))$. We also modify M_1 to include markers in its alphabet; the resulting machine M_5 behaves like M_1 and ignores markers in its input.

From these constructions we have $L(M) = g_1 g_2(L(M_5) \cap L(M_4)^R)$, and it is now trivial to construct from M_4 and M_5 a 1-input reversal 2DCM M' such that $L(M) = g_1 g_2(L(M'))$. Since the emptiness problem for Turing machines is undecidable, so is the emptiness problem for 1-input reversal 2DCM. ∎

For one-way machines, we have:

Theorem 2 ([13]). *The equivalence (hence also the emptiness) problem for* 1DCM's *is decidable, but the containment problem is undecidable.*

Now suppose we restrict the 2DCM so that its counter is finite-reversal (i.e., the number of alternations between increasing and decreasing modes is at most a fixed number, independent of the input) but its input is unrestricted. Such machines can accept fairly complex languages, even when the counter makes only 1-reversal. For example, such a machine can recognize the language consisting of strings of the form $a^i b^j$ where i divides j. A 1-reversal counter machine stores j in the counter and makes sweeps over the a-segment, subtracting i from the counter for each sweep.

In fact, it can be shown that finite-reversal 2DCM's can verify the validity of existential sentences in the first-order logic for integers with order and addition.

In contrast to Theorem 1, we have:

Theorem 3 ([14]). *The ECE problems for finite-reversal* 2DCM's *are decidable.*

We believe that the class of finite-reversal 2DCM's is one of the largest known classes of machines, which are a natural generalization of the two-way finite automaton, for which the ECE problems are decidable.

Consider now a 2DCM where the counter can become zero only at most a fixed number of times regardless of its input (call it *finite-reset* 2DCM). These machines are powerful. For example, there is a finite-reset 2DCM M that accepts the set $S = \{0^x\#(0^y\#)^x : x, y \geq 1\}$. M first checks that its input is of the form $0^x\#0^{y_1}\#0^{y_2}\#\ldots\#0^{y_n}$ and that $n = x$. This requires 1 reset of the counter. Next M verifies that $y_1 \geq y_2$ by adding $1+y_1-y_2$ to its counter; M rejects if the counter resets to zero during this operation. Otherwise, $y_1 \geq y_2$ and M adds $2y_2 - y_1$ to the counter so that its value now becomes $1 + y_2$. Note that the counter does not reset during this operation. M now checks that $y_2 \geq y_3$ and so on, until it has verified that $y_1 \geq y_2 \geq \ldots \geq y_n$. Finally, M verifies that $y_1 = y_n$. It is easy to see that the whole computation requires only 2 resets of the counter.

It follows from the above remark that the emptiness problem for finite-reset counter machines is undecidable, because they can multiply and hence compute the value of any polynomial $p(y, x_1, x_2, \ldots, x_n)$ (see [2] for the proof of an analogous result). However, over bounded languages, we can show that the emptiness problem is decidable.

Definition 1. A language L is *bounded* if it is a subset of $a_1^* \cdots a_k^*$, for some distinct symbols $a_1, \ldots, a_k$.

Actually, the notion can be relaxed. The symbols need not be distinct, provided there are at most a fixed number of *runs* of symbols in the strings. For instance, a language $\subseteq a_1^*a_2^*a_1^*a_2^*a_3^*a_1^*$ is also viewed as bounded since it has at most 6 runs of symbols. But $(a_1^*a_2^*)^*$ is not bounded. In this paper, we use the relaxed definition.

Each accepting computation of a finite-reset 2DCM M can be divided into a finite number of phases at the times the counter value becomes zero. Since the language is bounded, it is easy to see that the counter makes a reversal only at the boundary points, i.e., the endmarkers or the first symbols of each type. The computation in each phase can then be rewritten in the form uv^iw, such that the lengths of u, v, and w are bounded by some constant depending only M, and v starts and ends in the same state and on the same boundary point. Again, the number of different such v's is bounded by a constant. So we can construct a finite-reversal 2DCM M', whose input is padded with the net change to the counter value by each possible loop v, such that $L(M)$ is empty iff $L(M')$ is. Hence from Theorem 3, we have:

Theorem 4. *The ECE problems for finite-reset 2DCM's over bounded languages are decidable.*

It is interesting to note that there are bounded languages that can be recognized by finite-reset 2DCM's but not by finite-reversal 2DCM's. One example is $L = \{0^a1^b2^c : a = n(b-c) \text{ for some } n \geq 0\}$.

3 Multi-Counter Machines

When the input and counter of a one-counter machine are restricted, the decision problems become decidable. It is known that the ECE problems are decidable for counter machines with a finite-turn input and a finite-reversal counter. In fact, stronger results are known [8,9]. For nonnegative integers m, k and r, we define a finite-crossing finite-reversal multi-counter machine as a two-way finite automaton with input delimiters (end markers), augmented with m counters such that on any input: (i) Any (possibly nondeterministic) computation leads to a halting state with the input head on the right delimiter; (ii) No boundary between input symbols (including the delimiters) is *crossed* by the input head more than k times (note that the number of turns, i.e., changes in directions, the input head makes on the input may be unbounded); (iii) Each counter makes no more than r reversals. Note that each counter can be tested whether it is 0 or non-0 and incremented or decremented by 1. We count each alternation from increasing mode to decreasing mode or vice-versa as a *reversal*. We call such a machine a k-crossing r-reversal m-counter machine. When m, k, r are not specified, we refer to the machine as finite-crossing finite-reversal multi-counter machine. Note that 1-crossing corresponds to *one-way* input.

Finite-crossing finite-reversal multi-counter machines are quite powerful. For example, a deterministic 5-crossing 1-reversal 1-counter machine M can accept the language over the alphabet $\{a, b, c, d\}$ consisting of all strings such that the *sum* of the lengths of all runs of c's occurring between pairs of symbols a and b (in this order) equals the number of d's. For example, M accepts the string "*dacbacaccbdd*". M operates in the following manner. It computes the *sum* in its counter by looking at the input and whenever it sees an a, it first checks that there is a matching b to the right and that all symbols in-between are c's. It then moves left (to a), adding the length of the run of c's to the counter. The process is repeated until the whole string has been examined. (So far, M crosses any boundary between two input symbols at most 3 times.) M then moves the input head from the right delimiter to the left delimiter and checks that the number of d's is equal to the *sum* in the counter. Finally, the input head is moved to right delimiter and the machine accepts if and only if the string is in the language. Thus, M is 5-crossing, although its input head makes an unbounded number of (left-to-right and right-to-left) turns, i.e., it is not finite-turn. In fact, it can be shown that no deterministic finite-turn finite-reversal multi-counter machine can accept the language.

As another example, let L be the language consisting of all strings $x\#y\#z$, such that x, y, z are pair-wise distinct binary strings. A nondeterministic one-way 1-reversal 3-counter machine M can accept L. M uses one counter to check that x is different from y, a second counter to compare x and z, and a third counter to check that y is different from z. To verify that x is different from y, M "guesses" a position of discrepancy (within the string x). It does this by incrementing the first counter by 1 for every symbol it encounters while moving right on x, and nondeterministically terminating the counting at some point, guessing that a position of discrepancy has been reached. M records in it's finite-control the symbol in that position. M uses the value in the counter to arrive at the same location within y where a discrepancy was guessed to occur. The second and third counters are used in a similar way to compare x with z and y with z.

The following results are from [9]:

Theorem 5. *The emptiness problem for nondeterministic k-crossing r-reversal m-counter machines is decidable in nondeterministic* $(mkr \log n)$-*SPACE and in deterministic* n^{cmkr}-*TIME, where n is the length of the binary representation of the counter machine being tested and c a constant.*

Theorem 6. *The containment and equivalence problems for deterministic k-crossing r-reversal m-counter machines are decidable in nondeterministic* $(mkr \log n)$-*SPACE and in deterministic* n^{cmkr}-*TIME, where n is the sum of the lengths of the binary representations of the two counter machines involved and c a constant.*

Theorem 6 fails for nondeterministic machines. In fact, the containment and equivalence problems are undecidable for nondeterministic one-way 1-reversal 1-counter machines [4]. However, when the machines accept "bounded" languages, the problems become decidable [8,9].

Theorem 7. *Let k, r be fixed nonnegative integers. The containment and equivalence problems for nondeterministic k-crossing r-reversal m-counter machines accepting bounded languages are decidable in* $2^{2^{p(mn)}}$-*TIME for some polynomial p, where n is the sum of the lengths of the binary representations of the machines involved.*

Theorem 8. *Every nondeterministic finite-crossing finite-reversal multi-counter machine which accepts a bounded language can effectively be converted to an equivalent deterministic finite-crossing finite-reversal multi-counter machine. The result also holds for the case of "one-way" instead of "finite-crossing".*

In the results above, the restriction that the input head is finite-crossing is crucial in view of the following result in [8].

Theorem 9. *The emptiness problem for deterministic finite-reversal 2-counter machines (with unrestricted input, i.e., not finite-crossing) accepting bounded languages is undecidable.*

4 Some Applications

We give some applications of the results of the previous sections.

The first concerns finite-reversal two-way deterministic pushdown automata (2DPDA's) on bounded languages. A 2DPDA is finite-reversal if it makes at most a fixed number of alternations between "pushing" and "popping", independent of the input.

Theorem 10. *The emptiness problem for finite-reversal* 2DPDA's *accepting bounded languages is decidable.*

The proof follows from Theorem 3 and the following lemma.

Lemma 1. *Let M be a finite-reversal* 2DPDA *accepting a bounded language over $a_1^* \dots a_k^*$. We can effectively construct a finite-reversal* 2DCM *machine M' (not necessarily accepting the same language) such that $L(M)$ is empty iff $L(M')$ is empty.*

Proof. We may assume, without loss of generality, that on every step, M pushes exactly 1 symbol on top of the stack, does not change the top of the stack, or pops exactly 1 symbol, i.e., M is not allowed to rewrite the top of the stack. In the discussion that follows, we assume $M$is processing an input that is accepted, i.e., the computation is halting.

A writing phase is a sequence of steps which starts with a push and the stack is never popped (i.e., the stack height does not decrease) during the sequence. A writing phase is periodic if there are strings u, v, w with v nonnull such that for the entire writing phase, the string written on the stack is of the form uv^iw for some i (the multiplicity) , and the configuration of M (state, symbol, top of the stack) just before the first symbol of the first v is written is the same as the configuration just before the first symbol of the second v is written. Note that w is a prefix of v. Clearly, a writing phase can only end when the input head reaches an endmarker or a boundary between the a_i's. A writing phase can only be followed by a popping of the stack (i.e. reversal) or another writing phase with possibly different triple (u, v, w).

By enlarging the state set, we can easily modify M so that all writing phases are periodic. One can easily verify that because M is finite-reversal, there are at most a fixed number t of writing phases (in the computation), and t is effectively computable from the specification of M.

We now describe the construction of M'. Let $x = a_1^{i_1} \dots a_k^{i_k}$ be the input to M. The input to M' is of the form: $y\#c_1\#c_2\# \dots \#c_t$, where the c_i's are unary strings; y is x but certain positions are marked with markers $m_1, m_2, \dots, m_t$. Note that a position of y can have $0, 1, \dots$ at most t markers.

M' simulates M on the segment y ignoring the markers. The c_i's are used to remember the counter values. Informally, every time the counter enters a new writing phase, the machine "records" the current value of the counter by checking the c_i's.

M' begins by simulating M on y (ignoring the markers). When a writing phase is entered, M' records the triple (u_1, v_1, w_1) in its finite control and the multiplicity in the counter. Suppose M enters another writing phase. Then M' checks that the input head is on a symbol marked by marker m_1; hence M' can "remember" the input head position. (If it's not marked m_1, M' rejects.) Then it "records" the current value of the counter on the input by checking that the current value is equal to c_1. (If it's not, M' rejects.) M' restores the input head to the position marked m_1 and resets the counter to 0. It can then proceed with the simulation. Next time M' has to record the value of the counter, M' use the input marker m_2 and checks c_2, while storing the triple (u_2, v_2, w_2) in its finite control, etc. Popping of the stack is easily simulated using the appropriate triple (u, v, w) and the counter value. Note that if in the simulation of a sequence of pops, the counter becomes 0, M' must first retrieve the appropriate c_i (corresponding to the pushdown segment directly below the one that was just consumed) and restore it in the counter before it can continue with the simulation; retrieving and restoring the count in the counter requires the input head to leave the input position, but M' can remember the "new" position with a new "marker". If after a sequence of pops M enters a new writing phase before the counter becomes 0, the "residual" counter value is recorded as a new c_i like before. We leave the details to the reader. It is clear that M' is a finite-reversal 2DCM. ∎

Since finite-reversal 2DPDA's accepting bounded languages can be made halting [14], we have:

Corollary 1. *The containment and equivalence problems for finite-reversal* 2DPDA's *accepting bounded languages are decidable.*

Note that the above result (for equivalence) is a generalization over bounded languages of the well-known result [15] that the equivalence problem for one-way finite-reversal pushdown automata is decidable.

It is known that 2DPDA's are more powerful that 2DCM's [16]. Consider the set of palindromes $L = \{x\#x^R \mid x \in \{0,1\}^* \}$. This language cannot be accepted by any finite-reversal 2DCM [1]. On the other hand, L can easily be accepted by a 2DPDA (in fact, a one-way DPDA) whose stack reverses once, and by a 2DCM with an unrestricted counter. It follows that finite-reversal 2DCM's are weaker than 2DCM's and weaker than finite-reversal 2DPDA's, over general (i.e., unbounded) languages. In contrast, we can show:

Theorem 11. *Over bounded languages, every finite-reversal* 2DPDA *can effectively be converted to an equivalent finite-reversal* 2DCM.

Note that the above theorem implies Theorem 10 and Corollary 1 . Theorem 11 is best possible since: (1) the emptiness problem for 2DCM's (and hence also for 2DPDA's) over unary alphabet is undecidable [12]; (2) the emptiness problem for 2DPDA's over unbounded inputs whose input head makes at most one turn and whose stack makes at most one reversal is undecidable [8]. It remains an interesting open question whether 2DCM's and 2DPDA's over bounded languages are equivalent.

The emptiness, equivalence and containment problems for finite-turn 2DPDA's (with unrestricted stack) over bounded languages are decidable. In fact there is a stronger result. One can augment the pushdown automata with counters. The following interesting result was shown in [8] ("N" stands for nondeterministic).

Theorem 12. *The following problems are decidable.*

1. *The emptiness problem for one-way* NPDA's (1NPDA's) *augmented with finite-reversal counters (but the stack is unrestricted).*
2. *The emptiness, equivalence and containment problems for finite-turn* 2NPDA's *augmented with finite-reversal counters over bounded languages.*

Theorem 12 has some nice applications. An example is concerned with checking PCP (Post Correspondence Problem) - like properties. Given two n-tuples of nonnull words $(u_1, \ldots, u_n)$, $(v_1, \ldots, v_n)$ and two integer functions f, g whose range is [1, n]. Define the following languages:

- $L = \{ x \mid x = u_{f(1)} \ldots u_{f(k)} = v_{g(1)} \ldots v_{g(k)}$ for some k $\}$,
- $L' = \{ x \mid x = u_{f(1)} \ldots u_{f(k)} = v_{g(1)} \ldots v_{g(k)}$ for some k,
  where $g(1), \ldots, g(k)$ is a permutation of $f(1), \ldots, f(k)$ $\}$,

Let C be a context-free language specified by a 1NPDA or a CFG. Then one can effectively construct 1NPDA's augmented with finite-reversal counters to accept L, L', and $L' \cap C$. It follows that the properties defined by L, L', and $L' \cap C$ for arbitrary $\{u_1, \ldots, u_n\}$ and $\{v_1, \ldots, v_n\}$ and C are decidable. Note that these properties are almost like the PCP property which is undecidable.

In part 1 of Theorem 12, the one-way machine has one unrestricted pushdown and several finite-reversal counters. If instead there is only one unrestricted counter and one finite-reversal pushdown, the problem becomes undecidable:

Theorem 13. *The emptiness problem is undecidable for one-way deterministic one-counter machines augmented with 1-reversal pushdown (the counter is unrestricted).*

Theorem 13 follows from the fact that the emptiness problem is undecidable for 1-turn 2DCM's [8].

We now consider finite-reversal 2NPDAs over bounded languages. We shall show that the emptiness problem for this class is undecidable even when restricted to unary alphabets. We first make the following observation.

Observation:
Given a single-tape TM M, we can effectively construct a two-counter machine (without input tape) M' which simulates M on blank tape. M' is normalized in that its computation can be organized in phases starting with both counters zero. Each phase starts with one counter C1 decreasing its value until it becomes zero, while the other counter C2 increases its value (from zero). In the next phase, C2 decreases to zero while C1 increases from zero, etc. The construction of M' is a simple modification of Minsky's construction (see, e.g., [10]).

Theorem 14. *The emptiness problem is undecidable for*
1. *sweeping 1-reversal* 2NPDA's *over unary languages.*
2. *1-turn 1-reversal* 2DPDA's *over the language* $(0^+\#)^+$.

Proof. To prove 1), let M be a normalized 2-counter machine as described in the Observation. We construct a 1-reversal 2DPDA M' which simulates M as follows. Given an input a^n, M' first writes $(\#a^n)^i$ on the pushdown for some nondeterministically chosen i. Then M uses the pushdown to simulate the decreasing counter and the input to simulate the other counter. The simulation of a counter by the pushdown can be made by just popping the stack. When the simulation of a phase of the 2-counter machine M has been completed, M' continues moving its input head (in the same direction) and popping its stack (thus preserving the difference between the counter values) until the input head reaches an endmarker. By exchanging the roles of the input head and pushdown stack (with respect to the counters they are simulating), the simulation can then be continued. Clearly, M' is sweeping, almost deterministic, and makes exactly one reversal on the stack.

To prove 2), we use the fact that a 2-counter machine (without input) normalized as described in Observation above can simulate a TM on blank tape. So let M be such a 2-counter machine. We construct a 2DPDA M' which simulates M as follows. Given an input, M' copies the input on the pushdown and checks that it is of the form $\%0^{i_1}\#0^{i_2}\#\ldots\#0^{i_k}$. (% is the left endmarker.) At the end of this process, the input and pushdown heads are on \$. The simulation of M is done while M' moves its input head to the left and pops it stack towards %. M' simulates the 2-counter machine but when a phase of M is completed, M' continues moving its head and popping its stack until the input head and stack are on #. Then it simulates the next phase, etc. Note that the simulation can be carried out correctly if the i_j's are large enough and it is not necessary that they are the same. ∎

Another interesting application is in the theory of query languages for constraint databases. Counter machines provide a new technique to the study of *containment* and *equivalence* (CE) of queries with linear constraints (over $\mathbb{N}, \mathbb{Z}, \mathbb{Q}, \mathbb{R}$). In [11], it is shown that the CE problems are decidable for conjunctive queries and for a subclass of first-order queries. In the following, we give a short discussion on the problems.

We fix the domain $\mathbb{Z}$ (the domain $\mathbb{N}$ is similar; $\mathbb{Q}, \mathbb{R}$ are dealt with in [11]). A *(database) schema* is a set of *predicates* and a *database* is a mapping that maps each predicate to a finite subset of $\mathbb{Z}^n$ of an appropriate arity. A *conjunctive query* over a schema σ has the following form: "$ans(\bar{x}) \leftarrow q_1(\bar{y}_1), \ldots, q_n(\bar{y}_n), \varphi(\bar{z})$" where $n \in \mathbb{N}$, $\bar{x}, \bar{z}, \bar{y}_i$'s are sequences of variables such that $\{\bar{z}\} \subseteq \{\bar{y}_1, ..., \bar{y}_n\}$, q_i's are predicates in σ, and φ is a quantifier-free first-order formula involving linear constraints over $\mathbb{Z}$. The *answer* to a conjunctive query Q on a database I, $Q(I)$, is the result of applying the rule to I. A query Q *is contained in* another Q', $Q \sqsubseteq Q'$, if for each database I, $Q(I) \subseteq Q'(I)$. Two queries are *equivalent* if they contain each other.

Theorem 15. [11] *The CE problems of conjunctive queries are decidable.*

The proof consists of two steps. First, we show that the CE problems can be determined by examining only finite databases of a fixed size, i.e., the total number of elements occurring in the database (dependent on the queries).

Lemma 2. *For conjunctive queries Q, Q', there is $k \in \mathbb{N}$ ($\leq c \times$ size of Q for some constant c) such that $Q \sqsubseteq Q'$ iff for all databases I of size $\leq k$, $Q(I) \subseteq Q'(I)$.*

Proof. Since all relation names occur positively in a conjunctive query, the query is *monotonic* with respect to databases. The lemma follows directly. ∎

For the second step, let Q, Q' be two conjunctive queries. We construct two bounded languages L_Q and $L_{Q'}$ that are accepted by nondeterministic one-way 1-reversal m-counter machines for some $m \in \mathbb{N}$ (depending on the queries) such that $L_Q \subseteq L_{Q'}$ if and only if $Q \sqsubseteq Q'$. The construction is accomplished by Lemmas 3–5 in the following.

The languages L_Q and $L_{Q'}$ are defined using a unary encoding of databases. The alphabet includes the symbols $c_1, ..., c_m, +, -, \$$, where m is the largest arity of relations in the database. A tuple $t = (a_1, ..., a_k) \in \mathbb{Z}^k$ is encoded as $\text{ENC}(t)$ = "$s_1 c_1^{v_1} ... s_k c_k^{v_k}$", where s_i is the sign of a_i and v_i the absolute value of a_i. Let I be a database of the schema σ. If $t_1, ..., t_\ell$ is an enumeration of a relation p in I, the encoding of p is "$\text{ENC}(t_1)...\text{ENC}(t_\ell)\$$". The encoding of the database I, $\text{ENC}(I)$, is the concatenation of encodings of its relations.

Let $inst_k(\sigma)$ be the set of databases over σ with at most k elements. We define $L_Q^k = \{ \text{ENC}(t) \, \$ \, \text{ENC}(I) \mid I \in inst_k(\sigma), t \in Q(I)\}$. Since k is fixed, L_Q^k is a bounded language. It is easy to verify the following.

Lemma 3. *Let Q, Q' be two conjunctive queries over a schema σ and $k \in \mathbb{N}$ greater than the product of the size of Q and the largest arity of predicates in σ. Then $L_Q^k \subseteq L_{Q'}^k$ if and only if $Q \sqsubseteq Q'$.*

We now show that L_Q^k can be accepted by a nondeterministic one-way 1-reversal m-counter machine. Since databases have at most k elements, there are at most k^n assignments, where n is the size of Q. Let $\alpha_1, ..., \alpha_{k^n}$ be all assignments. Let L^k_{Q,α_i} be the sub-language of L^k_Q where the encoded answer tuple can be produced by α_i. We first need the following technical lemma.

Let φ be a formula with free variables $x_1, ..., x_k$ and linear constraints (e.g., $(5x_1 - 3x_2 - 4x_3 + 6 < 2x_4) \wedge \neg(-2x_1 + 7x_3 = x_4)$). Let $a_1, \ldots, a_k$ be distinct symbols and $L(\varphi)$ the language consisting of strings $s_1 a_1^{v_1} \cdots s_k a_k^{v_k}$, where each s_i is the sign (+ or −) and each $v_i \in \mathbb{N}$ such that the formula φ is true under the assignment $x_i \mapsto s_i v_i$ for each $1 \leq i \leq k$.

Lemma 4. *For each quantifier-free formula φ of size n we can effectively construct a deterministic one-way $f(n)$-reversal multi-counter machine M_φ that accepts $L(\varphi)$, where both $f(n)$ and the size of M_φ are polynomials in n.*

Proof. The states (finite control) of M_φ remember all coefficients including the signs and φ. For each variable x, if a is the largest absolute value of its coefficients, M_φ has the following counters: $c_{x,0}, c_{x,1}, ..., c_{x,\lceil \log a \rceil}, c'_x$, where $c_{x,i}$ will store the value $2^i \times x$. Initially $c_{x,0}$ gets the value from the (one-way) input. For each $0 \leq i \leq \lceil \log a \rceil - 1$, the value of $c_{x,i+1}$ is computed from $c_{x,i}$ using c'_x as an auxiliary counter. The value of $c_{x,i}$ is restored after computing the value of $c_{x,i+1}$. For each (occurrence of a) term t, M_φ includes a counter c_t that will hold the absolute value for the term (the sign is remembered in the control). The (absolute) values of the counters for terms "ax" are computed from the counters $c_{x,i}$, $0 \leq i \leq \lceil \log |a| \rceil$, possibly using the auxiliary space c'_x. The values of the counters $c_{x,i}$'s are again restored after their use. We note that the binary representation of $|a|$ and the sign are remembered in the finite control. The values of the counters for the other terms are then computed based on the syntax of the terms. Finally, M_φ performs comparison tests for $=$, $\leq$, and $<$, and then computes truth value of the formula φ with Boolean operations. Clearly M_φ has polynomially many counters and a polynomial size in the size of φ. As for the number of reversals, we note that each "doubling" in computing $c_{x,i}$ and each addition/subtraction operation may need a couple of reversals. It follows that the total number of reversals is polynomial in the size of φ. ■

Lemma 5. *For each conjunctive query Q of size n, each $k \in \mathbb{N}$, and each assignment α, $L^k_{Q,\alpha}$ is accepted by a deterministic one-way 1-reversal $f(n,k)$-counter machine M, where f is a function polynomial in both n and k. The size of M is also polynomial in n and k.*

Proof. For the assignment α, we construct first a deterministic one-way f-reversal multi-counter machine M' which, on an input "ENC(t)\$ENC(I)", verifies if $t \in Q(I)$ under the assignment α, where f is a polynomial in both n and k. The verification involves (1) checking if the tuple for each atom is in the corresponding relation of I, which can be done in the straightforward manner, and (2) checking if the formula is satisfied, which can be done by a one-way, polynomial-size, polynomial-reversal counter machine by Lemma 4.

Therefore, M' can be constructed, and it is easy to obtain an equivalent 1-reversal machine M. ∎

To establish Theorem 15, let Q, Q' be two conjunctive queries and $k \in \mathbb{N}$ such that $Q \sqsubseteq Q'$ coincides with their containment over databases with at most k-elements. Let $\alpha_1, ..., \alpha_{k^n}$ be all possible assignments. For each $1 \leq i \leq k^n$, let M_i (resp. M_i') be the counter machine constructed in Lemma 5 for Q (resp. Q'). Then, $L_Q^k = \bigcup_{1 \leq i \leq k^n} L(M_i)$ and $L_{Q'}^k = \bigcup_{1 \leq i \leq k^n} L(M_i')$. It follows that both L_Q^k and $L_{Q'}^k$ are accepted by nondeterministic one-way finite-reversal multi-counter machines. By Theorem 7, the problem of determining $Q \sqsubseteq Q'$ is decidable and so is the equivalence problem.

The above decision procedure has an exponential space complexity in the size of queries. With an improvement in the last step, it can be shown to be in polynomial space [11]. In [11], the decidability results were further extended to a larger class of conjunctive queries with linear arithmetic over integers or rational numbers, and to a syntactically restricted class of first-order queries over bounded degree databases.

References

1. M. CHROBAK, *A Note on Bounded-reversal Multipushdown Machines*, Info. Proc Letters, 19 (1984), pp. 179–180.
2. T.-H. CHAN, *On Two-way Weak Counter Machines*, Math. System Theory, 20 (1987), pp. 31–41.
3. M. DAVIS, Y. MATIJASEVIČ, AND J. ROBINSON, *Hilbert's Tenth Problem. Diophantine Equations: Positive Aspects of a Negative Solution*, Proc. Symp. Pure Math., 28 (1976), pp. 323–378.
4. B. BAKER AND R. BOOK, *Reversal-bounded Multipushdown Machines*, J. Comput. System Sci., 8 (1974), pp. 315–332.
5. E. M. GURARI AND O. H. IBARRA, *Simple Counter Machines and Number-theoretic Problems*, J. Comput. System Sci., 19 (1979), pp. 145–162.
6. E. M. GURARI AND O. H. IBARRA, *Two-way Counter Machines and Diophantine Equations*, J. Assoc. Comput. Mach., 29 (1982), pp. 863–873.
7. O. H. IBARRA, T. JIANG, N. TRAN, AND H. WANG, *On the Equivalence of Two-way Pushdown Automata and Counter Machines Over Bounded Languages*, in Proc. of the 10th Symposium on Theoretical Aspects of Computer Science, 1993.
8. O. H. IBARRA, *Reversal-bounded Multicounter Machines and Their Decision Problems*, J. Assoc. Comput. Mach., 25 (1978), pp. 116–133.
9. O. H. IBARRA AND E. GURARI, *The Complexity of Decision Problems for Finite-turn Multicounter Machines*, J. Comput. System Sci., 22 (1981), pp. 220–229.
10. O. H. IBARRA AND T. JIANG, *The Power of Alternating One-reversal Counters and Stacks*, SICOMP, 20 (1991), pp. 278–290.
11. O. H. IBARRA AND J. SU, *On the Containment and Equivalence of Database Queries with Linear Constraints*, in Proc. ACM Symp. on Principles of Database Systems, 1997.

12. M. Minsky, *Recursive Unsolvability of Post's Problem of Tag and Other Topics in the Theory of Turing Machines*, Ann. of Math., 74 (1961), pp. 437–455.
13. L. G. Valiant and M. S. Paterson, *Deterministic One-counter Automata*, J. Comput. System Sci., 10 (1975), pp. 340–350.
14. O. H. Ibarra, T. Jiang, N. Tran, and H. Wang, *New Decidability Results Concerning Two-way Counter Machines*, SICOMP, 24 (1995), pp. 123–137.
15. L. Valiant, *The Equivalence Problem for Deterministic Finite-turn Pushdown Automata*, Inform. and Contr., 25 (1974), pp. 123–133.
16. P. Duris and Z. Galil, *Fooling a Two Way Automaton or One Pushdown Store is Better Than One Counter for Two Way Machines*, Theoretical Computer Science, 21 (1982), pp. 39–53.

On the Equivalence of Finite Substitutions and Transducers

Juhani Karhumäki and Leonid P. Lisovik*

Summary. This paper discusses on several variants of a fascinating problem of deciding whether two finite substitutions are equivalent on a regular language, as well as its relations to the equivalence problems of sequential transducers. Among other things it is proved to be decidable whether for a regular language L and two substitutions φ and ψ, the latter one being a prefix substitution, the relation $\varphi(w) \subseteq \psi(w)$ holds for all w in L.

1 Introduction

Equivalence problems of automata have played a crucial role in the theory of automata from its very beginning. Two such problems, namely the equivalence problems of deterministic multitape automata and deterministic pushdown automata, were for several decades among the most important open problems of the theory, until they were solved in the 90's, cf. [10] and [21], respectively.

In its pure form the *equivalence problem of automata* asks to decide whether two automata of a certain type define the same language or relation. An interesting variant of this problem was introduced by Culik II and Salomaa in [5], where they proposed to study so-called *morphic equivalence problem for languages*, i.e. a problem of deciding whether two morphisms are equivalent on a given language. Here being equivalent means that the morphisms map each word of the language to the same word. The celebrated $D0L$ problem is an example of such a problem, cf. [4].

A natural generalization of the latter problem is to consider more general mappings than morphisms. Let us refer such decision problems to as the *equivalence problems of mappings on languages*. As an example of articles considering these problems we mention [15], where it was shown that the equivalence of mappings of the form $h^{-1}g$, with h and g morphisms, is decidable on regular languages while that of mappings of the form gh^{-1} is undecidable.

* This work was initiated when this author visited Turku University under the grant 14047 of the Academy of Finland.

This paper concentrates on a particular instance of the above problem, namely on the equivalence of finite substitutions on regular languages. Surprisingly, this problem was shown to be undecidable in [18]. Our goals here are as follows. First, we recall connections of the above result with the equivalence problems of transducers, cf. also [22]. Second, we prove some new results, mainly decidability ones, for certain restricted variants of the above problem. These, we believe, sharpen the borderline between the decidability and undecidability. And finally, we consider the remaining major open problem, namely the problem of deciding whether two finite substitutions are equivalent on the language ab^*c, and give both a decidable and undecidable variant of this problem.

In details this paper is organized as follows.

In Section 2 we fix our terminology and recall several major results related to our problems. In Section 3 we consider so-called prefix substitutions, and prove, as our most complicated technical result, that not only the equivalence but also the (pointwise) inclusion of such substitutions on regular languages is decidable. In Section 4 we translate our results to transducers. Finally, in Section 5 we discuss on the most challenging open problem, the equivalence of finite substitutions on the language ab^*c.

2 Preliminaries

In this section we fix our terminology, as well as recall a few important results we shall need in our considerations. We refer to [12] and [1] as general references.

We denote by Σ^+, Σ^* and $\Sigma^{(*)}$ the *free semigroup*, the *free monoid* and the *free group* generated by an alphabet Σ, respectively. Elements of Σ^* are called words and 1 denotes the neutral element of Σ^*, i.e. the empty word. The length of a word w is denoted by $|w|$. For two words u and v the word u is a prefix of v, in symbols $u \leq v$, if there exists a word t such that $v = ut$. We use the notation $\mathrm{Pref}(v)$ to denote the set of all prefixes of v.

We say that a subset $X \subseteq \Sigma^*$ is a *prefix set*, or briefly a *prefix*, if none of its words is a prefix of another of its words. Consequently, 1 is in a prefix X iff $X = \{1\}$. Further we say that $X \subseteq \Sigma^*$ has a *bounded deciphering delay* if there exists a natural number $p \geq 0$ such that

$$uX^p\Sigma^* \cap vX^* = \emptyset \text{ for all } u, v \in X \text{ with } u \neq v.$$

Obviously, X with a bounded deciphering delay is a code.

A finite *substitution* is a morphism $\varphi\colon \Sigma^* \to \mathcal{P}(\Delta^*)$, where $\mathcal{P}(\Delta^*)$ denotes the monoid of finite subsets of Δ^* under the operation of the product of subsets of Δ^*. We call φ *1-free* if $1 \notin \varphi(a)$ and *prefix* if $\varphi(a)$ is a prefix set for all $a \in \Sigma$. Consequently, a prefix substitution need not be 1-free, since it may be that $\varphi(a) = \{1\}$ for some $a \in \Sigma$. Similarly, we say that φ has a bounded deciphering delay if $\varphi(a)$ has such for each $a \in \Sigma$. Finally, we say that two substitutions $\varphi, \psi \colon \Sigma^* \to \mathcal{P}(\Delta^*)$ *are equivalent* on a language $L \subseteq \Sigma^*$ if

$$\varphi(w) = \psi(w) \text{ for all } w \in L.$$

Now, we are ready to formulate our crucial problems.

Problem 1. *Is it decidable for two finite substitutions $\varphi, \psi : \Sigma^* \to \mathcal{P}(\Delta^*)$ and a regular language $L \subseteq \Sigma^*$ whether or not φ and ψ are equivalent on L?*

Problem 2. *Is it decidable for two finite substitutions $\varphi, \psi : \{a, b, c\}^* \to \mathcal{P}(\Delta^*)$ whether or not φ and ψ are equivalent on ab^*c?*

Of course, Problem 2 is a very restricted variant of Problem 1 but we prefer to formulate it separately due to its importance. Note that the deterministic variant of Problem 1, i.e. when φ and ψ are morphisms, in other words when sets $\varphi(a)$ and $\psi(a)$ are singletons, is easily seen to be decidable.

Next we fix our terminology on transducers, or more precisely on sequential transducers. A *sequential transducer* is a sixtuple $\mathcal{M} = (Q, \Sigma, \Delta, T, q_0, F)$, where Q is a finite set of states, Σ and Δ are finite input and output alphabets, respectively, $T \subseteq Q \times \Sigma \times \Delta^* \times Q$ is a set of *transitions*, q_0 is an initial state and F is a set of final states. The relation *defined* or *computed* by $\mathcal{M}$ is denoted by $\mathcal{R}(\mathcal{M}) \subseteq \Sigma^* \times \Delta^*$, for details cf.[1]. The *underlying automaton* of $\mathcal{M}$ is obtained from it by omitting the third components in the transitions.

We shall need some special types of sequential transducers. We say that a sequential transducer $\mathcal{M}$

- is *1-free* if, for each $(p, a, v, q) \in T$, the word $v \neq 1$;
- is *input deterministic* if the underlying automaton is deterministic;
- satisfies the *prefix condition* if, for each $p \in Q$ and $a \in \Sigma$ we have

$$(p, a, v, q), (p, a, v', q') \in T \Rightarrow v \not\leq v' \text{ and } v' \not\leq v.$$

The above definitions deserve a few comments. First the prefix condition implies (but is not equivalent to) the property: the set of outputs associated to each triple (p, a, q), i.e.

$$O_{p,a,q} = \{v | (p, a, v, q) \cap T \neq \emptyset\} \tag{1}$$

is a prefix set. For input deterministic transducers the latter condition is equivalent to the prefix property. Finally, a transducer with the prefix property need not be 1-free, but cannot contain transitions $(p, a, 1, q)$ and $(p, a, 1, q')$ with $q \neq q'$.

The following connection between two types of problems we are considering was noticed in [3].

Theorem 1. *The equivalence problem for input deterministic sequential transducers is decidable if and only if the equivalence of finite substitutions on regular languages is decidable.*

The proof of Theorem 1 is based on the fact that the set of accepting computations of an automaton is a regular language. Moreover, the equivalence in Theorem 1 remains if the sets (1) of the input deterministic transducers are of the same type, for example prefixes, as the images of letters in the finite substitutions.

We conclude this section by recalling two remarkable results related to our problems, and used later in our considerations. The first one, proved simultaneously in [13] and [17], extended in the very nontrivial way the basic undecidability result of sequential transducers established in [6]. Recently this result was further extended in [19] by considering also the number of states as a parameter.

Theorem 2. *The equivalence problem for 1-free sequential transducers having a unary output (or input) alphabet is undecidable.*

The other result proved in [18], cf. also [8] or [9] for a more detailed proof, gives a surprising answer to our Problem 1. It also gives a partial answer why we introduced Problem 2 as a separate problem.

Theorem 3. *It is undecidable whether two finite 1-free substitutions* $\varphi, \psi : \{a, b, c, d\}^* \to \mathcal{P}(\Delta^*)$ *are equivalent on the language* $a\{b, c\}^*d$.

Using Theorem 1 the above can be interpreted as an undecidability result of sequential transducers.

Corollary 1. *The equivalence problem for 1-free input deterministic sequential tranducers is undecidable.*

3 Problems on finite substitutions

In this section we state two decidability results for a special type of substitutions, namely prefix substitutions. Our first result is noticed in [14], and should be compared to Theorem 3.

Theorem 4. *It is decidable whether two prefix substitutions are equivalent on a given regular language.*

In order to compare the proof techniques of different results of this paper we recall basic ideas of the proof of Theorem 4. It is based on three facts: The pumping lemma for regular languages, the freeness of the submonoid of the monoid $\mathcal{P}(\Delta^*)$ constituting of finite prefix languages, and the following implication of equations over free monoids:

$$\begin{cases} uv &= u'v' \\ uxv &= u'x'v' \\ uyv &= u'y'v' \end{cases} \Rightarrow uxyv = u'x'y'v'.$$

The second condition was first noticed in [20] while the third one is a simple exercise on word equations.

In our second result we consider instead of the equivalence the *inclusion* of two substitutions φ and ψ on a regular language L, i.e. the relation $\varphi(w) \subseteq \psi(w)$ for all $w \in L$. It seems that we need a completely new method than the one used in Theorem 4 to prove the decidability.

Theorem 5. *It is decidable whether for a given regular language L and two 1-free prefix substitutions φ and ψ the relation $\varphi(w) \subseteq \psi(w)$ holds for all w in L.*

Proof. Let φ and ψ be 1-free prefix substitutions defined on Σ^* and $L \subseteq \Sigma^*$ a regular language recognized by an automaton $\mathcal{A}_L$. The problem of the theorem is reduced to the emptiness problem of finite automaton, i.e. we construct a nondeterministic finite automaton $\mathcal{A}$ such that

$$L(\mathcal{A}) = \emptyset \quad \text{if and only if} \quad \varphi(w) \subseteq \psi(w) \text{ for all } w \in L. \qquad (*)$$

We may assume that the automaton $\mathcal{A}_L = (Q, \Sigma, \delta, q_0, F)$ is deterministic and reduced. Then define the constant

$$t = \max_{q \in Q}\{\min\{|v| \mid \delta(q, v) \in F\}\}.$$

Consequently, from any state of $\mathcal{A}_L$ there exists a word of length at most t leading to a final state. Two other constants are defined as

$$m_1 = \max_{a \in \Sigma}\{\max\{|v| \mid v \in \varphi(a)\}\},$$

$$m_2 = \max_{a \in \Sigma}\{\max\{|v| \mid v \in \psi(a)\}\}.$$

Now, we construct the nondeterministic $\mathcal{A}$ as follows.

Consider a constant N (fixed later). The automaton $\mathcal{A}$

(i) reads nondeterministically N symbols, i.e. a word $w = a_1 \dots a_N$, and checks that w is a prefix of a word in L;
(ii) quesses nondeterministically a word $u = u_1 \dots u_N$, with $u_i \in \varphi(a_i)$ for $i = 1, \dots, N$, and;
(iii) checks that ψ can respond to the above choice of *(ii)*, i.e. either finds words $v_i \in \psi(a_i)$, for $i = 1, \dots, N$ $(= n_1)$, such that

$$v\gamma = v_1 \dots v_N \gamma = u \quad \text{for some word } \gamma, \qquad (1)$$

or finds an $n_1 < N$ and words $v_1, \dots, v_{n_1}$ such that

$$v\gamma = v_1 \dots v_{n_1} \gamma = u \quad \text{for some } \gamma \in \mathrm{Pref}(\varphi(a_{n_1+1})). \qquad (2)$$

Now, a crucial observation is that, for a fixed w and u, either (1) or (2), but not both, can hold and, moreover, the v_i-sequence is unique. This is due to the fact that the sets $\psi(a_i)$ are prefix sets. Further, if neither (1) nor (2) in (*iii*) cannot be satisfied, then, by the choice of w and u, w can be extended to a word ww_1 such that $\varphi(ww_1) \not\subseteq \psi(ww_1)$, that is the automaton $\mathcal{A}$ can accept its input.

So it remains to analyse how the automaton $\mathcal{A}$ is constructed if (1) or (2) is satisfied.

Consider first the case (1). Now, the automaton $\mathcal{A}$ checks whether

$$|\gamma| > m_2 t, \tag{3}$$

and if the answer is affirmative, then, as above, we can extend the word w to a word ww' in L such that $\varphi(ww') \not\subseteq \psi(ww')$, and so the automaton $\mathcal{A}$ can accept its input. Intuitively, this means that ψ is *too much behind* in order to respond the choice of u done in (*ii*).

So assume the case (2) in (*iii*). Now, the automaton $\mathcal{A}$ checks whether

$$N - n_1 > m_1 t, \tag{4}$$

and if the answer is affirmative, then again, but due to a different reason, namely because ψ is 1-free, we can extend w to a word ww' in L such that $\varphi(ww') \not\subseteq \psi(ww')$, and so the automaton $\mathcal{A}$ accepts its input. Intuitively, this case means that ψ is *too much ahead* in order to respond to a minimal continuation of w.

If the answers to (3) and (4) are negative, i.e. the automaton $\mathcal{A}$ did not accept anything yet, then it remembers the word γ (from (1) or (2)) and the number $N - n_1$, and starts a new round of actions (*i*)-(*iii*): the automaton $\mathcal{A}$ reads new N symbols, say $a'_1, \ldots, a'_N$, chooses new N words, say $u'_i \in \varphi(a'_i)$, for $i = 1, \ldots, N$, and the v'_i words as in (*iii*). The differences from the step described above are that now the new u, say u', is

$$u' = \gamma u'_1 \ldots u'_N,$$

and that the new n_1, say n'_1, satisfies

$$n_1 + n'_1 < 2N,$$

and accordingly (4) gets the form

$$2N - n_1 - n'_1 > m_1 t.$$

The word γ' is defined as in the basic step.

During the second round the information $(\gamma, N - n_1)$ is changed to $(\gamma',$ $2N - n_1 - n'_1)$, where $|\gamma|, |\gamma'| \leq m_2 t$ and $N - n_1, 2N - n_1 - n'_1 \leq m_1 t$. Hence, the procedure goes into a cycle.

To complete the description of the construction of $\mathcal{A}$ we have to analyse what happens when a prefix of the n-letter word w in (*ii*) corresponds an element of L. In such a situation the automaton $\mathcal{A}$ has to check that ψ indeed can respond to the choice of the values of φ selected in (*ii*).

Consider in more details the situations when the word in L is of the form ww' with $|w| = N$ and $|w'| \leq N$. Say $w = a_1 \dots a_N a'_{N+1} \dots a'_{N+j}$. Now, after reading w the automaton $\mathcal{A}$ remembers the word γ and the number $N - n_1$ such that, for the values $u_i \in \varphi(a_i)$, for $i = 1, \dots, N$, there exists values $v_i \in \psi(a_i)$, for $i = 1, \dots, n_1$, such that

$$u_1 \dots u_N = v_1 \dots v_{n_1} \gamma.$$

Now, the automaton $\mathcal{A}$ guesses nondeterministically the values $u'_i \in \varphi(a'_i)$, for $i = N + 1, \dots, N + j$, and checks if for this choice there exist words $v'_i \in \psi(a'_i)$, for $i = n_1 + 1, \dots, N + j$ such that

$$\gamma u_{N+1} \dots u_{N+j} = v_{n_1+1} \dots v_{N+j}.$$

If the answer to this check is no, then the automata accepts the input, otherwise not.

The value of N can be chosen rather freely. Clearly, everything works if it would be larger than any of the important constants t, tm_1 and tm_2, say $N = (2m_1 + 2m_2 + 2)t$.

Now, we have completed the description of the automaton satisfying (∗), and also the proof of the theorem. □

The above result deserves two comments. First, no special properties of the substitution φ are needed, it need not be 1-free nor prefix. So the result can be generalized accordingly. Second, and more importantly, also ψ need not be 1-free:

Theorem 6. *It is decidable for a given regular language L and two prefix substitutions φ and ψ whether the relation $\varphi(w) \subseteq \psi(w)$ holds for all w in L.*

Proof. In the proof of Theorem 5 the 1-freeness of ψ was needed only in the case when ψ was ahead, i.e. in the case corresponding (2) of (*iii*). Here the essential point was when the relation $N - n_1 > m_1 t$ was satisfied. From that we concluded that $\varphi(w) \not\subseteq \psi(w)$. Now, the same conclusion can be drawn if, instead of the above inequality, the inequality

$$p > m_1 t$$

is satisfied, where p is the number of those letters a of the suffix of the input word $a_1 a_2 \dots a_{jN}$ for which $\psi(a) \neq 1$. In the proof of Theorem 5 the automaton $\mathcal{A}$ remembered the previous N block of letters, and hence also the required suffix of length $N - n_1$ of the input word read so far. In the current case the automaton have to remember a sparse subword of length p.

Otherwise the changes are very straightforward. □

4 Problems on transducers

In this section we modify the results of Section 3 for sequential transducers. We prove three decidability results each of those using completely different techniques. Actually, these results form a sequence of proper extensions of each other.

From Theorems 1 and 4, cf. also the discussion before Theorem 1, we directly derive:

Theorem 7. *The equivalence problem for input deterministic sequential transducers satisfying the prefix condition is decidable.*

We recall that the transducers considered in Theorem 7 are those input deterministic sequential transducers for which the sets

$$O_{p,a,q} = \{v|(p,a,v,q) \in T\}, \text{ for } p,q \in Q, a \in \Sigma,$$

are prefix sets. In this formulation we cannot omit the input determimism, since clearly for any sequential transducer we can construct equivalent one satisfying the condition that the sets $O_{p,a,q}$ are singletons, and thus prefix sets.

On the other hand, Theorem 7 can be extended to:

Theorem 8. *The equivalence problem for sequential transducers satisfying the prefix condition is decidable.*

Proof. Let $\mathcal{M}$ be a sequential transducer satisfying the prefix condition. Then the relation $\mathcal{R}(\mathcal{M})$ defined by $\mathcal{M}$ is accepted by a deterministic 2-tape automaton. Indeed, such an automaton $\mathcal{A}$ is constructed as follows: It reads in its first tape the input word, and in the second tape the output word. When $\mathcal{M}$ is in its state p so is $\mathcal{A}$, and it reads a symbol, say a, from its first tape, and moves either to state q if $(p,a,1,q)$ is a transition of $\mathcal{M}$, or moves to the state p_a where it reads the second tape until it finds a word v such that (p,a,v,q) is a transition of $\mathcal{M}$. Then $\mathcal{A}$ moves to state q and again reads a symbol from the first tape.

The above explains the basic simulation step, and $\mathcal{A}$ indeed is deterministic due to the prefix condition of $\mathcal{M}$. The other details of the simulation are straightforward.

Now, Theorem 8 follows from the decidability of the equivalence problem for 2-tape deterministic automata, cf. [2] or [10]. □

Our method of Section 3 allows to extend Theorem 8.

Theorem 9. *The inclusion problem for sequential transducers satisfying the prefix condition is decidable.*

Proof. The proof of Theorem 5, as well as that of Theorem 6, can be modified for Theorem 9. Let $\mathcal{M}_1$ and $\mathcal{M}_2$ be two transducers of Theorem 9. Then a finite nondeterministic automaton $\mathcal{A}$ is constructed such that

$$L(\mathcal{A}) = \emptyset \quad \text{if and only if} \quad \mathcal{R}(\mathcal{M}_1) \subseteq \mathcal{R}(\mathcal{M}_2).$$

The main difference from the proof of Theorem 5 is that now $\mathcal{A}$ has to remember also the corresponding states of $\mathcal{M}_1$ and $\mathcal{M}_2$. Moreover, the prefix property of $\mathcal{M}_2$ guaranties that, for a given input word $w = a_1 \ldots a_N$ and the u-sequence, the v-sequence defined in *(iii)* is unique. □

We conclude this section with two remarks. First, as in Theorem 6, the transducer $\mathcal{M}_1$ can be assumed to an arbitrary sequential transducer. Second, Theorem 9 can not be deduced using the method of Theorem 8, since the inclusion problems for deterministic 2-tape automata is undecidable, cf. e.g. [10].

5 Finite substitutions on ab^*c

In this section we consider the most challenging open problem of the topic of this paper, namely the problem of deciding whether two finite substitutions are equivalent on the language ab^*c. We introduce two variants of this problem, one being decidable and the other undecidable.

This can be seen as an indication of the hardness of this problem. Further evidence of that was proved already in [16] where it was shown that the language $L = ab^*c$ does not have a finite *test set* with respect to finite substitutions, i.e. finite subset F of L such that to test the equivalence of two finite substitutions on L it is enough to test that on F, cf. [11]. The proof of Theorem 4 shows that for prefix substitutions such a test set exists.

Our first result is a decidability one. Its proof uses two-way finite automata with a finite number of counters. Such an automaton is called *reversal bounded* if there is a constant t such that on any computation the automaton changes the direction of the head and the directions of the use of the counters at most t times. A computation of such an automaton is *stack increasing* (resp. *stack decreasing*) if the length of the stack is never decreased (resp. increased). Finally, a configuration of such an automaton, with two stacks, on input w is completely determined by the quadruple

$$(q, \overset{\downarrow}{w}, n_1, n_2)$$

where q is a state, n_1 and n_2 are the contents of the stacks, and $\downarrow$ points the position of the head on input w.

Theorem 10. *It is decidable whether two finite substitutions with bounded deciphering delay are equivalent on the language ab^*c.*

Proof. The problem is reduced to the emptiness problem of reversal bounded two-way nondeterministic counter automata, which was shown to be decidable in [7].

Let $\varphi, \psi\colon \{a,b,c\}^* \to \mathcal{P}(\Delta^*)$ be finite substitutions with a bounded deciphering delay. We may assume that each $\varphi(a)$ and $\psi(a)$ (resp. $\varphi(c)$ and $\psi(c)$) starts with a special symbol, say ¢ (resp. \$), which does not occur anywhere else in the images.

The following claims constitute the core of the proof.

Claim 1. *There exists a one-way nondeterministic counter automaton $\mathcal{M}_1$, with i and t initial and final states, respectively, such that $\mathcal{M}_1$ has a stack increasing computation*

$$\text{from } (i, \overset{\downarrow}{¢}w\$, 0) \text{ to } (t, ¢w\overset{\downarrow}{\$}, k)$$

if and only if

$$¢w\$ \in \varphi(ab^k c).$$

Claim 2. *There exists a reversal bounded two-way deterministic 2-counter automaton $\mathcal{M}_2$, with i and t initial and final states, respectively, such that $\mathcal{M}_2$ has a computation*

$$\text{from } (i, \overset{\downarrow}{¢}w\$, k, 0) \text{ to } (t, ¢w\overset{\downarrow}{\$}, 0, 0) \tag{1}$$

if and only if

$$¢w\$ \notin \psi(ab^k c).$$

Proofs of the Claims. The proof of Claim 1 is very obvious, and actually does not require any assumption on φ. The automaton $\mathcal{M}_2$ is constructed according to the following ideas. First, $\mathcal{M}_2$ copies k to the second counter. Then, assuming that the elements $\psi(a)$ are lexicographically ordered, $\mathcal{M}_2$ chooses the first element α_1 of $\psi(a)$, and checks whether $¢w\$ \in \alpha_1\psi(a^k c)$. This can be done deterministically since ψ has a bounded deciphering delay. If the answer is no, the stacks are regenerated and the procedure is continued with the second element of $\psi(a)$, and so on. If the answer to the above tests is always no, then $\mathcal{M}_2$ goes to the required configuration of (1). Clearly, $\mathcal{M}_2$ does several, but only a bounded number of reversals on the input, as well as on the counters. So the claims are proved. □

Now, it is not difficult – using automata from Claims 1 and 2 – to construct a reversal bounded two-way nondeterministic counter automaton $\mathcal{M}$ such that

$$L(\mathcal{M}) = \emptyset \quad \text{if and only if} \quad \varphi(ab^k c) \subseteq \psi(ab^k c) \text{ for all } k \geq 0.$$

Indeed, the automaton of Claim 1 is used to create a number k such that $¢w\$ = \varphi(ab^kc)$, and the automaton from Claim 2 is used to determine whether ψ can respond, i.e. whether $¢w\$ \in \psi(ab^kc)$. Moreover, $L(\mathcal{M}) \neq \emptyset$ if and only if ψ is not able to respond, i.e. if and only if $\varphi(ab^kc) \not\subseteq \psi(ab^kc)$ for some k.

Now, the result follows when we check the inclusions $\varphi(ab^*c) \subseteq \psi(ab^*c)$ and $\psi(ab^*c) \subseteq \varphi(ab^*c)$. □

As we already pointed out, we actually proved a stronger result than Theorem 10, namely the decidability of the condition

$$\varphi(ab^kc) \subseteq \psi(ab^kc), \text{ for all } k \geq 0,$$

where φ is arbitrary and ψ has a bounded deciphering delay. In particular, in the formulation of Theorem 10 it is not necessary to assume that φ and ψ has a bounded deciphering delay to the same direction.

Our second result of this section is an undecidability result, although rather special one. Now we consider finite substitutions from a free monoid into the free group that is morphisms $\varphi \colon \Sigma^* \to \mathcal{P}(\Delta^{(*)})$.

Theorem 11. *It is undecidable whether two finite substitutions $\varphi, \psi \colon \Sigma^* \to \mathcal{P}(\Delta^{(*)})$ are equivalent on the language ab^*c modulo a free submonoid of $\Delta^{(*)}$, i.e. whether the following holds for a free submonoid Γ^* of $\Delta^{(*)}$:*

$$\varphi(ab^kc) \cap \Gamma^* = \psi(ab^kc) \cap \Gamma^* \text{ for all } k \geq 0.$$

Proof. The result follows straightforwardly from Theorem 2. Let

$$\mathcal{M} = (Q, \{a\}, \Delta, T, q_0, F)$$

be a sequential transducer with a unary input alphabet. We define a finite substitution

$$\varphi_{\mathcal{M}} \colon \{a, b, c\}^* \to \mathcal{P}((Q \cup \Delta)^{(*)})$$

by setting

$$\begin{aligned} \varphi_{\mathcal{M}}(a) &= q_0^{-1}, \\ \varphi_{\mathcal{M}}(b) &= pvq^{-1} \text{ if } (p, b, v, q) \in T, \\ \varphi_{\mathcal{M}}(c) &= F. \end{aligned}$$

Now, for two sequential transducers $\mathcal{M}$ and $\mathcal{M}'$ we clearly have:

$$\mathcal{M} \text{ and } \mathcal{M}' \text{ are equivalent}$$

if and only if

$$\varphi_{\mathcal{M}}(ab^kc) \cap \Delta^* = \varphi'_{\mathcal{M}'}(ab^kc) \cap \Delta^* \text{ for all } k \geq 0.$$

This completes the proof. □

References

1. Berstel, J., Transductions and Context-Free Languages, Teubner, Stuttgart, 1979.
2. Bird, M., The equivalence problem for deterministic two-tape automata, J. Comput. System Sci. **7**, 1973, 218 - 236.
3. Culik II, K. and Karhumäki, J., The equivalence of finite valued transducers (on HDTOL languages) is decidable, Theoret. Comput. Sci. **47**, 1986, 71 - 84.
4. Culik II, K. and Fris, I., The decidability of the equivalence problem for D0L-systems, Inform. and Control **33**, 1976, 20 - 33.
5. Culik II, K. and Salomaa, A., On the decidability of homomorphism equivalence for languages, J. Comput. System. Sci **17**, 1978, 163 - 175.
6. Griffiths, T.V., The unsolvability of the equivalence problem for λ-free nondeterministic generalized machines, J. Assoc. Comput. Mach. **15**, 1968, 409 - 413.
7. Gurari, E. and Ibarra, O., The complexity of decision problems for finite-turn multicounter machines, J. Comput. System Sci. **22**, 1981, 220 - 229.
8. Halava, V., Finite substitutions and integer weighted finite automata, Turku Centre for Computer Science, TUCS Report **197**, 1998.
9. Halava, V. and Harju, T., Undecidability of the equivalence of finite substitutions on regular language, RAIRO Theoret. Informatics and Applications (to appear).
10. Harju, T. and Karhumäki, J., The equivalence problem of multitape finite automata, Theoret. Comp. Sci. **78**, 1991, 345 - 353.
11. Harju, T. and Karhumäki, J., Morphisms, in: G. Rozenberg and A. Salomaa (eds), Handbook of Formal Languages, Vol. I, Springer, 1997, 439 - 510.
12. Hopcroft, J.E. and Ullman J.D., Introduction to Automata Theory, Languages and Computation, Addison-Wesley, 1973.
13. Ibarra, O., The unsolvability of the equivalence problem for ε-free NGSM's with unary input (output) alphabet and applications, SIAM J. Comput. **7**, 1978, 524 - 532.
14. Karhumäki, J., Equations on finite sets of words and equivalence problems in automata theory, Theoret. Comput. Sci. **108**, 1993, 103 - 118.
15. Karhumäki, J. and Kleijn, H.C.M., On the equivalence of compositions of morphisms and inverse morphisms on regular languages, RAIRO Theor. Informatics **19**, 1985, 203 - 211.
16. Lawrence, J., The non-existence of finite test sets for set equivalence of finite substitutions, Bull. EATCS **28**, 1986, 34 - 37.
17. Lisovik, L.P., The identity problem of regular events over cartesian product of free and cyclic semigroups, Doklady of Academy of Sciences of Ukraina **6**, 1979, 480 - 413.
18. Lisovik, L.P., The equivalence problem for finite substitutions on regular language, Doklady of Academy of Sciences of Russia **357**, 1997, 299 - 301.
19. Lisovik, L.P., The equivalence problems for transducers with bounded number of states, Kibernetika and Sistemny Analiz **6**, 1997, 109 - 114.
20. Perrin, D., Codes conjegués, Inform. and Control **20**, 1972, 221 - 231.
21. Senizergues, G., $L(A) = L(B)$?, Manuscript, 1998.
22. Turakainen, P., The undecidability of some equivalence problems concerning ngsm's and finite substitutions, Theoret. Comput. Sci. **174**, 1997, 269 - 274.

Complementation of Büchi Automata Revisited

Wolfgang Thomas

Summary. As an alternative to the two classical proofs for complementation of Büchi automata, due to Büchi himself and to McNaughton, we outline a third approach, based on stratified alternating automata with a "weak" acceptance condition. Building on work by Muller, Saoudi, Schupp (1986) and Kupferman and Vardi (1997), we present a streamlined version of this complementation proof. An essential point is a determinacy result on infinite games with a weak winning condition. In a unifying logical setting, the three approaches are shown to correspond to three different types of second-order definitions of ω-languages.

1 Introduction

In his seminal paper [Bü62], Büchi introduced a framework for defining sets of sequences, today called the framework of Büchi automata, and showed that the class of properties definable therein is closed under complementation. This result was the key to establish a bridge between ω-automata theory and monadic second-order logic over infinite strings, which in turn opened a new chapter of automata theory with interesting applications in logic and computer science. Even after forty years this chapter is far from being closed, and it is worthwhile to reconsider the beginnings.

There are two "classical" approaches to the complementation of Büchi automata. The first, as found by Büchi himself, stays in the framework of Büchi automata and provides a transformation of a nondeterministic Büchi automaton into such an automaton for the complement language. The second proof, due to McNaughton [McN66] and later sharpened by Safra [Sa88], involves a transformation to deterministic automata with a more general acceptance condition, the Muller condition [Mu63], and uses the fact that for deterministic Muller automata the complementation step is obvious. Both approaches involve nontrivial arguments: In the first case, a combinatorial result is applied (Ramsey's Theorem B [Ra29]), and in the second a very intricate automaton construction is required.

In the present paper, we expose a third proof strategy which so far did not attract much attention in the literature but has some advantages. Instead of reducing the nondeterminism of Büchi automata to determinism with a more complex acceptance condition, this third approach is based on a more general transition mode than nondeterminism, namely alternation, but at the same time the acceptance condition is made simpler: Instead of Büchi acceptance the so-called "weak acceptance" condition is used. As in the deterministic case, the weak acceptance condition is closed under negation, whence complementation is relatively easy. Moreover, the task of connecting Büchi automata to weak alternating automata is not as complex as to show that Büchi automata and deterministic automata are equivalent.

The idea of weak alternating automata is due to Muller, Saoudi, and Schupp ([MSS86], [MS87], [MSS88]). However, in their work they emphasize complexity issues (especially regarding program logics and temporal logics) and not so much a reconsideration of the complementation problem. Recently, Vardi and Kupferman [KV97] have taken up the approach and supplied a self-contained complementation proof for Büchi automata. Their proof strategy does not make use, however, of the duality phenomenon which is characteristic for alternating automata. In the subsequent sections we outline such a more symmetric proof. The key ingredient is a result on determinacy of infinite games with a weak winning condition. An advantage of this proof architecture is its composition from rather elementary, easily verified "modules". Moreover, some conceptual points are clarified. For example, one sees that for defining regular ω-languages by automata, the use of liveness conditions (such as the Büchi acceptance condition) can be avoided, if one works with alternating automata. The complementation proof via infinite games also sheds some light on the relation between automata on infinite words and automata on infinite trees. In the game-theoretic framework, the proofs of complementation for ω-automata and for tree automata can be compared via the respective determinacy results. For Büchi automata complementation we shall need only a very simple determinacy proof based on a reachability analysis, whereas tree automata complementation requires the more complicated determinacy proof for parity games (cf. [Th97]). This pinpoints a characteristic difference between ω-automata theory and tree automata theory.

This paper provides an introduction to results obtained in collaboration with C. Löding (see his diploma thesis [Lö98]) on alternating ω-automata. We confine ourselves to the use of weak alternating automata in the complementation proof, mentioning only briefly how the transformation of Büchi automata into this model and conversely works, and leaving aside complexity issues and further applications of alternating automata. A joint paper with a more detailed exposition and further results is in preparation.

2 Review of the classical proofs

A Büchi automaton over the alphabet A is a finite automaton of the form $\mathcal{A} = (Q, A, q_0, \Delta, F)$ with finite set Q of states, initial state q_0, transition relation $\Delta \subseteq Q \times A \times Q$, and a set $F \subseteq Q$ of final states. It accepts an ω-word

$\alpha = \alpha(0)\alpha(1)\ldots$ from A^ω if there is a run $\rho = \rho(0)\rho(1)\ldots$ from Q^ω with $\rho(0) = q_0$ and $(\rho(i), \alpha(i), \rho(i+1)) \in \Delta$ for $i \geq 0$, which is *Büchi accepting*, i.e. such that $\rho(i) \in F$ for infinitely many i. Formally, we write

$$(*) \quad \exists \rho \; (\rho(0) = q_0 \;\wedge\; \forall i((\rho(i), \alpha(i), \alpha(i+1)) \in \Delta \;\wedge\; \forall i \exists j > i \; \rho(j) \in F)$$

The ω-language $L(\mathcal{A})$ recognized by $\mathcal{A}$ consists of all ω-words α for which $(*)$ holds.

In the logical classification of quantifier alternation hierarchies, one calls $(*)$ a Σ^1_1-formula, referring to the existential sequence quantifier in front (which in an arithmetical setting is captured by a tuple of existential second-order quantifiers ranging over sets of natural numbers). We shall also take into account the logical status of the acceptance condition, which in the case above is called a Π^0_2-condition, referring to its two first-order quantifiers, with a universal quantifier coming first. Taking both aspects together, we speak of a $\Sigma^1_1[\Pi^0_2]$-definition.

Büchi's complementation theorem says that the negation of the formula $(*)$ may again be written in this form, with different Q, Δ, F. Büchi stated the result in this "logical" form (see [Bü62, Lemma 9]). Let us sketch his proof. Given a Büchi automaton $\mathcal{A}$ as above, define a congruence $\sim_\mathcal{A}$ over A^+ by declaring two finite words u, v as equivalent iff the following holds: for any $p, q \in Q$, $\mathcal{A}$ can reach q from p via the input u iff this is possible via the input v, and furthermore $\mathcal{A}$ can reach q from p via the input u by passing through a state of F iff this is possible via the input v. It is easy to verify that $\sim_\mathcal{A}$ is a congruence with finitely many (regular) equivalence classes. Denoting the $\sim_\mathcal{A}$-class of the word u by $[u]$, one observes that an ω-language $[u] \cdot [v]^\omega$ is either contained in $L(\mathcal{A})$ or disjoint from $L(\mathcal{A})$. Invoking Ramsey's Theorem B ([Ra29]), one shows that any ω-word can be cut into a sequence $u_0 u_1 \ldots$ of finite words where all u_i for $i > 1$ belong to a fixed $\sim_\mathcal{A}$-class. Applying this decomposition to the ω-words outside $L(\mathcal{A})$, one sees that $A^\omega \setminus L(\mathcal{A})$ is representable as a union of sets $[u] \cdot [v]^\omega$, taking those pairs u, v where $u \cdot v^\omega$ is not in $L(\mathcal{A})$. The union is of course a finite one since there are only finitely many equivalence classes. This representation of $A^\omega \setminus L(\mathcal{A})$ is easily converted into a Büchi automaton.

The complementation theorem can also be shown via a transformation of Büchi automata into deterministic Muller automata. This is the content of McNaughton's Theorem ([McN66]). A Muller automaton is specified in the form $\mathcal{A} = (Q, A, q_0, \delta, \mathcal{F})$ where $\delta : Q \times A \to Q$ is the transition function and $\mathcal{F}$ is a subset of the powerset of Q. Acceptance of an ω-word α means that in the unique run ρ of $\mathcal{A}$ on α, the set $\mathrm{In}(\rho)$ of states visited infinitely often belongs to $\mathcal{F}$. Formally, we express this as follows:

$$(**) \qquad \exists \rho \; ((\rho(0) = q_0 \;\wedge\; \forall i \; \rho(i+1) = \delta(\rho(i), \alpha(i))) \;\wedge\; \mathrm{In}(\rho) \in \mathcal{F}).$$

We have $\mathrm{In}(\rho) \in \mathcal{F}$ iff for some $F \in \mathcal{F}$, precisely the states in F are infinitely often visited in ρ. A formalization of this condition reads as follows:

$$\bigvee_{F \in \mathcal{F}} \big(\bigwedge_{q \in F} \forall i\, \exists j > i\ \rho(j) = q\ \wedge \bigwedge_{q \in Q \setminus F} \neg\, \forall i\, \exists j > i\ \rho(j) = q \big)$$

Taking the set $2^Q \setminus \mathcal{F}$ instead of $\mathcal{F}$, one obtains an automaton recognizing the complement language.

Let us analyze the logical status of deterministic Muller automata. The acceptance condition is a Boolean combination of Π^0_2-formulas, and thus we shall call the formula (**) a $\Sigma^1_1[\mathrm{Bool}(\Pi^0_2)]$-formula. Since the run required in (**) is unique, one can also use a universal condition instead:

$$(**') \qquad \forall \rho\ (\rho(0) = q_0\ \wedge\ \forall i\ \rho(i+1) = \delta(\rho(i), \alpha(i))\ \ \rightarrow\ \ \mathrm{In}(\rho) \in \mathcal{F})$$

This condition (**') is a $\Pi^1_1[\mathrm{Bool}(\Pi^0_2)]$-formula, equivalent to the Σ^1_1-definition (**) above. Properties which are definable in Σ^1_k-form and also in Π^1_k-form are said to have a Δ^1_k-representation. (Note that this involves two separate definitions.) So a deterministic Muller automaton provides a $\Delta^1_1[\mathrm{Bool}(\Pi^0_2)]$-representation of the recognized ω-language.

In this logical setting, McNaughton's Theorem says that the $\Sigma^1_1[\Pi^0_2]$-definitions as provided by Büchi automata can be brought into $\Delta^1_1[\mathrm{Bool}(\Pi^0_2)]$-form, which means a "decrease" of the second-order quantifier complexity at the cost of a more complicated first-order kernel.

3 Alternating automata

The concept of alternating automaton combines the idea of existential branching, as found in nondeterministic automata, with its dual, universal branching. The two branching modes are specified by Boolean expressions over the state set Q. For example, $q_1 \vee (q_2 \wedge q_3)$ denotes the nondeterministic choice of going either to q_1 or to q_2, q_3 simultaneously. The set of such positive (i.e., negation-free) Boolean expressions is denoted by $\mathrm{B}_+(Q)$. We introduce alternating automata here with the so-called weak acceptance condition. It refers to a ranking r of the states by natural numbers.

So an alternating automaton is presented in the form $\mathcal{A} = (Q, A, q_0, \delta, r)$ with entries Q, A, q_0 as before for Büchi automata and functions $\delta : Q \times A \rightarrow \mathrm{B}_+(Q)$ and $r : Q \rightarrow \{0, \ldots, m\}$ for some m. Moreover, the ranking function r defines a stratification of Q in the sense that for any q occurring in an expression $\delta(p, a)$ we have $r(p) \geq r(q)$. This means that by applying transitions we can only keep or decrease the ranks of states.

The definition of acceptance by alternating automata is somewhat involved. Usually, it refers to the notion of run tree (or computation tree). We use here a different terminology which allows a more elegant logical comparison with the previous modes of acceptance (nondeterministic and deterministic).

A run of an alternating automaton is a dag (directed acyclic graph) whose elements are labelled with states from Q. The dag can be presented as a sequence of "slices" $S_0, S_1, \ldots$, where the states occurring in S_i are the simultaneously "active" ones at the i-th letter of the input word. Acceptance will mean that a run dag exists (which represents the existential branching in the automaton) such that for all paths through the dag (representing its universal branching) a condition regarding the ranks of the states occurring on this path is satisfied.

In the definition of run dags we refer to the models of Boolean expressions in $B_+(Q)$. We shall identify such a Boolean model with a subset S of Q, given by the assignment of states to truth values which sends the states in S to value 1 and the states in $Q \setminus S$ to value 0. Since our expressions are positive, a superset S' of a model S of an expression β is again a model of β. By a minimal model of β we mean a model S of which no proper subset is again a model. If the expression β from $B_+(Q)$ is presented in disjunctive normal form, the minimal models of β are given by the sets S which constitute the individual conjuncts of the disjunctive form.

Let us make precise how a run dag is built up. It is started with the slice S_0 consisting solely of the initial state q_0 (more precisely: of a node labelled with q_0). From a given S_i one obtains a slice S_{i+1} as follows, assuming that the input letter $\alpha(i)$ is the letter a: For each p from S_i one chooses a minimal model of the expression $\delta(p, a)$; the union of all these minimal models for $p \in S_i$ is taken to be S_{i+1}. One inserts an edge from $p \in S_i$ to $q \in S_{i+1}$ (more precisely: from the node labelled p in S_i to the node labelled q in S_{i+1}) if q belongs to the minimal model chosen for $\delta(p, a)$.

If the expressions $\delta(p, a)$ are given in disjunctive normal form, the formation of a run dag can be described more easily. Again S_0 consists of q_0 only. Starting from a slice S_i and assuming again $\alpha(i) = a$, pick one disjunction member from $\delta(p, a)$ for each p in S_i. Collect all states arising by these choices to form the slice S_{i+1}, and introduce an edge from p in S_i to q in S_{i+1} if q occurs in the chosen disjunction member of $\delta(p, a)$.

Example 1. Given $Q = \{q_0, q_1, q_2, q_3\}$, we may have, for a certain letter $a \in A$:

$$\begin{aligned}
\delta(q_0, a) &= q_1 \wedge q_2 \\
\delta(q_1, a) &= (q_1 \wedge q_3) \vee (q_2 \wedge q_3) \\
\delta(q_2, a) &= q_1 \\
\delta(q_3, a) &= (q_1 \wedge q_2) \vee q_3
\end{aligned}$$

With this function δ, a run dag on the input word $aaa\ldots$ may start with the following slices:

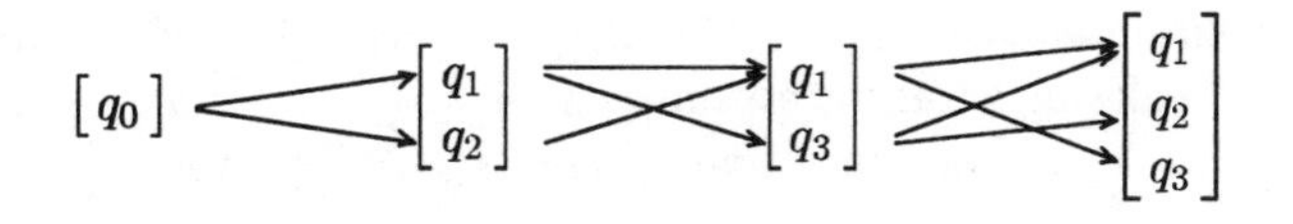

Now we introduce the weak acceptance condition. This condition refers to the infinite paths through a run dag. By the stratification property of the rank function on states, the ranks on such a path will stay constant from some point onwards. The path is accepting if this ultimately assumed rank is even. Equivalently, the set $\mathrm{Occ}(r(\pi))$ of r-ranks occurring in the path π under consideration has an even minimum. So the acceptance condition reads as follows:

$$\exists \text{ run dag } \rho \ \forall \text{ paths } \pi \text{ through } \rho : \min(\mathrm{Occ}(r(\pi)) \text{ is even}$$

A *weak alternating automaton* is an alternating automaton used with this "weak" acceptance condition.

In order to fix the quantifier complexity of weak acceptance, we rewrite the minimum condition in a more formal way. Denoting by $\pi(i)$ the i-th state of the path π, we can express it as follows:

$$\bigvee_{k \text{ even}} (\exists i \ r(\pi(i)) = k \ \wedge \ \bigwedge_{l<k} \neg \exists i \ r(\pi(i)) = l)$$

So weak acceptance for a given path π is a $\mathrm{Bool}(\Sigma_1^0)$-condition. In the quantifier prefix "$\exists$ run dag ρ $\forall$ paths π" of the acceptance clause, the existential quantifier on run dags and the universal quantifier on paths can both be coded as ranging over infinite sequences over appropriate finite alphabets: A run dag can be represented by a sequence of slices and pointers between adjacent slices, so the possible entries of the sequence are representable by the letters of a finite alphabet. Similarly, a path through a run dag is codable in this way. (At this point one sees an advantage of the notion of run dag over the standard approach involving computation trees. Their existence is not directly formalizable by sequence quantifiers.)

In summary, the definition of an ω-language by a weak alternating automaton is a $\Sigma_2^1[\mathrm{Bool}(\Sigma_1^0)]$-representation. In comparison to the $\Sigma_1^1[\Pi_2^0]$-definition given by Büchi automata, the second-order quantifier complexity is increased, while the acceptance condition is simpler: Instead of a requirement that certain states are visited infinitely often, we only ask for the visit of certain states and the avoidance of others.

In the next section we show that the class of ω-languages accepted by weak alternating automata is closed under complementation. As a consequence we can improve the $\Sigma_2^1[\mathrm{Bool}(\Sigma_1^0)]$-form of the definition to obtain even a $\Delta_2^1[\mathrm{Bool}(\Sigma_1^0)]$-representation.

4 Duality and weak determinacy

The purpose of this section is to show closure under complement for weak alternating automata. In this analysis, game-theoretic notions are useful. With each (weak) alternating automaton $\mathcal{A} = (Q, A, q_0, \delta, r)$ and each ω-word $\alpha \in A^\omega$ we associate an infinite game $G(\mathcal{A}, \alpha)$, played by two persons called Automaton and Pathfinder. The idea is that in the process of scanning the input word α, Automaton picks sets of simultaneously active states according to the transition function of $\mathcal{A}$, whereas Pathfinder picks, at each point, one of these momentary active states. Making such choices in alternation they build up a path through a run dag of $\mathcal{A}$ on α, and Automaton is declared the winner of the play if the acceptance condition of $\mathcal{A}$ is satisfied on this path.

Formally, a game position refers to the number i supplying the momentary input letter $\alpha(i)$. If Automaton has the next move, the game position is of the form (i, p) with $p \in Q$, and if Pathfinder has to make the next move, the game position has the form (i, S) with $S \subseteq Q$. The initial position is $(0, q_0)$; so Automaton starts. A move of Automaton from position (i, p) consists in the choice of a minimal model of $\delta(p, \alpha(i))$, i.e. a set $S \subseteq Q$, yielding the game position $(i + 1, S)$. Pathfinder reacts by picking a state s from S, producing the game position $(i + 1, s)$. The play determines a sequence of states (extracted from the positions of Automaton), and Automaton wins the play if the minimal rank occurring in this sequence of states is even.

A *local strategy* (also called memoryless strategy) for a player P is a function which associates with any game position of P a move which can be performed in this position. Such a function is called a winning strategy for player P from game position *pos* if its application will produce, when starting from *pos*, for any moves of the opponent, a play won by P.

Proposition 1. *The weak alternating automaton $\mathcal{A}$ accepts α iff in the game $G(\mathcal{A}, \alpha)$, Automaton has a local winning strategy from the initial position.*

Proof. First assume that there is an accepting run dag of $\mathcal{A}$ on α, say with slices $S_0, S_1, \ldots$. Define a strategy for Automaton by choosing, given game position (i, p), the set S of states from S_{i+1} which are reachable from p by an edge of the run dag. In this way, starting from position $(0, q_0)$, Automaton ensures that the play proceeds along a path through the run dag. Since the run dag is accepting, Automaton wins by this local strategy.

Conversely, a local strategy for Automaton defines an accepting run dag: For $i = 0, 1, \ldots$ the slices S_i are built up inductively, beginning with the singleton $S_0 = \{q_0\}$: For any game position (i, p) as picked by Pathfinder, Automaton's local strategy prescribes a set of states as next move; the union of these is taken to form S_{i+1}, and edges are inserted which allow to trace the connections to the different choices of p. It is clear that the constructed run dag is accepting. □

The complementation proof for weak alternating automata will be given in this game-theoretic setting. It has two steps: the dualization of alternating automata and a determinacy result for infinite games.

For the first step, we introduce the *dual automaton* $\widetilde{\mathcal{A}}$ of a given weak alternating automaton $\mathcal{A} = (Q, A, q_0, \delta, r)$. The definition uses the dualization of Boolean expressions: Given an expression $\beta \in \mathrm{B}_+(Q)$, let its dual $\widetilde{\beta}$ arise from β by exchanging $\vee$ and $\wedge$. Now the dualized transition function $\widetilde{\delta}$ is defined by $\widetilde{\delta}(p,a) = \widetilde{\delta(p,a)}$. The dual automaton $\widetilde{\mathcal{A}}$ is obtained as $(Q, A, q_0, \widetilde{\delta}, r)$ with the convention that in an accepting run dag, on each path the minimal rank of visited states should be *odd*.

We need a remark on models of the dual of an expression in $\mathrm{B}_+(Q)$. Recall that a model of an expression β is considered as a subset S of Q.

Remark 1. A set S is a model of $\widetilde{\beta}$ iff every minimal model R of β contains a state from S.

Proof. Let $MM(\beta)$ be the set of minimal models of β. We have the logical equivalence

$$\beta \equiv \bigvee_{R \in MM(\beta)} \bigwedge_{q \in R} q$$

By duality, we have

$$\widetilde{\beta} \equiv \bigwedge_{R \in MM(\beta)} \bigvee_{q \in R} q$$

This shows the claim. □

Now we are able to connect local winning strategies in the two games $G(\mathcal{A},\alpha)$ and $G(\widetilde{\mathcal{A}},\alpha)$:

Proposition 2. *Automaton has a local winning strategy in $G(\mathcal{A},\alpha)$ from the initial position iff Pathfinder has a local winning strategy in $G(\widetilde{\mathcal{A}},\alpha)$ from the initial position.*

Proof. We show how to transform a local winning strategy of Automaton in $G(\mathcal{A},\alpha)$ into a local Pathfinder strategy for the dual game. The desired strategy in $G(\widetilde{\mathcal{A}},\alpha)$ has to tell Pathfinder which state to take for any game position $(i+1, S)$ (where $i \geq 0$). Note that in fixing the strategy it suffices to consider only game positions $(i+1, S)$ which are reachable, i.e. for which a sequence of moves in $G(\widetilde{\mathcal{A}},\alpha)$ exists starting in the initial position $(0, q_0)$ and ending in $(i+1, S)$. The set S of the game position $(i+1, S)$ is produced by Automaton from a game position (i, s), such that S is a minimal model of $\widetilde{\delta(s, \alpha(i))}$. Pathfinder chooses such a state s which could produce S via $\alpha(i)$. Now in the game $G(\mathcal{A},\alpha)$ at position (i, s), the given local winning strategy of Automaton picks a minimal model R of $\delta(s, \alpha(i))$. By the remark above, there is a state in $R \cap S$. For his move from the game position $(i+1, S)$, Pathfinder chooses such a state. Then in $G(\widetilde{\mathcal{A}},\alpha)$ a state sequence is built up which is compatible with Automaton's winning strategy in $G(\mathcal{A},\alpha)$ and hence is won in that game by Automaton. In the game $G(\widetilde{\mathcal{A}},\alpha)$, where the roles of even and odd ranks are exchanged, this state sequence gives a play won by Pathfinder. So the described strategy is a winning strategy for Pathfinder in $G(\widetilde{\mathcal{A}},\alpha)$, and its specification shows that it is local.

The other direction is shown analogously, by exchanging the roles of $\mathcal{A}$ and $\widetilde{\mathcal{A}}$. □

Proposition 3. *Let $\mathcal{A}$ be a weak alternating automaton. From any game position in $G(\mathcal{A}, \alpha)$, either Automaton or Pathfinder has a local winning strategy.*

Proof. Let $\mathcal{A} = (Q, A, q_0, \delta, r)$ be a weak alternating automaton, where $r : Q \to \{0, \ldots, m\}$ and $Q_i := \{q \in Q \mid r(q) = i\}$. Let Pos_0 be the set of game positions of the game $G(\mathcal{A}, \alpha)$. It is divided into the sets Pos_A and Pos_P where Automaton, respectively Pathfinder has the next move.

As a preparation we need the definition of "attractor set of a set T of game positions". This attractor set (for player Automaton, say) is denoted $\mathrm{Attr}_A(T)$; it contains all game positions from which Automaton can force in finitely many moves a visit in the target set T. The set is constructed inductively by collecting, for $i \geq 0$, the positions from where a visit to T can be forced within i moves: Let $\mathrm{Attr}_A^0 := T$ and set

$$\begin{aligned}\mathrm{Attr}_A^{i+1}(T) := \mathrm{Attr}_A^i(T) & \\ \cup \{p \in Pos_A \mid & \text{ there is a move from } p \text{ to } \mathrm{Attr}_A^i(T)\} \\ \cup \{p \in Pos_P \mid & \text{ all moves from } p \text{ lead to } \mathrm{Attr}_A^i(T)\}\end{aligned}$$

Now let $\mathrm{Attr}_A(T) = \bigcup_{i \geq 0} \mathrm{Attr}_A^i(T)$. From the positions in this set player Automaton can force a decrease of distance to T in each step (which defines a local strategy). Also note that for the game positions pos outside $\mathrm{Attr}_A(T)$, Pathfinder will be able to avoid entering this set. (If at pos it is Pathfinder's turn, one move to the complement of $\mathrm{Attr}_A(T)$ is possible, and if it is Automaton's turn, all moves lead to the complement of $\mathrm{Attr}_A(T)$; otherwise pos would be already in $\mathrm{Attr}_A(T)$ itself.) So from outside $\mathrm{Attr}_A(T)$, Pathfinder can avoid, by a local strategy, to enter this set and hence can avoid the visit of T.

The set $\mathrm{Attr}_P(T)$ for player Pathfinder is constructed analogously.

Using the notion of attractor set and corresponding local strategies we determine inductively the game positions in $G(\mathcal{A}, \alpha)$ from where player Automaton, respectively Pathfinder, wins.

Clearly, from the positions in $A_0 := \mathrm{Attr}_A(Q_0)$, Automaton can force, by a local strategy, to reach states of rank 0 and thus win. Consider the subgame whose set of positions is $Pos_1 := Pos_0 \setminus A_0$ (all of which have rank ≥ 1). From the positions in $A_1 := \mathrm{Attr}_P(Pos_1 \cap Q_1)$, Pathfinder can force, again by a local strategy, to reach (and stay in) states of rank 1 and hence win. (Note that Pathfinder can avoid to enter A_0, as explained above.) In this way we continue: In the game with position set $Pos_2 := Pos_1 \setminus A_1$ (containing only states of rank ≥ 2) we form the attractor set $A_2 := \mathrm{Attr}_A(Pos_2 \cap Q_2)$, etc. Then the positions from which Automaton wins (by the local attractor strategies) are those in the sets A_i with even $i \leq m$. Similarly, Pathfinder wins from the positions in the sets A_i with odd $i \leq m$ (again by his local attractor strategies). □

Now we have all prerequisites for the complementation of weak alternating automata:

Theorem 1. *For any weak alternating automaton $\mathcal{A}$ over the alphabet A, we have $A^\omega \setminus L(\mathcal{A}) = L(\widetilde{\mathcal{A}})$.*

Proof. By Proposition 2, the automaton $\mathcal{A}$ does not accept the input word α iff Automaton does not have a local winning strategy in $G(\mathcal{A}, \alpha)$ from the initial position. By Proposition 4, this means that Pathfinder does not have a local winning strategy in $G(\widetilde{\mathcal{A}}, \alpha)$ from the initial position. By the determinacy result (Proposition 5), this holds iff Automaton has a local winning strategy in $G(\widetilde{\mathcal{A}}, \alpha)$ from the initial position, which in turn means that $\widetilde{\mathcal{A}}$ accepts α. □

The present game-theoretic complementation proof for Büchi automata has the same general structure as the complementation of nondeterministic tree automata in the framework of parity games (or "Rabin chain games"; see for example [Th97]). So it is possible to compare the two proofs. (Note that the two classical proofs of Büchi automata complementation, via deterministic automata or via a finite congruence saturating the given ω-language, do not extend – as far as we know – to tree automata, which makes a direct comparison difficult.) The parity games associated with tree automata have a winning condition defined by a Boolean combination of Σ_2^0-formulas, and the corresponding determinacy proof, also by induction on the ranks of game positions, involves a nontrivial combination of attractor strategies with strategies given by the inductive hypothesis. In the "weak" games considered in the present paper, where Boolean combinations of Σ_1^0-conditions serve as winning conditions, a straightforward reachability analysis, yielding the attractor sets $A_0, \dots, A_m$, suffices for the determinacy proof. This direct comparison in the game-theoretical setting shows in which sense complementation is easier for ω-automata than for tree automata and thus reveals a characteristic difference between ω-automata theory and tree automata theory.

5 Equivalence of Büchi automata and weak alternating automata

In order to complete the complementation proof for Büchi automata, we have to supply transformations from Büchi automata to weak alternating automata and conversely. These transformations, as developed by C. Löding in [Lö98], are described here very briefly.

Proposition 4. *For any Büchi automaton $\mathcal{A}$ there is a weak alternating automaton $\mathcal{A}'$ with $L(\mathcal{A}) = L(\mathcal{A}')$.*

Proof. Let $\mathcal{A} = (Q, A, q_0, \Delta, F)$ be a Büchi automaton with n states. The desired weak alternating automaton is constructed over the state set $Q \times \{0, \dots, 2n\}$, taking $(q_0, 2n)$ as initial state, and defining the rank function

r by $r((q,i)) = i$. The transition function δ associates to a state $(p,0)$ of rank 0 and letter a simply the disjunction over all $(q,0)$ with $(p,a,q) \in \Delta$. For even ranks $i > 0$, however, we take the disjunction over all expressions $(q,i) \wedge (q,i-1)$ with $(p,a,q) \in \Delta$. This will open a track of states of odd rank until some final state of the Büchi automaton is reached: Namely, as long as $p \notin F$ we set, for odd i, $\delta((p,i),a)$ to be the disjunction over all (q,i), again of rank i, with $(p,a,q) \in \Delta$, while for $p \in F$ we take the disjunction over all $(q,i-1)$ with $(p,a,q) \in \Delta$.

From an accepting run ρ of $\mathcal{A}$ one obtains an accepting run dag ρ' of $\mathcal{A}'$ and conversely. The construction of ρ' from ρ is straightforward: With the states of even rank, one simulates the given run ρ, and by the definition of the transition function δ of $\mathcal{A}$ no path eventually stays on an odd level. The converse direction requires to compose an accepting Büchi run ρ from an accepting run dag ρ'. This composition is achieved by concatenating run segments leading from a state $(p,2i)$ to a state $(q,2j)$ with $i > j$, for in this case an intermediate visit in F is ensured by the construction of δ. In order to be able to do this infinitely often, one has to reset the (even) rank to a higher level infinitely often. This in turn is made possible by the presence of $n+1$ even ranks (from 0 to $2n$), which means that once rank 0 is present, some state occurs on two ranks. □

Proposition 5. *For any weak alternating automaton $\mathcal{A}$, there is a Büchi automaton $\mathcal{A}'$ with $L(\mathcal{A}) = L(\mathcal{A}')$.*

Proof. Apply a subset construction as given by Miyano and Hayashi [MH84]: The desired Büchi automaton $\mathcal{A}'$ has states (S,R) where S,R are subsets of the state set Q of the given weak alternating automaton $\mathcal{A}$. In the first component S, $\mathcal{A}'$ guesses the slices of a run dag of $\mathcal{A}$ and thus a run dag, while in the second component R keeps track of those states from S which are of odd rank. If they eventually vanish, R is reset to the whole set S again. Then infinitely many such reset operations (captured by the Büchi acceptance condition) signal that no path in the run dag finally stays in states of odd rank. □

6 Discussion

We have outlined a complementation proof for Büchi automata by invoking a determinacy theorem on weak infinite games. Referring to the logical classification of automata theoretic definability explained in Section 2, we passed from $\Sigma^1_1[\Pi^0_2]$-representations of ω-languages, as given by Büchi automata, to the level of $\Delta^1_2[\text{Bool}(\Sigma^0_1)]$-representations, as given by weak alternating automata. This is an alternative to the option to pass to $\Delta^1_1[\text{Bool}(\Sigma^0_2)]$-representations, as provided by deterministic automata.

None of the steps as described in the propositions above is very difficult; so the complementation proof via weak alternation is composed of simple "modules", in some contrast to the two classical proofs (which rely on a nontrivial combinatorial result or a complicated automaton construction). Of course, there is a price to be paid in using the more involved definition of acceptance of alternating automata.

Another conceptual advantage of weak alternating automata may be the fact that acceptance is defined without resorting to liveness conditions; in the kernel of the second-order definition of acceptance one finds here only conditions on mere reachability or non-reachability of states. This phenomenon may be helpful in the investigation of still unsolved problems of ω-automata theory, for instance in the question of finding a good framework for the minimization of ω-automata.

References

[Bü62] J.R. Büchi, On a decision method in restricted second order arithmetic, in: *Proc. 1960 Int. Congr. on Logic, Methodology and Philosophy of Science*, Stanford Univ. Press, Stanford, 1962, pp. 1-11.

[KV97] O. Kupferman, M.Y. Vardi, Weak alternating automata are not so weak, in: *Proc. 5th Israeli Symp. on Theory of Computing and Systems*, IEEE Computer Soc. Press, 1997, pp. 147-158.

[Lö98] C. Löding, *Methods for the Transformation of ω-Automata: Complexity and Connection to Second Order Logic* Diploma Thesis, Universität Kiel, Inst. f. Informatik u. Prakt. Math., 1998.

[McN66] R. McNaughton, Testing and generating infinite sequences by a finite automaton, *Inf. Contr.* **9** (1966), 521-530.

[MH84] S. Miyano, T. Hayashi, Alternating finite automata on ω-words, *Theor. Comput. Sci.* **32** (1984), 321-330.

[MS87] D.E. Muller, P.E. Schupp, Alternating automata on infinite trees, *Theor. Comput. Sci.* **54** (1987), 267-276.

[MSS86] D.E. Muller, A. Saoudi, P.E. Schupp, Alternating automata, the weak monadic theory of the tree and its complexity, in: *Proc. 13th ICALP*, Lecture Notes in Computer Science **226**, Springer-Verlag, Berlin 1986, pp. 275-283.

[MSS88] D.E. Muller, A. Saoudi, P.E. Schupp, Weak alternating automata give a simple explanation of why most temporal and dynamic logics are decidable in exponential time, in: *Proc. 3rd IEEE Symp. on Logic in Computer Science*, 1988, pp. 422-427.

[Mu63] D.E. Muller, Infinite sequences and finite machines, in: *Proc. 4th Ann. Symp. on Switching Circuit Theory and Logical Design*, IEEE, 1963, pp. 3-16.

[Ra29] F.P. Ramsey, On a problem of formal logic, *Proc. London Math. Soc.* (2) **30** (1929), 264-286.

[Sa88] S. Safra, On the complexity of omega-automata, in: *Proc. 29th Ann. Symp. on Foundations of Computer Science*, IEEE Computer Society, 1988, pp. 319-327.

[Th97] W. Thomas, Languages, automata, and logic, in *Handbook of Formal Languages* (G. Rozenberg, A. Salomaa, Eds.), Vol. 3, Springer-Verlag, Berlin 1997, pp. 389-455.

Part III

Automata with Multiplicities

Languages Accepted by Integer Weighted Finite Automata

Vesa Halava and Tero Harju

Summary. We study the family of languages accepted by the integer weighted finite automata. Especially the closure properties of this family are investigated.

1 Introduction

The integer weighted automata, as were studied in [4], [5] and [6], are closely related to the 1-turn one–counter automata as considered by Baker and Book [1], Greibach [3], and especially by Ibarra [7]. In our model the counter is replaced by a weight function of the transitions, and while doing so, the finite automaton becomes independent of the counter. The differences between one–counter automata and integer weighted finite automata as well as the deterministic integer weighted automata are considered in [4].

We shall first give the definition of the *weighted finite automata.*

Consider a (nondeterministic) finite automaton $\mathcal{A} = (Q, A, \delta, q_0)$ *without final states* with the states Q, the alphabet A, the set of transitions $\delta \subseteq Q \times A \times Q$ and the initial state q_0.

We shall redefine the transitions using a *set of edges* $T = \{t_1, t_2, \ldots, t_m\}$ and a transition function $\sigma: T \to \delta$. We do this to allow transitions t_i and t_j, where $i \neq j$, but $\sigma(t_i) = \sigma(t_j)$. In other words, there may exists many copies of one transition in this new set of edges. Clearly this new definition of the transitions does not affect the language accepted by the automaton.

Let G, or $(G, \cdot, \iota)$, be a group with identity ι. A (G-)*weighted finite automaton* $\mathcal{A}^\gamma$ consists of a finite automaton $\mathcal{A} = (Q, A, \sigma, q_0)$ as above, and a *weight function* γ: $\{t_1, \ldots, t_n\} \to G$ of the edges. To simplify the notation, we shall write the edges in the form

$$t = \langle q, a, p, z \rangle$$

if $\sigma(t) = (q, a, p)$ and $\gamma(t) = z$. Similarly, we shall write the transition function σ as a set, $\sigma \subseteq Q \times A \times Q \times \mathbb{Z}$, where

$$\sigma = \{\langle q, a, p, z \rangle \mid \exists t \in T \colon \sigma(t) = (q, a, p) \text{ and } \gamma(t) = z\}.$$

In the figures we shall denote such an edge t by

$$q \xrightarrow{(a,z)} p.$$

But note that the function γ does not affect the computations of the finite automaton $\mathcal{A}$. It will be used only in the acceptance of words.

Let $\pi = t_{i_0}t_{i_1}\dots t_{i_n}$ be a path of $\mathcal{A}$, where $\sigma(t_{i_j}) = (q_{i_j}, a_j, q_{i_{j+1}})$ for $0 \leq j \leq n-1$. Define a morphism $\|\cdot\|\colon \{t_1,\dots,t_m\}^* \to A^*$ by setting $\|t\| = a$ if $\sigma(t) = (q,a,p)$. The *weight of the path* π is the element

$$\gamma(\pi) = \gamma(t_{i_0})\gamma(t_{i_1})\cdots\gamma(t_{i_n}) \in G.$$

Further, we let $L(\mathcal{A}^\gamma) = \|\gamma^{-1}(\iota)\|$, that is,

$$L(\mathcal{A}^\gamma) = \{w \in A^* \mid w = \|\pi\|,\ \gamma(\pi) = \iota\},$$

be the language *accepted by* $\mathcal{A}^\gamma$.

A *configuration* of $\mathcal{A}^\gamma$ is any triple $(q,w,g) \in Q \times A^* \times G$.

A configuration (q, aw, g_1) is said to *yield* in a configuration (p, w, g_1g_2), denoted by

$$(q, aw, g_1) \models_{\mathcal{A}^\gamma} (p, w, g_1g_2),$$

if there is an edge t such that $\sigma(t) = (q,a,p)$ with $\gamma(t) = g_2$. Let $\models^*_{\mathcal{A}^\gamma}$, or simply $\models^*$ if $\mathcal{A}^\gamma$ is clear from the context, be the reflexive and transitive closure of the relation $\models_{\mathcal{A}^\gamma}$.

We shall restrict to the case, where the group of the automaton is the additive group of integers, namely $(\mathbb{Z},+,0)$. Such automata are called *integer weighted finite automata* and denoted by $FA(\mathbb{Z})$.

Note that our definition of the integer weighted finite automata is a restricted case of the extended finite automata of Mitrana and Stiebe [8]. In the extended finite automata the underlying automata has final states $F \subseteq Q$ and the transitions reading the empty word are allowed.

Let $\mathcal{A}^\gamma$ be an $FA(\mathbb{Z})$. The empty word is always included in $L(\mathcal{A}^\gamma)$, since in an integer weighted finite automaton all states, including the initial state, are final. Therefore we known that not all regular languages can be accepted by an $FA(\mathbb{Z})$. On the other hand,

Theorem 1. *For each regular language L, there exists an $FA(\mathbb{Z})$ $\mathcal{A}^\gamma$ such that $L(\mathcal{A}^\gamma) = L \cup \{\varepsilon\}$.*

Proof. Let L be a regular language and let $\mathcal{A} = (Q, A, \delta, q_0, F)$ be a (non-deterministic) finite automaton such that $L(\mathcal{A}) = L$, where $\delta \subseteq Q \times A \times Q$. We may assume that $\mathcal{A}$ has one initial state q_0 and one final state q_f (or two final states q_0 and q_f, if $\varepsilon \in L$) such that there are no transitions to q_0 and no transitions from q_f. It is well known that such an $\mathcal{A}$ exists for all regular languages L.

We shall define an $FA(\mathbb{Z})$ $\mathcal{A}^\gamma = (Q, A, \sigma, q_0)$, where σ is a bijection and therefore we may define $T = \delta$. One such required weight function γ is defined, for $(p, a, q) \in \delta$, by

$$\gamma(p, a, q) = \begin{cases} 0 & \text{if } p = q_0 \text{ and } q = q_f, \\ 1 & \text{if } p = q_0 \text{ and } q \neq q_f, \\ -1 & \text{if } p \neq q_0 \text{ and } q = q_f, \\ 0 & \text{if } p \neq q_0 \text{ and } q \neq q_f. \end{cases} \quad (1)$$

It is obvious that $\gamma(\pi) = 0$ only in the case where π is an accepting path of $\mathcal{A}$. This proves our claim.

As we mentioned, not all regular languages are accepted by a $FA(\mathbb{Z})$. On the other hand, it is easy to show that not all languages accepted by integer weighted finite automata are regular, since the language

$$L = \{a^n b^n \mid n \geq 0\}$$

is accepted by an $FA(\mathbb{Z})$ of Figure 1, but L is not regular.

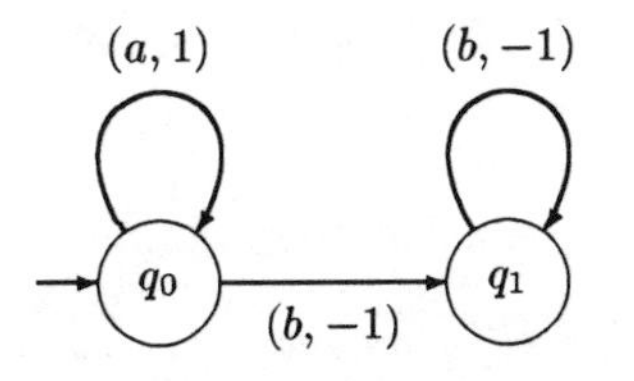

Fig. 1. An $FA(\mathbb{Z})$ accepting the language $\{a^n b^n \mid n \geq 0\}$.

It was proved in [5] that the universe problem, asking whether all input words are accepted by an integer weighted automata, is undecidable. Actually, it was proved there that the universe problem is undecidable for a rather restricted type of 4-state integer weighted finite automata. In this type the edges on every possible path have first positive weights, then zero weights, then negative weights and then again zero weights. Any of these parts may be trivial. Such a restricted type of integer weighted finite automaton is called *unimodal*.

2 Closure properties

In this section we consider the closure properties of the family of $FA(\mathbb{Z})$ languages, denoted by $\mathcal{L}_{FA(\mathbb{Z})}$, under various operations. We shall begin with the union.

Theorem 2. *The family of $FA(\mathbb{Z})$ languages is closed under unions.*

Proof. Let L_1 and L_2 be two $FA(\mathbb{Z})$ languages accepted by the $FA(\mathbb{Z})$'s $\mathcal{A}^\gamma$, where $\mathcal{A} = (Q_1, A, \sigma_1, q_1)$, and $\mathcal{B}^\alpha$, where $\mathcal{B} = (Q_2, A, \sigma_2, q_2)$, respectively. Assume that $Q_1 \cap Q_2 = \emptyset$. We define an $FA(\mathbb{Z})$ $\mathcal{C}^\beta$, where

$$\mathcal{C} = (Q_1 \cup Q_2 \cup \{q_0\}, A, \sigma, q_0),$$

and $q_0 \notin Q_1 \cup Q_2$ is the new initial state. The edges in $\mathcal{C}^\beta$ are as in $\mathcal{A}^\gamma$ and $\mathcal{B}^\alpha$, except that there is an edge $\langle q_0, a, p, z\rangle$ for $p \in Q_1 \cup Q_2$ in $\mathcal{C}^\beta$ if the edge $\langle q_1, a, p, z\rangle$, where $p \in Q_1$, is in $\mathcal{A}^\gamma$ or the edge $\langle q_2, a, p, z\rangle$, where $p \in Q_2$, is in $\mathcal{B}$. Clearly $L(\mathcal{C}^\beta) = L(\mathcal{A}^\gamma) \cup L(\mathcal{B}^\alpha) = L_1 \cup L_2$. This proves the claim.

Next we consider the intersection. For this, let $L_1 = a^* \{b^n c^n \mid n \geq 0\}$ and $L_2 = \{a^n b^n \mid n \geq 0\} c^*$. These languages can be accepted by integer weighted finite automata, but

$$L = L_1 \cap L_2 = \{a^n b^n c^n \mid n \geq 0\}$$

cannot be, since L is not even a context-free language and it was proved in [4] that $\mathcal{L}_{FA(\mathbb{Z})} \subset \mathcal{L}_{CF}$, where $\mathcal{L}_{CF}$ denotes the family of context-free languages.

Theorem 3. *The family of $FA(\mathbb{Z})$ languages is not closed under intersections.*

Next we consider the concatenation operation on languages.

Theorem 4. *The family of $FA(\mathbb{Z})$ languages is not closed under concatenations. In fact, the square L^2 of a language $L \in \mathcal{L}_{FA(\mathbb{Z})}$ need not be in $\mathcal{L}_{FA(\mathbb{Z})}$.*

Proof. Let $L = \{a^n b^n \mid n \geq 0\}$. We shall show that the (one-counter) language

$$L^2 = L \cdot L = \{a^n b^n a^m b^m \mid n, m \geq 0\}$$

cannot be accepted by an $FA(\mathbb{Z})$.

Assume on the contrary that there is an $FA(\mathbb{Z})$ $\mathcal{A}^\gamma$ with states Q and weight function γ such that it accepts L^2. Since L^2 is an infinite language and it contains words that have length greater than $|Q|$, necessarily there is a cycle in $\mathcal{A}$. We have now two cases to consider:

1) Suppose $\mathcal{A}$ has only one cycle. Assume that the weight of the cycle is d. To accept all words in L^2 with only one cycle, necessarily $d = 0$, since L^2 is infinite. Let $uv \in L^2$, $u, v \in L$ and $|uv| > |Q|$. Then we have a factorization $uv = xyz$, where y is read during the cycle. Since $uv = xyz$ gives a path of weight 0, we get that $xy^k z \in L(\mathcal{A}^\gamma)$ for all $k \in \mathbb{N}$. But L^2 does not contain any words $xy^k z$ for $k > 2$ and $y \neq \varepsilon$.

2) Suppose that $\mathcal{A}^\gamma$ has more than one cycle. If there exists a cycle of weight zero in any accepting path, then we get a contradiction as in the previous case. On the other hand, the weights of all cycles cannot be of the

same sign, since L^2 is infinite. It follows that there must be $u, v \in L$ such that $uv = xyzrs$, and, in an accepting path $\pi = \pi_x \pi_y \pi_z \pi_r \pi_s$ of uv, y is read during the cycle π_y and r is read during the cycle π_r, and $\gamma(\pi_y)$ and $\gamma(\pi_r)$ are of different sign. But now the word

$$xy^{k|\gamma(\pi_r)|+1} zr^{k|\gamma(\pi_y)|+1} s \in L(\mathcal{A}^\gamma)$$

for each $k > 0$ is accepted by the path

$$\pi' = \pi_x \pi_y^{k|\gamma(\pi_r)|+1} \pi_z \pi_r^{k|\gamma(\pi_y)|+1} \pi_s,$$

since

$$\begin{aligned}\gamma(\pi') &= \gamma(\pi_x) + \gamma(\pi_y)(k|\gamma(\pi_r)| + 1) + \gamma(\pi_z) + \gamma(\pi_r)(k|\gamma(\pi_y)| + 1) + \gamma(\pi_s) \\ &= \gamma(\pi_x) + \gamma(\pi_y) + \gamma(\pi_z) + \gamma(\pi_r) + \gamma(\pi_s) \\ &\quad + k(\gamma(\pi_y)|\gamma(\pi_r)| + \gamma(\pi_r)|\gamma(\pi_y)|) = 0.\end{aligned}$$

Since these new words are not in L^2, we get a contradiction.

Note that although the family of $FA(\mathbb{Z})$ languages is not closed under intersections and concatenation, it is closed under these operations with regular languages (that contain the empty word). This is stated in the next theorem.

Theorem 5. *Let R be a regular language with $\varepsilon \in R$ and L be an $FA(\mathbb{Z})$ language. Then $R \cap L$, RL and LR are $FA(\mathbb{Z})$ languages.*

Proof. Assume that R is accepted by the FA $\mathcal{B} = (Q, A, \delta, q_0, F)$ and L with the $FA(\mathbb{Z})$ $\mathcal{A}^\gamma$, where $\mathcal{A} = (P, A, \sigma_1, p_0)$. Assume also that $Q \cap P = \emptyset$. We can transform $\mathcal{B}$ into a $\mathcal{B}^\alpha$ accepting R, where α is defined as in 1. We may also assume that in $\mathcal{A}^\gamma$ all edges have even weight, since we can multiply each weight by 2, and the accepted language remains the same.

For $R \cap L$, we define $\mathcal{C}^\beta$, where $\mathcal{C} = (Q \times P, A, \sigma, (q_0, p_0))$, and there is an edge

$$\langle (q,p), a, (r,s), (z_1 + z_2) \rangle$$

in $\mathcal{C}^\beta$, for all edges $\langle q, a, r, z_1 \rangle$ in $\mathcal{B}^\alpha$ and $\langle p, a, s, z_2 \rangle$ in $\mathcal{A}^\gamma$. By the definition in 1 of α, the weight of a path starting from (q_0, p_0) and ending in a (q, p) is $0 \pmod 2$ if and only if $q \in F$, which is equivalent to the fact that the word is in R. It follows that $w \in A^*$ is in $R \cap L$ if and only if it is in $L(\mathcal{C}^\beta)$.

For RL, we construct an $FA(\mathbb{Z})$ $\mathcal{C}^\beta$, where $\mathcal{C} = (Q \cup P, A, \sigma, q_0)$, by connecting the two $FA(\mathbb{Z})$'s by introducing new edges

$$\langle f, a, p, z \rangle,$$

for $f \in F$ and $a \in A$, if $\langle p_0, a, p, z \rangle$ is an edge in $\mathcal{A}^\gamma$. It is then clear that $L(\mathcal{C}^\beta) = RL$.

For LR, the same construction can be used, but this time we connect the two $FA(\mathbb{Z})$ in the opposite order.

Next we shall consider the star operation. Let K be a language and

$$K^* = \bigcup_{i=0}^{\infty} K^i.$$

By using again the language $L = \{a^n b^n \mid n \geq 0\}$ and the proof of Theorem 4, we can show

Theorem 6. *The family of $FA(\mathbb{Z})$ languages is not closed under star.*

Proof. Assume that there is an $FA(\mathbb{Z})$ $\mathcal{A}^\gamma$ accepting L^* for $L = \{a^n b^n \mid n \geq 0\}$. It follows by Theorem 5 that the language

$$L^* \cap a^*b^*a^*b^* = L^2$$

can be accepted by an $FA(\mathbb{Z})$, since $a^*b^*a^*b^*$ is a regular language. But this is a contradiction by the proof of Theorem 4.

The family of $FA(\mathbb{Z})$ languages is not closed under complement, since each $FA(\mathbb{Z})$ language contains the empty word, and therefore it is not in the complement. But let us consider the complement modulo ε. For any language $L \subseteq A^*$, the *complement modulo* ε of L is

$$\bar{L}_\varepsilon = (A^* \setminus L) \cup \{\varepsilon\}.$$

For example, let $A = \{a, b\}$ and $L = \{a^n b^n \mid n \geq 0\}$. Now there is a partition

$$\begin{aligned} \bar{L}_\varepsilon = {} & a^* \cup bA^* \cup aA^*baA^* \\ & \cup \{a^n b^m \mid 0 < m < n\} \cup \{a^m b^n \mid m > n > 0\}, \end{aligned}$$

and this language can be accepted by the $FA(\mathbb{Z})$ in Figure 2. In other words, to prove that the family of $FA(\mathbb{Z})$ languages is not closed under complement modulo ε, we have to use some other language than L. It is also clear that this new language cannot be regular, since the family of regular languages is closed under complement.

Let $A = \{a, b, c\}$ and $S = \{a^n b^n c^n \mid n \geq 0\}$. Now S is not an $FA(\mathbb{Z})$ language, since it is not a context–free language. But

$$\begin{aligned} \bar{S}_\varepsilon = {} & a^* \cup \{b, c\} A^* \cup A^*baA^* \cup A^*c\{a, b\} A^* \\ & \cup \left\{a^n b^m c^k \mid n \neq k,\ k, m, n \in \mathbb{N}\right\} \\ & \cup \left\{a^n b^m c^k \mid n \neq m,\ k, m, n \in \mathbb{N}\right\}. \end{aligned}$$

is a $FA(\mathbb{Z})$ language, since the languages on the first row are regular and the last two languages are easily seen to be $FA(\mathbb{Z})$ languages. Since

$$\overline{(\bar{S}_\varepsilon)}_\varepsilon = S,$$

we may write

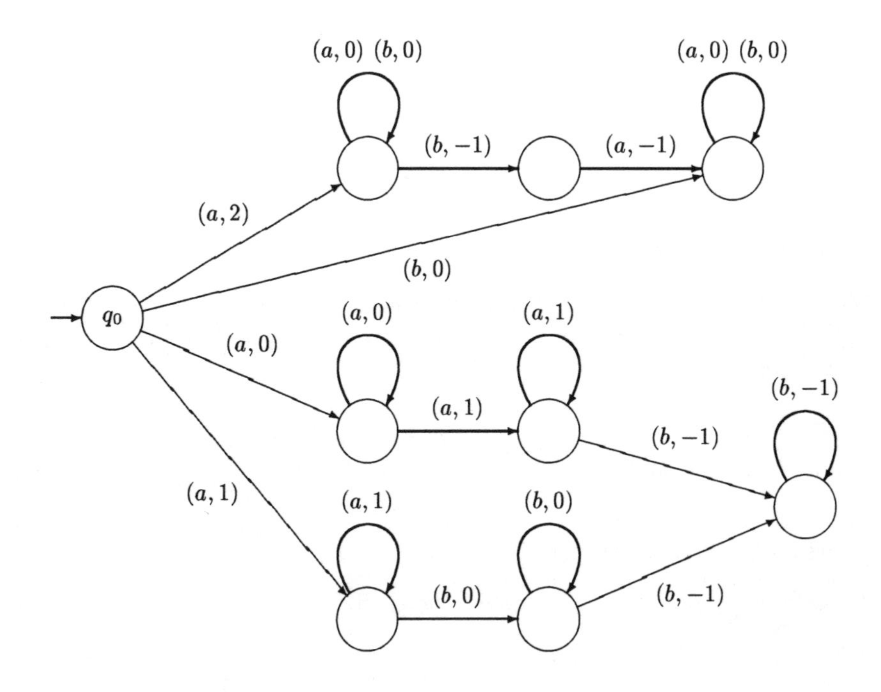

Fig. 2. An $FA(\mathbb{Z})$ accepting the complement of $\{a^n b^n \mid n \geq 0\}$ modulo ε.

Theorem 7. *The family of $FA(\mathbb{Z})$ languages is not closed under complement modulo ε.*

It is obvious that the family of $FA(\mathbb{Z})$ languages is closed under taking the image of a nonerasing morphism $h : A^* \to B^*$, since each transition reading a letter $a \in A$ can be replaced by a new path, which reads the image $h(a)$ and has the same weight as the original transition. Note that the morphism must be nonerasing, since otherwise we would get ε–transitions.

Next we consider inverse morphisms.

Lemma 1. *Let $h : B^+ \to A^+$ be a morphism and $\mathcal{A}^\gamma$, where $\mathcal{A} = (Q, A, \sigma, q_0)$, be an $FA(\mathbb{Z})$. Then there exists an $FA(\mathbb{Z})$ $\mathcal{B}^\alpha$ such that*

$$L(\mathcal{B}^\alpha) = h^{-1}(L(\mathcal{A}^\gamma)) = \left\{h^{-1}(w) \mid w \in L(\mathcal{A}^\gamma)\right\}.$$

Proof. Let $\mathcal{B}^\alpha$, where $\mathcal{B} = (Q, B, \sigma', q_0)$, and the edges of $\mathcal{B}^\alpha$, for all $q \in Q$ and $b \in B$, are defined by

$$\langle q, b, p, z\rangle \quad \text{if} \quad (q, h(b), 0) \models_{\mathcal{A}^\gamma} (p, \varepsilon, z),$$

where $p \in Q$ and $z \in \mathbb{Z}$. Note that if $h(b) = \varepsilon$, then there is a loop

$$\langle q, b, q, 0 \rangle ,$$

for all $q \in Q$. Now it is straightforward to show that $(q_0, v, 0) \models_{\mathcal{B}^\alpha} (q, \varepsilon, 0)$ if and only if $(q_0, h(v), 0) \models_{\mathcal{A}^\gamma} (q, \varepsilon, 0)$. This proves the claim.

We have proved

Theorem 8. *The family of $FA(\mathbb{Z})$ languages is closed under taking the images of nonerasing morphisms and under taking the image of arbitrary inverse morphisms.*

Next we shall consider the *shuffle* operation, denoted by $\sqcup\!\sqcup$. For $u, v \in A^*$ and $a, b \in A$, the shuffle is defined recursively by

1. $(au \sqcup\!\sqcup bv) = a(u \sqcup\!\sqcup bv) \cup b(au \sqcup\!\sqcup v)$, and
2. $(\varepsilon \sqcup\!\sqcup u) = (u \sqcup\!\sqcup \varepsilon) = \{u\}$.

For two languages $K, L \subseteq A^*$, define the shuffle by

$$K \sqcup\!\sqcup L = \bigcup_{u \in K,\ v \in L} (u \sqcup\!\sqcup v).$$

Theorem 9. *The family of $FA(\mathbb{Z})$ languages is not closed under shuffle.*

Proof. Let $L_1 = \{a^n b^n \mid n \geq 0\}$ and $L_2 = \{c^m d^m \mid m \geq 0\}$. We shall prove that $L_1 \sqcup\!\sqcup L_2$ is not an $FA(\mathbb{Z})$ language.

Assume on the contrary that $L_1 \sqcup\!\sqcup L_2$ is an $FA(\mathbb{Z})$ language. Then by Theorem 5 the language

$$(L_1 \sqcup\!\sqcup L_2) \cap (a^+b^+c^+d^+ \cup \{\varepsilon\}) = L_1 \cdot L_2$$

is also an $FA(\mathbb{Z})$ language, since $a^+b^+c^+d^+$ is regular. Since $L_1 \cdot L_2$ has a morphic image L_1^2, also L_1^2 is an $FA(\mathbb{Z})$ language. But this contradicts the proof of Theorem 4.

Note that all these closure properties hold also for unimodal $FA(\mathbb{Z})$ languages, except that in the case of inverse image of a morphism we have to assume that the morphism is nonerasing. Closure properties for unimodal and deterministic integer weighted finite automata are studied in [4].

3 Final states

In this section we shall consider the integer weighted finite automata with final states. This is done for the purposes of further closure properties of $FA(\mathbb{Z})$ languages.

An integer weighted finite automata with final states is defined as the integer weighted finite automata in Section 1 except that the underlying automata has final states, i.e. the underlying automata is $\mathcal{A} = (Q, A, \delta, q_0, F)$, where $F \subseteq Q$ is the *set of final states.* Let σ be an edge set and γ be a weight function for $\mathcal{A}$. The accepting of $\mathcal{A}^\gamma$ is now defined by the final states, in other words, we let

$$L(\mathcal{A}^\gamma) = \{w \in A^* \mid (q_0, w, 0) \models_{\mathcal{A}^\gamma} (q, \varepsilon, 0) \text{ for some } q \in F\},$$

be the language *accepted by* $\mathcal{A}^\gamma$.

We shall denote the family of languages accepted by integer weighted finite automata with final states by $\mathcal{L}^f_{FA(\mathbb{Z})}$.

Theorem 10. $\mathcal{L}^f_{FA(\mathbb{Z})} = \mathcal{L}_{FA(\mathbb{Z})} \cup \left(\bigcup_{L \in \mathcal{L}_{FA(\mathbb{Z})}} (L \setminus \{\varepsilon\})\right)$.

Proof. It is obvious that $\mathcal{L}_{FA(\mathbb{Z})} \subseteq \mathcal{L}^f_{FA(\mathbb{Z})}$, since in an $FA(\mathbb{Z})$ all states are final. We need to show that each language $L \in \mathcal{L}^f_{FA(\mathbb{Z})}$ can be accepted by an $FA(\mathbb{Z})$ or $L \cup \varepsilon$ is accepted by an $FA(\mathbb{Z})$.

Consider now an arbitrary integer weighted finite automaton $\mathcal{A}^\gamma$ with final states, where $\mathcal{A} = (Q, A, \sigma, q_0, F)$. We have two cases depending on whether or not ε is in $L(\mathcal{A}^\gamma)$:

(i) Assume that $\varepsilon \in L(\mathcal{A}^\gamma)$. This implies that $q_0 \in F$. The only thing we shall change in this case is the weight function γ. For $\langle q, a, p, z\rangle \in \sigma$, let $\langle q, a, p, z'\rangle$ be in σ', where

$$z' = \begin{cases} 2z + 1 & \text{if } q = q_0 \text{ and } p \notin F, \\ 2z & \text{if } p, q \notin F \text{ or } p, q \in F, \\ 2z + 1 & \text{if } q \in F \text{ and } p \notin F, \\ 2z - 1 & \text{if } q \notin F \text{ and } p \in F. \end{cases}$$

We define an $FA(\mathbb{Z})$ $\mathcal{B}^\alpha$, where $\mathcal{B} = (Q, A, \sigma', q_0)$. By the definition of the edges in $\mathcal{B}^\alpha$, the weight of a path is $0 \pmod 2$ only when we end in a state of F. If we are in the state in F then the weight of the path is $2w$, where $w \in \mathbb{Z}$ is the weight of the same path in $\mathcal{A}^\gamma$. This proves that $L(\mathcal{B}^\alpha) = L(\mathcal{A}^\gamma)$.

(ii) Assume now that $\varepsilon \notin L(\mathcal{A}^\gamma)$. Since q_0 is not a final state, we must assure that the only word accepted in the initial state is ε. Therefore we define a new initial state $q_0' \notin Q$. The edges from q_0' are defined in the following way. Let

$$\sigma' = \sigma \cup \{\langle q_0', a, p, z\rangle \mid \langle q_0, a, p, z\rangle \in \sigma\}.$$

Now the automaton $\mathcal{B} = (Q \cup \{q_0'\}, A, \sigma', q_0', F \cup \{q_0'\})$ accepts the language $L(\mathcal{A}^\gamma) \cup \{\varepsilon\}$. The claim follows then by the case (i).

We proved that each nonempty language in $\mathcal{L}^f_{FA(\mathbb{Z})}$ can be accepted with integer weighted finite automaton where all states are final or all but the initial state are final. In an integer weighted finite automata with final states the number of final states does not make any difference, since it can be proved that each language in $\mathcal{L}^f_{FA(\mathbb{Z})}$ can be accepted by an integer weighted finite automaton with zero, one or two final states depending on whether ε is in the language or not. This follows, since we may define a new final state and make a copy of each edge ending in the final state to this new final state. No outgoing edges are defined for this new state. Clearly all accepted nonempty words are accepted also in this new final state. To have ε accepted we must also have the initial state as a final state. Naturally, the empty language can be accepted only by zero final states.

It is obvious that the closure properties proved for the family $\mathcal{L}_{FA(\mathbb{Z})}$ in Section 2 also hold for the family $\mathcal{L}^f_{FA(\mathbb{Z})}$, with the difference that Theorem 5 can be stated as follows:

Theorem 11. *Let R be a regular language and L be in $\mathcal{L}^f_{FA(\mathbb{Z})}$. Then $R \cap L$, RL and LR are $\mathcal{L}^f_{FA(\mathbb{Z})}$ languages.*

The reason for studying the integer weighted finite automata with final states is that, by Theorem 11, we get that $\mathcal{L}^f_{FA(\mathbb{Z})}$ is a *semi-AFL*, where AFL stands for *abstract family of languages.* Indeed, $\mathcal{L}^f_{FA(\mathbb{Z})}$ is *trio*, meaning that it is closed under ε-free morphisms and inverse morphisms, and intersections with regular languages. Furthermore, a trio is a semi-AFL, if it is closed under union. For these definitions and for properties of a semi-AFL, we refer to [2].

A family of languages is called a *AFL* if it is a semi-AFL and closed under concatenation and Kleene plus. Obviously $\mathcal{L}^f_{FA(\mathbb{Z})}$ is not an AFL, since by Theorem 4 it is not closed under concatenation. Neither can it be closed under Kleene plus, since by Theorem 3.1.2 in [2] this would imply that $\mathcal{L}^f_{FA(\mathbb{Z})}$ is an AFL. This also follows by the proof of Theorem 6.

A *finite transducer with accepting states* is a 6-tuple $\mathcal{M} = (K, A, B, H, p_0, F)$, where K is a finite set of states, A is the input alphabet and B is the output alphabet, H is a finite subset of $K \times A^* \times B^* \times K$, $p_0 \in K$ is the initial state and $F \subseteq K$ is the set of final states. The elements in H are called *moves.* A finite transducer is a finite automaton with output, i.e. a move (p, a, b, q) means that when we are in the state p and read $a \in A^*$ as input, we move to the state q and output $b \in B^*$.

As usual, we define the relation $\models_{\mathcal{M}}$ on $K \times A^* \times B^*$ by $(p, xw, z) \models_{\mathcal{M}} (q, w, zy)$ for each $w \in A^*$ if (p, x, y, q) is in H. The triple (p, w, z) represents the fact that $\mathcal{M}$ is in state p, w is the input still to be read and z is the output so far.

For each word $w \in A^*$, we define the set

$$\mathcal{M}(w) = \{z \in B^* \mid (p_0, w, \varepsilon) \models^*_{\mathcal{M}} (q, \varepsilon, z), \text{ and } q \in F\},$$

where $\models^*_{\mathcal{M}}$ denotes the reflexive and transitive closure of $\models_{\mathcal{M}}$. The mapping $\mathcal{M}$ from 2^{A^*} into 2^{B^*} is called a *rational transduction.*

Note that $\mathcal{M}$ is ε-free if $\mathcal{M}(w)$ is ε-free for all $w \neq \varepsilon$ and $\mathcal{M}(\varepsilon)$ contains ε if and only if $p_0 \in F$.

In [2], Corollary 2 of Theorem 3.2.1 states that each trio is closed under ε-free rational transduction, and therefore

Theorem 12. *The family $\mathcal{L}^f_{FA(\mathbb{Z})}$ is closed under ε-free rational transductions, i.e. for all $L \in \mathcal{L}^f_{FA(\mathbb{Z})}$ and ε-free finite transducers $\mathcal{M}$*

$$\mathcal{M}(L) \in \mathcal{L}^f_{FA(\mathbb{Z})}.$$

A finite transducer $\mathcal{G} = (K, A, B, H, p_0, F)$ is called a *generalized sequential machine* (*gsm* for short), if $H \subseteq K \times A \times B^* \times K$. $\mathcal{G}$ is ε-free, if $(p, a, \varepsilon, q) \notin H$ for all $p, q \in K$ and $a \in A$. The *gsm mapping* is defined as the rational transduction, and the inverse of a gsm mapping is called an *inverse gsm mapping.*

By Corollary 3 of Theorem 3.2.2 in [2], each trio is closed under inverse gsm mappings. Therefore

Theorem 13. *The family $\mathcal{L}^f_{FA(\mathbb{Z})}$ is closed under ε-free gsm mappings and arbitrary inverse gsm mappings.*

The closure under ε-free gsm mappings follows from Theorem 12.

Let A and B be alphabets. A function $\tau: A^* \to 2^{B^*}$ is called a *substitution,* if

1. $\tau(\varepsilon) = \varepsilon$, and
2. $\tau(xy) = \tau(x)\tau(y)$ for all $x, y \in A^*$.

A substitution τ is extended to languages by defining $\tau(L) = \cup_{w \in L}\tau(w)$. A substitution is said ε-free, if $\tau(a)$ is ε-free.

By Theorem 3.3.1 in [2], we have the next theorem.

Theorem 14. *Let $\tau: A^* \to 2^{B^*}$ be a substitution such that, for all $a \in A$, $\tau(a)$ is an ε-free regular language, and $L \in \mathcal{L}^f_{FA}$. Then $\tau(L) \in \mathcal{L}^f_{FA}$.*

Finally, we give a negative result on the closure under ε-free substitution. By Proposition 3.3.3 in [2], each trio closed under ε-free substitutions is an AFL.

Theorem 15. *The family $\mathcal{L}^f_{FA}$ is not closed under ε-free substitutions.*

References

1. B. Baker and R. Book, *Reversal-bounded multipushdown machines*, J. Comput. System Sci. **8** (1974), 315–332.

2. S. Ginsburg, *Algebraic and automata-theoretic properties of formal languages*, North-Holland Publishing Co., Amsterdam, 1975, Fundamental Studies in Computer Science, Vol. 2.
3. S. A. Greibach, *An infinite hierarchy of context-free languages*, J. Assoc. Comput. Mach. **16** (1969), 91–106.
4. V. Halava, *Finite Substitutions and Integer Weighted Finite Automata*, Tech. Report 197, Turku Centre for Computer Science, August 1998.
5. V. Halava and T. Harju, *Undecidability in integer weighted finite automata*, Fund. Inform., to appear.
6. V. Halava and T. Harju, *Undecidability of the equivalence of finite substitutions on regular language*, Tech. Report 160, Turku Centre for Computer Science, February 1998.
7. O. H. Ibarra, *Restricted one-counter machines with undecidable universe problems*, Math. Systems Theory **13** (1979), 181–186.
8. V. Mitrana and R. Stiebe, *The accepting power of finite automata over groups*, New Trends in Formal Language (G. Păun and A. Salomaa, eds.), Lecture Notes in Comput. Sci., vol. 1218, Springer-Verlag, 1997, pp. 39–48.

A Power Series Approach to Bounded Languages

Juha Honkala

Summary. We use the framework of the monograph "Semirings, Automata, Languages" by Kuich and Salomaa to study bounded languages and algebraic series having bounded supports. We characterize A-algebraic series having bounded supports in the case A is a subring of the real numbers or a positive semiring. This result implies the wellknown characterization of bounded context-free languages and shows that the characterization holds true for languages of finite degree of ambiguity even if multiplicities are counted.

1 Introduction

Classical language theory can be viewed as a study of formal power series over the Boolean semiring. In their monograph "Semirings, Automata, Languages" Kuich and Salomaa give a presentation of automata and language theory over an arbitrary semiring. Kuich and Salomaa also use formal power series as an efficient tool to obtain deep decidability results (see also Salomaa [21]).

In this paper we use the framework of Kuich and Salomaa to study bounded languages and algebraic power series with bounded supports. We first show that for a commutative semiring A the A-algebraic series having bounded supports form a semiring and characterize this semiring in the case A is a positive semiring or a subring of the real numbers. This result is used to give a simple proof of the characterization of bounded context-free languages due to Ginsburg and Spanier [5]. It is also deduced that the result of Ginsburg and Spanier holds true for languages having finite degree of ambiguity even if multiplicities are taken into account.

For further background and motivation see Kuich and Salomaa [15], Rozenberg and Salomaa [20] and Ginsburg [4]. Also all unexplained terminology and notation is from these references.

2 Definitions and earlier results

Suppose A is a commutative semiring and X is an alphabet. The set of *formal power series* (resp. *polynomials*) with noncommuting variables in X and coefficients in A is denoted by $A \ll X^* \gg$ (resp. $A < X^* >$). The set of A-*algebraic* (resp. A-*rational*) series with noncommuting variables in X

is denoted by $A^{\mathrm{alg}} \ll X^* \gg$ (resp. $A^{\mathrm{rat}} \ll X^* \gg$). If $w \in X^*$, the series $r \in A \ll X^* \gg$ is called *w-unary* if

$$\mathrm{supp}(r) \subseteq w^*.$$

A series $r \in A \ll X^* \gg$ is called *unary* if there is a word $w \in X^*$ such that r is w-unary.

In this paper we discuss A-algebraic series having bounded supports. By definition, a series $r \in A \ll X^* \gg$ is *s-bounded* if supp(r) is a bounded set, i.e., there exist an integer $m \geq 1$ and nonempty words $w_1, \ldots, w_m \in X^*$ such that

$$\mathrm{supp}(r) \subseteq w_1^* w_2^* \ldots w_m^*.$$

Clearly, unary series are s-bounded.

Lemma 1. *Suppose A is a commutative semiring. If $r_1, r_2 \in A \ll X^* \gg$ are s-bounded, so are $r_1 + r_2$ and $r_1 r_2$. Consequently, the set of s-bounded A-algebraic series is a semiring.*

Proof. Suppose r_1 and r_2 are s-bounded. Then there exist integers $m, n \geq 1$ and nonempty words $u_i, v_j \in X^*$, $1 \leq i \leq m$, $1 \leq j \leq n$, such that

$$\mathrm{supp}(r_1) \subseteq u_1^* u_2^* \ldots u_m^*$$

and

$$\mathrm{supp}(r_2) \subseteq v_1^* v_2^* \ldots v_n^*.$$

Therefore

$$\mathrm{supp}(r_1 + r_2) \subseteq \mathrm{supp}(r_1) \cup \mathrm{supp}(r_2) \subseteq u_1^* u_2^* \ldots u_m^* v_1^* v_2^* \ldots v_n^*$$

and

$$\mathrm{supp}(r_1 r_2) \subseteq \mathrm{supp}(r_1) \cdot \mathrm{supp}(r_2) \subseteq u_1^* u_2^* \ldots u_m^* v_1^* v_2^* \ldots v_n^*.$$

Hence $r_1 + r_2$ and $r_1 r_2$ are s-bounded, and the set of s-bounded A-algebraic series is a subsemiring of $A \ll X^* \gg$. □

Let $\mathbf{B} = \{0, 1\}$ be the Boolean semiring. The semiring of s-bounded $\mathbf{B}$-algebraic series is isomorphic to the semiring of bounded context-free languages. This semiring plays an important role in classical language theory, see Ginsburg [4]. It is also the natural framework to study thin and slender languages introduced in Andraşiu, Dassow, Păun and Salomaa [1]. (For thin and slender languages see also Dassow, Păun and Salomaa [2], Păun and Salomaa [17-19], Nishida and Salomaa [16], Ilie [13-14] and Honkala [6,8,10-12].)

The following characterization of bounded context-free languages is due to Ginsburg and Spanier [5].

Theorem 1. *The family of bounded context-free languages is the smallest family of sets containing all finite sets and closed with respect to the following operations:*
(i) Finite union.
(ii) Finite product.
(iii) $(x, y) * Z$ *where* x *and* y *are words.*
(Here $(x, y) * Z = \bigcup_{i=0}^{\infty} x^i Z y^i$*.)*

A new proof of Theorem 1 will be given in Section 4. Next, we define the class of power series corresponding to the class of languages considered in Theorem 1.

Let X_∞ be a fixed countably infinite alphabet and denote (see Kuich and Salomaa [15])

$$A\{\{X_\infty^*\}\} = \{r \in A \ll X_\infty^* \gg \mid \text{there exists a finite alphabet } X \subseteq X_\infty \text{ such that } r \in A \ll X^* \gg\},$$

$$A\{X_\infty^*\} = \{r \in A\{\{X_\infty^*\}\} \mid \operatorname{supp}(r) \text{ is finite}\},$$

$$A^{\mathrm{alg}}\{\{X_\infty^*\}\} = \{r \in A \ll X_\infty^* \gg \mid \text{there exists a finite alphabet } X \subseteq X_\infty \text{ such that } r \in A^{\mathrm{alg}} \ll X^* \gg\}.$$

Now, let $\mathcal{B}_0$ be the least set $\mathcal{S} \subseteq A\{\{X_\infty^*\}\}$ having the following properties:
(i) $A\{X_\infty^*\} \subseteq \mathcal{S}$.
(ii) If $r_1, r_2 \in \mathcal{S}$, then $r_1 + r_2 \in \mathcal{S}$ and $r_1 r_2 \in \mathcal{S}$.
(iii) If $r \in \mathcal{S}$ and w_1, w_2 are words of X_∞^*, at least one of which is nonempty, then

$$\sum_{i=0}^{\infty} w_1^i r w_2^i \in \mathcal{S}.$$

Note that the class $\mathcal{B}_0$ depends upon the basic semiring A.

The following theorem is equivalent to Theorem 1.

Theorem 2. *A series* $r \in \mathbf{B}\{\{X_\infty^*\}\}$ *is* $\mathbf{B}$*-algebraic and s-bounded if and only if* $r \in \mathcal{B}_0$.

The following results due to Honkala [9] show that s-bounded algebraic series have nice decision properties. The proofs apply the theory of Parikh simplifying mappings and rely heavily on deep decision methods developed in Kuich and Salomaa [15].

Theorem 3. *Suppose* E *is a computable field and* X *is a finite alphabet. It is decidable whether or not two given s-bounded series* $r, s \in E^{alg} \ll X^* \gg$ *are equal.*

Theorem 4. *Suppose* X *is a finite alphabet and* $A = \mathbf{Z}$ *or* $A = \mathbf{Q}$*. It is decidable whether or not a given s-bounded series* $r \in \mathbf{Q}^{alg} \ll X^* \gg$ *is* A*-rational.*

3 Characterization of s-bounded algebraic series

Suppose A is a commutative semiring. To characterize s-bounded A-algebraic series we first define generic linear series.

Suppose n is a positive integer and let $X_n = \{c_i, d_{ij}, e_{ij} \mid 1 \leq i, j \leq n\}$ be an alphabet with $n + 2n^2$ letters. Then the proper $A \ll X_n^* \gg$-algebraic system

$$y_i = c_i + \sum_{j=1}^{n} d_{ij} y_j e_{ji}, \quad 1 \leq i \leq n, \tag{1}$$

is called a *generic linear system* of order n. The series $r \in A \ll X_n^* \gg$ is called a *generic linear series* of order n if r equals the first component of the strong solution of (1).

Let $(r_1, \ldots, r_n)$ be the strong solution of (1). The approximation sequence $(r^t)_{t \geq 0}$ associated to (1) is given by

$$r_i^{t+1} = c_i + \sum_{u=1}^{t} \sum_{1 \leq k_1, \ldots, k_u \leq n} d_{ik_1} d_{k_1 k_2} \ldots d_{k_{u-1} k_u} c_{k_u} e_{k_u k_{u-1}} \ldots e_{k_2 k_1} e_{k_1 i},$$

$1 \leq i \leq n$, $t \geq 0$. Hence the generic linear series r_1 defined by (1) is given by

$$r_1 = c_1 + \sum_{u=1}^{\infty} \sum_{1 \leq k_1, \ldots, k_u \leq n} d_{1k_1} d_{k_1 k_2} \ldots d_{k_{u-1} k_u} c_{k_u} e_{k_u k_{u-1}} \ldots e_{k_2 k_1} e_{k_1 1}.$$

It is often useful to rewrite r_1 in the following way. For $1 \leq j \leq n$, denote

$$L_j = \{d_{1k_1} d_{k_1 k_2} \ldots d_{k_{u-1} k_u} \mid k_u = j, 1 \leq k_1, \ldots, k_{u-1} \leq n, u \geq 1\}.$$

Clearly, L_j is a regular language. Now, let $\alpha : X_n^* \longrightarrow X_n^*$ be the isomorphism defined by $\alpha(d_{ij}) = e_{ji}$, $\alpha(e_{ji}) = d_{ij}$, $\alpha(c_i) = c_i$, $1 \leq i, j \leq n$. Then

$$r_1 = c_1 + \sum_{j=1}^{n} \sum_{w \in L_j} w c_j \alpha(w)^T \tag{2}$$

where $\alpha(w)^T$ stands for the mirror image of $\alpha(w)$.

To define closure under linear image we first define compatible morphisms. We say that the morphism $h : A < X_n^* > \longrightarrow A^{\mathrm{alg}}\{\{X_\infty^*\}\}$ is *compatible* with the generic linear series of order n given by (2) if the following conditions hold:

(i) There exists a word $w_1 \in X_\infty^*$ such that $h(d_{ij})$ is w_1-unary for all $1 \leq i, j \leq n$.

(ii) There exists a word $w_2 \in X_\infty^*$ such that $h(e_{ij})$ is w_2-unary for all $1 \leq i, j \leq n$.

(iii) For each i, $1 \leq i \leq n$, the series $h(c_i)$ is quasiregular. For each pair (i, j), $1 \leq i, j \leq n$, the product

$$h(d_{ij}) h(e_{ji})$$

is quasiregular. Furthermore, if $h(d_{ij})$ (resp. $h(e_{ji})$) is not quasiregular then $h(d_{ij}) = \varepsilon$ (resp. $h(e_{ji}) = \varepsilon$).

Note that if r is the generic linear series of order n and h is compatible with r, then $h(r) \in A^{\text{alg}}\{\{X_\infty^*\}\}$. Now, a set $\mathcal{S} \subseteq A\{\{X_\infty^*\}\}$ is said to be *closed under linear image* if $h(r) \in \mathcal{S}$ whenever $n \geq 1$, r is the generic linear series of order n given by (2), h is compatible with r and, finally, $h(c_i) \in \mathcal{S}$ for all $1 \leq i \leq n$.

Next, let $\mathcal{B}$ be the least class $\mathcal{S} \subseteq A\{\{X_\infty^*\}\}$ such that $\mathcal{S}$ contains unary A-algebraic series and is closed under sum, product and linear image. Note that the class $\mathcal{B}$ depends upon the basic semiring A.

We proceed to show that in many interesting cases $\mathcal{B}$ equals the semiring of s-bounded A-algebraic series.

Theorem 5. *Suppose A is a positive commutative semiring. Then $r \in A\{\{X_\infty^*\}\}$ is A-algebraic and has a bounded support if and only if $r \in \mathcal{B}$.*

Proof. Suppose first that $r \in \mathcal{B}$. By Lemma 1, the set of A-algebraic series having bounded supports is closed under sum and product. Clearly, this set contains also unary A-algebraic series. Suppose then that n is a positive integer, r_1 is the generic linear series of order n and h is a morphism compatible with r_1 such that $h(c_i)$, $1 \leq i \leq n$, have bounded supports. Because $h(r_1) \in A^{\text{alg}}\{\{X_\infty^*\}\}$, it suffices to show that $h(r_1)$ has a bounded support. This follows by equation (2).

Suppose then that $r \in A^{\text{alg}}\{\{X_\infty^*\}\}$ has a bounded support. Then there exist a finite alphabet X, a positive integer m and nonempty words $w_1, \ldots, w_m \in X^*$ such that

$$\text{supp}(r) \subseteq w_1^* w_2^* \ldots w_m^*.$$

Let $\Delta = \{a_1, \ldots, a_m\}$ be a new alphabet with m letters and define the morphism $g : \Delta^* \longrightarrow X^*$ by $g(a_i) = w_i$, $1 \leq i \leq m$. By the Cross-Section Theorem due to Eilenberg [3] there exists a rational language $R \subseteq a_1^* a_2^* \ldots a_m^*$ such that g maps R bijectively onto $w_1^* w_2^* \ldots w_m^*$. Define the series $s \in A \ll \Delta^* \gg$ by

$$s = g^{-1}(r) \odot \text{char}(R).$$

Then $s \in A^{\text{alg}} \ll \Delta^* \gg$ and $\text{supp}(s) \subseteq a_1^* a_2^* \ldots a_m^*$. Furthermore, $g(s) = r$. Because g is nonerasing we have $g(\mathcal{B}) \subseteq \mathcal{B}$. Therefore it suffices to show that $s \in \mathcal{B}$. We induct on m.

If $m = 1$, the series s is unary and hence belongs to $\mathcal{B}$. Suppose that $m \geq 2$. Without restriction we assume that s is quasiregular. Let

$$y_i = p_i, \qquad 1 \leq i \leq n, \tag{3}$$

be a proper $A \ll \Delta^* \gg$-algebraic system such that s equals the first component of the strong solution $(s_1, \ldots, s_n)$ of (3). Without loss of generality we assume that

$$\emptyset \neq \text{supp}(s_i) \subseteq a_1^* a_2^* \ldots a_m^*$$

for all $1 \leq i \leq n$. Without restriction we assume also that in (3) for any $1 \leq i \leq n$, each variable in $Y = \{y_1, \ldots, y_n\}$ has at most one occurrence in p_i. (If necessary, we add new variables and equations to (3).)

Next, choose α, $1 \leq \alpha \leq n$, such that exactly α components of the strong solution of (3) have minimal alphabet Δ. We may assume that $s_1, \ldots, s_\alpha$ have minimal alphabet Δ and $s_{\alpha+1}, \ldots, s_n$ have minimal alphabets smaller than Δ. By induction, $s_{\alpha+1}, \ldots, s_n \in \mathcal{B}$. Now, let

$$y_i = p_i(y_1, \ldots, y_\alpha, s_{\alpha+1}, \ldots, s_n), \quad 1 \leq i \leq \alpha, \tag{4}$$

be the system obtained from (3) by deleting the last $n - \alpha$ equations and replacing each y_β, $\beta > \alpha$, by s_β in the first α equations. Then the righthand side of (4) is linear in $Y' = \{y_1, \ldots, y_\alpha\}$. More precisely, there exist series $t_i, u_{ij}, v_{ij} \in A \ll \Delta^* \gg$, $1 \leq i, j \leq \alpha$, such that (4) equals

$$y_i = t_i + \sum_{j=1}^{\alpha} u_{ij} y_j v_{ji}, \quad 1 \leq i \leq \alpha.$$

Here $t_i \in \mathcal{B}$, $u_{ij} \in A^{\text{alg}} \ll a_1^* \gg$ and $v_{ij} \in A^{\text{alg}} \ll a_m^* \gg$ for $1 \leq i, j \leq \alpha$. Furthermore, the series t_i and $u_{ij} v_{ji}$ are quasiregular and, if u_{ij} (resp. v_{ij}) is not quasiregular, then $u_{ij} = \varepsilon$ (resp. $v_{ij} = \varepsilon$), $1 \leq i, j \leq \alpha$. Let now

$$y_i = c_i + \sum_{j=1}^{\alpha} d_{ij} y_j e_{ji}, \quad 1 \leq i \leq \alpha,$$

be the generic linear system of order α defining the generic linear series $\overline{r}$ of order α. Define the morphism $h : A < X_\alpha^* > \longrightarrow A^{\text{alg}} \ll \Delta^* \gg$ by

$$h(c_i) = t_i, h(d_{ij}) = u_{ij}, h(e_{ij}) = v_{ij}, 1 \leq i, j \leq \alpha.$$

Then h is compatible with $\overline{r}$ and

$$s_1 = h(\overline{r}).$$

Hence $s = s_1 \in \mathcal{B}$. □

A similar characterization now follows for all subrings of the real numbers. If $A \subseteq \mathbf{R}$, denote $A_+ = \{x \in A \mid x \geq 0\}$.

Theorem 6. *Suppose $A \subseteq \mathbf{R}$ is a ring. Then $r \in A\{\{X_\infty^*\}\}$ is A-algebraic and has a bounded support if and only if $r \in \mathcal{B}$.*

Proof. If $r \in \mathcal{B}$ it follows as in the proof of Theorem 5 that r is A-algebraic and has a bounded support. Suppose then that $r \in A^{\text{alg}} \ll X^* \gg$ has a bounded support where X is a finite alphabet. Suppose

$$\text{supp}(r) \subseteq w_1^* w_2^* \ldots w_m^*$$

where m is a positive integer and $w_1, \ldots, w_m \in X^*$. By Theorem IV 2.4 in Salomaa and Soittola [22] there exist series $r_1, r_2 \in A_+^{\mathrm{alg}} \ll X^* \gg$ such that

$$r = r_1 - r_2.$$

Then

$$\begin{aligned} r &= r \odot \mathrm{char}(w_1^* w_2^* \ldots w_m^*) \\ &= r_1 \odot \mathrm{char}(w_1^* w_2^* \ldots w_m^*) - r_2 \odot \mathrm{char}(w_1^* w_2^* \ldots w_m^*). \end{aligned}$$

Because A_+ is a positive semiring, Theorem 5 implies that the series $r_1 \odot \mathrm{char}(w_1^* w_2^* \ldots w_m^*)$ and $r_2 \odot \mathrm{char}(w_1^* w_2^* \ldots w_m^*)$ belong to the class $\mathcal{B}$ when the basic semiring in the definition of $\mathcal{B}$ equals A_+. Hence $r \in \mathcal{B}$ where we now have A as the basic semiring. □

4 Language-theoretic applications

Consider the classes $\mathcal{B}_0$ and $\mathcal{B}$ introduced in Sections 2 and 3, respectively. In general, $\mathcal{B}_0$ is a proper subclass of $\mathcal{B}$. Indeed, all series of $\mathcal{B}_0$ are commutatively rational whereas $\mathcal{B}$ contains nonrational unary series.

Next we give a new proof of the characterization of bounded context-free languages given in Theorems 1 and 2 due to Ginsburg and Spanier [5]. By Theorem 5 it suffices to prove the next result.

Theorem 7. *If the basic semiring A equals the Boolean semiring, then $\mathcal{B}_0 = \mathcal{B}$.*

Proof. We have to show that $\mathcal{B} \subseteq \mathcal{B}_0$. First, a unary series in $\mathbf{B}^{\mathrm{alg}}\{\{X_\infty^*\}\}$ is in fact in $\mathbf{B}^{\mathrm{rat}}\{\{X_\infty^*\}\}$ and hence in $\mathcal{B}_0$. Because $\mathcal{B}_0$ is a semiring it remains to show that $\mathcal{B}_0$ is closed under linear image. Let n be a positive integer, r be the generic linear series of order n given by (2) and h be a morphism compatible with r_1. Suppose $h(c_i) \in \mathcal{B}_0$ for $1 \leq i \leq n$. Factorize h as a product $h = h_2 h_1$ by defining

$$h_1(c_i) = c_i, h_1(d_{ij}) = h(d_{ij}), h_1(e_{ij}) = h(e_{ij}), 1 \leq i, j \leq n,$$

and

$$h_2(c_i) = h(c_i) \text{ for } 1 \leq i \leq n, h_2(x) = x, \text{ otherwise.}$$

Then there exist sets $I_i \subseteq \mathbf{N}^2$, $1 \leq i \leq n$, and words w_1, w_2 such that

$$h_1(r) = \sum_{i=1}^{n} \sum_{(j,k) \in I_i} w_1^j c_i w_2^k.$$

Because $h_1(r)$ is $\mathbf{B}$-algebraic, the sets I_i are semilinear. Therefore

$$\sum_{(j,k) \in I_i} w_1^j h(c_i) w_2^k \in \mathcal{B}_0, \tag{5}$$

because $h(c_i) \in \mathcal{B}_0$, $1 \le i \le n$. Indeed, if

$$I_i = (m_0, n_0) + \mathbf{N}(m_1, n_1) + \ldots + \mathbf{N}(m_t, n_t)$$

where $m_\beta, n_\beta \in \mathbf{N}$ for $0 \le \beta \le t$, then

$$\sum_{(j,k)\in I_i} w_1^j h(c_i) w_2^k =$$

$$\sum_{j_t=0}^{\infty} w_1^{m_t j_t}(\ldots(\sum_{j_2=0}^{\infty} w_1^{m_2 j_2}(\sum_{j_1=0}^{\infty} w_1^{m_1 j_1}(w_1^{m_0} h(c_i) w_2^{n_0}) w_2^{n_1 j_1}) w_2^{n_2 j_2})\ldots) w_2^{n_t j_t}.$$

Hence (5) holds true and, consequently, $h(r) = h_2 h_1(r) \in \mathcal{B}_0$. □

As another interesting consequence of Theorem 5 we show that the characterization of Ginsburg and Spanier holds true even if multiplicities are considered in the case of bounded context-free languages with *finite* degree of ambiguity. The following theorem states this in a precise way.

Theorem 8. *Suppose the basic semiring* $A = \mathbf{N}$. *If* $s \in \mathbf{N}^{alg}\{\{X_\infty^*\}\}$ *has a bounded support and finite image then* $s \in \mathcal{B}_0$.

In the proof of Theorem 8 we will use results concerning formal power series with commuting variables. Let $X^\oplus$ be the free commutative monoid generated by X and $c : A \ll X^* \gg \longrightarrow A \ll X^\oplus \gg$ be the canonical morphism. Denote

$$A^{\text{rat}} \ll X^\oplus \gg = \{c(r) \mid r \in A^{\text{rat}} \ll X^* \gg\}$$

and

$$A^{\text{alg}} \ll X^\oplus \gg = \{c(r) \mid r \in A^{\text{alg}} \ll X^* \gg\}.$$

The following lemma is due to Honkala [7].

Lemma 2. *If* $r \in \mathbf{N}^{alg} \ll X^\oplus \gg$ *has a finite image then* r *is a finite* $\mathbf{N}$-*linear combination of series in* $\mathbf{N}^{rat} \ll X^\oplus \gg$ *of the form* $uv_1^* \ldots v_p^*$ *with pairwise disjoint supports. Here* $u, v_1, \ldots, v_p \in X^\oplus$ *and the Parikh vectors* $\psi(v_1), \ldots, \psi(v_p)$ *of the words* $v_1, \ldots, v_p$, *respectively, are linearly independent over* $\mathbf{Q}$.

The proof of Lemma 2 relies heavily on deep results due to Kuich and Salomaa [15] and Semenov [23].

Proof of Theorem 8. Suppose $s \in \mathbf{N}^{\text{alg}}\{\{X_\infty^*\}\}$ has a bounded support. By Theorem 5, $s \in \mathcal{B}$. We show inductively that all elements of $\mathcal{B}$ having finite images belong to $\mathcal{B}_0$. First, if s is unary and has a finite image, Lemma 2 implies that $s \in \mathcal{B}_0$. If $s = s_1 + s_2$ or $s = s_1 s_2$, and s has a finite image, so do s_1 and s_2. Hence we may assume inductively that $s_1, s_2 \in \mathcal{B}_0$. Consequently, $s \in \mathcal{B}_0$.

In what follows we use the notation of the proof of Theorem 7. Suppose that $s = h(r)$ and s has a finite image. Then $h(c_i)$, $1 \leq i \leq n$, have finite images and we may assume inductively that $h(c_i) \in \mathcal{B}_0$, $1 \leq i \leq n$. Let

$$h_1(r) = \sum_{i=1}^{n} t_i \tag{6}$$

where $\text{supp}(t_i) \subseteq w_1^* c_i w_2^*$ for $1 \leq i \leq n$.

We want to show that $h_2(t_i) \in \mathcal{B}_0$ for $1 \leq i \leq n$. We consider first the series t_1. If $g : X_\infty^* \longrightarrow X_\infty^*$ is a nonerasing morphism then $g(\mathcal{B}_0) \subseteq \mathcal{B}_0$. Therefore we may assume that w_1 and w_2 are distinct letters. If $h(c_1) = 0$, trivially $h_2(t_1) \in \mathcal{B}_0$. Otherwise, by (6), t_1 has a finite image. Furthermore, no two words in $\text{supp}(t_1)$ are commutatively equal. Hence also $c(t_1)$ has a finite image. Because $c(t_1)$ is $\mathbf{N}$-algebraic, Lemma 2 implies that $c(t_1)$ is a finite $\mathbf{N}$-linear combination of series of the form $uv_1^* \ldots v_p^*$ where $u, v_1, \ldots, v_p \in X_\infty^\oplus$ and the Parikh vectors $\psi(v_1), \ldots, \psi(v_p)$ are linearly independent over $\mathbf{Q}$. Furthermore, u contains the letter c_1 and $v_1, \ldots, v_p$ do not contain c_1. Fix $u, v_1, \ldots, v_p$. Let $u = w_1^{i_0} c_1 w_2^{j_0}$ and $v_\beta = w_1^{i_\beta} w_2^{j_\beta}$, $1 \leq \beta \leq p$. Then $h_2(t_1)$ is a finite $\mathbf{N}$-linear combination of series of the form

$$\sum_{k_p=0}^{\infty} w_1^{i_p k_p} (\ldots (\sum_{k_2=0}^{\infty} w_1^{i_2 k_2} (\sum_{k_1=0}^{\infty} w_1^{i_1 k_1} (w_1^{i_0} h(c_1) w_2^{j_0}) w_2^{j_1 k_1}) w_2^{j_2 k_2}) \ldots) w_2^{j_p k_p}.$$

Hence, $h_2(t_1) \in \mathcal{B}_0$. It is similarly seen that $h_2(t_i) \in \mathcal{B}_0$ for $2 \leq i \leq n$. Consequently

$$h(r) = h_2 h_1(r) = \sum_{i=1}^{n} h_2(t_i) \in \mathcal{B}_0. \qquad \square$$

References

[1] M. Andraşiu, J. Dassow, G. Păun and A. Salomaa, Language-theoretic problems arising from Richelieu cryptosystems, *Theoret. Comput. Sci.* **116** (1993), 339-357.

[2] J. Dassow, G. Păun and A. Salomaa, On thinness and slenderness of L languages, *EATCS Bulletin* **49** (1993), 152-158.

[3] S. Eilenberg, "Automata, Languages and Machines," Vol. A, Academic Press, New York, 1974.

[4] S. Ginsburg, "The Mathematical Theory of Context-Free Languages," McGraw-Hill, New York, 1966.

[5] S. Ginsburg and E. H. Spanier, Bounded ALGOL-like languages, *Trans. Amer. Math. Soc.* **113** (1964), 333-368.

[6] J. Honkala, On Parikh slender languages and power series, *J. Comput. System Sci.* **52** (1996), 185-190.

[7] J. Honkala, On images of algebraic series, *J. Universal Comput. Sci.* **2** (1996), 217-223.

[8] J. Honkala, A decision method for Parikh slenderness of context-free languages, *Discrete Appl. Math.* **73** (1997), 1-4.

[9] J. Honkala, Decision problems concerning algebraic series with noncommuting variables, in: J. Mycielski, G. Rozenberg and A. Salomaa (eds.): Structures in Logic and Computer Science, Springer, Berlin, 1997, pp. 281-290.

[10] J. Honkala, On N-algebraic Parikh slender power series, *J. Universal Comput. Sci.* **3** (1997), 1114-1120.

[11] J. Honkala, Decision problems concerning thinness and slenderness of formal languages, *Acta Inform.* **35** (1998), 625-636.

[12] J. Honkala, On Parikh slender context-free languages, submitted

[13] L. Ilie, On a conjecture about slender context-free languages, *Theoret. Comput. Sci.* **132** (1994), 427-434.

[14] L. Ilie, On lengths of words in context-free languages, *Theoret. Comput. Sci.*, to appear.

[15] W. Kuich and A. Salomaa, "Semirings, Automata, Languages," Springer, Berlin, 1986.

[16] T. Nishida and A. Salomaa, Slender 0L languages, *Theoret. Comput. Sci.* **158** (1996), 161-176.

[17] G. Păun and A. Salomaa, Decision problems concerning the thinness of D0L languages, *EATCS Bulletin* **46** (1992), 171-181.

[18] G. Păun and A. Salomaa, Closure properties of slender languages, *Theoret. Comput. Sci.* **120** (1993), 293-301.

[19] G. Păun and A. Salomaa, Thin and slender languages, *Discrete Appl. Math.* **61** (1995), 257-270.

[20] G. Rozenberg and A. Salomaa (eds.), "Handbook of Formal Languages," Vol. 1-3, Springer, Berlin, 1997.

[21] A. Salomaa, Formal languages and power series, in: J. van Leeuwen, ed., Handbook of Theoretical Computer Science, Vol. B., North-Holland, Amsterdam, 1990, pp. 103-132.

[22] A. Salomaa and M. Soittola, "Automata-Theoretic Aspects of Formal Power Series," Springer, Berlin, 1978.

[23] A. L. Semenov, Presburgerness of predicates regular in two number systems (in Russian), *Sibirsk. Mat. Zh.* **18** (1977), 403-418. English translation: *Siberian Math J.* **18** (1977), 289-299.

Full Abstract Families of Tree Series I

Werner Kuich*

Summary. We introduce full abstract families of tree series and show that their yields form full abstract families of power series that are closed under algebraic transductions. In the language case, the yield of a full abstract family of tree languages forms a full abstract family of languages that is closed under context-free substitutions.

1 Introduction and preliminaries

Full abstract families of tree series (briefly, full AFTs) are families of tree series with certain closure properties. They are defined analogous to full abstract families of power series (briefly, full AFPs; see Kuich, Salomaa [19], Kuich [14]). These are a generalization of the full abstract families of languages (briefly, full AFLs) as introduced by Ginsburg, Greibach [13] (see also Salomaa [20], Ginsburg [12] and Berstel [2]). Bozapalidis, Rahonis [7] introduced families of tree languages called sheaves. These sheaves are defined to have certain closure properties and are a generalization of the full AFLs in the framework of tree languages. The main difference of full AFTs in the language case and of sheaves is that the former are closed under linear nondeleting recognizable tree transductions (see Kuich [17]) while the latter are closed under alphabetic tree transductions (see Bozapalidis [4]). The relation of linear nondeleting recognizable tree transductions and of alphabetic tree transductions is yet unclear.

The paper is divided in this and another two sections. The main result of Section 2 is that linear nondeleting recognizable tree transductions are closed under functional composition. In Section 3 we introduce recognizable tree cones and rec-closed families of tree series. Here a recognizable tree cone is a family of tree series closed under linear nondeleting recognizable tree transductions while a family of tree series closed under addition, topcatenation and least solutions of equations is called rec-closed. A family of tree series that is a recognizable tree cone and is rec-closed is called a full AFT. The main result of Section 3 considers the yields of the tree series of a full AFT: they form a full AFP.

* Partially supported by the "Stiftung Aktion Österreich-Ungarn".

It is assumed that the reader is familiar with the basics of semiring theory (see Kuich, Salomaa [19] and Kuich [14], Section 2). Throughout the paper, $\langle A, +, \cdot, 0, 1 \rangle$ denotes a *commutative continuous* semiring. This means:

(o) the multiplication $\cdot$ is commutative;
(i) A is partially ordered by the relation $\sqsubseteq$: $a \sqsubseteq b$ iff there exists a c such that $a + c = b$,
(ii) $\langle A, +, \cdot, 0, 1 \rangle$ is a complete semiring,
(iii) $\sum_{i \in I} a_i = \sup(\sum_{i \in E} a_i \mid E \subseteq I,\ E \text{ finite})$, $a_i \in A$, $i \in I$, for an arbitrary index set I, where sup denotes the least upper bound with respect to $\sqsubseteq$.

In the sequel, we denote $\langle A, +, \cdot, 0, 1 \rangle$ briefly by A.

Furthermore, Σ will always denote a *ranked alphabet*, where Σ_k, $k \geq 0$, contains the symbols of rank k and X will denote an *alphabet of leaf symbols*. By $T_\Sigma(X)$ we denote the set of *trees* formed by $\Sigma \cup X$. If $\Sigma_0 \neq \emptyset$ then X may be the empty set. By $A\langle\langle T_\Sigma(X) \rangle\rangle$ we denote the set of *formal tree series* over $T_\Sigma(X)$ and by $A\langle T_\Sigma(X) \rangle$ the set of *polynomials*. (For more definitions see Kuich [15].)

Formal tree series induce continuous mappings called *substitutions*. Let $Y = \{y_1, \ldots, y_n\}$ denote a set of variables, and $s = s(y_1, \ldots, y_n)$ and s_i, $1 \leq i \leq n$, be tree series in $A\langle\langle T_\Sigma(X \cup Y) \rangle\rangle$. Then the formal tree series s *induces a mapping* $s : (A\langle\langle T_\Sigma(X \cup Y) \rangle\rangle)^n \to A\langle\langle T_\Sigma(X \cup Y) \rangle\rangle$, whose value at $s_1, \ldots, s_n$ is defined as the substitution of the tree series s_i for the variable y_i, $1 \leq i \leq n$, into the tree series s. It is denoted by $s(s_1, \ldots, s_n)$ or $s[s_i/y_i,\ 1 \leq i \leq n]$. (For a more detailed definiton see Kuich [15].)

Our tree transducers will be defined by transition matrices. Let $Y_k = \{y_1, \ldots, y_k\}$, $k \geq 1$, and Y be sets of variables. A *matrix* $M \in (A\langle\langle T_\Sigma(X \cup Y_k) \rangle\rangle)^{I' \times I^k}$, $k \geq 1$, I' and I arbitrary index sets, *induces a mapping*

$$M : (A\langle\langle T_\Sigma(X \cup Y) \rangle\rangle)^{I \times 1} \times \ldots \times (A\langle\langle T_\Sigma(X \cup Y) \rangle\rangle)^{I \times 1} \to (A\langle\langle T_\Sigma(X \cup Y) \rangle\rangle)^{I' \times 1}$$

(there are k argument vectors), defined by the entries of the resulting vector as follows: For $P_1, \ldots, P_k \in (A\langle\langle T_\Sigma(X \cup Y) \rangle\rangle)^{I \times 1}$ we define, for all $i \in I'$,

$$M(P_1, \ldots, P_k)_i = \sum_{i_1, \ldots, i_k \in I} M_{i,(i_1, \ldots, i_k)}((P_1)_{i_1}, \ldots, (P_k)_{i_k}) = \sum_{i_1, \ldots, i_k \in I} \sum_{t \in T_\Sigma(X \cup Y_k)} (M_{i,(i_1, \ldots, i_k)}, t) t((P_1)_{i_1}, \ldots, (P_k)_{i_k}) .$$

Throughout the whole paper, I (resp. Q) will denote an arbitrary (resp. a finite) index set.

2 Linear nondeleting recognizable tree transductions

In this section we consider the linear nondeleting recognizable tree transducers of Kuich [17] and show that the mappings defined by them are closed under functional composition.

In the sequel, Σ, Σ' and Σ'' denote finite ranked alphabets, X, X' and X'' denote leaf alphabets and $Z = \{z_i \mid i \geq 1\}$ denotes an alphabet of variables. We denote $Z_k = \{z_i \mid 1 \leq i \leq k\}$, $k \geq 1$, and $Z_0 = \emptyset$.

A *recognizable tree representation* (with state set Q) is a mapping μ from $\Sigma \cup X$ into matrices with entries in $A^{\mathrm{rec}}\langle\langle T_{\Sigma'}(X' \cup Z)\rangle\rangle$ such that

$$\mu : \Sigma_k \to (A^{\mathrm{rec}}\langle\langle T_{\Sigma'}(X' \cup Z_k)\rangle\rangle)^{Q\times Q^k}, \; k \geq 1,$$
$$\mu : \Sigma_0 \cup X \to (A^{\mathrm{rec}}\langle\langle T_{\Sigma'}(X')\rangle\rangle)^{Q\times 1}.$$

Observe that, for $f \in \Sigma_k$, $k \geq 1$, $\mu(f)$ induces a mapping and

$$\langle (A\langle\langle T_{\Sigma'}(X' \cup Z)\rangle\rangle)^{Q\times 1}, (\mu(f) \mid f \in \Sigma)\rangle, \; \langle (A\langle\langle T_{\Sigma'}(X')\rangle\rangle)^{Q\times 1}, (\mu(f) \mid f \in \Sigma)\rangle$$

are Σ-algebras. Hence, the mapping $\mu : X \to (A\langle\langle T_{\Sigma'}(X')\rangle\rangle)^{Q\times 1}$ can be uniquely extended to a morphism $\mu : T_\Sigma(X) \to (A\langle\langle T_{\Sigma'}(X')\rangle\rangle)^{Q\times 1}$. This morphic extension is defined by $\mu(f(t_1, \ldots, t_k)) = \mu(f)(\mu(t_1), \ldots, \mu(t_k))$ for $f \in \Sigma_k$, $t_1, \ldots, t_k \in T_\Sigma(X)$. For $s \in A\langle\langle T_\Sigma(X)\rangle\rangle$ we define $\mu(s) = \sum_{t \in T_\Sigma(X)} (s, t) \otimes \mu(t)$, where $\otimes$ denotes the Kronecker product (see Kuich, Salomaa [19], Section 4).

A tree $t \in T_\Sigma(X \cup Z_k)$, $k \geq 1$, is called *linear* iff the variable z_j appears at most once in t, $1 \leq j \leq k$. A tree $t \in T_\Sigma(X \cup Z_k)$, $k \geq 1$, is called *nondeleting* iff the variable z_j appears at least once in t, $1 \leq j \leq k$. A tree series $s \in A\langle\langle T_\Sigma(X \cup Z_k)\rangle\rangle$, $k \geq 1$, is called *linear* or *nondeleting* iff all $t \in \mathrm{supp}(s)$ are linear or nondeleting, respectively. A recognizable tree representation μ is called *linear* or *nondeleting* iff all entries of $\mu(f)$, $f \in \Sigma_k$, $k \geq 1$, are linear or nondeleting tree series, respectively.

A *linear nondeleting recognizable tree transducer* (with input alphabet Σ, input leaf alphabet X, output alphabet Σ', output leaf alphabet X')

$$\mathfrak{T} = (Q, \mu, S)$$

is given by

(i) a nonempty finite set Q of *states*,
(ii) a *linear nondeleting recognizable tree representation* μ with state set Q,
(iii) $S \in (A\langle T_{\Sigma'}(Z_1)\rangle)^{1\times Q}$, where $S_q = a_q z_1$, $a_q \in A$, $q \in Q$, called the *initial state vector.*

The *mapping* $||\mathfrak{T}|| : A\langle\langle T_\Sigma(X)\rangle\rangle \to A\langle\langle T_{\Sigma'}(X')\rangle\rangle$ *realized by a tree transducer* $\mathfrak{T} = (Q, \mu, S)$ is defined by

$$||\mathfrak{T}||(s) = S(\mu(s)) = \sum_{q \in Q} a_q \sum_{t \in T_\Sigma(X)} (s, t)\mu(t)_q \,.$$

Before we show the main result of this section we need three technical lemmas. For these technical lemmas we have to extend the domain of a linear nondeleting recognizable tree representation μ from $\Sigma \cup X$ to $\Sigma \cup X \cup Z$; μ is now a mapping from $\Sigma \cup X \cup Z$ into matrices with entries in $A^{\mathrm{rec}}\langle\langle T_{\Sigma'}(X' \cup Z \cup Z_Q)\rangle\rangle$, where $Z_Q = \{(z_i)_q \mid i \geq 1,\ q \in Q\}$. The extension

$$\mu : Z_k \to (A\langle T_{\Sigma'}(Z_Q^k)\rangle)^{Q \times 1}, \quad k \geq 1,$$

is defined by

$$\mu(z_i)_q = (z_i)_q, \quad 1 \leq i \leq k,\ q \in Q\,.$$

Here $Z_Q^k = \{(z_i)_q \mid 1 \leq i \leq k,\ q \in Q\}$, $k \geq 1$, and $Z_Q^0 = \emptyset$. We call μ *extended* linear nondeleting recognizable tree representation.

Let $t \in T_\Sigma(X \cup Z_k)$ (resp. $t \in T_{\Sigma'}(X' \cup Z_Q^k)$), $k \geq 0$. Then we say that the variables $z_{i_1}, \ldots, z_{i_m} \in Z_k$ (resp. $(z_{i_1})_{r_1}, \ldots, (z_{i_m})_{r_m} \in Z_Q^k$), $m \geq 0$, appear *exactly* once in t iff $z_{i_1}, \ldots, z_{i_m}$ (resp. $(z_{i_1})_{r_1}, \ldots, (z_{i_m})_{r_m}$) do appear once in t and the variables of $Z - \{z_{i_1}, \ldots, z_{i_m}\}$ (resp. $Z_Q^k - \{(z_{i_1})_{r_1}, \ldots, (z_{i_m})_{r_m}\}$) do not appear in t.

Lemma 2.1 *Let μ be an extended linear nondeleting recognizable tree representation with state set Q, mapping $\Sigma \cup X \cup Z$ into matrices with entries in $A^{\mathrm{rec}}\langle\langle T_{\Sigma'}(X' \cup Z \cup Z_Q)\rangle\rangle$. Let $t \in T_\Sigma(X \cup Z_k)$, $k \geq 0$, be a linear nondeleting tree. Consider a tree $t' \in T_{\Sigma'}(X' \cup Z_Q^k)$ that is in* $\mathrm{supp}(\mu(t)_q)$, *$q \in Q$. Then there exist states $r_1, \ldots, r_k \in Q$ such that the variables $(z_1)_{r_1}, \ldots, (z_k)_{r_k}$ appear exactly once in t'.*

Proof. Consider first a tree $t \in T_\Sigma(X \cup Z_k)$ such that the variables $z_{i_1}, \ldots, z_{i_m} \in Z_k$ appear exactly once in t. Then we show by induction on the structure of these trees that there exist states $r_1, \ldots, r_m \in Q$ such that the variables $(z_{i_1})_{r_1}, \ldots, (z_{i_m})_{r_m}$ appear exactly once in t'.

(i) If $t = x$, the lemma is true. (Clearly, if no variable of Z_k appears in t, the lemma is true.)

(ii) If $t = z_j$ for some $z_j \in Z_k$ we have $\mu(t)_q = (z_j)_q$ and the lemma is true.

(iii) Let now $t = f(t_1, \ldots, t_n)$, $f \in \Sigma_n$, $t_1, \ldots, t_n \in T_\Sigma(X \cup Z_k)$, $n \geq 1$. Then there exist mutually disjoint sets $J_1, \ldots, J_n$ with $J_1 \cup \ldots \cup J_n = \{i_1, \ldots, i_m\}$ such that the variables z_j, $j \in J_i$, appear exactly once in t_i, $1 \leq i \leq n$. By induction hypothesis there exist, for each $1 \leq i \leq n$, each $q_i \in Q$ and each $t'_i \in \mathrm{supp}(\mu(t_i)_q)$, states r_j, $j \in J_i$, such that the variables $(z_j)_{r_j}$, $j \in J_i$, appear exactly once in t'_i. By Lemma 11 of Kuich [17], we obtain

$$\begin{aligned}
&\mu(f(t_1, \ldots, t_n))_q = \mu(f(z_1, \ldots, z_n))_q[\mu(t_i)/\mu(z_i),\ 1 \leq i \leq n] \\
&= \textstyle\sum_{q_1,\ldots,q_n \in Q}(\mu(f)_{q,(q_1,\ldots,q_n)}[\mu(z_i)_{q_i}/z_i,\ 1 \leq i \leq n])[\mu(t_i)/\mu(z_i),\ 1 \leq i \leq n] \\
&= \textstyle\sum_{q_1,\ldots,q_n \in Q} \mu(f)_{q,(q_1,\ldots,q_n)}[\mu(t_i)_{q_i}/z_i,\ 1 \leq i \leq n]\,.
\end{aligned}$$

Hence, there exist states $s_1, \ldots, s_n \in Q$ such that $t' \in \mathrm{supp}(\mu(f)_{q,(s_1,\ldots,s_n)}$ $[\mu(t_i)_{s_i}/z_i,\ 1 \leq i \leq n])$. Since $\mu(f)_{q,(s_1,\ldots,s_n)}$ is linear and nondeleting, we have completed the proof by induction and the lemma is proved. □

We now define, for $r_1, \ldots, r_k \in Q$, the operator

$$\varphi_{r_1,\ldots,r_k} : A\langle\langle T_\Sigma(X \cup Z_Q^k)\rangle\rangle \to A\langle\langle T_\Sigma(X \cup \{(z_1)_{r_1}, \ldots, (z_k)_{r_k}\})\rangle\rangle$$

as follows: For $s \in A\langle\langle T_\Sigma(X \cup Z_Q^k)\rangle\rangle$ and $t \in A\langle\langle T_\Sigma(X \cup \{(z_1)_{r_1}, \ldots, (z_k)_{r_k}\})\rangle\rangle$,

$$(\varphi_{r_1,\ldots,r_k}(s), t) = \begin{cases} (s,t) & \text{iff the variables } (z_1)_{r_1}, \ldots, (z_k)_{r_k} \\ & \text{appear exactly once in } t, \\ 0 & \text{otherwise.} \end{cases}$$

Lemma 2.2 *Let μ be an extended linear nondeleting recognizable tree representation with state set Q and $s \in A\langle\langle T_\Sigma(X \cup Z_k)\rangle\rangle$, $k \geq 1$, be a linear nondeleting tree series. Then, for $q \in Q$,*

$$\mu(s)_q = \sum_{r_1,\ldots,r_k \in Q} \varphi_{r_1,\ldots,r_k}(\mu(s)_q).$$

Proof. By Lemma 2.1, we have, for $t \in T_\Sigma(X \cup Z_k)$ and $q \in Q$,

$$\mu(t)_q = \sum_{r_1,\ldots,r_k \in Q} \varphi_{r_1,\ldots,r_k}(\mu(t)_q).$$

Hence, we obtain

$$\mu(s)_q = \textstyle\sum_{t \in T_\Sigma(X \cup Z_k)} (s,t)\mu(t)_q =$$
$$\textstyle\sum_{t \in T_\Sigma(X \cup Z_k)} \sum_{r_1,\ldots,r_k \in Q} (s,t)\varphi_{r_1,\ldots,r_k}(\mu(t)_q) =$$
$$\textstyle\sum_{r_1,\ldots,r_k \in Q} \varphi_{r_1,\ldots,r_k}(\sum_{t \in T_\Sigma(X \cup Z_k)} (s,t)\mu(t)_q) = \sum_{r_1,\ldots,r_k \in Q} \varphi_{r_1,\ldots,r_k}(\mu(s)_q).$$

□

Compare the construction in the next lemma with the constructions in the proofs of Lemma 4.2 of Engelfriet [8] and of Theorem IV.3.15 of Gécseg, Steinby [10].

Lemma 2.3 *Let μ_1 be a linear nondeleting recognizable tree representation with state set Q_1 mapping $\Sigma \cup X$ into matrices with entries in $A^{\mathrm{rec}}\langle\langle T_{\Sigma'}(X' \cup Z)\rangle\rangle$. Furthermore, let μ_2 be an extended linear nondeleting recognizable tree representation with state set Q_2 mapping $\Sigma' \cup X' \cup Z$ into matrices with entries in $A^{\mathrm{rec}}\langle\langle T_{\Sigma''}(X'' \cup Z \cup Z_{Q_2})\rangle\rangle$. Define the recognizable tree representation μ with state set $Q_1 \times Q_2$ mapping $\Sigma \cup X$ into matrices with entries in $A^{\mathrm{rec}}\langle\langle T_{\Sigma''}(X'' \cup Z)\rangle\rangle$ by*

$$\mu(x)_{(q_1,q_2)} = \mu_2(\mu_1(x)_{q_1})_{q_2}, \quad \textit{for } x \in \Sigma_0 \cup X,\ q_1 \in Q_1,\ q_2 \in Q_2,$$
$$\mu(f)_{(q_1,q_2),((r_1,s_1),\ldots,(r_k,s_k))} =$$
$$\varphi_{s_1,\ldots,s_k}(\mu_2(\mu_1(f)_{q_1,(r_1,\ldots,r_k)})_{q_2})[z_1/(z_1)_{s_1}, \ldots, z_k/(z_k)_{s_k}],$$
$$\textit{for } f \in \Sigma_k,\ k \geq 1,\ q_1, r_1, \ldots, r_k \in Q_1,\ q_2, s_1, \ldots, s_k \in Q_2.$$

Then μ is a linear nondeleting recognizable tree representation and, for $t \in T_\Sigma(X)$ and $q_1 \in Q_1$, $q_2 \in Q_2$,

$$\mu(t)_{(q_1,q_2)} = \mu_2(\mu_1(t)_{q_1})_{q_2}.$$

Proof. By Corollary 13 of Kuich [17] and by the definition of $\varphi_{s_1,\ldots,s_k}$, $s_1, \ldots, s_k \in Q$, μ is a linear nondeleting recognizable tree representation. The proof is now by induction on the structure of the trees t in $T_\Sigma(X)$.

(i) For $t \in \Sigma_0 \cup X$, the lemma is true by the definiton of μ.

(ii) Let $t = f(t_1, \ldots, t_n)$, $f \in \Sigma_n$, $t_1, \ldots, t_n \in T_\Sigma(X)$, $n \geq 1$. Then we obtain, for $q_1 \in Q_1$, $q_2 \in Q_2$,

$$\begin{aligned}
&\mu_2(\mu_1(t)_{q_1})_{q_2} = \\
&\mu_2(\textstyle\sum_{r_1,\ldots,r_n \in Q_1} \mu_1(f)_{q_1,(r_1,\ldots,r_n)}(\mu_1(t_1)_{r_1}, \ldots, \mu_1(t_n)_{r_n}))_{q_2} = \\
&\textstyle\sum_{r_1,\ldots,r_n \in Q_1} \mu_2(\mu_1(f)_{q_1,(r_1,\ldots,r_n)})_{q_2}[\mu_2(\mu_1(t_i)_{r_i})/\mu_2(z_i),\ 1 \leq i \leq n] = \\
&\textstyle\sum_{r_1,\ldots,r_n \in Q_1} \mu_2(\mu_1(f)_{q_1,(r_1,\ldots,r_n)})_{q_2}[\mu(t_i)_{(r_i,s)}/(z_i)_s,\ 1 \leq i \leq n,\ s \in Q] = \\
&\textstyle\sum_{r_1,\ldots,r_n \in Q_1} \sum_{s_1,\ldots,s_n \in Q_2} \varphi_{s_1,\ldots,s_n}(\mu_2(\mu_1(f)_{q_1,(r_1,\ldots,r_n)})_{q_2}) \\
&\qquad\qquad [\mu(t_i)_{(r_i,s)}/(z_i)_s,\ 1 \leq i \leq n,\ s \in Q] = \\
&\textstyle\sum_{r_1,\ldots,r_n \in Q_1} \sum_{s_1,\ldots,s_n \in Q_2} (\varphi_{s_1,\ldots,s_n}(\mu_2(\mu_1(f)_{q_1,(r_1,\ldots,r_n)})_{q_2}) \\
&\qquad\qquad [z_i/(z_i)_{s_i},\ 1 \leq i \leq n])[\mu(t_i)_{(r_i,s_i)}/z_i,\ 1 \leq i \leq n] = \\
&\textstyle\sum_{r_1,\ldots,r_n \in Q_1} \sum_{s_1,\ldots,s_n \in Q_2} \mu(f)_{(q_1,q_2),((r_1,s_1),\ldots,(r_n,s_n))} \\
&\qquad\qquad [\mu(t_i)_{(r_i,s_i)}/z_i,\ 1 \leq i \leq n] = \\
&\mu(f(t_1, \ldots, t_n))_{(q_1,q_2)} = \mu(t)_{(q_1,q_2)}\,.
\end{aligned}$$

Here the second equality follows by Lemma 11 of Kuich [17], the third equality by the induction hypothesis and the fourth equality by Lemma 2.2. □

We are now ready for the main result of Section 2.

Theorem 2.4 *Let μ_1 (resp. μ_2) be a linear nondeleting recognizable tree representation with state set Q_1 (resp. Q_2) mapping $\Sigma \cup X$ (resp. $\Sigma' \cup X'$) into matrices with entries in $A^{\mathrm{rec}}\langle\langle T_{\Sigma'}(X' \cup Z)\rangle\rangle$ (resp. $A^{\mathrm{rec}}\langle\langle T_{\Sigma''}(X'' \cup Z)\rangle\rangle$). Let $\mathfrak{T}_1 = (Q_1, \mu_1, S_1)$ and $\mathfrak{T}_2 = (Q_2, \mu_2, S_2)$ be linear nondeleting recognizable tree transducers. Then there exists a linear nondeleting recognizable tree transducer $\mathfrak{T}$ such that $||\mathfrak{T}||(s) = ||\mathfrak{T}_2||(||\mathfrak{T}_1||(s))$ for all $s \in A\langle\langle T_\Sigma(X)\rangle\rangle$.*

Proof. The linear nondeleting recognizable tree transducer $\mathfrak{T} = (Q_1 \times Q_2, \mu, S_1 \odot S_2)$ is defined by the linear nondeleting recognizable tree representation μ of Lemma 2.3.

Let $(S_1)_{q_1} = a_{q_1} z_1$, $a_{q_1} \in A$, $q_1 \in Q_1$, and $(S_2)_{q_2} = b_{q_2} z_1$, $b_{q_2} \in A$, $q_2 \in Q_2$. Then $(S_1 \odot S_2)_{(q_1,q_2)} = a_{q_1} b_{q_2} z_1$ for $q_1 \in Q_1$, $q_2 \in Q_2$. We now obtain, for $s \in A\langle\langle T_\Sigma(X)\rangle\rangle$,

$$\begin{aligned}
&||\mathfrak{T}_2||(||\mathfrak{T}_1||(s)) = \textstyle\sum_{q_2 \in Q_2} b_{q_2} \sum_{t_2 \in T_{\Sigma'}(X')} (||\mathfrak{T}_1||(s), t_2)\mu_2(t_2)_{q_2} = \\
&\textstyle\sum_{q_2 \in Q_2} b_{q_2} \sum_{t_2 \in T_{\Sigma'}(X')} (\sum_{q_1 \in Q_1} a_{q_1} \sum_{t_1 \in T_\Sigma(X)} (s, t_1)\mu_1(t_1)_{q_1}, t_2)\mu_2(t_2)_{q_2} = \\
&\textstyle\sum_{q_1 \in Q_1} \sum_{q_2 \in Q_2} a_{q_1} b_{q_2} \sum_{t_1 \in T_\Sigma(X)} (s, t_1) \sum_{t_2 \in T_{\Sigma'}(X')} (\mu_1(t_1)_{q_1}, t_2)\mu_2(t_2)_{q_2} = \\
&\textstyle\sum_{(q_1,q_2) \in Q_1 \times Q_2} a_{q_1} b_{q_2} \sum_{t_1 \in T_\Sigma(X)} (s, t_1)\mu(t_1)_{(q_1,q_2)} = \\
&||\mathfrak{T}||(s)\,.
\end{aligned}$$

□

3 Full abstract families of tree series and their yield

In this section we introduce recognizable tree cones, rec-closed families of tree series and full AFTs, and consider their yields: The yield of a rec-closed family of tree series is a rationally closed family of power series; and the yield of a full AFT is a full AFP (closed under algebraic transductions). We consider this result on full AFTs to be a necessary condition that the definition of the notion of a full AFT makes sense.

We make the following convention for the remainder of this paper: The set Σ_∞ (resp. X_∞) is a fixed *infinite* ranked alphabet (resp. *infinite* alphabet) and Σ (resp. X), possibly provided with indices, is a *finite* subalphabet of Σ_∞ (resp. X_∞); moreover, Σ_∞ contains the symbols $\bullet$ and e of ranks 2 and 0, respectively. Our basic semiring will be $A\langle\langle T_{\Sigma_\infty}(X_\infty)\rangle\rangle$ and we define

$$A\{\{T_{\Sigma_\infty}(X_\infty)\}\} = \{s \in A\langle\langle T_{\Sigma_\infty}(X_\infty)\rangle\rangle \mid \text{there exist finite alphabets } \Sigma \subset \Sigma_\infty \text{ and } X \subset X_\infty \text{ such that } s \in A\langle\langle T_\Sigma(X)\rangle\rangle\}\,.$$

Any nonempty subset of $A\{\{T_{\Sigma_\infty}(X_\infty)\}\}$ is called *family of tree series*. Furthermore, we define three subsemirings of $A\{\{T_{\Sigma_\infty}(X_\infty)\}\}$, namely the families of *algebraic tree series* (see Bozapalidis [6], Kuich [16]), of *recognizable tree series* (see Berstel, Reutenauer [3]) and of *polynomials* by

$$A^{\mathrm{alg}}\{\{T_{\Sigma_\infty}(X_\infty)\}\} = \{s \in A\{\{T_{\Sigma_\infty}(X_\infty)\}\} \mid \text{there exist finite alphabets } \Sigma \subset \Sigma_\infty \text{ and } X \subset X_\infty \text{ such that } s \in A^{\mathrm{alg}}\langle\langle T_\Sigma(X)\rangle\rangle\}\,,$$
$$A^{\mathrm{rec}}\{\{T_{\Sigma_\infty}(X_\infty)\}\} = \{s \in A\{\{T_{\Sigma_\infty}(X_\infty)\}\} \mid \text{there exist finite alphabets } \Sigma \subset \Sigma_\infty \text{ and } X \subset X_\infty \text{ such that } s \in A^{\mathrm{rec}}\langle\langle T_\Sigma(X)\rangle\rangle\}\,,$$
$$A\{\Sigma_\infty\} = \{s \in A\{\{T_{\Sigma_\infty}(X_\infty)\}\} \mid \mathrm{supp}(s) \text{ is finite}\}\,.$$

A mapping $\tau : A\{\{T_{\Sigma_\infty}(X_\infty)\}\} \to A\{\{T_{\Sigma_\infty}(X_\infty)\}\}$ is called *linear nondeleting recognizable tree transduction* iff, for $s \in A\{\{T_{\Sigma_\infty}(X_\infty)\}\} \cap A\langle\langle T_\Sigma(X)\rangle\rangle$, there exists a linear nondeleting recognizable tree transducer $\mathfrak{T}$ with input alphabet Σ and input leaf alphabet X, such that $\tau(s) = ||\mathfrak{T}||(s)$.

For a family $\mathfrak{L}$ of tree series, we define

$$\mathfrak{M}(\mathfrak{L}) = \{\tau(s) \mid s \in \mathfrak{L} \text{ and } \tau \text{ is a linear nondeleting recognizable tree transduction}\}\,.$$

Observe that, by Theorem 2.4, $\mathfrak{M}(\mathfrak{M}(\mathfrak{L})) = \mathfrak{M}(\mathfrak{L})$.

A family $\mathfrak{L}$ of tree series is said to be *closed under linear nondeleting recognizable tree transductions* iff $\mathfrak{L} = \mathfrak{M}(\mathfrak{L})$.

A family $\mathfrak{L}$ of tree series is called a *recognizable tree cone* iff it satisfies the following conditions:

(i) $\mathfrak{L}$ contains a tree series s such that $(s, x) = 1$ for some $x \in X_\infty$;
(ii) $\mathfrak{L} = \mathfrak{M}(\mathfrak{L})$.

Condition (i) is needed to assure that the forthcoming Theorem 3.2 is valid. In the case of $A = \mathbb{B}$, it can be replaced by the condition that $\mathfrak{L} \neq \{0\}$. This definition and Corollary 14 of Kuich [17] yield the first result of this section.

Theorem 3.1 *$A^{\mathrm{rec}}\{\{T_{\Sigma_\infty}(X_\infty)\}\}$ is a recognizable tree cone.*

The next theorem shows that $A^{\mathrm{rec}}\{\{T_{\Sigma_\infty}(X_\infty)\}\}$ is the smallest recognizable tree cone.

Theorem 3.2 *Let $\mathfrak{L}$ be a recognizable tree cone. Then $A^{\mathrm{rec}}\{\{T_{\Sigma_\infty}(X_\infty)\}\} \subseteq \mathfrak{L}$.*

Proof. By definition, $\mathfrak{L}$ contains a tree series s such that $(s, x) = 1$ for some $x \in X_\infty$. Consider a recognizable tree series r and the linear nondeleting recognizable tree transducer $\mathfrak{T} = (\{q\}, \mu, z_1)$, where $\mu(x) = r$, $\mu(x') = 0$ for $x' \neq x$, $x' \in X_\infty$, and $\mu(f) = 0$, $f \in \Sigma_\infty$. Then $||\mathfrak{T}||(s) = r$. □

For a family $\mathfrak{L}$ of tree series, we define the *yield of* $\mathfrak{L}$ to be

$$\mathrm{yield}(\mathfrak{L}) = \{\mathrm{yd}(s) \mid s \in \mathfrak{L}\}.$$

Here yd is the yield mapping. Observe that $\mathrm{yield}(\mathfrak{L}) \subseteq A\{\{X_\infty^*\}\}$, i. e., $\mathrm{yield}(\mathfrak{L})$ is a family of power series (see Kuich, Salomaa [19], Kuich [14]). Corollary 7.3 of Kuich [15] (see also Bozapalidis [5]) yields the following result.

Theorem 3.3 $\mathrm{yield}(A^{\mathrm{rec}}\{\{T_{\Sigma_\infty}(X_\infty)\}\}) = A^{\mathrm{alg}}\{\{X_\infty^*\}\}$.

Analogous to Bozapalidis, Rahonis [7] we introduce rec-closed families of tree series. A family $\mathfrak{L}$ of tree series is called *rec-closed* whenever the following conditions are satisfied:

(i) $0 \in \mathfrak{L}$.
(ii) If $s_1, s_2 \in \mathfrak{L}$ then $s_1 + s_2 \in \mathfrak{L}$.
(iii) If $f \in \Sigma_\infty$ is of rank $k \geq 0$ and $s_1, \ldots, s_k \in \mathfrak{L}$ then $f(z_1, \ldots, z_k)[s_1/z_1, \ldots$ $\ldots, s_k/z_k] \in \mathfrak{L}$; if $x \in X_\infty$ then $x \in \mathfrak{L}$.
(iv) If $s \in \mathfrak{L}$ and $x \in X_\infty$ then the least solution $\mu x.s$ of the equation $x = s$ is in $\mathfrak{L}$.

Hence, a family $\mathfrak{L}$ of tree series is rec-closed iff $\langle \mathfrak{L}, +, 0, (\bar{f} \mid f \in \Sigma_\infty \cup X_\infty)\rangle$ is a distributive multioperator monoid (see Kuich [15], Theorem 2.9) that satisfies condition (iv), i. e., our "rational" operations are 0, addition, topcatenation and least solutions of equations. (Observe that we do not ask for the closure under substitution as Bozapalidis, Rahonis [7] do for their REC-closed families of tree languages.)

Theorem 3.4 *$A^{\mathrm{rec}}\{\{T_{\Sigma_\infty}(X_\infty)\}\}$ is a rec-closed family of tree series.*

Proof. By Bozapalidis [5] and Kuich [15]. □

We are now ready to introduce full AFTs. We use the notation $\mathfrak{F}(\mathfrak{L})$, $\mathfrak{L}$ a family of tree series, for the smallest rec-closed family of tree series that is closed under linear nondeleting recognizable tree transductions and contains $\mathfrak{L}$. A family $\mathfrak{L}$ of tree series is called *full AFT* iff $\mathfrak{L} = \mathfrak{F}(\mathfrak{L})$.

Theorem 3.5 *$A^{\text{rec}}\{\{T_{\Sigma_\infty}(X_\infty)\}\}$ is a full AFT.*

Theorem 3.6 *$A^{\text{alg}}\{\{T_{\Sigma_\infty}(X_\infty)\}\}$ is a rec-closed family of tree series.*

Proof. By Corollary 28 of Bozapalidis [6]. □

We now show that the yield of a rec-closed family of tree series is a rationally closed family of power series.

Theorem 3.7 *Let $\mathfrak{L}$ be a rec-closed family of tree series. Then* yield$(\mathfrak{L})$ *is closed under addition, multiplication and star and contains 0 and 1.*

Proof. (i) Let $r_1, r_2 \in \text{yield}(\mathfrak{L})$. Then there exist $s_1, s_2 \in \mathfrak{L}$ auch that $\text{yd}(s_i) = r_i$, $i = 1, 2$. Sincer $\mathfrak{L}$ is closed under addition, $s = s_1 + s_2 \in \mathfrak{L}$ and $\text{yd}(s) = r_1 + r_2 \in \text{yield}(\mathfrak{L})$. Since $\mathfrak{L}$ is closed under top-catenation, $s' = \bullet(s_1, s_2) \in \mathfrak{L}$ and $\text{yd}(s') = r_1 r_2 \in \text{yield}(\mathfrak{L})$.

(ii) Let $s \in \mathfrak{L}$ and assume that $x \in X_\infty$ does not appear in s. Consider the equation $x = \bullet(s, x) + e$. Its least solution $\mu x.(\bullet(s, x) + e)$ is in $\mathfrak{L}$. Hence, the least solution $\mu x.\text{yd}(\bullet(s, x) + e) = \mu x.(\text{yd}(s)x + \varepsilon) = \text{yd}(s)^*$ of $x = \text{yd}(\bullet(s, x) + e) = \text{yd}(s)x + \varepsilon$ is in $\text{yield}(\mathfrak{L})$. Moreover, $\text{yd}(0) = 0$ and $0^* = 1$ are in $\text{yield}(\mathfrak{L})$. □

We will show that, for a full AFT $\mathfrak{L}$, $\text{yield}(\mathfrak{L})$ is a full AFP. Here a full AFP is defined as follows. Let $\bar{\mathfrak{F}}(\bar{\mathfrak{L}})$, $\bar{\mathfrak{L}}$ a family of power series, be the smallest rationally closed family of power series that is closed under rational transductions and contains $\bar{\mathfrak{L}}$. A family $\bar{\mathfrak{L}}$ of power series is called *full AFP* iff $\bar{\mathfrak{L}} = \bar{\mathfrak{F}}(\bar{\mathfrak{L}})$. (See Kuich, Salomaa [19] and Kuich [14].) A lemma on algebraic representations and a theorem on algebraic cones are needed before the the main result of this section. (Kuich [14], below Theorem 7.10, gives the definition of an algebraic representation.)

Lemma 3.8 *Let ν be an algebraic representation defined by*

$$\nu : X \to (A^{\text{alg}}\langle\langle X'^* \rangle\rangle)^{Q \times Q}.$$

Then there exists a linear nondeleting recognizable tree representation μ with state set $Q \times Q$ mapping $\Sigma \cup X$ into matrices with entries in $A^{\text{rec}}\langle\langle T_{\Sigma'}(X' \cup Z)\rangle\rangle$, $\Sigma' = \{\bullet, e\}$, such that, for all $s \in A\langle\langle T_\Sigma(X)\rangle\rangle$ and $q_1, q_2 \in Q$,

$$\text{yd}(\mu(s)_{(q_1,q_2)}) = \nu(\text{yd}(s))_{q_1,q_2}.$$

Proof. We construct μ:

(i) For $x \in X$ and $q_1, q_2 \in Q$ we construct $\mu(x)_{(q_1,q_2)}$ according to Theorem 3.3 with the property that $\text{yd}(\mu(x)_{(q_1,q_2)}) = \nu(x)_{q_1,q_2}$.

(ii) For $f \in \Sigma_0$ and $q_1, q_2 \in Q$ we define $\mu(f)_{(q_1,q_2)} = \delta_{q_1,q_2} e$, where δ is the Kronecker symbol; hence, $\text{yd}(\mu(f)_{(q_1,q_2)}) = \delta_{q_1,q_2}\varepsilon = \nu(\varepsilon)_{q_1,q_2}$.

(iii) For $f \in \Sigma_k$, $k \geq 1$, and $q_1, q_2, r_1, \ldots, r_k, s_1, \ldots, s_k \in Q$, we define

$$\mu(f)_{(q_1,q_2),((r_1,s_1),\ldots,(r_k,s_k))} = \delta_{q_1,r_1}\delta_{s_1,r_2}\cdots\delta_{s_{k-1},r_k}\delta_{s_k,q_2} \bullet (z_1, \bullet(z_2, \bullet(\ldots \bullet (z_{k-1}, z_k)\ldots))).$$

First, for a tree $t \in T_\Sigma(X)$ we show that $\mathrm{yd}(\mu(t)_{(q_1,q_2)}) = \nu(\mathrm{yd}(t))_{q_1,q_2}$, $q_1, q_2 \in Q$. The proof is by induction on the structure of trees in $T_\Sigma(X)$. The induction basis is true by (i) and (ii). Let now $t = f(t_1, \ldots, t_k)$, $f \in \Sigma_k$, $k \geq 1$, $t_1, \ldots, t_k \in T_\Sigma(X)$. Then we obtain, for $q_1, q_2 \in Q$,

$$\begin{aligned}
&\mathrm{yd}(\mu(t)_{(q_1,q_2)}) = \mathrm{yd}(\mu(f(t_1,\ldots,t_k))_{(q_1,q_2)}) = \\
&\mathrm{yd}(\textstyle\sum_{r_1,\ldots,r_k\in Q}\sum_{s_1,\ldots,s_k\in Q}\mu(f)_{(q_1,q_2),((r_1,s_1),\ldots,(r_k,s_k))} \\
&\qquad [\mu(t_1)_{(r_1,s_1)}/z_1,\ldots,\mu(t_k)_{(r_k,s_k)}/z_k]) = \\
&\mathrm{yd}(\textstyle\sum_{r_1,\ldots,r_k\in Q}\sum_{s_1,\ldots,s_k\in Q}\delta_{q_1,r_1}\delta_{s_1,r_2}\ldots\delta_{s_{k-1},r_k}\delta_{s_k,q_2} \\
&\quad \bullet(\mu(t_1)_{(r_1,s_1)}, \bullet(\mu(t_2)_{(r_2,s_2)}, \bullet(\ldots, \bullet(\mu(t_{k-1})_{(r_{k-1},s_{k-1})}, \mu(t_k)_{(r_k,s_k)})\ldots)))) = \\
&\textstyle\sum_{s_1,\ldots,s_{k-1}\in Q}\mathrm{yd}(\mu(t_1)_{(q_1,s_1)})\mathrm{yd}(\mu(t_2)_{(s_1,s_2)})\ldots \\
&\qquad \ldots\mathrm{yd}(\mu(t_{k-1})_{(s_{k-2},s_{k-1})})\mathrm{yd}(\mu(t_k)_{(s_{k-1},q_2)}) = \\
&\textstyle\sum_{s_1,\ldots,s_{k-1}\in Q}\nu(\mathrm{yd}(t_1))_{q_1,s_1}\nu(\mathrm{yd}(t_2))_{s_1,s_2}\ldots \\
&\qquad \ldots\nu(\mathrm{yd}(t_{k-1}))_{s_{k-2},s_{k-1}}\nu(\mathrm{yd}(t_k))_{s_{k-1},q_2} = \\
&\nu(\mathrm{yd}(t_1)\ldots\mathrm{yd}(t_k))_{q_1,q_2} = \nu(\mathrm{yd}(t))_{q_1,q_2}\,.
\end{aligned}$$

Hence, for $s \in A\langle\langle T_\Sigma(X)\rangle\rangle$ and $q_1, q_2 \in Q$,

$$\begin{aligned}
&\mathrm{yd}(\mu(s)_{(q_1,q_2)}) = \textstyle\sum_{t\in T_\Sigma(X)}(s,t)\mathrm{yd}(\mu(t)_{(q_1,q_2)}) = \\
&\textstyle\sum_{t\in T_\Sigma(X)}(s,t)\nu(\mathrm{yd}(t))_{q_1,q_2} = \nu(\mathrm{yd}(s))_{q_1,q_2}\,.
\end{aligned}$$

□

A nonempty family of power series is called *algebraic cone* iff it is closed under algebraic transductions. Observe that each algebraic cone is a (rational) cone, i.e., a family of power series closed under rational transductions.

Theorem 3.9 *Let $\mathfrak{L}$ be a full AFT. Then* yield($\mathfrak{L}$) *is an algebraic cone.*

Proof. Let $s \in \mathfrak{L}$, $s \in A\langle\langle T_\Sigma(X)\rangle\rangle$, $r = \mathrm{yd}(s)$, and $\mathfrak{Z} = (Q, \nu, S, P)$ be an algebraic transducer. We will show that $||\mathfrak{Z}||(r) \in A\langle\langle X'^*\rangle\rangle$ is again in yield($\mathfrak{L}$). Observe that $||\mathfrak{Z}||(r) = S\nu(r)P = \sum_{q_1,q_2\in Q} S_{q_1}\nu(r)_{q_1,q_2}P_{q_2}$, where $S_q, P_q \in A^{\mathrm{alg}}\langle\langle X'^*\rangle\rangle$, $q \in Q$. By Theorem 3.3 there exist $s_q, p_q \in A^{\mathrm{rec}}\langle\langle T_{\Sigma'}(X')\rangle\rangle$, $\bullet, e \in \Sigma'$, such that $\mathrm{yd}(s_q) = S_q$, $\mathrm{yd}(p_q) = P_q$, $q \in Q$. By Lemma 3.8 there exists a linear nondeleting recognizable tree representation μ with state set $Q \times Q$ such that $\mathrm{yd}(\mu(s)_{(q_1,q_2)}) = \nu(r)_{q_1,q_2}$ for all q_1, q_2. Since $\mathfrak{L}$ is rec-closed, $\sum_{q_1,q_2\in Q}\bullet(s_{q_1}, \bullet(\mu(s)_{(q_1,q_2)}, p_{q_2}))$ is in $\mathfrak{L}$. Hence,

$$\begin{aligned}
&\mathrm{yd}(\textstyle\sum_{q_1,q_2\in Q}\bullet(s_{q_1}, \bullet(\mu(s)_{(q_1,q_2)}, p_{q_2}))) = \\
&\textstyle\sum_{q_1,q_2\in Q}\mathrm{yd}(s_{q_1})\mathrm{yd}(\mu(s)_{(q_1,q_2)})\mathrm{yd}(p_{q_2}) = \\
&\textstyle\sum_{q_1,q_2\in Q}S_{q_1}\nu(r)_{q_1,q_2}P_{q_2} = ||\mathfrak{Z}||(r)
\end{aligned}$$

is in yield($\mathfrak{L}$). □

Corollary 3.10 *Let $\mathfrak{L}$ be a full AFT. Then* yield($\mathfrak{L}$) *is a full AFP that is closed under algebraic transductions.*

Consider the macro power series introduced in Kuich [16], Section 6. They are the generalization of the OI languages of Fischer [9] and the indexed languages of Aho [1] to power series. By Theorem 5.8 of Kuich [16], the family of macro power series is the yield of the family of algebraic tree series. In a forthcoming paper, Kuich [18], we will show that $A^{\mathrm{alg}}\{\{T_{\Sigma_\infty}(X_\infty)\}\}$ is closed under linear nondeleting recognizable tree transducers, i. e., is a full AFT. This result will imply that the family of macro power series is a full AFP that is closed under algebraic transductions.

We now turn to the language case, i. e., our basic semiring is now given by $\mathfrak{P}(T_{\Sigma_\infty}(X_\infty))$ ($\mathfrak{P}$ denotes the power set). We use without mentioning the isomorphism between $\mathfrak{P}(T_{\Sigma_\infty}(X_\infty))$ and $\mathbb{B}\langle\langle T_{\Sigma_\infty}(X_\infty)\rangle\rangle$.

A family $\mathfrak{L} \neq \{\emptyset\}$ of tree languages is called *rec-closed* iff $\langle \mathfrak{L}, \cup, \emptyset, (\bar{f} \mid f \in \Sigma_\infty \cup X_\infty)\rangle$ is a distribute multioperator monoid (see Kuich [15]) that satisfies the following condition:

> If $L \in \mathfrak{L}$ and $x \in X_\infty$ then the least solution $\mu x.L$ of the tree language equation $x = L$ is in $\mathfrak{L}$.

Define $\hat{\mathfrak{F}}(\mathfrak{L})$ to be the smallest rec-closed family of tree languages that is closed under linear nondeleting recognizable tree transductions and contains $\mathfrak{L}$. A family $\mathfrak{L}$ of tree languages is called *full abstract family of tree languages* iff $\mathfrak{L} = \hat{\mathfrak{F}}(\mathfrak{L})$.

We now connect our full abstract families of tree languages with full AFLs (see Salomaa [20], Ginsburg [12] and Berstel [2]).

Theorem 3.11 *Let $\mathfrak{L}$ be a full abstract family of tree languages. Then* yield($\mathfrak{L}$) *is a full AFL that is closed under algebraic transductions.*

A substitution σ is called *context-free* iff $\sigma(x)$ is a= context-free language for each $x \in X$.

Corollary 3.12 *Let $\mathfrak{L}$ be a full abstract family of tree languages. Then* yield($\mathfrak{L}$) *is a full AFL that is closed under context-free substitutions.*

Open question: What is the relation of full abstract families of tree languages and sheaves (Bozapalidis, Rahonis [7])?

References

1. Aho, A. V.: Indexed grammars—an extension of context-free grammars. JACM 15(1968) 647–671.
2. Berstel, J.: Transductions and Context-Free Languages. Teubner, 1979.
3. Berstel, J., Reutenauer, C.: Recognizable formal power series on trees. Theor. Comput. Sci. 18(1982) 115–148.
4. Bozapalidis, S.: Alphabetic tree relations. Theoret. Comput. Sci. 99(1992) 177–211.

5. Bozapalidis, S.: Equational elements in additive algebras. Technical Report, Aristotle University of Thessaloniki, 1997.
6. Bozapalidis, S.: Context-free series on trees. Preprint, Aristotle University of Thessaloniki, February 1998.
7. Bozapalidis, S., Rahonis, G.: On two families of forests. Acta Informatica 31(1994) 235–260.
8. Engelfriet, J.: Bottom-up and top-down tree transformations— a comparison. Math. Systems Theory 9(1975) 198–231.
9. Fischer M. J.: Grammars with macro-like productions. 9th Annual Symposium on Switching and Automata Theory, 1968, 131–142.
10. Gécseg, F., Steinby, M.: Tree Automata. Akademiai Kiado, 1984.
11. Gécseg, F., Steinby, M.: Tree Languages. In: Handbook of Formal Languages (Eds.: G. Rozenberg and A. Salomaa), Springer, 1997, Vol. 3, Chapter 1, 1–68.
12. Ginsburg, S.: Algebraic and Automata-Theoretic Properties of Formal Languages. North-Holland, 1975.
13. Ginsburg, S., Greibach, S.: Abstract families of languages. In: Studies in Abstract Families of Languages. Mem. Am. Math. Soc. 87(1969) 1–32.
14. Kuich, W.: Semirings and formal power series: Their relevance to formal languages and automata theory. In: Handbook of Formal Languages (Eds.: G. Rozenberg and A. Salomaa), Springer, 1997, Vol. 1, Chapter 9, 609–677.
15. Kuich, W.: Formal power series over trees. In: Proceedings of the 3rd International Conference Developments in Language Theory (S. Bozapalidis, ed.), Aristotle University of Thessaloniki, 1998, 61–101.
16. Kuich, W.: Pushdown tree automata, algebraic tree systems, and algebraic tree series. Preprint, Technische Universität Wien, 1998.
17. Kuich, W.: Tree transducers and formal tree series. Acta Cybernetica, to appear.
18. Kuich, W.: Full abstract families of tree series II. In preparation.
19. Kuich, W., Salomaa, A.: Semirings, Automata, Languages. EATCS Monographs on Theoretical Computer Science, Vol. 5. Springer, 1986.
20. Salomaa, A.: Formal Languages. Academic Press, 1973.

Linear Automata, Rational Series and a Theorem of Fine and Wilf

Giovanna Melideo, Cesidia Pasquarelli, and Stefano Varricchio

Summary. We consider partial functions from a free monoid A^* to a field K and define a natural notion of periodicity in terms of linear semiautomata. We introduce the notion of greatest common divisor of two linear semiautomata and prove its existence and uniqueness. If a partial function is periodic with respect to two linear semiautomata $\mathcal{A}_1$ and $\mathcal{A}_2$ then, under suitable conditions on its domain, it is also periodic with respect to the greatest common divisor of $\mathcal{A}_1$ and $\mathcal{A}_2$. As a corollary of this result one obtains a generalization of Fine and Wilf's periodicity theorem and a stronger version of Eilenberg Equality Theorem for rational series.

1 Introduction

In this paper we shall present some recent results [12] that give a generalization of the Fine and Wilf periodicity theorem in terms of linear automata and (partial) formal power series.

A word is periodic of period p if it presents the same letter in any two positions which differ by p. The Fine and Wilf theorem states that if a word w has two periods p and q and $|w| \geq p + q - \gcd(p, q)$, then w has period $\gcd(p, q)$ (cf. [11]. A weaker version of the Fine and Wilf [4] theorem can be stated as follows. Let m and t be two infinite words having periods p and q, respectively. If m and t have a common prefix of length $\geq p + q - \gcd(p, q)$, then $m = t$.

The notion of periodicity can be defined for k-ary tree in terms of finite semiautomata. In fact, a finite semiautomaton, with an initial state q_0, induces a periodicity on a tree requiring that any two nodes u, v (u and v are words over the branching alphabet of the tree), with $q_0 u = q_0 v$, must have the same label.

Giammarresi et al. [5–7] introduced a natural notion of a greatest common divisor of two semiautomata, and proved that any tree which has two semiautomata as periodicities, under certain conditions on its domain, has also a periodicity in terms of the greatest common divisor of the two semiautomata.

We will extend the previous results and notions to a different context in which one considers labels which are not letters but rational numbers or, more in general, elements of a field; in this way one can consider (partial) functions from a free monoid into a field K, i.e. formal series in non-commutative variables [1,14].

As it will become clear in the paper, our notion of periodicity will lead us to consider rational series that have been widely investigated in Algebra and Formal Language Theory. Usually rational series are represented using multiplicity automata [3] or linear representations [1].

In this paper we consider linear automata and semiautomata [8,9,13] that give another representation of rational series.

Considering the natural notion of morphism between linear semiautomata, we give a natural notion of divisibility between linear semiautomata. This notion allows us to define the greatest common divisor (gcd) of two linear semiautomata and prove its existence and uniqueness. We then give a notion of periodicity in terms of linear semiautomata using linear dependencies and we prove that if a function has periodicities in terms of two linear semiautomata then, under suitable conditions on its domain, it also has a periodicity in terms of the greatest common divisor of them.

This result extends that of Fine and Wilf for words [4,11] and that of Giammarresi et al. for k-ary labeled trees [5–7].

As a consequence of our result we can easily prove classical theorems of rational series theory, in particular the uniqueness (up to isomorphisms) of the minimal linear automaton of a rational series. Moreover, we give an improvement of the celebrated Eilenberg Equality Theorem [3].

In the last part of the paper we show how our main result can be seen as an effective generalization of Fine and Wilf periodicity theorem. In fact we will obtain a stronger version of this result that will be expressed in terms of linear recurrence relations.

2 Rational series and linear automata

We briefly present rational series [1,3,14]. Let K be a field and A^* be the free monoid over the finite alphabet A. An application $S : A^* \to K$ is called a *formal series* with (non-commuting) variables in A and coefficients in K. We recall that a formal series S is called *recognizable* or *rational* if there exists a positive integer n, a row-vector $\lambda \in K^{1\times n}$, a column-vector $\gamma \in K^{n\times 1}$, and a morphism $\mu : A^* \to K^{n\times n}$ w.r.t. the row-by-column product such that, for any $w \in A^*$, $S(w) = \lambda\mu(w)\gamma$. The triplet (λ, μ, γ) is called a *linear representation* of S of dimension n.

Now we introduce some notions on linear automata [8,9,13]. From now on, we will always consider vector spaces over K.

A *linear semiautomaton* $\mathcal{A}$ of dimension n is a 4-tuple (Q, A, δ, q_0) where:

- Q is a vector space of dimension n;
- A is a finite alphabet;
- $\delta : Q \times A \to Q$ is a transition function such that for any $a \in A$, the map $\delta_a : Q \to Q$, with $\delta_a(q) = \delta(q, a)$, is a linear application;
- $q_0 \in Q$ is the initial vector.

Now, a *linear automaton* of dimension n is a 5-tuple $(Q, A, \delta, q_0, \psi)$ where (Q, A, δ, q_0) is a linear semiautomaton of dimension n and $\psi : Q \to K$ is a linear application.

We will sometimes denote a linear automaton as $(\mathcal{A}, \psi)$, where $\mathcal{A}$ is a linear semiautomaton. We extend the application δ on words in the usual way: $\delta(q, \epsilon) = q$, and for any $w \in A^*$ and $a \in A$ $\delta(q, wa) = \delta(\delta(q, w), a)$. Moreover, we shall always write qw instead of $\delta(q, w)$ and $q\psi$ instead of $\psi(q)$. We can associate a formal series with any linear automaton. In fact, given $\mathcal{A} = (Q, A, \delta, q_0, \psi)$, we define the *behavior* of $\mathcal{A}$ as the formal series $S_{\mathcal{A}} : A^* \to K$ such that for any $w \in A^*$, $S_{\mathcal{A}}(w) = (q_0 w)\psi$. Thus, we say that two linear automata on the same alphabet are *equivalent* if their behaviors coincide. Moreover, we say that a linear semiautomaton (Q, A, δ, q_0) is *complete* if the vector space Q is generated by the set $\{q_0 v \mid v \in A^*\}$, and then extend this concept to the corresponding linear automaton $(\mathcal{A}, \psi)$. We remark that, given a linear automaton, it is always possible to construct an equivalent complete one. Consequently, from this point on, we will only concern ourselves with complete linear automata and semiautomata. Any linear representation of dimension n can be regarded as a linear automaton of the same dimension over the vector space K^n. Conversely, one can easily prove that to any linear automaton $\mathcal{A}$ one can associate a linear representation (λ, μ, γ) of the same dimension such that the formal series $S(w) = \lambda\mu(w)\gamma$ coincides with the behavior of $\mathcal{A}$.

3 Greatest common divisor of linear semiautomata

In this section we define the notion of greatest common divisor of two linear semiautomata and prove its existence and uniqueness. First of all we need to introduce a suitable divisibility relation in the class of linear semiautomata.

Let $\mathcal{A} = (Q, A, \delta, q_0)$ be a linear semiautomaton and Q' a vector subspace of Q. We say that Q' is *stable* or *invariant* if

$$\text{for any } a \in A \text{ and } q \in Q', \; qa \in Q'. \tag{1}$$

We denote the *quotient* linear semiautomaton of $\mathcal{A}$ with respect to Q' as $(Q/Q', A, \delta', [q_0])$ and define the transition function in the following way: for any $[q] \in Q/Q'$ and $a \in A$, $\delta'([q], a) = [q]a = [qa]$. The stability condition assures that the function δ' is well defined. Let us consider two linear semiautomata $\mathcal{A} = (Q, A, \delta, q_0)$ and $\mathcal{B} = (\bar{Q}, A, \bar{\delta}, p_0)$. We can define a *morphism* from $\mathcal{A}$ to $\mathcal{B}$ as a homomorphism from Q to $\bar{Q}$ such that $q_0\varphi = p_0$, and for any $q \in Q$ and $a \in A$, $(qa)\varphi = (q\varphi)a$. If such a homomorphism is both injective and surjective, then it is called an *isomorphism* and we say that the two linear semiautomata are *isomorphic*. We note that the inverse of an isomorphism between linear semiautomata is an isomorphism too. We can extend the notion of morphism from linear semiautomata to linear automata in a natural way: if $\mathcal{B}_1 = (\mathcal{A}_1, \psi_1) = (Q_1, A, \delta_1, q_1, \psi_1)$ and $\mathcal{B}_2 = (\mathcal{A}_2, \psi_2) = (Q_2, A, \delta_2, q_2, \psi_2)$

are two linear automata, we say that $\varphi : Q_1 \to Q_2$ is a *morphism* from $\mathcal{B}_1$ to $\mathcal{B}_2$ if φ is a morphism from $\mathcal{A}_1$ to $\mathcal{A}_2$ such that $q\psi_1 = (q\varphi)\psi_2$, for any $q \in Q_1$. If such morphism is both injective and surjective, then it is called an *isomorphism* and we say that the two linear automata are *isomorphic*. Also in this case, we remark that the inverse of an isomorphism between linear automata is an isomorphism too.

We introduce now the notion of divisibility between semiautomata. Given two linear semiautomata $\mathcal{A}$ and $\mathcal{B}$, we say that $\mathcal{B}$ *divides* $\mathcal{A}$, and we write $\mathcal{B}/\mathcal{A}$, if there exists a surjective morphism from $\mathcal{A}$ to $\mathcal{B}$. We remark that if $\mathcal{B}/\mathcal{A}$ and φ is an epimorphism from $\mathcal{B}$ to $\mathcal{A}$, then $\mathcal{B}$ is isomorphic to the quotient of $\mathcal{A}$ through the invariant subspace $Q' = \ker(\varphi)$. Let $\mathcal{A}_1$, $\mathcal{A}_2$ be two linear semiautomata. We say that a linear semiautomaton $\mathcal{B}$ is a *greatest common divisor* of $\mathcal{A}_1$ and $\mathcal{A}_2$, and we write $\mathcal{B} = \gcd(\mathcal{A}_1, \mathcal{A}_2)$, if $\mathcal{B}/\mathcal{A}_1$, $\mathcal{B}/\mathcal{A}_2$ and for any linear semiautomaton $\mathcal{B}'$ such that $\mathcal{B}'/\mathcal{A}_1$ and $\mathcal{B}'/\mathcal{A}_2$, one has $\mathcal{B}'/\mathcal{B}$.

Proposition 1. *The greatest common divisor of two linear semiautomata is unique up to isomorphisms.*

Let $\mathcal{A}_1 = (Q_1, A, \delta_1, q_1)$, $\mathcal{A}_2 = (Q_2, A, \delta_2, q_2)$ be two linear semiautomata, with Q_1 and Q_2 vector spaces of dimensions n and m, respectively. Let us consider the semiautomaton $(Q_1 \times Q_2, A, \delta_1 \times \delta_2, (q_1, \underline{0}))$ where the transition function is defined as follows: $(p, q)a = (pa, qa)$, for any $(p, q) \in Q_1 \times Q_2$ and $a \in A$. Let Q' be the subspace of $Q_1 \times Q_2$ generated by the set $\{(q_1, -q_2)w \mid w \in A^*\}$. Obviously, Q' is stable. We can then consider the quotient linear semiautomaton

$$\mathcal{B} = ((Q_1 \times Q_2)/Q', A, \bar{\delta}, [(q_1, \underline{0})]).$$

We remark that for any $[(p, q)] \in (Q_1 \times Q_2)/Q'$ and $w \in A^*$, $[(p, q)]w = [(pw, qw)]$. Furthermore, we note that $[(q_1, \underline{0})] = [(\underline{0}, q_2)]$. In fact, $(q_1, \underline{0}) - (\underline{0}, q_2) = (q_1, -q_2) \in Q'$, that is $(q_1, \underline{0})$ and $(\underline{0}, q_2)$ are equivalent vectors. This fact trivially implies that

$$[(q_1 w, \underline{0})] = [(\underline{0}, q_2 w)], \ \forall w \in A^*. \tag{2}$$

Proposition 2. *Let $\mathcal{A}_1 = (Q_1, A, \delta_1, q_1)$, $\mathcal{A}_2 = (Q_2, A, \delta_2, q_2)$ be two linear semiautomata, with Q_1, Q_2 vector spaces of dimensions n and m, respectively. Then the linear semiautomaton $\mathcal{B}$ defined above is the greatest common divisor of $\mathcal{A}_1$ and $\mathcal{A}_2$.*

Proof. We will show that $\mathcal{B}/\mathcal{A}_1$, omitting the proof for $\mathcal{B}/\mathcal{A}_2$ which is symmetric. In order to show that $\mathcal{B}/\mathcal{A}_1$, we will prove the existence of an epimorphism between the two linear semiautomata. Let $\psi : Q_1 \to (Q_1 \times Q_2)/Q'$ be the function such that $q\psi = [(q, \underline{0})]$ for any $q \in Q_1$. The function ψ is trivially a linear application between the two vector spaces. Moreover, $q_1\psi = [(q_1, \underline{0})]$, and for any $q \in Q_1$ and $a \in A$, $(q\psi)a = [(q, \underline{0})]a = [(qa, \underline{0})] = (qa)\psi$. Therefore, ψ is a morphism from $\mathcal{A}_1$ to $\mathcal{B}$. The function ψ is surjective, i.e. every element in $(Q_1 \times Q_2)/Q'$ has an inverse image through ψ in Q_1. In fact, since $\mathcal{A}_1$ and $\mathcal{A}_2$ are complete linear semiautomata, there exist two bases for Q_1 and Q_2 such that all the vectors of the base of Q_1 can be written as $q_1w_1, \ldots, q_1w_n$, and those of the base of Q_2 as $q_2u_1, \ldots, q_2u_m$. Then, by using Eq. (2), one can prove that for any $[(p,q)] \in (Q_1 \times Q_2)/Q'$, there exists $\bar{p} \in Q_1$ such that $[(p,q)] = [(\bar{p}, \underline{0})] = \bar{p}\psi$. It remains to show that $\mathcal{B}$ is the greatest common divisor of $\mathcal{A}_1$ and $\mathcal{A}_2$. Then, let $\mathcal{B}' = (\tilde{Q}, A, \tilde{\delta}, \tilde{q})$ be a linear semiautomaton such that $\mathcal{B}'/\mathcal{A}_1$ and $\mathcal{B}'/\mathcal{A}_2$. That means that there exist two epimorphisms φ, ψ with $\varphi : Q_1 \to \tilde{Q}$, $\psi : Q_2 \to \tilde{Q}$ such that $q_1\varphi = \tilde{q} = q_2\psi$. We note that the last condition trivially implies that for any $w \in A^*$, $(q_1\varphi)w = (q_2\psi)w$, that is

$$\text{for any } w \in A^*,\ (q_1w)\varphi = (q_2w)\psi. \tag{3}$$

If we show that $\mathcal{B}'/\mathcal{B}$, the thesis follows. Consider $\tilde{\chi} : Q_1 \times Q_2 \to \tilde{Q}$ such that $(p,q)\tilde{\chi} = p\varphi + q\psi$. The function $\tilde{\chi}$ is a linear application since it is a linear combination of linear applications. We show that $Q' \subseteq \ker(\tilde{\chi})$. Since any generator $(q_1w, -q_2w)$ of Q' is such that $(q_1w, -q_2w)\tilde{\chi} = (q_1w)\varphi - (q_2w)\psi = \underline{0}$ (by Eq. (3)), then we can conclude that all the linear combinations of the generators (that is the elements of Q') are in $\ker(\tilde{\chi})$. We define now $\chi : (Q_1 \times Q_2)/Q' \to \tilde{Q}$ such that $[(p,q)]\chi = (p,q)\tilde{\chi}$. The function χ is well defined. In fact, let (p,q) and (p',q') be vectors in the same class; then $(p,q)-(p',q') \in Q' \subseteq \ker(\tilde{\chi})$. Thus, $[(p,q)]\chi - [(p',q')]\chi = (p,q)\tilde{\chi} - (p',q')\tilde{\chi} = \underline{0}$, that is $[(p,q)]\chi = [(p',q')]\chi$. We have shown that χ does not depend on the particular element of the class. Moreover, one can check that χ is a homomorphism of the two vector spaces $(Q_1 \times Q_2)/Q'$ and $\tilde{Q}$. We can conclude that χ is a morphism between the two linear semiautomata, in fact: $[(q_1, \underline{0})]\chi = \tilde{q}$, trivially; furtheremore $([(p,q)]a)\chi = [(pa, qa)]\chi = (pa)\varphi + (qa)\psi = (p\varphi)a + (q\psi)a = ([(p,q)]\chi)a$. We prove now that χ is a surjective morphism. It suffices to show that any vector $\tilde{p}$ of a base of $\tilde{Q}$ has an inverse image by χ. By the completeness assumption on $\tilde{Q}$, we can choose a base in which each element $\tilde{p}$ is such that $\tilde{p} = \tilde{q}u$ for some $u \in A^*$. Therefore $[(q_1u, \underline{0})]\chi = (q_1u)\varphi + \underline{0} = (q_1\varphi)u = \tilde{q}u = \tilde{p}$.

4 The periodicity theorem

A natural notion of periodicity has been introduced for k-ary labelled trees [6]. A tree t is represented as a pair $t = (T, f)$, where T is a prefix-closed set of words over A and $f : T \to \Gamma$ is a map (Γ is the alphabet of the labels). The tree $t = (T, f)$ has periodicity in terms of a finite semiautomaton with initial state q_0 if the following condition is satisfied: for any $u, v \in T$, $q_0u = q_0v \Rightarrow f(u) = f(v)$, i.e. two nodes that reach the same state must have the same label.

Since we are considering linear semiautomata and partial functions from A^* to K, it is quite natural to express periodicity in terms of linear dependencies. Thus, if $\mathcal{A} = (Q, A, \delta, q_0)$ is a linear semiautomaton and $S : T \to K$

is a partial function ($T \subseteq A^*$), we say that S is $\mathcal{A}$*-linear* or $\mathcal{A}$*-periodic* if for any $u_0, \ldots, u_{n-1}$ in T and $\alpha_0, \ldots, \alpha_{n-1}$ in K, $\sum_{i=0}^{n-1} \alpha_i q_0 u_i = \underline{0}$ implies that $\sum_{i=0}^{n-1} \alpha_i S(u_i) = 0$.

Proposition 3. *Let $\mathcal{A} = (Q, A, \delta, q_0)$ be a linear semiautomaton and $S : T \rightarrow K$ be a map , with $T \subseteq A^*$. The function S is $\mathcal{A}$-periodic if and only if there exists a linear application $\psi : Q \rightarrow K$ such that S coincides in T with the behavior of the linear automaton $(\mathcal{A}, \psi)$, i.e. for any $u \in T$, $S(u) = (q_0 u)\psi$.*

We remark that a formal series $S : A^* \rightarrow K$ is rational if and only if there exists a linear semiautomaton $\mathcal{A}$ such that S is $\mathcal{A}$-periodic. We give now the notion of a complete set of words for two linear semiautomata by adapting an analogous notion for finite semiautomata [5–7].

We say that a subset P of A^* is a *complete* set with respect to two linear semiautomata $\mathcal{A}_1 = (Q_1, A, \delta_1, q_1)$ and $\mathcal{A}_2 = (Q_2, A, \delta_2, q_2)$ if the subspace Q' of $Q_1 \times Q_2$ generated by the set $\{(q_1, -q_2)w \mid w \in A^*\}$ is generated by the set $\{(q_1, -q_2)w \mid w \in P\}$. We remark that, given two linear semiautomata $\mathcal{A}_1$ and $\mathcal{A}_2$, one can effectively construct a subset P of A^* which is complete with respect to them. In fact it can be obtained by setting $P = \{\epsilon\}$ and upgrading it as follows: for all w in P and for all a in A, if $(q_1, -q_2)wa$ is not a linear combination of the vectors $\{(q_1, -q_2)u \mid u \in P\}$, then wa must be inserted in P. One can prove that the set P, so constructed, is complete. We remark that P is a prefix-closed finite set by construction. Moreover, from the definition of Q' and the fact that the elements in V_P are linearly independent, we know that $\dim(Q') = \mathrm{card}(P)$. Since $\dim((Q_1 \times Q_2)/Q') = \dim(Q_1) + \dim(Q_2) - \dim(Q')$, one has

$$\mathrm{card}(P) = \dim(Q_1) + \dim(Q_2) - \dim((Q_1 \times Q_2)/Q').$$

Lemma 1. *Let $\mathcal{A}_1 = (Q_1, A, \delta_1, q_1)$ and $\mathcal{A}_2 = (Q_2, A, \delta_2, q_2)$ be two linear semiautomata. If any $\mathcal{A}_1$-periodic function is also $\mathcal{A}_2$-periodic, then linear dependencies are preserved, i.e. for any $v, v_i \in A^*, \alpha_i \in K, i = 1, \ldots, n$:*

$$q_2 v = \alpha_1 q_2 v_1 + \cdots + \alpha_n q_2 v_n \ \Rightarrow q_1 v = \alpha_1 q_1 v_1 + \cdots + \alpha_n q_1 v_n.$$

Proposition 4. *Let $\mathcal{A}_1$ and $\mathcal{A}_2$ be two linear semiautomata. Then $\mathcal{A}_1/\mathcal{A}_2$ if and only if any $\mathcal{A}_1$-linear function is also $\mathcal{A}_2$-linear.*

Theorem 1. *Let $S : T \rightarrow K$, with $T \subseteq A^*$ and let $\mathcal{A}_1$ and $\mathcal{A}_2$ be two linear semiautomata. If S is $\mathcal{A}_1$-periodic and $\mathcal{A}_2$-periodic and if T contains a complete set of words with respect to $\mathcal{A}_1$ and $\mathcal{A}_2$, then S is also $\mathcal{B}$-periodic, where $\mathcal{B} = \gcd(\mathcal{A}_1, \mathcal{A}_2)$.*

Proof. Let $\mathcal{A}_1 = (Q_1, A, \delta_1, q_1)$ and $\mathcal{A}_2 = (Q_2, A, \delta_2, q_2)$ with Q_1 and Q_2 vector spaces of dimension n and m, respectively. In order to show that $\mathcal{B} = ((Q_1 \times Q_2)/Q', A, \bar{\delta}, [(q_1, \underline{0})])$ determines a new periodicity for the function S, we have to define a linear application $\psi : (Q_1 \times Q_2)/Q' \rightarrow K$ such that $S(w) =$

$([(q_1, \underline{0})]w)\psi$ for any $w \in T$. Since S is $\mathcal{A}_1$-periodic and $\mathcal{A}_2$-periodic, there exist $\psi_1 : Q_1 \to K$, $\psi_2 : Q_2 \to K$ such that for any $w \in T$, $S(w) = (q_1 w)\psi_1$ and $S(w) = (q_2 w)\psi_2$. Then we define a morphism $\tilde{\psi} : Q_1 \times Q_2 \to K$ such that for any $(p, q) \in Q_1 \times Q_2$, $(p, q)\tilde{\psi} = p\psi_1 + q\psi_2$. We show that $Q' \subseteq \ker(\tilde{\psi})$, that is for any $(p, q) \in Q'$, $(p, q)\tilde{\psi} = 0$. It suffices to show that this is true for the vectors of a base of Q'. Since T contains a complete set by hypothesis, we can choose a base such that each of its vectors has the shape $(q_1 w, -q_2 w)$, for some $w \in T$. Then, $(q_1 w, -q_2 w)\tilde{\psi} = (q_1 w)\psi_1 - (q_2 w)\psi_2 = S(w) - S(w) = 0$. This allows us to define $\psi : (Q_1 \times Q_2)/Q' \to K$ such that $[(p, q)]\psi = (p, q)\tilde{\psi}$. The function ψ is well defined because it does not depend on the representative of the class. In fact, if $(p, q), (p', q')$ are in the same class, we have that $(p - p', q - q') \in Q' \Rightarrow (p - p', q - q') \in \ker(\tilde{\psi}) \Rightarrow (p - p', q - q')\tilde{\psi} = 0$. Thus $[(p, q)]\psi - [(p', q')]\psi = (p, q)\tilde{\psi} - (p', q')\tilde{\psi} = (p - p', q - q')\tilde{\psi} = 0$. Since for any $w \in T$ $([(q_1, \underline{0})]w)\psi = [(q_1 w, \underline{0})]\psi = (q_1 w)\psi_1 = S(w)$, we can conclude that S is also $\mathcal{B}$-periodic.

Corollary 1. *Let $\mathcal{B}_1 = (\mathcal{A}_1, \psi_1)$ and $\mathcal{B}_2 = (\mathcal{A}_2, \psi_2)$ be two linear automata. Then they are equivalent if and only if their behaviors coincide on all the strings of length* $\leq \dim(\mathcal{A}_1) + \dim(\mathcal{A}_2) - \dim(\gcd(\mathcal{A}_1, \mathcal{A}_2)) - 1$.

Proof. If $\mathcal{B}_1$ and $\mathcal{B}_2$ are equivalent linear automata, their behaviors coincide for any $w \in A^*$ and then the thesis trivially follows. Conversely, let $\mathcal{A} = (Q, A, \delta, q) = \gcd(\mathcal{A}_1, \mathcal{A}_2)$ and suppose that the behaviors of the two automata coincide on all the strings of length $\leq \dim(\mathcal{A}_1) + \dim(\mathcal{A}_2) - \dim(\mathcal{A}) - 1$. Let T be the set of all the words of length $\leq \dim(\mathcal{A}_1) + \dim(\mathcal{A}_2) - \dim(\mathcal{A}) - 1$ and let S be the restriction of $S_{\mathcal{B}_1}$ and then of $S_{\mathcal{B}_2}$ to T. The function S is, by hypothesis, $\mathcal{A}_1$-linear and $\mathcal{A}_2$-linear. Then, by Theorem 1, S is $\mathcal{A}$-linear, since T must contain a complete set with respect to $\mathcal{A}_1$ and $\mathcal{A}_2$.

Therefore, there exists a linear application $\psi : Q \to K$, such that for any $u \in T$, $(qu)\psi = S(u) = (q_1 u)\psi_1 = (q_2 u)\psi_2$. We observe that, since $\dim(\mathcal{A}_1) \leq \dim(\mathcal{A}_1) + \dim(\mathcal{A}_2) - \dim(\mathcal{A})$, the set T contains a complete set of words w.r.t. to $\mathcal{A}_1$. This means that for any word $u \in A^*$ one has $q_1 u = \sum_{v \in T} \lambda_v q_1 v$, for suitable λ_v's in K. Since $\mathcal{A}$ divides $\mathcal{A}_1$, by Lemma 1 and Proposition 4 one also has $qu = \sum_{v \in T} \lambda_v qv$. From the previous relations one derives that

$$S_{\mathcal{B}_1}(u) = (q_1 u)\psi_1 = \sum_{v \in T} \lambda_v (q_1 v)\psi_1 = \sum_{v \in T} \lambda_v (qv)\psi = (qu)\psi.$$

Similarly one can prove that for any $u \in A^*$ one has $S_{\mathcal{B}_2}(u) = (qu)\psi$ and then $S_{\mathcal{B}_1}(u) = S_{\mathcal{B}_2}(u)$.

The previous result can be seen as a stronger version of Eilenberg's Equality Theorem. In fact, this theorem states that two linear automata $\mathcal{A}_1$ and $\mathcal{A}_2$ are equivalent if and only if their behaviors coincide on all the strings of length less than or equal to $\dim(\mathcal{A}_1) + \dim(\mathcal{A}_2) - 1$. The following result is well known. It is the analogous for linear automata of the theorem on the uniqueness of the minimal linear representation of a rational series (cf. [1]).

Corollary 2. *Let S be a rational series. The minimal linear automaton of S is unique up to isomorphisms.*

Proof. Let $\mathcal{B}_1 = (\mathcal{A}_1, \psi_1)$ and $\mathcal{B}_2 = (\mathcal{A}_2, \psi_2)$ be two minimal linear automata of S. Let $\mathcal{A}_1 = (Q_1, A, \delta_1, q_1)$ and $\mathcal{A}_2 = (Q_2, A, \delta_2, q_2)$ and $\mathcal{A} = (Q, A, \delta, q) = \gcd(\mathcal{A}_1, \mathcal{A}_2)$. By the Theorem 1 the formal series S is $\mathcal{A}$-linear, i.e. there exists a linear application $\psi : Q \to K$ such that S is the behavior of the linear automaton $(\mathcal{A}, \psi)$. From the fact that $\dim(\mathcal{A}) \leq \dim(\mathcal{A}_1)$, $\dim(\mathcal{A}) \leq \dim(\mathcal{A}_2)$ and the hypothesis of minimality of $\mathcal{B}_1$ and $\mathcal{B}_2$, we can derive that $\dim(\mathcal{A}) = \dim(\mathcal{A}_1) = \dim(\mathcal{A}_2)$. Moreover, we know that there exist an epimorphism φ from $\mathcal{A}_1$ to $\mathcal{A}$ and an epimorphism φ' from $\mathcal{A}_2$ to $\mathcal{A}$; since $\mathcal{A}$, $\mathcal{A}_1$, $\mathcal{A}_2$ have the same dimension, φ and φ' are isomorphisms, and then $\mathcal{A}_1$ and $\mathcal{A}_2$ are isomorphic. It remains to show that, if ϕ denotes such isomorphism, $q'\psi_1 = (q'\phi)\psi_2$, for any $q' \in Q_1$.

It suffices to prove this condition on a base for Q_1. To this aim, one can consider a set P of words such that $\{q_1 w \mid w \in P\}$ is a base for the vector space. Then, as S is the behavior of the linear automaton $(\mathcal{A}, \psi)$, for any $u \in A^*$ $(qu)\psi = (q_1 u)\psi_1 = (q_2 u)\psi_2$; this implies that $(q_1 w)\psi_1 = (q_2 w)\psi_2$ for any $w \in P$, and then $(q_1 w)\psi_1 = ((q_1 w)\phi)\psi_2$ for any $w \in P$, where $q_2 w = (q_1 w)\phi$ by definition of morphism between semiautomata. Thus, our claim holds.

5 Linear semiautomata on polynomials

In this section we deal with linear semiautomata in which the transition functions are extended so to allow processing polynomials in non-commuting variables over an alphabet A, that is the formal series $S : A^* \to K$ whose support $\mathrm{Supp}(S) = \{w \in A^* \mid S(w) \neq 0\}$ is a finite set. Let us denote the set of all such polynomials as $K\langle A\rangle$ and consider a linear semiautomaton $\mathcal{A} = (Q, A, \delta, q_0)$. We can extend by linearity the transition function over $K\langle A\rangle$ in the following way: for any $q \in Q$ and $\sum_{i=1}^{n} \lambda_i u_i \in K\langle A\rangle$, $q(\sum_{i=1}^{n} \lambda_i u_i) = \sum_{i=1}^{n} \lambda_i q u_i$. Also in this case we are going to consider complete semiautomata. We remark that, if $\mathcal{A}$ is a complete semiautomaton, then the vector space Q is equal to $\{q_0 p \mid p \in K\langle A\rangle\}$. We note that for any $p, r \in K\langle A\rangle$, one has $q_0(p \cdot r) = (q_0 p)r$. We can consider the following set of polynomials associated with the semiautomaton $\mathcal{A}$:

$$I_{\mathcal{A}} = \{p \mid q_0 p = \underline{0}\}.$$

The following result holds.

Proposition 5. *Given a linear semiautomaton $\mathcal{A}$, the set $I_{\mathcal{A}}$ is a right-ideal for $K\langle A\rangle$.*

One can prove that the right-ideal associated with a linear semiautomaton characterizes its class of isomorphism.

Proposition 6. *Let $\mathcal{A}$ and $\mathcal{B}$ be two linear semiautomata. Then, $\mathcal{A}$ and $\mathcal{B}$ are isomorphic if and only if $I_{\mathcal{A}} = I_{\mathcal{B}}$.*

We try to express the notion of greatest common divisor in terms of right-ideals. At this purpose, we recall that the sum of two right-ideals I_1 and I_2 is the set $I_1 + I_2 = \{p_1 + p_2 \mid p_1 \in I_1, p_2 \in I_2\}$. Trivially, the sum $I_1 + I_2$ is the smallest right-ideal containing both I_1 and I_2.

Proposition 7. *Let $\mathcal{A} = (Q_{\mathcal{A}}, A, q_0, \delta)$ and $\mathcal{B} = (Q_{\mathcal{B}}, A, q'_0, \delta')$ be two linear semiautomata and let $\mathcal{C}$ =gcd$(\mathcal{A}, \mathcal{B})$. Then, $I_{\mathcal{C}} = I_{\mathcal{A}} + I_{\mathcal{B}}$.*

Proof. By Proposition 2, the semiautomaton $\mathcal{C}$ is of the form $((Q_{\mathcal{A}} \times Q_{\mathcal{B}})/Q'$, A,$\bar{\delta}$, $[(q_0, \underline{0})] = [(\underline{0}, q'_0)])$. First we show that $I_{\mathcal{C}} \subseteq I_{\mathcal{A}} + I_{\mathcal{B}}$. By definition, for any $p \in I_{\mathcal{C}}$ we have that $[(q_0, \underline{0})]p = [(q_0 p, \underline{0})] = [(\underline{0}, \underline{0})]$, that is $(q_0 p, \underline{0}) \in Q'$. Since $Q' = \{(q_0, -q'_0)t \mid t \in K\langle A\rangle\}$, there exists a polynomial $\tilde{p}$ such that $(q_0 p, \underline{0}) = (q_0 \tilde{p}, -q'_0 \tilde{p})$. By this equation one has that $q_0 p = q_0 \tilde{p}$, that is $p - \tilde{p} = \hat{p} \in I_{\mathcal{A}}$ and that $\underline{0} = -q'_0 \tilde{p}$, that is $\tilde{p} \in I_{\mathcal{B}}$. This implies that $p = \hat{p} + \tilde{p} \in I_{\mathcal{A}} + I_{\mathcal{B}}$. Conversely, we show that $I_{\mathcal{A}} + I_{\mathcal{B}} \subseteq I_{\mathcal{C}}$. For any $p_1 \in I_{\mathcal{A}}$ and $p_2 \in I_{\mathcal{B}}$, it holds that $[(q_0, \underline{0})](p_1 + p_2) = [(q_0(p_1 + p_2), \underline{0})] = [(q_0 p_1, \underline{0})] + [(q_0 p_2, \underline{0})] = [(q_0 p_2, \underline{0})] =$ (by Eq. (2) generalized over $K\langle A\rangle$) $= [(\underline{0}, q'_0 p_2)] = [(\underline{0}, \underline{0})]$, that is $p_1 + p_2 \in I_{\mathcal{C}}$.

6 Linear recurrence relations

In this section, as application of the previous results, we give a proof of a theorem on finite sequences satisfying two linear recurrence relations (cf. Theorem 3). This result, which is a generalization of Fine and Wilf's theorem, seems to be known to people working on the theory of Padé approximants.

We can transpose on linear recurrence relations the results obtained in the previous sections. This is achieved by showing that we can always identify a linear recurrence relation with a linear semiautomaton over a one-letter alphabet $\{x\}$ and conversely.

Given a semiautomaton over a one-letter alphabet $\mathcal{A} = (Q, \{x\}, \delta, q_0)$, we will represent $\mathcal{A}$ by the triplet (Q, q_0, φ), where $\varphi : Q \to Q$ is the linear application defined as follows: for any $q \in Q, q\varphi = qx$. A *linear recurrence relation of degree* n is defined as a sequence of n coefficients $(\alpha_0, \ldots, \alpha_{n-1})$ in K that we will denote by $\underline{\alpha}$, setting also $\deg(\underline{\alpha}) = n$. Let I be either the set of natural numbers $\mathbf{N}$ or an interval $[0, h]$ of $\mathbf{N}$. We say that a sequence $(t_i)_{i \in I}$ of elements of K satisfies the linear recurrence relation $\underline{\alpha}$ if

$$t_{n+l} = \sum_{j=0}^{n-1} \alpha_j t_{j+l}, \text{ for any } l \geq 0 \text{ and } n + l \in I.$$

We show that a linear semiautomaton $\mathcal{A} = (Q, q_0, \varphi)$ of dimension n can be associated with a linear recurrence relation $(\alpha_0, \ldots, \alpha_{n-1})$ of degree n and conversely. In fact, given the linear semiautomaton $\mathcal{A}$ of dimension n over a one-letter alphabet $\{x\}$, since $\mathcal{A}$ is complete, one can prove that the set $\{q_0, q_0x, \ldots, q_0x^{n-1}\}$ is a base for the vector space Q; hence, since the vector q_0x^n is a linear combination of the elements of the base, there exist coefficients α_j's in K such that $q_0x^n = \sum_{j=0}^{n-1} \alpha_j q_0 x^j$. Consequently, the linear recurrence relation $(\alpha_0, \ldots, \alpha_{n-1})$ can be associated with the linear semiautomaton $\mathcal{A}$.

On the other hand, given a linear recurrence relation $(\alpha_0, \ldots, \alpha_{n-1})$ of degree n we can define the associated linear semiautomaton as (K^n, q_0, φ) where $q_0 = (1, 0, \ldots, 0) \in K^n$, and for any $q \in K^n$, $q\varphi = qM$ with

$$M = \begin{pmatrix} 0 & 1 & 0 & \cdots & 0 \\ 0 & 0 & 1 & \cdots & 0 \\ . & . & . & \cdots & . \\ 0 & 0 & 0 & \cdots & 1 \\ \alpha_0 & \alpha_1 & \alpha_2 & \cdots & \alpha_{n-1} \end{pmatrix} \in K^{n \times n}.$$

Furthermore, it is easy to see that $q_0x^n = \sum_{j=0}^{n-1} \alpha_j q_0 x^j$.

Proposition 8. *Let $\mathcal{A}_1$ and $\mathcal{A}_2$ be two linear semiautomata over a one-letter alphabet. Then $\mathcal{A}_1$ and $\mathcal{A}_2$ are isomorphic if and only if they can be associated with the same linear recurrence relation.*

One can easily prove the following result.

Proposition 9. *Let $\underline{\alpha} = (\alpha_0, \ldots, \alpha_{n-1})$ be the linear recurrence relation associated with a linear semiautomaton $\mathcal{A}$. Then for any linear map $\psi : Q_{\mathcal{A}} \to K$, the behavior of the linear automaton $(\mathcal{A}, \psi)$ is a rational sequence $(a_n)_{n\geq 0}$ of elements in K that satisfies the linear recurrence relation $(\alpha_0, \ldots, \alpha_{n-1})$.*

In the sequel we will denote by $\mathcal{A}_{\underline{\alpha}}$ the linear semiautomaton associated with a linear recurrence relation $\underline{\alpha}$. We remark that if we consider the function $S : T \to K$ where $T = \{\epsilon, x, \ldots, x^{k-1}\}$, that is a prefix-closed subset of $\{x\}^*$, the sequence $(S(x^i))_{i=0,\ldots,k-1}$ satisfies the linear recurrence relation $\underline{\alpha}$ if and only if S is $\mathcal{A}_{\underline{\alpha}}$-periodic.

From the results of the previous section linear semiautomata on one-letter alphabet $\{x\}$ process polynomials in the commutative ring $K[x]$. We remark that in this context the distinction between right-ideals and left-ideals falls, so that we will always deal with (two-sided) ideals. From polynomials rings theory, we know that every ideal I of $K[x]$ is principal, that is there exists a polynomial $p \in I$ such that $I = \langle p \rangle = \{pr \mid r \in K[x]\}$, and that such a polynomial is of minimal degree in the ideal. We will show that this polynomial has a special form in the context of linear recurrence relations. We first recall the notion of ***characteristic polynomial*** of a given linear recurrence relation $\lambda = (\lambda_0, \ldots, \lambda_{p-1})$, that is

$$p_\lambda(x) = x^p - \lambda_{p-1}x^{p-1} - \ldots - \lambda_0.$$

Obviously $\deg(\lambda) = \deg(p_\lambda)$. Moreover, we remark that, if q_0 is the initial state of $\mathcal{A}_\lambda$, then, by definition of linear recurrence relation associated with a linear semiautomaton, it holds that $q_0 p_\lambda = q_0 x^p - q_0\lambda_{p-1}x^{p-1} - \ldots - q_0\lambda_0 = \underline{0}$, i.e. the characteristic polynomial of a linear recurrence relation always belongs to the ideal associated with the relative semiautomaton.

Proposition 10. *Let λ be a linear recurrence relation and $\mathcal{A}_\lambda$ its associated linear semiautomaton. The ideal $I_{\mathcal{A}_\lambda}$ is generated by the characteristic polynomial p_λ of λ.*

Proof. As well known, every ideal I of $K[x]$ is principal and is generated by a polynomial of I of minimal degree. By construction λ is such that p_λ is a polynomial of minimal degree satisfying $q_0 p_\lambda = \underline{0}$. Then p_λ is a polynomial of minimal degree in the ideal $I_{\mathcal{A}_\lambda}$, and thus the thesis follows.

Theorem 2. *If $\mathcal{A}_\gamma = \gcd(\mathcal{A}_\lambda, \mathcal{A}_\mu)$, then $p_\gamma = \gcd(p_\lambda, p_\mu)$.*

Proof. By Proposition 10 we know that p_γ, p_λ, p_μ generate the ideals $I_{\mathcal{A}_\gamma}$, $I_{\mathcal{A}_\lambda}$, $I_{\mathcal{A}_\mu}$ respectively. Therefore, by Proposition 7, we can state that $< p_\gamma >= I_{\mathcal{A}_\gamma} = I_{\mathcal{A}_\lambda} + I_{\mathcal{A}_\mu} =< p_\lambda > + < p_\mu >=< p_\lambda, p_\mu >$. From a well known result about Euclidean rings, we can state that $p_\gamma = \gcd(p_\lambda, p_\mu)$.

Thus we obtain a generalization of Fine and Wilf's periodicity theorem.

Theorem 3. *Let λ and μ be two linear recurrence relations with associated characteristic polynomials p_λ and p_μ respectively, and let p_γ =$\gcd(p_\lambda, p_\mu)$. If a sequence of elements in K satisfies λ and μ and is longer than or equal to $\deg(p_\lambda) + \deg(p_\mu) - \deg(p_\gamma)$, then it satisfies also the linear recurrence relation γ relative to p_γ.*

Proof. Let $\lambda = (\lambda_0, \ldots, \lambda_{n-1})$, $\mu = (\mu_0, \ldots, \mu_{m-1})$ and $\gamma = (\gamma_0, \ldots, \gamma_{d-1})$. One can consider a sequence $\{t_0, \ldots, t_{k-1}\}$ satisfying the two linear recurrence relations λ and μ, with $k \geq n + m - d$. Let $T = \{\epsilon, x, \ldots, x^{k-1}\}$ and let $S : T \to K$ be a partial function such that $S(x^i) = t_i$, for $i = 0, \ldots, k-1$. The set of words T contains a complete set with respect to $\mathcal{A}_\lambda$ and $\mathcal{A}_\mu$ and then, by Theorem 1, S is $\mathcal{A}_\gamma$-linear with $\mathcal{A}_\gamma = \gcd(\mathcal{A}_\lambda, \mathcal{A}_\mu)$. By Propositions 3 and 9, one has that S satisfies the linear recurrence relation γ. Moreover, by Theorem 2, it holds that γ is associated with the polynomial p_γ that is equal to $\gcd(p_\lambda, p_\mu)$. Thus, one can conclude that S is $\mathcal{A}_\gamma$-linear, with $\gamma = \gcd(\lambda, \mu)$, that is the sequence $\{t_0, \ldots, t_{k-1}\}$ satisfies γ.

Remark 1. We observe that the previous result is well known if one considers an infinite sequence of elements of K satisfying two linear recurrence equations. This follows from the fact that the characteristic polynomials of the linear recurrence relations satisfied by a given infinite sequence form a right ideal of $K[x]$ (cf. [2]). However Theorem 3 gives an extension of this result to finite sequences in the spirit of Fine and Wilf's periodicity theorem.

Now, suppose that a sequence is p-periodic and q-periodic; such a sequence satisfies the two linear recurrence relations $\lambda = (1, 0, \ldots, 0)$ with $\deg(\lambda) = p$ and $\mu = (1, 0, \ldots, 0)$ with $\deg(\mu) = q$. We have that $p_\lambda = x^p - 1$ and $p_\mu = x^q - 1$. Since $\gcd(x^p - 1, x^q - 1) = x^d - 1$, where $d = \gcd(p, q)$, Fine and Wilf's periodicity theorem is a particular case of Theorem 3.

References

1. J. Berstel and C. Reutenauer, *Rational series and their languages*, EACTS Monographs on Theoretical Computer Science, Vol. 12, Springer-Verlag, Berlin, 1988.
2. L. Cerlienco, M. Mignotte and F. Piras, *Suites récurrentes linéaires, propriétés algébriques et arithmétiques*, L'Enseignement Mathématique, 33 (1987) 67–108.
3. S. Eilenberg, *Semirings, Automata, Languages and Machines*, Vol.A, Academic Press, New York, 1974.
4. N.J. Fine and H.S. Wilf, *Uniqueness Theorem for Periodic Functions*, in Proc. Am. Mathematical Society, 16 (1965) 109–114.
5. D. Giammarresi, S. Mantaci, F. Mignosi and A. Restivo, *Periodicities on Trees*, *Theoretical Computer Science*, 205 (1998) 145–181.
6. D. Giammarresi, S. Mantaci, F. Mignosi and A. Restivo, *A periodicity theorem for trees*, in Proc. 13th World Computer Congress - IFIP '94, Hamburg, Germany, 1994; vol. A-51, pp. 473–478. Elsevier Science B.V. (North-Holland), 1994.
7. D. Giammarresi, S. Mantaci, F. Mignosi and A. Restivo, *Congruences, automata and periodicities*, in Proc. of the workshop *Semigroups, Automata and Languages*, Porto, June 1994. J. Almeida and P. Silva, Eds., pp. 125–135. World Scientific Publishing Co., 1995.
8. A. Gill, *The Reduced Form of a Linear Automaton*, in Automata Theory, E.R. Caianiello, Ed., Academic Press, New York, 1966.
9. A. Gill, *State Graphs of Autonomous Linear Automata*, in Automata Theory, E.R. Caianiello, Ed., Academic Press, New York, 1966.
10. W. Kuich and A. Salomaa, *Semirings, Automata, Languages*, EACTS Monographs on Theoretical Computer Science, Vol. 5, Springer-Verlag, Berlin, 1986.
11. M. Lothaire, *Combinatorics on Words*, Addison-Wesley, Reading, MA 1983.
12. G. Melideo, C. Pasquarelli and S. Varricchio, *On the generalization of Fine and Wilf's theorem to linear automata*, in *Proc.s of the "Sixth Italian Conference on Theoretical Computer Science"*, P. Degano, U. Vaccaro, G. Pirillo Eds., World Scientific, pp. 371–383; Prato, November 1998.
13. I.B. Plotkin, L.J. Greenglaz and A.A. Gvaramija, *Algebraic structures in automata and database theory*, World Scientific Publishing Co., 1992.
14. A. Salomaa and M. Soittola, *Automata theoretic aspects of formal power series*, Springer-Verlag, New York, 1978.

Part IV

Formal Languages

Numerical Parameters of Evolutionary Grammars

Jürgen Dassow

Summary. Evolutionary grammars form a grammatical model for the evolution of genomes where the basic derivation steps correspond to mutation in genomes. The generated language consists of all words which can be generated from the axioms (a finite set of words) by iterated derivation steps.

In this paper we study the number of axioms, the number of mutations and the size of the mutations which are necessary in order to generate some languages. Especially we show that these parameters cannot be bounded in order to generate all possible languages.

Furthermore we study the function which gives the number of words which can be generated by a certain number of derivation steps. We present some bounds and examples for such functions.

1 Introduction and Definitions

Grammars can be considered as language generating devices. The classical operation iteratively performed in order to obtain the words of the language from the axiom is the (sequential or parallel) substitution of some subword(s) by (an)other word(s).

In the last years there have been introduced some other types of grammars (or systems) motivated by considerations from molecular genetics. The starting point is the fact that DNA and RNA structures, genomes and chromosomes can be described to a certain extend as words, e.g. a DNA strand can be presented as a word over the alphabet of the compound symbols $(A,T),(T,A),(C,G),(G,C)$. Further, the basic operations in a derivation step model mutations occuring in the evolution of genomes and/or chromosomes. Thus the grammars (or systems) obtained can be considered as a model for parts of the evolution.

In the evolution the genome of an organism mutates at the level of individual genes by point mutations and at the genome level by some large-scale rearrangements in one evolutionary event. The most well-known rearrangements are deletion, inversion, translocation and duplication.

- Deletion cancels a segment of a chromosome.
- Inversion replaces a segment of a chromosome with its reverse DNA sequence.

- Translocation moves a segment to a new location in the genome.
- Duplication copies a segment of a chromosome.

A formalization of these mutations in terms of formal language theory was first given in [11]. Results on the power of these operations and relations between them can be found in [13], [2] and [6].

On the basis of the above large-scale reaarrangements a grammatical model for the evolution of genomes was presented in in [4] and further studied in [5]; a context-sensitive variant was considered in [3]. Instead of rules for replacements (as in CHOMSKY grammars or LINDENMAYER systems) finite sets C, I, T and D of words are given; these sets contain the words which can be deleted (cancelled), reversed, translocated or duplicated, respectively. Any derivation step corresponds to an application of a deletion of a word from C or of a reversal of a word from I or of a translocation of a word from T or of a duplication of a word from D. The language consists of all words which can be generated by a finite sequence of derivation steps starting with a word from the a given set A of axioms. In biological terms, the language consists of all genomes which can be obtained from a given set A of genomes by a finite sequence of certain given mutations.

We now give the formal definition of a context-free evolutionary grammar and start with some notations. Throughout the paper we assume that the reader is familiar with basic notions of the theory of formal languages. We refer to [9] and [10].

For an alphabet V, by V^* (V^+) we denote the set of all (non-empty) words over V. The empty word is denoted by λ. For a word $w \in V^*$ and a letter $a \in V$, $\#_a(w)$ denotes the number of occurrences of a in w. The reversal w^R of a word $w \in V^*$ is inductively defined by

$$\lambda^R = \lambda \quad \text{and} \quad (vx)^R = xv^R \text{ for } x \in V, v \in V^* .$$

The cardinality of a (finite) set M is denoted by $\#(M)$. For two functions f and g from the set $\mathbf{N}$ of natural numbers into $\mathbf{N}$, we write $f(n) = \Theta(g(n))$ if f and g are asmptotically equal.

A *context-free evolutionary grammar* is a system

$$G = (V, A, C, I, T, D) ,$$

where V is an alphabet and A, C, I, T and D are finite languages over V.

For $x, y \in V^*$ we write:

$$x \underset{C}{\Longrightarrow} y \text{ iff } x = x_1 z x_2,\ y = x_1 x_2 \text{ for some } x_1, x_2 \in V^*, z \in C,$$

$$x \underset{I}{\Longrightarrow} y \text{ iff } x = x_1 z x_2,\ y = x_1 z^R x_2 \text{ for some } x_1, x_2 \in V^*, z \in I,$$

$$\begin{aligned} x \underset{T}{\Longrightarrow} y \text{ iff } & x = x_1 z x_2 x_3,\ y = x_1 x_2 z x_3 \text{ for some } x_1, x_2, x_3 \in V^* \text{ or} \\ & x = x_1 x_2 z x_3,\ y = x_1 z x_2 x_3 \text{ for some } x_1, x_2, x_3 \in V^*, z \in T, \end{aligned}$$

$$x \underset{D}{\Longrightarrow} y \text{ iff } x = x_1 z x_2,\ y = x_1 z z x_2 \text{ for some } x_1, x_2 \in V^*, z \in D.$$

We write $x \Longrightarrow y$ if and only if $x \underset{X}{\Longrightarrow} y$ for some $X \in \{C, I, T, D\}$; the reflexive and transitive closure of $\Longrightarrow$ is denoted by $\overset{*}{\Longrightarrow}$.

The language $L(G)$ generated by G is defined by

$$L(G) = \{w \mid x \overset{*}{\Longrightarrow} w \text{ for some } x \in A\}.$$

The language contains all words which can be generated by iterated derivation steps.

In this paper we study the number of axioms, the number of operations and the maximal length of the words in $C \cup I \cup T \cup D$ which are necessary in order to generate some languages. Especially we show that these parameters cannot be bounded in order to generate all possible languages.

Furthermore we study the differentiation function which gives the number of words which can be generated in a certain number of derivation steps. We present some bounds and examples for such functions.

2 Descriptional Complexity

In this section we consider the descriptional (syntactic) complexity of languages generated by context-free evolutionary grammars. We are interested in the minimal number of axioms and operations, respectively, and the maximal length of the words associated with an operation. Formally, for a context-free evolutionary grammar $G = (V, A, C, I, T, D)$, we set

$$\begin{aligned} a(G) &= \#(A), \\ o(G) &= \#(C) + \#(I) + \#(T) + \#(D), \\ l(G) &= \max\{|w| \mid w \in C \cup I \cup T \cup D\} \end{aligned}$$

and extend these measures to a language L generated by a context-free evolutionary grammar by

$$\begin{aligned} a(L) &= \min\{a(G) \mid L = L(G), G \text{is a context-free evolutionary grammar}\}, \\ o(L) &= \min\{o(G) \mid L = L(G), G \text{is a context-free evolutionary grammar}\}, \\ l(L) &= \min\{l(G) \mid L = L(G), G \text{is a context-free evolutionary grammar}\}. \end{aligned}$$

Theorem 1. *A language L is finite if and only if $o(L) = 0$.*

Proof. Let L be a finite language, and let V be the set of symbols occurring in at least one word of L. Then $L = L(G)$ for the context-free evolutionary grammar $G = (V, L, \emptyset, \emptyset, \emptyset, \emptyset)$. Since $o(G) = 0$ we obtain $o(L) = 0$.

If $o(L) = 0$ for some language L, then there is a context-free evolutionary grammar $G = (V, A, \emptyset, \emptyset, \emptyset, \emptyset)$ with $L = L(G)$. By $L(G) = A$, G generates a finite language which proves that L is finite.

The measure $o(G)$ corresponds to the number of productions in a (usual) CHOMSKY grammar. The context-free languages form an infinite hierarchy with respect to the number of productions (see [7]). Furthermore, the measure $l(G)$ corresponds to the radius of a H system which is grammatical device based on splicing as another mutation. With respect to the radius the languages generated by H systems form an infinite hierarchy, too (see [8]). In this section we shall prove analogous assertions for the measures for context-free evolutionary grammars introduced above.

Theorem 2. *For any measure $d \in \{a, o, l\}$ and any natural number $r \geq 1$, there is a language L generated by a context-free evolutionary grammar such that $d(L) = r$.*

Proof. We consider the language

$$L = \bigcup_{i=1}^{n} L_i \quad \text{where} \quad L_i = \{(ba^ib)^m \mid m \geq 0\} \text{ for } 1 \leq i \leq n\,.$$

Because L is generated by the context-free evolutionary grammar

$$G = (\{a, b\}, \{ba^ib \mid 1 \leq i \leq n\}, \emptyset, \emptyset, \emptyset, \{ba^ib \mid 1 \leq i \leq n\})$$

with $a(G) = n$, $o(G) = n$ and $l(G) = n + 2$, we obtain

$$a(L) \leq n, \quad o(L) \leq n \quad \text{and} \quad l(L) \leq n + 2. \tag{1}$$

Now let us assume that $H = (V, A, C, I, T, D)$ is a context-free evolutionary grammar with $L(H) = L$. If there is a derivation $w' \Longrightarrow w$ in H with $w' \in L_i$, $w \in L_j$, $1 \leq i, j \leq n$ and $i \neq j$, then there also is a derivation $w'ba^ib \Longrightarrow wba^ib$. Since $w'ba^ib \in L_i \subset L$ and $wba^ib \notin L$, we get a contradiction. Thus, for any i, $1 \leq i \leq n$, $A \cap L_i$ has to be a non-empty set. Therefore $a(H) \geq n$ for any context-free evolutionary grammar H with $L(H) = L$ which implies $a(L) \geq n$. By (1), we obtain $a(L) = n$.

Let $a = \max\{|z| \mid z \in A\}$. We consider a word $w \in L_i$, $1 \leq i \leq n$, with $|w| \geq a + 1$. Let $w = (ba^ib)^l$. By the length of w there is a word $w' \in L$ with $w' \Longrightarrow w$ and $w' \neq w$. By the above considerations $w' \in L_i$, too, say $w' = (ba^ib)^k$ for some k.

If $w' \underset{I}{\Longrightarrow} w$ or $w' \underset{T}{\Longrightarrow} w$, then $|w'| = |w|$ and hence $w' = w$ in contrast to the choice of w'.

Let us assume that $w' \underset{D}{\Longrightarrow} w$. Then $w' = w_1 x w_2$ and $w = w_1 x x w_2$ for some $w_1, w_2 \in V^*$, $x \in V^+$. Thus

$$\#_a(x) = \#_a(w) - \#_a(w') = i(l - k) \quad \text{and} \quad \#_b(x) = 2(l - k)\,.$$

Therefore

$$x \in \{a^r b (ba^ib)^{l-k-1} ba^s \mid r, s \geq 0, r + s = i\} \cup \{(ba^ib)^{l-k}\}\,.$$

If $x = a^r bba^s$, $r, s \geq 0$, $r + s = i$, i.e. $l = k + 1$, then we can apply $x \in D$ to $ba^n bba^n b$ which yields $ba^{2n-i}b$ and hence $n = i$. If, in addition, $r > 0$ and $s > 0$, we can apply $x \in D$ to $ba^{n-1}bba^{n-1}bba^{n-1}b$ and obtain $ba^{2n-2-i}bba^{n-1}b$ from which $i = n-1$ follows. This contradicts $i = n$. Thus $x = a^r bba^s$ implies $r + s = n$ and $r = 0$ or $s = 0$.

We now define

$$\begin{aligned} M_j &= \{a^r b(ba^j b)^t ba^s \mid r, s \geq 0, r + s = j, t \geq 0\} \cup \{(ba^j b)^t \mid t \geq 0\} \\ &\quad \text{for } 1 \leq j < n, \\ M_n &= \{a^r b(ba^n b)^t ba^s \mid r, s \geq 0, r + s = n, t \geq 0\} \cup \{a^n bb, bba^n\} \\ &\quad \cup \{(ba^n b)^t \mid t \geq 0\}. \end{aligned}$$

By the considerations above, we get $x \in M_i$.

Let $w' \underset{C}{\Longrightarrow} w$. Then $w' = w_1 x w_2$ and $w = w_1 w_2$. By analogous arguments we can show that $x \in M_i$, again.

Thus $(C \cup D) \cap M_i \neq \emptyset$ for $1 \leq i \leq n$. Furthermore, $M_i \cap M_j = \emptyset$ for $1 \leq i, j \leq n$ and $i \neq j$. Therefore $C \cup D$ contains at least n elements and $o(H) \geq n$ holds for any context-free evolutionary grammar H with $L(H) = L$. Hence $o(L) \geq n$. By (1), $o(L) = n$.

Moreover, for $1 \leq i \leq n$, $|x| \geq 2 + i$ holds for any $x \in (C \cup D) \cap M_i$. Thus $l(H) \geq n + 2$ for any context-free evolutionary grammar H with $L(H) = L$. Therefore $l(L) \geq n + 2$ and, by (1), $l(L) = n + 2$.

Hence the statement holds for $d \in \{a, o\}$, $r \geq 1$ and $d = l$, $r \geq 3$. It is easy to see that $l(\{a\}^+) = 1$ and $l(\{a^2\}^+) = 2$, and therefore the statement holds in the remaining cases, too.

In order to get a finer measure than the number of all operations given by $o(G) = \#(C) + \#(I) + \#(T) + \#(D)$ one can consider the quadruple $O(G) = (\#(C), \#(I), \#(T), \#(D))$. The investigation of this measure is left as an open problem.

3 The differentiation function

The notion of a differentiation function of a grammar was firstly introduced in [1] for deterministic tabled Lindenmayer systems. It presents a measure for the number of objects which can be derived in a given grammar by a given number of derivation steps. Formally we obtain the following notion for context-free evolutionary grammars.

Let $G = (V, A, C, I, T, D)$ be an evolutionary grammar. Then we define the its *differentiation function*

$$f_G : \mathbf{N} \to \mathbf{N} \quad \text{by} \quad f_G(k) = card(L_k(G)).$$

Example 1. We consider the context-free evolutionary grammars

$$G_1 = (\{a, b\}, \{aa\}, \emptyset, \emptyset, \emptyset, \{aa\}),$$
$$G_2 = (\{a, b\}, \{aa\}, \emptyset, \{aa\}, \emptyset, \{aa\}),$$
$$G_3 = (\{a, b\}, \{aab\}, \emptyset, \{aa\}, \{b\}, \{aa\}).$$

Then, for $k \geq 1$,

$$L_k(G_1) = \{a^{2k+2}\},$$
$$L_k(G_2) = \{a^2, a^4, \ldots, a^{2k+2}\},$$
$$L_k(G_3) = \{a^r b a^s \mid r + s = 2i, 1 \leq i \leq k\} \cup \{a^{2k+2}b\},$$

and thus

$$f_{G_1}(k) = 1 \quad \textit{for} \quad k \geq 1,$$
$$f_{G_2}(k) = k + 1 \quad \textit{for} \quad k \geq 1,$$
$$f_{G_3}(k) = 3 + 5 + \ldots (2k + 1) + 1 = (k + 1)^2 \quad \textit{for} \quad k \geq 1.$$

We only show the statement concerning $L_k(G_3)$, *the modifications for the other cases are obvious.*

From the axiom aab *of* G_3 *we can generate by inversion of* aa *the word* aab, *again, by translocation of* b *the words* baa, aba, aab *and by duplication of* aa *the word* $aaaab$. *Thus the statement holds for* $k = 1$.

Now let $w \in L_k(G_3)$. *By induction hypothesis,* $w = a^r b a^s$ *with* $r + s = 2i$ *for some* i, $1 \leq i \leq k$ *or* $w = a^{2k+2}b$.

We first consider the former case. Since the translocation and inversion does not change the number of occurrences of a *and* b, *we obtain by these operation a word* $a^{r'} b a^{s'}$ *with* $r' + s' = 2i$. *If we applied the duplication of* aa *we get* $a^{r+2}ba^s$ *or* $a^r b a^{s+2}$. *Because* $r + s + 2 = 2(i + 1)$, *in all cases the generated words have the desired form.*

In the former case we generate from w *a word of the* $a^r b a^s$ *with* $r + s = 2k + 2 = 2(k + 1)$ *or* $a^{2k+4}b = a^{2(k+1)+2}b$, *and all words have the desired form, again.*

Moreover, these considerations also show that all words of the desired form are contained in $L_{k+1}(G_3)$.

We now give an upper bound for differentiation functions of context-free evolutionary grammars.

Theorem 3. *For any context-free evolutionary grammar* G, *there are constants* c_1 *and* c_2 *such that* $f_G(k) \leq c_1 \cdot c_2^k$ *for* $k \geq 1$.

Proof. Let $G = (V, A, C, I, T, D)$ be an evolutionary grammar. We set

$$d = \max\{|u| \mid u \in D\}, \quad a = \#(A), \quad b = \max\{|v| \mid v \in A\}.$$

Then, for any $k \geq 0$ and any word in $z \in L_k(G)$, $|z| \leq b + k \cdot d$. Thus

$$\begin{aligned} f_G(k+1) &\leq \#\{w \mid |w| \leq b + kd\} \\ &= \sum_{i=0}^{b+kd} (\#(V))^i = \frac{1}{\#(V)-1} \cdot ((\#(V))^{b+kd+1} - 1) \\ &= \frac{(\#(V))^{b+1}}{\#(V)-1} ((\#(V))^d)^k . \end{aligned}$$

By setting $c_1 = \frac{(\#(V))^{b+1}}{\#(V)-1}$ and $c_2 = (\#(V))^d)$ the assertion follows.

The following shows that the exponentially upper bound is obtained for some context-free evolutionary grammars.

Theorem 4. *For any natural number c, there is a context-free evolutionary grammar G such that $f_G(k) = c^k$ for $k \geq 0$.*

Proof. Let c be given. We consider the context-free evolutionary grammar

$$G = (V, \{a^c ba^c\}, \emptyset, \emptyset, \emptyset, \{a^c ba^i \mid 1 \leq i \leq c\}.$$

By induction on k we prove that

$$L_k(G) = \{a^c ba^{i_1+c} ba^{i_2+c} b \dots a^{i_k+c} ba^c \mid i_1, i_2, \dots i_k \in \{1, 2 \dots c\}\}$$

which implies

$$f_G(k) = \#(L_k(G)) = c^k$$

for $k \geq 0$.

By definition, $L_0(G) = \{a^c ba^c\}$ and therefore the statement holds for $k = 0$.

Now let $w \in L_{k+1}(G)$. Then $w' \Longrightarrow w$ for some $w' \in L_k(G)$. By induction hypothesis $w' = a^c ba^{i_1+c} ba^{i_2+c} b \dots a^{i_k+c} ba^c$. Let w be derived from w' by duplikation of some $a^c ba^i$, $1 \leq i \leq c$, where the duplication involves the j-th occurrence of b. Then

$$\begin{aligned} w' &= a^c ba^{i_1+c} ba^{i_2+c} b \dots a^{i_j+c} ba^{i_{j+1}+c} b \dots a^{i_k+c} ba^c \\ &= a^c ba^{i_1+c} ba^{i_2+c} b \dots a^{i_j} a^c ba^i a^{i_{j+1}+c-i} b \dots a^{i_k+c} ba^c \\ &\Longrightarrow a^c ba^{i_1+c} ba^{i_2+c} b \dots a^{i_j} a^c ba^i a^n ba^i a^{i_{j+1}+c-i} b \dots a^{i_k+c} ba^c \\ &= a^c ba^{i_1+c} ba^{i_2+c} b \dots a^{i_j+c} ba^{i+c} ba^{i_{j+1}+c} b \dots a^{i_k+c} ba^c \end{aligned}$$

which proves that w has the desired form.

If we apply in succession the duplications of $a^c ba^{i_1}, a^c ba^{i_2}, \dots, a^c ba^{i_{k+1}}$ such that in any step the last b is involved, then we get $a^c ba^{i_1+c} ba^{i_2+c} b \dots ba^{i_{k+1}+c} ba^n \in L_{k+1}(G)$. Hence $L_{k+1}(G)$ contains all words of the considered form.

Thus the induction statement is shown for $k + 1$.

We now prove two assertion concerning the closure of the family of differentiation functions under operations.

Lemma 1. *Let f and g be two differentiation functions of context-free evolutionary grammars. Then their sum $f+g$ is a differentiation function, too.*

Proof. Let $f = f_G$ and $g = f_H$ for two context-free evolutionary grammars $G = (V_G, A_G, C_G, I_G, T_G, D_G)$ and $H = (V_H, A_H, C_H, I_H, T_H, D_H)$. Without loss of generality we can assume that $V_G \cap V_H = \emptyset$. Then no operation from H can be applied to an element of $L(G)$, and conversely, no operation from G can be applied to an element of $L(H)$. Moreover, $L(G) \cup L(H) = \emptyset$. We consider the context-free evolutionary grammar

$$G' = (V_G \cup V_H, A_G \cup A_H, C_G \cup C_H, I_G \cup I_H, T_G \cup T_H, D_G \cup D_H).$$

By induction we can show that $L_k(G') = L_k(G) \cup L_k(H)$ which implies

$$f_{G'}(k) = f_G(k) + f_H(k) = (f+g)(k)$$

by the disjointness of the languages generated by G and H.

Let f and g be two function from $\mathbf{N}$ into $\mathbf{N}$, then their convolution $f * g$ is defined by

$$(f * g)(k) = \sum_{i=0}^{k} f(i) \cdot g(k-i).$$

We now show an analogon of Lemma 1 for convolutions instead of sums.

Lemma 2. *Let f and g be two differentiation functions of context-free evolutionary grammars. Then their convolution $f * g$ is a differentiation function, too.*

Proof. Let $f = f_G$ and $g = f_H$ for two context-free evolutionary grammars $G = (V_G, A_G, C_G, I_G, T_G, D_G)$ and $H = (V_H, A_H, C_H, I_H, T_H, D_H)$. Without loss of generality we can assume that $V_G \cap V_H = \emptyset$. We consider the context-free evolutionary grammar

$$G' = (V_G \cup V_H, A_G \cdot A_H, C_G \cup C_H, I_G \cup I_H, T_G \cup T_H, D_G \cup D_H).$$

By induction we show that

$$L_k(G') = L_0(G) \cdot L_k(H) \cup L_1(G) \cdot L_{k-1}(H) \cup \cdots \cup L_k(G) \cdot L_0(H). \quad (2)$$

By the construction of G', the statement holds for $k = 0$. Let $w \in L_k(G')$. Then there is an integer i, $1 \leq i \leq k$, such that $w = w_1 w_2$ for two words $w_1 \in L_i(G)$ and $w_2 \in L_{k-i}(H)$. Let us apply an element $x \in C_G \cup C_H$ to w. If $x \in C_G$, we have to apply x to w_1 and get $w = w_1 w_2 \Longrightarrow w_1' w_2 = w'$ where $w_1' \in L_{k+1}(G)$. Hence

$$w' \in L_{i+1}(G) L_{k-i}(H) = L_{i+1}(G) L_{(k+1)-(i+1)}(H) \subseteq L_{k+1}(G').$$

Analogously, $x \in C_H$ implies

$$w' \in L_i(G)L_{k-i+1}(H) = L_i(G)L_{(k+1)-i}(H) \subseteq L_{k+1}(G').$$

Furthermore, it is easy to see that any word from $L_i(G)L_{k-i}(H)$ can be generated in k steps, more precisely, by i steps according to G and $k-i$ steps according to H.

By the supposition on disjointness and (2),

$$\begin{aligned} f_{G'}(k) = \#(L_k(G')) &= \sum_{i=0}^{k} \#(L_i(G)L_{k-i}(H) = \sum_{i=0}^{k} f_G(i)f_H(k-i) \\ &= \sum_{i=0}^{k} f(i)g(k-i) = (f * g)(k). \end{aligned}$$

We use the Lemma 2 to show that asymtotically any polynomial can be obtained as a differentiation function.

Theorem 5. *For any natural number n, there is a cgeg G such that $f_G(n) = \Theta(k^n)$.*

Proof. For $n \in \{0,1,2\}$ the statement follows from Example 1. Let us assume that there is already a context-free evolutionary grammar G with $f_G(n) = \Theta(k^n)$. Then we construct the context-free evolutionary grammar G' with $f_{G'} = f_G * f_{G_1}$, where G_1 is the context-free evolutionary grammar from Example 1 (see the proof of Lemma 2. Then

$$f_{G'}(k) = \sum_{i=0}^{k} f_G(i) = \Theta(k^{n+1}).$$

Thus the statement follows by induction.

The presented results give some upper and lower bounds for and some examples of differentiation functions; the characterization of the family of differentiation functions of context-free evolutionary grammars is left as an open problem.

Finally we present two classes of context-free evolutionary grammars with differentiation functions bounded by a constant or linear function.

Lemma 3. *If $G = (V, A, C, I, T, \emptyset)$, then there is a constant c such that $f_G(k) \leq c$ for $k \geq 0$.*

Proof. Obviously, because the set of duplications is empty, $L(G)$ is finite. Let c be the cardinality of $L(G)$. Then $f_G(k) = \#(L_k(G)) \leq \#(L(G)) = c$.

Example 1 shows that there are context-free evolutionary grammars with a non-empty set of duplications which also have a differentiation function bounded by constant.

Before we present the other class we give the definition and a property of a number-theoretic function. For a set A of natural numbers we define the function

$$f_A(k) = \#(\{\sum_{j=1}^{k} a_{i_j} \mid a_{i_j} \in A \text{ for } 1 \leq j \leq k\}).$$

Lemma 4. *Let $A = \{a_1, a_2, \ldots, a_s\}$ with $0 \leq a_1 < a_2 < a_3 < \ldots < a_s$ and $m = \max\{a_{i+1} - a_i \mid 1 \leq i \leq s-1\}$. Then*

$$\lfloor \frac{a_s - a_1}{m} \cdot k \rfloor + 1 \leq f_A(k) \leq (a_s - a_1) \cdot k + 1.$$

Moreover, both bounds are optimal.

Proof. Obviously $a_1 \cdot k$ and $a_s \cdot k$ are the minimal and maximal number which can be obtained by addition of k numbers of A. Hence any sum S of interest satisfies $a_1 k \leq S \leq a_s k$. This implies the upper bound.

We now prove that any interval $I_i = [a_1 k + im, a_1 k + (i+1)m)$, $0 \leq i \leq \lfloor \frac{(a_s - a_1)k}{m} \rfloor - 1$, contains at least one sum of k numbers of A. Obviously, this holds for $i = 0$ by $a_1 k \in I_0$. Now let $c = \sum_{j=1}^{k} a_{i_j}$ be the maximal number in I_i which can be represented by sum of k numbers of A. Since $i \leq \lfloor \frac{(a_s - a_1)k}{m} \rfloor - 1$ we obtain $c \leq a_1 k + (\lfloor \frac{(a_s - a_1)k}{m} \rfloor - 1)m < a_s k$. Thus there is a r such that $a_{i_r} < a_s$. Let $a_{i_r} = a_l$. Then we consider the sum

$$c' = (\sum_{j=1}^{r-1} a_{i_j}) + a_{l+1} + (\sum_{j=r+1}^{k} a_{i_j})$$

of k numbers of A. Because $c' = c + (a_{l+1} - a_l) \leq c + m$ and c is maximal in I_i, we obtain $c' \in I_{i+1}$. Since we have $\lfloor \frac{(a_s - a_1)k}{m} \rfloor$ intervals and the additional sum $a_s k$ (which belongs to no interval), the lower bound follows.

The optimality of both bounds follows by considering $A = \{1, 2, 3, \ldots, s\}$ or $A = \{m, 2m, 3m, \ldots, sm\}$ (for some m).

Lemma 5. *For any context-free evolutionary grammar $G = (\{a\}, \{a^n\}, \emptyset, I, T, D)$ where D contains a non-empty word (i.e. the underlying alphabet of the grammar is unary, there is only one axiom, no deletion and at least one non-empty duplication), $f_G(k) = \Theta(k)$.*

Proof. First we assume that $I \cup T = \emptyset$. The application of $v \in D$ to a^m leads to $a^{m+|v|}$. Thus taking $A = \{|v| \mid v \in D\}$, we obtain $f_A(k) = f_G(k)$ for $k \geq 1$. Hence the statement follows from Lemma 4.

Now let $I \cup T \neq \emptyset$ and $\lambda \notin D$. We set $r = \min\{|z| \mid z \in I \cup T\}$. If we apply an inversion or a translocation to a word w, then w is not changed by this operation because the underlying alphabet is unary. To any word $w \in L_k(G)$ with $|w| \geq r$ we can apply an inversion or a translocation, and thus w is also contained in $L_l(G)$ for $l \geq k$. Moreover, if D does not contain the empty word, all words from $L_r(G)$ have a length $\geq r$. It is easy to see that

$$f_{A \cup \{0\}}(k) \leq f_G(k+r) \leq \#(L_r(G)) f_{A \cup \{0\}}(k) \quad \text{for} \quad k \geq 1 .$$

Now the stement follows from Lemma 4.

The proof for the case $I \cup T \neq \emptyset$ and $\lambda \in D$ follows analogous arguments and is left to the reader.

References

1. J. Dassow, Eine neue Funktion für Lindenmayer-Systeme. *EIK* **12** (1976) 515–521.
2. J. Dassow and V. Mitrana, On some operations suggested by genome evolution. In: *Proc. Second Pacific Symposium on Biocomputing* (Eds.: R. B. Altman, A. K. Dunker, L. Hunter, T. Klein), World Scientific, 1997, 97–108.
3. J. Dassow and V. Mitrana, Evolutionary grammars: a grammatical model for genome evolution. In: *Bioinformatics*, Proc. German Conf. on Bioinformatics, Lecture Notes in Computer Science 1278 (Eds.: R. Hofestädt et al.), Springer-Verlag, 1997, 199–209.
4. J. Dassow, V. Mitrana and A. Salomaa, Context-free evolutionary grammars and the structural language of nucleic acids, *BioSystems* **43** (1997) 169–177.
5. J. Dassow and Gh. Păun, On the regularity of languages generated by context-free evolutionary grammars. To appear in *Applied Discrete Math.*
6. J. Dassow and Gh. Păun, Remarks on operations suggested by mutations in genomes. To appear in *Fund. Informaticae.*
7. J. Gruska, Descriptional complexity of context-free languages. In: *Proc. Math. Found. Comp. Science*, 1973, 71–83.
8. T. Head, Gh. Păun and D. Pixton, Language theory and molecular genetics: generative mechanisms suggested by DNA recombination. In: *Handbook of Formal Languages* (Eds.: G. Rozenberg and A. Salomaa), Springer-Verlag, Berlin, 1998, 295–360.
9. J. E. Hopcroft and J. D. Ullman, *Introduction to Automata Theory, Languages, and Computation.* Addison-Wesley Publ. Co., Reading, 1979.
10. G. Rozenberg and A. Salomaa, *The Mathematical Theory of L Systems.* Academic Press, New York, 1980.
11. D. B. Searls, The computational linguistics of biological sequences. In: *Artificial Intelligence and Molecular Biology* (Ed.: L. Hunter), AAAI Press, The MIT Press, 1993, 47–120.
12. H. J. Shyr, *Free Monoids and Languages.* Inst. Applied Math., Univ. Taichung and Hon Min Book Co., Taichung, Taiwan.
13. T. Yokomori and S. Kobayashi, DNA evolutionary linguistics and RNA structure modeling: a computational approach. In: *Proc. Internat. IEEE Symposium on Intelligence in Neural and Biological Systems*, IEEE Computer Science Press, 1995, 38–45.

Iterated GSM Mappings: A Collapsing Hierarchy

Vincenzo Manca, Carlos Martín-Vide, and Gheorghe Păun

Summary. With motivations from various areas, in several places mechanisms based on iterated (non-deterministic) finite state sequential transducers were considered. It is known that such mechanisms can characterize the family of recursively enumerable languages. We continue here the study of such devices, investigating the hierarchy on the number of states. We find that this hierarchy collapses: four states are enough in order to characterize the recursively enumerable languages, three states cover the ET0L languages, while two states can cover the E0L (hence also context-free) languages. The case of deterministic transducers remains open.

1 Introduction

Iterated gsm (generalized sequential machine) mappings were already investigated in seventies (see, e.g., [11], [5], [9]), especially in relation with L systems area. Because, as it is easy to see, such a device can simulate the work of a type-0 Chomsky grammar, in this way we obtain characterizations of recursively enumerable languages. (Depending on the precise mode of defining the accepted language, we need or not an intersection with T^*, where T is the alphabet of the desired language, as a final squeezing operation.)

The phenomenon of repeated translations of a text appears in many areas, especially in linguistics and in literary studies. We refer to [3] for bibliographical details.

A recent area where the need of iterated gsm's has appeared is DNA computing, in relation with the so-called *computing by carving*, [6]. We will briefly present this idea in the next section. In short, we "compute" a language by identifying strings in its complement and "carving" them from a given superlanguage; the removed strings are obtained by iterating a gsm on a starting regular language.

In this framework, it appears as a natural problem to consider simple gsm mappings, for instance, with a bounded number of states. The problem was formulated in [6] whether or not the number of states induces an infinite hierarchy of the languages computed by carving. We solve this problem here. Somewhat surprisingly, this hierarchy collapses at a rather low level: three states suffice. We place this question in a more general framework, that of languages obtained by an iterated use of a sequential transducer. We find that (non-deterministic) iterated transducers with four states characterize the recursively enumerable languages (the extra state is needed in order to select strings which are terminal with respect to a given alphabet; in the case of carving this is accomplished in a different way – see details in the sections below). Several connected results are also given (three states suffice in order to cover the ET0L languages, while two states suffice in order to obtain all E0L languages).

These results deal with non-deterministic gsm's. The case of iterated deterministic gsm's remains almost completely *open*.

2 Computing by Carving

DNA computing is a new field of research, trying to make use of the huge parallellism and other nice features of the structure and behavior of DNA molecules. The first practical step in this respect was the experiment reported in [1], where a small instance of the Hamiltonian Path Problem was solved by biochemical means. Another successful experiment was reported in [4]: finding the size of the maximal clique in a graph. Information about further developments in this area can be found in [7] and [8].

The procedures in [1] and [4] follow the same phases (on similar strategies are based other experiments of DNA computing): (1) one first generates a large set of candidate solutions to the problem, (2) one identifies sets of non-solutions, and (3) one removes such non-solutions from the candidate set, repeatedly, until only the solutions remain. In [4] one explicitly speaks about the *complete data pool* from which, by specific *filtering* procedures, one removes the non-solutions.

Inspired by these observations and by the (theoretically) easy way of implementing the difference of languages in DNA terms, the general strategy of *computing by carving* was proposed in [6], proceeding along the three steps mentioned above. In short, in order to construct the set of the solutions to a problem, we first construct a superset of it (one which is "easy" to be constructed) and then we filter it iteratively (by "carving" sets of non-solutions, also "easy" to be constructed) until only the solutions of the problem remain.

In some sense, this idea is not new at all: this is exactly the style of the Erathostene's sieve for constructing the prime numbers, where the "complete data pool" is the set of natural numbers (which is given "for free" in advance).

For the formal language theory, the computing by carving leads to a rather new way of identifying a language L: start from a superlanguage of L, say M, large and easy to be obtained, then remove from M strings or sets of strings, iteratively, until obtaining L. Contrast this strategy with the usual "positivistic" grammatical approach, where one produces the strings in L at the end of successful derivations, discarding the "wrong derivations" (derivations not ending with a terminal string), and constantly ignoring/avoiding the complement of L (for elements of formal language theory which we will use here we refer to [10]).

In particular, M above can be V^*, the total language over the alphabet V of the language L. Therefore, as a particular case, we can try to produce $V^* - L$, the complement of L, and to extract it from V^*. As one knows, the family of recursively enumerable languages is not closed under complementation. Therefore, *by carving we can "compute" non-recursively enumerable languages !*

However, identifying a language L by generating first its complement, $H = V^* - L$, and then computing the difference $V^* - H$ is a rough procedure. A more subtle one, present also in the DNA computing experiments mentioned above, is that where L is obtained at the end of several difference operations, possibly at the end of an infinite sequence of such operations: we construct V^*, as well as certain languages $L_1, L_2, \ldots$, and we iteratively compute $V^* - L_1$, $(V^* - L_1) - L_2, \ldots$, such that L is obtained at the limit.

Of course, the sequence of languages $L_i, i \geq 1$, must be defined in a regular, finite, way. The proposal in [6] is to consider *regular sequences of languages*:

A sequence $L_1, L_2, \ldots$ of languages over an alphabet V is called *regular* if L_1 is a regular language and there is a gsm g such that $L_{i+1} = g(L_i), i \geq 1$.

Then, a language $L \subseteq V^*$ is said to be *C-REG computable* if there is a regular sequence of languages, $L_1, L_2, \ldots$ (that is, a pair (L_1, g) as specified above), and a regular language $M \subseteq V^*$ such that $L = M - (\bigcup_{i \geq 1} L_i)$.

We can denote by g^i the ith iteration of $g, i \geq 0$, and then $L_{i+1} = g^i(L_1), i \geq 0$ (taking $g^0(H) = H$ for all $H \subseteq V^*$). Denoting by $g^*(L_1)$ the union $\bigcup_{i \geq 0} g^i(L_1)$, we can write $L = M - g^*(L_1)$. We are again back to the complement with respect to M of a single language, $g^*(L_1)$, but this language is the union of the languages in a sequence defined by finite tools: a regular grammar – or a finite automaton – for the language L_1, and the gsm g).

The following result is proved in [6]:

Theorem 1. *A language is C-REG computable if and only if it can be written as the complement of a recursively enumerable language.*

Consequently, the family of C-REG computable languages contains all context-sensitive languages and it is incomparable with the family of recursively enumerable languages (CS is closed, but RE is not closed under complementation).

The following necessary condition is proved in [6].

Lemma 1. *If (L_1, g) identifies an infinite language $L = g^*(L_1)$, then there exists a constant k such that for each $x \in L$ we can find $y \in L$ such that $|x| < |y| \leq k|x|$.*

For instance, the language $L = \{a^{2^{2^n}} \mid n \geq 1\}$ does not have this property. Therefore, RE includes strictly the family of languages of the form $g^*(L)$, for L regular. However, we have the next result (proofs can be found in [11], [5], [9]):

Theorem 2. *Every recursively enumerable language $L \subseteq T^*$ can be written in the form $L = g^*(\{a_0\}) \cap T^*$, where g is a gsm – depending on L – and a_0 is a fixed symbol not in T.*

At the end of [6] one asks the question whether or not the number of states in a gsm induces an infinite hierarchy of C-REG computable languages. We shall solve this problem in the next sections, by proving that the answer is negative.

3 Iterated Sequential Transducers

We first introduce a general set-up for our investigations, by defining the notion of an *iterated finite state sequential transducer* (in short, an IFT), as a language generating device. Among the reasons for considering this notion are the facts that, according to Theorem 2.2, in order to characterize RE by iterated gsm's we can start from a single symbol, but we also need an intersection with a language T^*. Such an intersection can easily be done by a gsm (for instance, using a special state for this purpose), so we introduce this feature "inside" the gsm work. Because we explicitly provide the starting symbol a_0 in the machinery, we prefer to use a new terminology, speaking of IFT's and not of gsm's of a special form and with a special functioning.

An IFT is a construct $\gamma = (K, V, s_0, a_0, F, P)$, where K, V are disjoint alphabets (the set of *states* and the *alphabet* of γ), $s_0 \in K$ (the *initial state*), $a_0 \in V$ (the *starting symbol*), $F \subseteq K$ (the set of *final states*), and P is a finite set of *transition rules* of the form $sa \to xs'$, for $s, s' \in K, a \in V, x \in V^*$ (in state s, the device reads the symbol a, passes to state s', and produces the string x).

For $s, s' \in K$ and $u, v, x \in V^*, a \in V$ we define:

$$usav \vdash uxs'v \text{ iff } sa \to xs' \in P.$$

This is a *direct transition step* with respect to γ. We denote by $\vdash^*$ the reflexive and transitive closure of the relation $\vdash$.

Then, for $w, w' \in V^*$ we write

$$w \Longrightarrow w' \text{ iff } s_0 w \vdash^* w's, \text{ for some } s \in K.$$

We say that w *derives* w'; note that this means that w' is obtained by translating the string w, starting from the initial state of γ and ending in any state of γ, not necessarily a final one. We denote by $\Longrightarrow^*$ the reflexive and transitive closure of the relation $\Longrightarrow$.

If in the writing above we have $s \in F$ (we stop in a final state), then we write $\stackrel{f}{\Longrightarrow}$ instead of $\Longrightarrow$; that is, $w \stackrel{f}{\Longrightarrow} w'$ iff $s_0 w \vdash^* w' s$, for some $s \in F$.

The *language generated* by γ is

$$L(\gamma) = \{w \in V^* \mid a_0 \Longrightarrow^* w' \stackrel{f}{\Longrightarrow} w, \text{ for some } w' \in V^*\}.$$

Therefore, we iteratively translate the strings obtained by starting from a_0, without care about the states we reach at the end of each translation, but at the last step we necessarily stop in a final state. This makes possible the avoiding of the intersection with a language of the form of T^*, as in the statement of Theorem 2.2.

The IFT's as defined above are *non-deterministic*. If for each pair $(s, a) \in K \times V$ there is at most one transition $sa \rightarrow xs'$ in P, then we say that γ is *deterministic*.

We denote by $IFT_n, n \geq 1$, the family of languages of the form $L(\gamma)$, for non-deterministic γ with at most n states; when using deterministic IFT's we write $DIFT_n$ instead of IFT_n. The union of all families $IFT_n, DIFT_n, n \geq 1$, is denoted by $IFT, DIFT$, respectively.

We will compare these families with the families *REG, CF, CS, RE* (of regular, context-free, context-sensitive, recursively enumerable languages, respectively), and with $0L, E0L, ET0L$ (of languages generated by interactionless, by extended interactionless, and by extended tabled interactionless L systems, respectively); $D0L, DE0L$ are the families corresponding to $0L, E0L$ generated by deterministic L systems.

Convention. When comparing the power of two generative devices, G_1, G_2, the empty string is ignored: G_1 is equivalent to G_2 if and only if $L(G_1) - \{\lambda\} = L(G_2) - \{\lambda\}$.

4 The Hierarchy IFT_n Collapses

The following relations are direct consequences of the definitions:

Lemma 2. (i) $IFT_n \subseteq IFT_{n+1}, DIFT_n \subseteq DIFT_{n+1}$, $n \geq 1$. (ii) $DIFT_n \subseteq IFT_n$, $n \geq 1$, $DIFT \subseteq IFT$. (iii) $IFT \subseteq RE$. (iv) $IFT_1 = 0L$, $DIFT_1 = D0L$.

Lemma 3. $E0L \subseteq IFT_2$.

Proof. Consider an E0L system $G = (V, T, w, P)$ (the total alphabet, the terminal alphabet, the axiom, and the set of rewriting rules). Let a_0, a_1 be two new symbols. We construct the IFT

$$\begin{aligned}\gamma &= (\{s_0, s_1\}, V \cup \{a_0, a_1\}, s_0, a_0, \{s_1\}, P'),\\ P' &= \{s_0 a_0 \rightarrow a_1 w s_0,\ s_0 a_1 \rightarrow a_1 s_0\} \cup \{s_0 a \rightarrow x s_0 \mid a \rightarrow x \in P\}\\ &\quad \cup \{s_0 a_1 \rightarrow s_1\} \cup \{s_1 a \rightarrow a s_1 \mid a \in T\}.\end{aligned}$$

The first step, $s_0a_0 \vdash a_1ws_0$, introduces the axiom of G and the left marker a_1. Each subsequent translation $s_0a_1z \vdash^* a_1z's_0$ precisely corresponds to an equivalent derivation step $z \Longrightarrow z'$ in G. The final state s_1 can be reached only by a transition $s_0a_1z \vdash s_1z$; the work of γ can be finished only when $z \in T^*$. After removing the symbol a_1, the final state s_1 cannot be reached again, hence no further iteration can modify the string. Consequently, $L(G) = L(\gamma)$. □

As a consequence of this proof we get the inclusion $CF \subset IFT_2$. This result is also proved in [9] (Theorem 1.7) for iterated gsm's (using an intersection with T^* as a squeezing mechanism). A characterization of context-free languages is also obtained in [9]: two-state iterated gsm's such that a copying cycle exists for each state generate only context-free languages (a copying cycle is a cycle using a rule of the form $sa \to as$, that is leaving unchanged the symbol a). In view of the previous lemma, the copying cycle property is crucial in this characterization.

The IFT γ constructed above is not deterministic, even when starting from a deterministic E0L system G: we have the rules $s_0a_1 \to a_1s_0$ and $s_0a_1 \to s_1$ with the same left hand member. We do not know whether or not $DE0L \subseteq DIFT_2$, but still we have:

Lemma 4. $DIFT_2 - CF \neq \emptyset$, $DIFT_2 - 0L \neq \emptyset$.

Proof. Consider the DIFT $\gamma = (\{s_0, s_1\}, \{a_0, a, b\}, s_0, a_0, \{s_1\}, P)$, with

$$P = \{s_0a_0 \to abas_1,\ s_0a \to as_0,\ s_0b \to bs_1,\ s_1a \to aas_1\}.$$

We obtain

$$L(\gamma) = \{aba^{2^n} \mid n \geq 0\}.$$

This is obviously a non-context-free language. It is also non-0L: in order to generate strings with arbitrarily large suffixes a^{2^n} we need a rule of the form $a \to a^i, i \geq 2$; using this rule for rewriting the leftmost occurrence of a in any string from $L(\gamma)$ we get a string not in $L(\gamma)$. □

The hierarchy $IFT_n, n \geq 1$, is not an infinite one. This fact is not unexpected: we know from [11], [5], [9] that iterated gsm's characterize the family RE, which can be written as $RE \subseteq IFT$; start the proof of this inclusion from a *universal type-0 grammar*; in this way we obtain an IFT with a fixed number of states; by slightly modifying it, we can obtain an IFT generating any given recursively enumerable language (at the first step of the work of our IFT we introduce the code of the particular grammar which is used by the universal one in order to simulate the particular grammar; then, our IFT works as the universal grammar, hence it generates the language of the particular grammar). This argument only shows that the hierarchy $IFT_n, n \geq 1$, collapses, without any indication on the hight of this hierarchy.

The hierarchy on the number of states collapses also in the case of iterated gsm's in the form considered in [5], [9], [11]: the previous argument holds for this case as well. In fact, the result is stated explicitly in [9] (Theorem 2.12), but the argument is much more complex: one first gives a normal form for iterated gsm's, one proves that state-complexity families with respect to such normal form iterated gsm's are full AFL's, and then one uses the fact that the family RE is a principal AFL. No estimation of the number of levels of the obtained hierarchy is given in [9]. (The question is not considered in [5], [11].)

Lemma 5. $RE \subseteq IFT_4$.

Proof. Consider a language $L \in RE, L \subseteq T^*$, and take a grammar $G = (N, T, S, P)$ in the Geffert normal form generating it, [2]. That is, we have $N = \{S, A, B, C\}$ and the rules in P are of the following forms: $S \to x$, for $x \in (N \cup T)^+$, and $ABC \to \lambda$.

Consider three new symbols, a_0, a_1, a_2, and construct the IFT

$$\begin{aligned}\gamma &= (\{s_0, s_1, s_A, s_B\}, V \cup \{a_0, a_1, a_2\}, s_0, a_0, \{s_1\}, P'),\\ P' &= \{s_0 a_0 \to a_1 S a_2 s_0\} \cup \{s_0 \alpha \to \alpha s_0 \mid \alpha \in N \cup T \cup \{a_1, a_2\}\}\\ &\cup \{s_0 S \to x s_0 \mid S \to x \in P\} \cup \{s_0 A \to s_A,\ s_A B \to s_B,\ s_B C \to s_0\}\\ &\cup \{s_0 a_1 \to s_1\} \cup \{s_1 a \to a s_1 \mid a \in T\} \cup \{s_1 a_2 \to s_1\}.\end{aligned}$$

At the first step we introduce the axiom of G, marked to the left with the symbol a_1 and to the right with the symbol a_2. At the subsequent translations, the context-free rules $S \to x$ of P are simulated by using the state s_0. The deletion of a substring ABC is performed as follows:

$$s_0 x_1 ABC x_2 \vdash^* x_1 s_0 ABC x_2 \vdash x_1 s_A BC x_2 \vdash x_1 s_B C x_2 \vdash x_1 s_0 x_2 \vdash^* x_1 x_2 s_0.$$

Of course, none or several applications of rules in P, irrespective whether or not they are context-free or non-context-free, can be simulated at the same translation, but this does not change the generated language.

We can reach the final state s_1 only by erasing the symbol a_1 by a step $s_0 a_1 w \vdash s_1 w$; after that, only terminal symbols can be parsed, hence the string w obtained in this way is a terminal one. After erasing the symbol a_1, the final state cannot be reached once again, hence new rules in P cannot be simulated in a successful way.

The presence of the symbol a_2 ensures the fact that we do not reach the end of a string in a state s_A or s_B (note that a_2 cannot be scanned in states s_A, s_B).

Consequently, $L(\gamma) = L(G)$, that is $RE \subseteq IFT_4$. □

We do not know whether or not the result above is optimal. Anyway, IFT's with three states are very powerful:

Lemma 6. *The membership problem with respect to IFT's with three states is not decidable.*

Proof. Consider two n-tuples of non-empty strings, $x = (x_1, \ldots, x_n), y = (y_1, \ldots, y_n)$, over the alphabet $\{a, b\}$. Let a', b' be two new symbols associated with a, b, respectively. Denote by h the coding defined by $h(a) = a', h(b) = b'$, and by $mi(x)$ the mirror image of x, for any string x. We construct the IFT

$$\gamma = (\{s_0, s_1, s_2\}, \{a, b, a', b', a_0, c, d\}, s_0, a_0, \{s_0\}, P),$$

with the rules:

1. $s_0 a_0 \to c x_i d\ mi(h(y_i)) c s_0$, for $1 \leq i \leq n$,
2. $s_0 \alpha \to \alpha s_0$, for $\alpha \in \{a, b, a', b', c\}$,
3. $s_0 d \to x_i d\ mi(h(y_i)) s_0$, for $1 \leq i \leq n$,
4. $s_0 d \to s_0$,
5. $s_0 a \to s_1$, $s_1 a' \to s_0$,
6. $s_0 b \to s_2$, $s_2 b' \to s_0$.

It is easy to see that after replacing a_0, at the first step, with a string $c x_i d\ mi(h(y_i)) c$, for some $1 \leq i \leq n$, at the next iterations one replaces the occurrence of d by a string of the form $x_j d\ mi(h(y_j))$, for $1 \leq j \leq n$. In this way, we obtain all strings of the form $c x_{i_1} x_{i_2} \ldots x_{i_r} d\ mi(h(y_{i_r})) \ldots mi(h(y_{i_2}))\ mi(h(y_{i_1})) c$, for some $r \geq 1, 1 \leq i_j \leq n$, $1 \leq j \leq r$.

Then, after removing the symbol d, by the rule of type 4, the following iterations either leave the string unchanged (by using rules of type 2), or a pair aa' or bb' is deleted, via states s_1, s_2, respectively (by using rules of types 5 and 6). If we arrive in s_1 (after erasing an occurrence of the symbol a) and the next symbol is not a', then the work of γ is blocked (at least the rightmost occurrence of the symbol c has remained non-scanned). The same happens if we arrive in s_2 (after deleting an occurrence of the symbol b) and the next symbol is not b'.

Therefore, we can erase all symbols in between the two occurrences of the marker c if and only if $h^{-1}(mi(h(y_{i_r})) \ldots mi(h(y_{i_2}))\ mi(h(y_{i_1}))) = mi(x_{i_1} x_{i_2} \ldots x_{i_r})$. This means that $x_{i_1} x_{i_2} \ldots x_{i_r} = h^{-1}(mi(mi(h(y_{i_r})) \ldots mi(h(y_{i_2}))\ mi(h(y_{i_1}))))$. Because $mi(mi(h(y_{i_r})) \ldots mi(h(y_{i_2})) mi(h(y_{i_1}))) = h(y_{i_1})\ h(y_{i_2}) \ldots h(y_{i_r}) = h(y_{i_1} y_{i_2} \ldots y_{i_r})$, and h is one-to-one, this is equivalent to $x_{i_1} x_{i_2} \ldots x_{i_r} = y_{i_1} y_{i_2} \ldots y_{i_r}$, that is, to the fact that the Post Correspondence Problem for the n-tuples x, y (abbreviated, $PCP(x, y)$) has a solution. Consequently, the string cc belongs to the language $L(\gamma)$ if and only if $PCP(x, y)$ has a solution, which is not decidable. □

Lemma 7. $ET0L \subseteq IFT_3$.

Proof. For every language $L \in ET0L$ there is an ET0L system which generates L and has only two tables. Let $G = (V, T, w, P_1, P_2)$ be such a system. Let a_0, c be new symbols and construct the IFT:

$$\gamma = (\{s_0, s_1, s_2\}, V \cup \{a_0, c\}, s_0, a_0, \{s_1\}, P),$$

$$\begin{aligned} P = \{&s_0a_0 \to cws_0\} \\ &\cup \{s_0c \to cs_0\} \cup \{s_0a \to xs_0 \mid a \to x \in P_1\} \\ &\cup \{s_0c \to cs_2\} \cup \{s_2a \to xs_2 \mid a \to x \in P_2\} \\ &\cup \{s_0c \to s_1\} \cup \{s_1a \to as_1 \mid a \in T\}. \end{aligned}$$

It is easy to see that $L(G) = L(\gamma)$. If one stays in state s_0 when scanning c, then one has to completely rewrite the string according to the rules in P_1 (the passing to s_1 or to s_2 is possible only when one scans c); if one passes to s_2 after scanning c, then one has to use the rules in P_2. The rules in the two tables cannot be mixed. Finally, if one passes to s_1 after scanning (and removing) c, then the work of γ should be finished: even if the rules in P_1 can be used again and again, the final state s_1 cannot be reached any more. The use of s_1 also ensures that the obtained string is terminal. □

Synthesizing the results above, and using the known relations among Chomsky and Lindenmayer families, we obtain:

Theorem 3. (i) $IFT_1 = 0L \subset E0L \subseteq IFT_2 \subseteq IFT_3 \subseteq IFT_4 = RE$. (ii) $CF \subset IFT_2$, $ET0L \subset IFT_3$.

We do not know which of the non-proper inclusions in this theorem are proper. Also the relationships between the family CS and the families IFT_2, IFT_3 are *open*. According to the next result, if $CS \subseteq IFT_2$, then $RE \subseteq IFT_3$.

Theorem 4. *If $CS \subseteq IFT_n$, for some $n \geq 1$, then $RE \subseteq IFT_{n+1}$.*

Proof. Consider a recursively enumerable language $L \subseteq T^*$. Take a type-0 grammar $G = (N, T, S, P)$ generating L and a new symbol, c. We construct the length-increasing grammar $G' = (N, T \cup \{c\}, S, P')$, with

$$\begin{aligned} P' = \{&u \to v \mid u \to v \in P, |u| \leq |v|\} \\ &\cup \{u \to vc^s \mid u \to v \in P, |u| = |v| + s, s \geq 1\} \\ &\cup \{c\alpha \to \alpha c,\ \alpha c \to c\alpha \mid \alpha \in N \cup T\}. \end{aligned}$$

It is easy to see that $L(G') \subseteq L ⧢ \{c\}^*$ and that for each string $w \in L$ there is a string $w' \in L(G'), w' \in \{w\} ⧢ \{c\}^*$. (We have denoted by ⧢ the shuffle operation, defined by $x ⧢ y = \{x_1y_1x_2y_2 \ldots x_ny_n \mid x = x_1x_2 \ldots x_n, y = y_1y_2 \ldots y_n, n \geq 1, x_i, y_i \in V^*, 1 \leq i \leq n\}$, for $x, y \in V^*$, and extended in the natural way to languages.) Indeed, the length-decreasing rules of P are replaced by monotonous rules which introduce occurrences of the symbol c; this symbol is moved freely to right and to left, making room for left hand sides of rules in P.

Consider now an IFT $\gamma = (K, V, s_0, a_0, F, P'')$ generating the context-sensitive language $L(G')$ and having $card(K) = n$. Let a_0', c_1, c_2, c_3 be new symbols and s_1 a new state, and construct the IFT

$$\gamma' = (K \cup \{s_1\}, V \cup \{a_0', c_1, c_2, c_3\}, s_0, a_0', \{s_1\}, P'''),$$

where P''' contains all transition rules in P'' as well as the following transitions:

1. $s_0 a_0' \to c_1 x c_2 s$, for all $s_0 a_0 \to xs \in P''$ with $x \in V^*, s \in K$,
2. $s_0 c_1 \to c_1 s_0$,
   $sc_2 \to c_2 s$, for all $s \in K$,
3. $sc_2 \to c_3 s$, for all $s \in F$,
4. $s_0 c_1 \to s_1$, $s_1 c \to s_1$, $s_1 c_3 \to s_1$,
   $s_1 a \to a s_1$, for all $a \in T$.

After replacing a_0' by a string introduced by γ when scanning a_0, bounded by the markers c_1, c_2, the IFT γ' simulates the work of γ, modulo the passing over c_1 and c_2 and leaving them unchanged. The symbol c_3 cannot be scanned in a state of K. This symbol can be introduced only from a state in F, by a rule $sc_2 \to c_3 s, s \in F$. If such a rule is used, then the obtained string, of the form $c_1 w c_3$, cannot be further translated by the rules in P''. This means that a rule of the form $sc_2 \to c_3 s, s \in F$, is used only once, at the end of simulating in γ' a sequence of translations in γ.

On the other hand, the symbol c_2 cannot be scanned in the state s_1, hence the work of γ' cannot be finished as long as the symbol c_2 is present in the string. Thus, a rule $sc_2 \to c_3 s, s \in F$, must be used. This ensures the fact that the work of γ ends in a final state, as imposed by F (note that the states in F are no longer final states in γ').

A string of the form $c_1 w c_3$ can be processed on the path from s_0 to s_1; all symbols c_1, c_3, c are erased, all symbols in T are left unchanged. This is the only way of reaching the end of a string in the final state s_1. Therefore, if w above is a string from $L(G')$ such that $w \in \{z\} ⧢ \{c\}^*$, for $z \in L$, then the string generated by γ' is exactly z. In conclusion, $L(\gamma') = L$, that is, $L \in IFT_{n+1}$. □

We close this section by mentioning some closure properties of families in the hierarchy $IFT_n, n \geq 1$; proofs can be found in [3].

Theorem 5. *If $L, L_1, L_2 \in IFT_n$ and h is a morphism, then $h(L)$, L^+, $L_1 \cup L_2$, and $L_1 L_2$ are in $IFT_{n+1}, n \geq 1$.*

5 Back to Computing by Carving

The main result of this paper is the fact that non-deterministic iterated finite state sequential transducers with four states characterize the recursively enumerable languages. This implies that the number of states does not induce an infinite hierarchy of languages which are C-REG computable: in view of Theorem 2.1, a language is C-REG computable if and only if it is the complement of a language in $RE = IFT_4$. Thus, if we take into account the number of states of an IFT generating the complement of a C-REG computable language, then we can have at most four levels in the hierarchy of

C-REG computable languages. If we take into account the number of states of the gsm g describing a regular sequence of languages, identified by a pair (L_1, g), then the hierarchy is still lower: there are at most three levels. To this aim, we will repeat the proof from [6] of one implication in Theorem 2.1, making use of the basic idea used here in the proof of Lemma 4.5.

Let us denote by $CREG_n, n \geq 1$, the family of languages of the form $M - g^*(L_1)$, for M, L_1 regular languages and g a gsm with at most n states; by $CREG$ we denote the union of all these families, that is, the family of all C-REG computable languages.

Theorem 6. *$CREG_1 \subset CREG_2 \subseteq CREG_3 = CREG$.*

Proof. The inclusions $\subseteq$ are obvious. According to Theorem 2.1, each language $L \subseteq T^*$, $L \in CREG$, has the complement in RE. Take such a language L and consider a grammar $G = (N, T, S, P)$ for the language $T^* - L$; take G in the Geffert normal form, that is with $N = \{S, A, B, C\}$ and with P containing context-free rules $S \to x, x \in (N \cup T)^*$, and the non-context-free rule $ABC \to \lambda$.

We construct the regular sequence of languages starting with $L_1 = \{S\}$ and using the gsm

$$\begin{aligned} g &= (\{s_0, s_A, s_B\}, N \cup T, N \cup T, s_0, \{s_0\}, R), \\ R &= \{s_0\alpha \to \alpha s_0 \mid \alpha \in N \cup T\} \cup \{s_0 S \to x s_0 \mid S \to x \in P\} \\ &\cup \{s_0 A \to s_A,\ s_A B \to s_B,\ s_B C \to s_0\}. \end{aligned}$$

It is easy to see that $g^*(L_1) \cap T^* = L(G)$ (at each iteration of g one can simulate the application of a rule in P; if several rules are simulated at the same iteration, then this does not change the generated language).

Therefore, for the regular language $M = T^*$ we obtain $M - g^*(L_1) = T^* - L(G) = T^* - (T^* - L) = L$. In conclusion, $L \in CREG_3$.

The inclusion $CREG_1 \subset CREG_2$ is clearly proper, because by iterating a gsm with only one state we get a language in $IFT_1 = 0L$. □

The proof before does not imply that $RE = IFT_3$, because one state has been saved here by using the language M, of a specific form, which makes unnecessary the use of a special (final) state in order to check whether or not the string produced by iterating the gsm is terminal.

The previous theorem shows that in order to classify the C-REG computable languages we need other parameters describing the complexity of the regular sequences of languages, the number of states is not sufficiently sensitive. Such more sensitive parameters remain to be found.

The case of deterministic iterated transducers remains to be investigated.

Note. We are much indebted to H. Fernau for useful remarks on a previous version of this paper.

References

1. Adleman, L.M. (1994) Molecular computation of solutions to combinatorial problems. *Science* **226**, 1021–1024
2. Geffert, V. (1991) Normal forms for phrase-structure grammars. *RAIRO. Th. Inform. and Appl.* **25**, 473–496
3. Manca, V., Martin-Vide, C., and Păun, Gh. (1998) *Iterated GSM Mappings: A Collapsing Hierarchy.* TUCS Report No. 206, Turku Centre for Computer Science
4. Ouyang, Q., Kaplan, P.D., Liu, S., and Libchaber, A. (1997) DNA solution of the maximal clique problem. *Science* **278**, 446–449
5. Păun, Gh. (1978) On the iteration of gsm mappings. *Revue Roum. Math. Pures Appl.* **23**, 921–937
6. Păun, Gh. (1998) (DNA) Computing by carving. *Soft Computing*, in press
7. Păun, Gh., ed. (1998) *Computing with Bio-Molecules. Theory and Experiments.* Springer
8. Păun, Gh., Rozenberg, G., and Salomaa, A. (1998) *DNA Computing. New Computing Paradigms.* Springer
9. Rovan, B. (1981) A framework for studying grammars. *Proc. MFCS 81, Lect. Notes in Computer Sci.* **118**, Springer, 473–482
10. Rozenberg, G. and Salomaa, A., eds. (1997) *Handbook of Formal Languages.* Springer
11. Wood, D. (1976) Iterated a-NGSM maps and Γ-systems. *Inform. Control* **32**, 1–26

On the Length of Words

Solomon Marcus

Summary. Some first steps are proposed for a general approach to the structure and typology of formal languages in respect to the length of their words. A special attention is paid to the asymptotic behavior of various parameters involved in this problem.

1 Introduction

Given a language L, how many words of a certain length are in L? Related to this question, some interesting concepts and problems were considered in formal language theory and perhaps among the first results was the famous pumping lemma due to Bar-Hillel, Perles and Shamir [1] and having as one of its consequences the fact that the length set of an infinite context-free language contains an infinite arithmetic progression.

Let us denote by $f_L(n)$ the number of words in L whose length is equal to n; this is the *density function* associated with L. Various (explicit or implicit) results related to this function, mainly in the particular case when L is regular or context-free, are due to Schützenberger [12], Chomsky and Schützenberger [3], Eilenberg [4], Salomaa and Soittola [11], Berstel and Reutenauer [2]. The "generating function" associated with a language L under the form of an infinite series $g_L(z) = \sum_{n>0} a_n z^n$, where $a_n = f_L(n)$, opened a new road by the theorem asserting that, when L is an unambiguous context-free language, the analytic function $g_L(z)$ is an algebraic function over the field of rationals (Chomsky and Schützenberger [3]). However, this very attractive approach failed to lead to a general theory of length sets in the field of formal languages.

Our aim is to stimulate the development of a general framework in the study of the length of words and to accomplish some first steps in this direction.

2 A Characterization of the Density Function

Given an arbitrary infinite sequence (a_n) of positive integers (zero included), under what conditions does there exist a language L over the alphabet A, whose density function is given just by $f_L(n) = a_n$? Denoting by p the number of elements in A, it is easy to see that a necessary condition for the existence of L is $a_n \leq p^n$ for $n = 1, 2, \ldots$ Is this condition also sufficient for the existence of L? The answer is affirmative, because we can choose in various ways a_n words of length equal to n among the total number p^n of such words on A (for $n = 1, 2, \ldots$) and define L as the language containing the words selected in this way. We conclude:

Proposition 1. *A necessary and sufficient condition for an infinite sequence* (a_n) *of positive integers to be the density function of a language* L *on the alphabet* A *(i.e.,* $f_L(n) = a_n$ *for* $n = 1, 2, \ldots$*) is* $a_n \leq p^n$*, where* p *is the cardinality of* A*.*

We also have:

Proposition 2. *If card* $A = 1$*, then the language* L *in Proposition 1 is uniquely determined. If card* $A > 1$*, then there are infinitely many languages on* A *having the same given density function.*

Remark 1. Proposition 2 suggests the consideration of the following equivalence relation: two languages L and M are *equivalent in length* if $f_L(n) = f_M(n)$ for $n = 1, 2, \ldots$

We may ask: *how much different can be two languages equivalent in length?* The question is relevant only when card $A > 1$ and in this case it is easy to see that *there always exist disjoint languages equivalent in length.* For instance, if $A = \{a, b\}$, then $L = \{a^n \mid n = 1, 2, \ldots\}$ and $M = \{b^n \mid n = 1, 2, \ldots\}$ are equivalent in length, because $f_L(n) = f_M(n) = 1$ for $n = 1, 2, \ldots$ We can also take $L = \{a, a^n b^k \mid n = 1, 2, \ldots, k = 1, 2, \ldots\}$, $M = \{b, b^n a^k \mid n = 1, 2, \ldots, k = 1, 2, \ldots\}$. L and M are disjoint and equivalent in length: $f_L(1) = f_M(1) = 1$, because a in L and b in M are the only words of length one; $f_L(2) = f_M(2) = 1$, because ab in L and ba in M are the only words of length 2; $f_L(3) = f_M(3) = 2$, because ab^2, a^2b in L and ba^2, b^2a in M are the only words of length 3, etc. This example shows that we need a more general equivalence relation: *two languages* L *and* M *are said to be almost-equivalent in length if there exists a positive integer* p *such that* $f_L(n) = f_M(n)$ *for any* $n > p$. An interesting problem is to determine what classes of languages (for instance, those in Chomsky hierarchy) are invariant through almost-equivalence in length.

So far, we paid no attention to the generative nature of the language L. In view of the pumping lemma [1], we can bring the following information:

A necessary condition for the language L *in Proposition 2 to be context-free is that the sequence* (a_n) *in Proposition 2 contains an infinite arithmetic progression.*

It would be interesting to obtain more information about the behavior of the density function for various classes of languages. The particular case of regular languages was investigated by Szilard, Yu, Zhang and Shallit [13], who proved the existence of some gaps for their possible density functions. For instance, there is no regular language that has a density of the order $\sqrt{n}$, $n \log n$, or $2^{\sqrt{n}}$; see, for more information, Yu [14].

3 Full Intervals and Lacunary Intervals

Given a language L on an alphabet A, a *full interval for* L is a sequence of consecutive positive integers, such that: 1) for each term k in the sequence there exists a word in L whose length is equal to k; 2) if i is the first term in the sequence, then there is no word in L whose length is equal to $i - 1$; 3) if j is the last term in the sequence, then there exists no word in L whose length is equal to $j + 1$. We will denote this interval by $[i, j]$. The length of this interval is the number of positive integers k such that $i \leq k \leq j$. It may happen that $i = j$ (intervals of length one), $i = 0$ (in this case condition 2 above is devoid of object), or $j = \infty$ (in this case we should write $[i, \infty)$ and condition 3 above is devoid of object).

Given two full intervals $[i, j]$ and $[p, r]$ for L, $j < p$, we say they are *consecutive* if there is no positive integer m, $j < m < p$, which is the length of a word in L. In this case, we say that the sequence $j+1, j+2, \ldots, p-2, p-1$ is a *lacunary interval* for L and we denote it by $[j + 1, p - 1]$.

Proposition 3. *Given two arbitrary sequences (m_k) and (n_k) of strictly positive integers, there exists a language L on the alphabet $\{a\}$ such that m_k is the length of the k-th full interval of L, while n_k is the length of the k-th lacunary interval of L $(k = 1, 2, \ldots)$.*

Proof. A language L with the required properties is $\{a^p \mid p \in S\}$, where S is the union of the following full intervals for L: $[1, m_1]$, $[m_1 + n_1 + 1, m_1 + n_1 + m_2]$, $[m_1+n_1+m_2+n_2+1, m_1+n_1+m_2+n_2+m_3], \ldots$, $[m_1+n_1+m_2+n_2+\ldots+m_k+n_k+1, m_1+n_1+m_2+n_2+\ldots+m_k+n_k+m_{k+1}], \ldots$, while the lacunary intervals for L are: $[m_1 + 1, m_1 + n_1]$, $[m_1 + n_1 + m_2 + 1, m_1 + n_1 + m_2 + n_2]$, $[m_1 + n_1 + m_2 + n_2 + m_3 + 1, m_1 + n_1 + m_2 + n_2 + m_3 + n_3], \ldots$

Remark 2. Proposition 3 remains valid when the alphabet $\{a\}$ is replaced by an arbitrary finite non-empty alphabet A. In order to avoid the trivial case when some letters in A are not used in the considered language on A, we define *complete words* on A as those words that use all letters in A.

Proposition 4. *Proposition 3 remains valid when $\{a\}$ is replaced by an arbitrary finite non-empty alphabet.*

Proof. The non-trivial case refers to languages whose words are complete. Let us proceed by induction. The first step was accomplished by Proposition 3. For the second step, let us suppose that Proposition 3 is valid for an alphabet A_{n-1} of cardinality $n-1$. This means that given the sequences (m_k) and (n_k) of strictly positive integers, there exists a language L_{n-1} such that m_k is the length of the k-th full interval of L_{n-1}, while n_k is the length of the k-th lacunary interval of L_{n-1}. Consider the k-th full interval $[i-1, j-1]$ for L_{n-1}. The language L_n we are looking for will be obtained by adding one more letter to the words in L_{n-1}, so the length of each word in L_n will increase with one in respect to words in L_{n-1}. The interval $[i-1, j-1]$ will become the full interval $[i, j]$ for L_n, the length of $[i, j]$ being the same as the length of $[i-1, j-1]$, i.e., m_k. A similar phenomenon will occur with the lacunary intervals.

4 Asymptotic Behavior

Let us consider, for any language L, the ratio $f_L(n)/n$, where f_L is the density function of L. A relevant parameter of the asymptotic behavior of L in respect to the length of its words will be the behavior of the above ratio, when n is increasing. Since $f_L(n) \leq p^n$ (see Proposition 1), we have $0 \leq \frac{f_L(n)}{n} \leq \frac{p^n}{n}$, where $p = \text{card } A$, so $\lim_{n\to\infty}(p^n/n) = \infty$. Obviously, we will have some times $\lim_{n\to\infty}(f_L(n)/n) = 0$, for instance, when $f_L(n) \leq k$, for a given constant k; the language L is called in this case *slender*; see Păun and Salomaa [8], [9], [10]. On the other hand, we may have $f_L(n) = p^n$ for all n enough large (when we introduce in L all words of length n, i.e., L is almost universal), so $\lim_{n\to\infty}(f_L(n)/n) = \infty$.

More generally, we can assert:

Proposition 5. *Given $0 \leq \alpha \leq \infty$, there exists a language L with*

$$\limsup_{n\to\infty} \frac{f_L(n)}{n} = \alpha.$$

Proof. The cases $\alpha = 0$ and $\alpha = \infty$ were already settled, so let us suppose $0 < \alpha < \infty$. If α is rational, $\alpha = m/r$ (m, r positive integers such that m/r is irreducible), then we consider an alphabet A of cardinality $p \geq 2$ and we choose L over A such that there are km words of length kr in L, for $k = 1, 2, \ldots$ Thus, $f_L(kr)/kr = km/kr = m/r$ for any positive integer k, while $f_L(n) = 0$ for any n which is not a multiple of r. This is possible because, for k enough large, $km < p^{kr}$; indeed, $m < (p^{kr})/k$ for k enough large, due to the fact that $\lim_{k\to\infty}(p^{kr})/k = \infty$.

If α is irrational, then there exists an increasing sequence of rational numbers (p_n/q_n) converging to α. We may assume that (q_n) is strictly increasing (so (p_n) is increasing, too). Define L by the following rule: for each n, L contains p_n words of length equal to q_n. We have $f_L(q_n) = p_n$ for $n = 1, 2, \ldots$ It follows that

$$\frac{f_L(q_n)}{q_n} = \frac{p_n}{q_n}, \text{ so } \lim_{n\to\infty} \frac{f_L(q_n)}{q_n} = \alpha.$$

For any n wich is not a term of (q_n) we put $f_L(n) = 0$, so

$$\limsup_{n\to\infty} \frac{f_L(n)}{n} = \alpha.$$

Remark 3. The converse of Proposition 5 is also true: to any language L we can associate a non-negative number $\alpha(L)$ such that $\lim\sup(f_L(n)/n) = \alpha(L)$ when $n \to \infty$. We can call L rational (irrational, algebraic, transcendental) if $\alpha(L)$ is rational (irrational, algebraic, transcendental, respectively), and it would be interesting how is this typology of $\alpha(L)$ represented for various classes of languages.

5 A Generalization and a Typology of Slender Languages

The considerations above offer a new framework to look at a slender language L, defined by the existence of a natural number k such that $f_L(n) \leq k$ for $n = 1, 2, \ldots$, [8], [9], [10], [6], [7]. Obviously, for any slender language L the associated parameter $\alpha(L)$ (see Remark 3) is equal to zero. A natural question arises: is the converse of this assertion true? The answer is negative, as it follows from

Proposition 6. *There exists on an alphabet A with card $A > 1$ a language L which is not slender, but*

$$\lim_{n\to\infty} \frac{f_L(n)}{n} = 0.$$

Proof. Define a sequence (a_n) by a_n = the number of prime numbers not larger than n. In view of the fundamental theorem of prime numbers, there exists $k > 1$ such that $a_n < k(n/\ln n)$. On the other hand, for n enough large and because the cardinality p of A is strictly larger than 1,

$$a_n < k(n/\ln n) < p^n, \text{ so } a_n < p^n.$$

We can apply Proposition 1. It follows that there exists a language L on A such that, for n enough large, $f_L(n) = a_n$. On the other hand, we have

$$\frac{a_n}{n} < k \cdot \frac{n}{n(\ln n)} = \frac{k}{\ln n}, \text{ so } \lim_{n\to\infty} \frac{a_n}{n} = 0,$$

and the proposition is proved.

Let us call a language L *asymptotically zero* if $\lim_{n\to\infty}(f_L(n)/n) = 0$. It follows from Proposition 6 that such languages are an effective generalization of slender languages. (Note that this extension is in a direction different from that considered by Honkala [5].) Moreover, in respect to the speed of the convergence to zero of the sequence $(f_L(n)/n)$ *we get an infinite hierarchy of asymptotically zero languages.* Slender languages are at the bottom of this hierarchy, because for these languages, for n enough large, all terms of the considered sequence are equal to zero. But slender languages, in their turn, are disposed in an infinite hierarchy: we say that a slender language is of rank k if the number of the full intervals of the associated set $\{f_L(n) \mid n = 1, 2, \ldots\}$ (this number is in any case finite for any slender language) is equal to k. There exist, for any strictly positive integer k, some languages of rank equal to k and having all their k full intervals of finite length. These languages are obligatory slender.

There is another hierarchy of slender languages, proposed by Păun and Salomaa: a language L is k-slender if $f_L(n) \leq k$ for any $n = 1, 2, \ldots$ For every positive integer k there are k-slender languages that are not $(k-1)$-slender. Obviously, a k-slender language may have different ranks. If we call a language *essentially k-slender* when it is k-slender, but not $(k-1)$-slender, then we can assert that the rank of a language tells us almost nothing about its essential slenderness. So, *rank and essential slenderness are rather incomparable.* Perhaps some restrictions appear when we consider some specific classes of languages.

6 Fullness Versus Lacunarity

Propositions 3 and 4 suggest the introduction of some relevant parameters measuring the asymptotic behavior of the fullness-lacunarity ratio. A first idea is to measure this behavior by the parameter

$$\lambda = \lim_{k \to \infty} \frac{m_k}{n_k}.$$

When $\lambda = \infty$, the language is said to be *rich in length.* When $\lambda = 0$, the language is *poor in length.* A language which is poor in length may not be slender, as it can be seen by taking $m_k =$ the largest positive integer not larger than $\sqrt{n_k}$ (this is possible in view of Propositions 3 and 4). Languages poor in length form an infinite hierarchy, in respect to the speed of convergence to zero of the ratio m_k/n_k. It would be interesting to relate this hierarchy to the hierarchy of asymptotically zero languages, the extension of slender languages considered in Section 5.

Another, more relevant, parameter measuring the ratio between fullness and lacunarity is

$$\mu = \lim_{n \to \infty} \frac{M_n}{N_n},$$

where $M_n = \sum m_k$ and $N_n = \sum n_k$, the sum being considered for all full intervals and for all lacunary intervals contained in the interval $[0, n]$.

For both λ and μ it may happen that the respective limit does not exist; in such cases we replace it by lim sup. The language L will be considered *balanced* if there exists a strictly positive number K such that $\frac{1}{K} < \frac{m_k}{n_k} < K$ for any k enough large or, better, $\frac{1}{K} < \frac{M_n}{N_n} < K$ for any n enough large. L is *predominantly full* if $\lambda > 1$ and *predominantly lacunary* if $\lambda < 1$. If $\lambda = 1$, then L is *perfectly balanced.* If $m_k = Cn_k$ for any k enough large, then the language is predominantly full when the constant C is > 1, predominantly lacunary if $C < 1$ and perfectly balanced if $C = 1$.

Since, in view of Propositions 3 and 4, m_k and n_k are arbitrary positive integers, λ can be an arbitrary positive real number and it can also be infinite. We get a potential typology of languages, in respect to the nature of λ. If the sequence (m_k/n_k) is convergent, then we will say that the corresponding language is *convergent in length*; otherwise it is divergent. Divergence can be of two types: if the limit λ exists, but it is infinite, then we say that the language is *divergent to infinity*; otherwise, it is *oscillating* and we direct our attention to the superior limit of (m_k/n_k). Similar distinctions can be made in respect to the parameter μ. Shortly, we can assert that the typology of languages in respect to λ and μ is reflected by the typology of real numbers. The language L will be considered *scattered in length* if all its full intervals in length are of length equal to one. If such a language is convergent in length, then it converges to a number between zero and one; conversely, any real number between zero and one (0 and 1 included) is the limit λ associated with a suitable language scattered in length.

An interesting situation occurs when $\lambda = 1$. In this case both L and its complementary are scattered in length. However, this is a very restrictive condition; it corresponds to one of the following two cases: 1) L contains all words whose length is an odd number; 2) L contains all words whose length is an even number.

7 Pairs of Languages and the Fullness-Lacunarity Duality

Two languages L and M are said to be *heterogeneous in length* if the length of a word in L (M) is never the length of a word in M (L, respectively). For instance, $L = \{a^n b^n \mid n = 1, 2, \ldots\}$ and $M = \{a^m b^n \mid m + n \text{ odd and prime}\}$ are in this situation, because any word in L has as its length an even integer, while any word in M has as its length an odd integer. Denoting by $s(L)$ the length set of L, for an arbitrary language L, heterogeneity in length means $s(L) \cap s(M) = \emptyset$. If $s(L) = s(M)$, then L and M are *weakly equivalent in length*. Obviously, *equivalence in length implies weakly equivalence in length, but the converse is not true* (excepting the case of an alphabet with only one element). For instance, if we take $A = \{a, b\}$, the universal language A^* and the language $\{a^m b^n \mid m, n \geq 0\}$ are not equivalent in length, but they are weakly equivalent in length. Weak equivalence raises problems similar to those related to equivalence in length. The main question is: how different can be two languages weakly equivalent in length? Taking into consideration that we are mainly interested in the behavior of words with an increasing length, two more equivalence relations have to be considered: *almost equivalence in length* and *almost weakly equivalence in length*. L and M are almost equivalent in length if there exists a positive integer p such that $f_L(n) = f_M(n)$ for any

$n \geq p$; L and M are almost weakly equivalent in length if the symmetric difference $s(L)\Delta s(M)$ is finite.

L is *poorer in length than* M if $s(L) \subseteq s(M)$; equivalently, M is *richer in length than* L. It may also happen that L and M are not comparable, without being disjoint. The "almost" variant of "disjoint", "poorer in length than" is obtained when some suitable finite subsets of $s(L)$ and $s(M)$ have to be ignored in order to obtain the respective relations. Examples of all these situations can be easily obtained.

Two languages L and M are said to be *complementary in length* if they are heterogeneous in length and if $s(L) \cup s(M) = \{0, 1, 2, \ldots\}$.

Proposition 7. *If card $A > 1$, then complementarity and complementarity in length are not comparable; for card $A = 1$, they are equivalent.*

Proof. Let $L = \{a^n b^n \mid n = 0, 1, 2, \ldots\}$ and $M = \{a^m b^n \mid m$ odd, n even, $n, m \geq 0\}$. We have $s(L) = \{0, 2, 4, \ldots, 2n, \ldots\}$ and $s(M) = \{1, 3, 5, \ldots, 2n+1, \ldots\}$, so $s(L) \cup s(M) =$ the set of positive integers, while L and M are heterogeneous in length. However M is only a part of the complementary set of L; so, complementarity in length does not imply complementarity. Conversely, let L be the same as above and $M = \{a^m b^n \mid m \neq n,\ m, n \geq 0\}$. L and M are complementary, but not complementary in length. The second part follows from Proposition 2.

We will now give a characterization of languages heterogeneous in length and of languages complementary in length.

Proposition 8. *L and M are heterogeneous in length if and only if any interval full in length for L (M) is lacunary in length for M (L, respectively). L and M are complementary in length if and only if any interval lacunary in length for L (M) is full in length for M (L, respectively).*

Proof. Suppose that L and M are heterogeneous in length, i.e., $s(L) \cap s(M) = \emptyset$ and let I be an interval full in length for L. In order to prove that I is a lacunary interval for M, let us accept, by contradiction, the existence of $a \in I \cap s(M)$. From $a \in I$ and $I \subseteq s(L)$ it follows that $a \in s(L)$, so $a \in s(L) \cap s(M)$, in contradiction with the hypothesis. The validity of the statement obtained by changing L with M and M with L follows from symmetry considerations (heterogeneity is a symmetric relation): if the hypothesis is symmetric in respect to L and M, then the conclusion should be symmetric too.

Conversely, if any interval full for L (M) is lacunary for M (L, respectively), then any a in $s(L)$ belongs to a full interval I for L, which should be, by hypothesis, lacunary for M, so a cannot be in $s(M)$; this means that $s(L) \cap s(M) = \emptyset$, so L and M are heterogeneous in length.

Suppose now that L and M are complementary in length; this means that $s(L) \cap s(M) = \emptyset$ and, moreover, that $s(L) \cup s(M) = \{0, 1, 2, \ldots\}$. If I is a lacunary interval for L and $a \in I$, then a is absent from $s(L)$, so, in view of the complementarity in length, it follows $a \in s(M)$, so $I \subseteq s(M)$ and I is full in length for M. For symmetry considerations, things behave similarly when L is replaced by M and M by L.

The use of the operator $s(L)$ requires the explicit status of its behavior, as it is given by

Proposition 9. *For any pair of languages L, M we have: a) $s(L \cup M) = s(L) \cup s(M)$; b) $s(L \cap M) \subseteq s(L) \cap s(M)$; c) If $L \subseteq M$, then $s(L) \subseteq s(M)$, but the converse is not true; d) $s(LM) = s(L) + s(M)$.*

Proof. If $\alpha \in s(L \cup M)$, then there exists $x \in L \cup M$ with $\alpha = s(x)$. If $x \in L$, then $a \in s(L)$; if $x \in M$, then $\alpha \in s(M)$; in both cases $\alpha \in s(L) \cup s(M)$. Starting now with the last relation as hypothesis, we have $\alpha \in s(L)$ or $\alpha \in s(M)$, but in both cases $\alpha \in s(L \cup M)$ and a) is proved.

If $\alpha \in s(L \cap M)$, then $\alpha \in s(L)$ and $\alpha \in s(M)$, so $\alpha \in s(L) \cap s(M)$. The converse is not true, as it can be seen when L and M are complementary, but not complementary in length, a possibility assured by Proposition 7. So, b) is proved. If $L \subseteq M$, then from $\alpha \in s(L)$ follows the existence of $x \in L$ with $s(x) = \alpha$ and, because $x \in M$ too, we have $\alpha \in s(M)$. The converse is not true, as it can be seen by taking $L = \{a^n b^n \mid n \geq 1\}$, $M = \{a^m b^n \mid m \neq n,\ m, n \geq o\}$. Indeed, $s(L) = \{2n \mid n = 1, 2, \ldots\}, s(M) = \{1, 2, \ldots\}$, so $s(L) \subseteq s(M)$, but L is not contained in M; moreover, $L \cap M = \emptyset$. So, c) is proved.

Let us now take $\alpha \in s(LM)$. This means the existence of $u \in L, v \in M$ with $\alpha = s(uv)$. But $s(uv) = s(u) + s(v)$, so $\alpha \in s(L) + s(M)$. Conversely, if $\alpha \in s(L) + s(M)$, then there exist $u \in L$ and $v \in M$ with $\alpha = s(u) + s(v) = s(uv)$, where $uv \in LM$, so $\alpha \in s(LM)$ and d) is proved.

All the above concepts and properties can be extended to classes of languages. Given a class F of languages and a binary relation r between languages, we may say that the *class F satisfies the relation r* if for any two languages L, M in F we have $L\ r\ M$. The relation r can mean here heterogeneity, almost heterogeneity, complementarity in length, etc.

References

1. Bar-Hillel, Y., Perles, M., and Shamir, E. (1961) On formal properties of phrase structure grammars. *Zeitschrift für Phonetik, Sprachwissenschaft und Kommunikationsforschung* **14**, 143–172
2. Berstel, J. and Reutenauer, C. (1978) *Relational Series and Their Languages*, Springer, Berlin, New York
3. Chomsky, N. and Schützenberger, M.P. (1963) The algebraic theory of context-free languages. In: *Computer Programs and Formal Systems*, North-Holland, Amsterdam, 118–161
4. Eilenberg, S. (1974) *Automata, Languages and Machines*, vol. A. Academic Press, New York

5. Honkala, J. (1996) Parikh slender languages and power series. *Journal of Computer and System Sciences* **52**, 185–190
6. Ilie, I. (1994) On a conjecture about slender context-free languages. *Theoretical Computer Science* **132**, 427–434
7. Ilie, L. (1998) *Decision Problems on Orders of Words.* Turku Centre for Computer Science, TUCS Dissertations No. 12, 160 pp.
8. Păun, Gh. and Salomaa, A. (1992) Decision problems concerning the thinness of D0L languages. *EATCS Bulletin* **46**, 171–181
9. Păun, Gh. and Salomaa, A. (1993) Closure properties of slender languages. *Theoretical Computer Science* **120**, 293–301
10. Păun, Gh. and Salomaa, A. (1995) Thin and slender languages. *Discrete Applied Mathematics* **61**, 257–270
11. Salomaa, A. and Soittola, M. (1978) *Automata Theoretic Aspects of Formal Power Series.* Springer, Berlin, New York
12. Schützenberger, M.P. (1962) Finite counting automata. *Information and Control* **5**, 91–107
13. Szilard, A., Yu, S., Zhang, K., and Shallit, J. (1992) Characterizing regular languages with polynomial densities. *Proc. of the 170th Intern. Symposium on Mathematical Foundations of Computer Science, Lecture Notes in Computer Science* **629**, Springer, Berlin, New York
14. Yu, S. (1997) Regular languages. In: *Handbook of Formal Languages* (Rozenberg, G. and Salomaa, A., eds.), vol. I, Springer, 41–110

An Insertion into the Chomsky Hierarchy?

Robert McNaughton*

Summary. This review paper will report on some recent discoveries in the area of Formal Languages, chiefly by F. Otto, G. Buntrock and G. Niemann. These discoveries have pointed out certain break-throughs connected with the concept of growing context-sensitive languages, which originated in the 1980's with a paper by E. Dahlhaus and M.K. Warmuth. One important result is that the deterministic growing context-sensitive languages turn out to be identical to an interesting family of formal languages definable in a certain way by confluent reduction systems.

1 Growing context-sensitive languages

There are several reasons for proposing that the family of GrCSL's (growing context-sensitive languages) be considered as a new level in the Chomsky hierarchy of languages. The insertion would be between the CFL's (context-free languages) and the CSL's (context-sensitive languages), in effect making a new family designated as the "type one-and-a-half languages". This family began to receive attention in 1986, when Dahlhaus and Warmuth published their result [5] that the complexity of its membership problem had a polynomial-time algorithm, in contrast to the P-space-complete membership problem for the larger family of context-sensitive languages.

A GrCSG (growing context-sensitive grammar) is one in which $|\alpha| < |\beta|$ for every rule $\alpha \to \beta$; a GrCSL is the language of a GrCSG. This class of grammars is a proper subclass of the class of CSG's (context-sensitive grammars) where rules are also permitted in which $|\alpha| = |\beta|$. (See pp. 200–214 of Arto Salomaa's 1969 book [3] for an elucidating treatment of context sensitive grammars.)

The GrCSL's are just one of many studied families properly between the CFL's and the CSL's. Another was the family of CSL's with linear-bounded derivations (see, e.g., Ron Book's dissertation [2]), i.e., languages having a CSG with a bound B such that every $w \in \Sigma^*$ derived in the grammar has a derivation whose length is $\leq B|w|$. In 1964 Gladkij proved in [1] that the CSL

$$\{wcw^R cw | w \in \{a, b\}^*\}$$

* Paljon onnea, Arto. In other words, Onnellista syntymäpäivää.

does not have linear-bounded derivations (see also the appendix to [2]). (w^R means w written backwards.)

More recently it has been proved that the GrCSL's are a proper subfamily of the family of CSL's with linear bounded derivations (see [12], and also [13], Corollary 5.4). However, the latter family will no longer be of concern in this paper.

From the work of Lautemann [6] and Buntrock [12] it follows that the language $\{ww | w \in \{a,b\}^*\}$ is not a GrCSL. (See also [13], especially the penultimate paragraph of Section 1.) Thus we have an improvement on the Gladkij language for a paradigm CSL that is not a GrCSL.

All CFL's consisting of words of length ≥ 2 are GrCSL's. The easy proof is by a constructive modification of the Chomsky normal form. But not all GrCSL's are context-free, e.g, $\{ba^{2^n} | n \geq 1\}$, a GrCSG for which is

$$S \to SK | baa$$

$$baaK \to baaaa$$

$$aK \to Kaa$$

The concept GrCSG is based on the length of strings. It is convenient to allow as GrCSG's the grammars that satisfy a variant of the definition based on the weighted length of strings. A *weighting function* ϕ on the words over an alphabet Σ maps each word to an integer satisfying the following: (1) $\phi(x) > 0$ for all $x \in \Sigma$, (2) $\phi(\lambda) = 0$ (λ is the null string) and (3) $\phi(xy) = \phi(x) + \phi(y)$ for all words x and y. We define a grammar to be a *GrCSG in the new sense* if there is a weighting function ϕ on the words over the total alphabet such that $\phi(\alpha) < \phi(\beta)$ for every rule $\alpha \to \beta$. Following [8] and [11] we can prove that, if G is a GrCSG in the new sense then there is a G' that is a GrCSG in the original sense such that $L(G') = L(G) \cap \Sigma\Sigma\Sigma^*$.

Note that a context-free grammar in Chomsky normal form whose language does not have the null word is a GrCSG in the new sense: take $\phi(x) = 1$ for x a variable and $\phi(x) = 2$ for x a terminal.

It will be convenient to adopt this new definition of GrCSG for the remainder of this paper, yielding the slight change in the definition of GrCSL.

As is well known, a language is context-sensitive if and only if it is recognized by a nondeterministic LBA (linear bounded automaton). We can get a corresponding result for GrCSL's by modifying this automaton to one whose tape decreases in its weighted length at every move. The best way to work out this idea precisely is to follow Buntrock and Otto in Section 3 of [13], stipulating an automaton with two pushdown tapes, representing the portions of the LBA tape to the left of the head and the right of the head, respectively. An elaborate weighting function is defined for configurations of the automaton, satisfying the condition that if one configuration is followed by another then the weight of the latter is less than the weight of the former. This weighting function is based on a weighting function of words over the alphabet, but it is far too complicated to be described here. (One trick is to get the effect of weighing each character of Σ more heavily on one pushdown than on the other.)

We thus have a nondeterministic shrinking two-pushdown automaton and the result that a language is a GrCSL if and only if it is recognized by such a device. The proof given by Buntrock and Otto is somewhat similar to the proof that a language is a CSL if and only if it is accepted by a nondeterministic LBA.

The question naturally arises as to which CSL's are GrCSL's and which are not. No broad answer has been given to this question. Some well known CSG's using length-preserving rules have languages that turn out to be GrCSL's, an example being the grammar:

$$S \to SABC|dABC$$

$$BA \to AB \qquad CB \to BC \qquad CA \to AC$$

$$dA \to da \qquad aA \to aa \qquad aB \to ab$$

$$bB \to bb \qquad bC \to bc \qquad cC \to cc$$

Its language $\{da^nb^nc^n|n \geq 1\}$ also has the GrCSG:

$$S \to SK|SL|dabc$$

$$cK \to Kcc \qquad bK \to Kbb$$

$$aK \to Kaa \qquad daK \to daa$$

$$cL \to Mccc \qquad cM \to Mcc \qquad bM \to Nbbb$$

$$bN \to Nbb \qquad aN \to Kaaa \qquad daN \to daaa$$

The weighting function is $\phi(x) = 1$ if x is a variable, $\phi(x) = 2$ if x is a terminal. (To derive $da^{32}b^{32}c^{32}$ in this grammar we would begin by deriving $dabcK^5$. But to derive $da^{37}b^{37}c^{37}$ we would note that $37 = 2^5 + 2^2 + 2^0$, and accordingly we would begin by deriving $dabcKKLKL$.)

With a bit more trouble we could get a GrCSG for $\{da^nb^nc^n\}$ in the original sense, i.e., one in which the weighting function is $\phi(x) = 1$ for all variables and terminals x. Also, if we wished to get rid of the d, which acts as a left-end marker, we could do so at the expense of further complication in the grammar.

GrCSG's have the advantage over other CSG's in that, in each such grammar, the length of a derivation has an upper bound that is linear in the length of the word derived. Moreover, as mentioned, the membership problem for every GrCSL has a polynomial-time algorithm. (The proof in [5] goes over to our new definition of GrCSL.)

Dahlhaus and Warmuth [5] prove that every GrCSL is log-tape reducible to some CFL. Buntrock and Otto [13] improve on this result by showing that this reduction can be done as a one-way log-space reduction; that is to say, the GrCSL is (quoting Section 1 of [13]) "accepted by an auxiliary pushdown automaton with logarithmic space bound and polynomial time bound that uses its input tape in a one-way fashion."

As pointed out in [8], the family of GrCSL's is an abstract family of languages, that is to say, this family is closed under union, concatenation, star iteration, intersection with regular languages, null-word-free homomorphisms and inverse homomorphisms.

A persistent problem for theoretical computer scientists has been to find a family of formal languages that will include all programming languages or most programming languages. The family must be reasonably simple conceptually and must not be so broad as to include languages that have properties that no reasonable programming language could have. This objective is necessarily vague and this is no place to attempt to refine it or even to discuss it, except to make a brief negative point about the family of GrCSL's: If the family of formal languages must include a language of programs whose variables occurring in executable statements must also occur in declaration statements, and if the family includes variables of unlimited length, then the family of GrCSL's is not a suitable family for this purpose. The argument for this negative assertion is based on the piece of evidence that $\{ww|w \in \{a,b\}^*\}$ is not a GrCSL, which indicates that any programming language in which a defined program may have arbitrarily long variables, but only those that are declared, is probably not a GrCSL.

The application of ideas from Theoretical Computer Science to actual computing is difficult to predict with any precision. But it helps to have some general idea of the possibility of some application, even if the exact nature of this application is vague. There will be more to say about the applicability of these ideas on GrCSL's when we investigate the deterministic variety of them in the next section.

2 The deterministic variety

A GrCSL is *deterministic* if it is recognized by a deterministic shrinking two-pushdown automaton. The name for this automaton is rather long; let us call it a "D-shrink" for short. As it is an important concept, it deserves a formal definition. The following is adapted from the paper by Buntrock and Otto [13]:

A *D-shrink* is a 7-tuple

$$(Q, \Sigma, \Gamma, \delta, q_0, B, F).$$

Here Q is the set of states, Σ the input alphabet, Γ the tape alphabet ($\Sigma \subset \Gamma$ and $\Gamma \cap Q = \emptyset$), q_0 is the initial state, B is the bottom marker of the two pushdown stores ($B \in \Gamma - \Sigma$), F is the set of accepting states, and δ is the transition function:

$$\delta : Q \times \Gamma^2 \to Q \times \Gamma^* \times \Gamma^* \cup \{\emptyset\}.$$

A configuration in a computation of a D-shrink is given as uqv where $u, v \in \Gamma^*$ and $q \in Q$. The idea is that u and v are the words on the left and right pushdown tapes, respectively. Normally B is the leftmost character of u and the rightmost character of v and occurs nowhere else. Where $u = u'a$, $v = bv'$, $a, b \in \Gamma$ and $\delta(q, a, b) = (q', w_1, w_2)$, the word $u'w_1q'w_2v'$ is the next configuration in that computation. If $\delta(q, a, b) = \emptyset$ then uqv is a halting configuration. An initial configuration is of the form Bq_0vB where v is the input. The bottom marker B is never created or destroyed in a computation but may be sensed. Acceptance of v is either by final state (a configuration $u'qv'$ where $q \in F$) or by empty store (a configuration BqB for $q \in Q$). The *language* of a D-shrink is the set of all accepted inputs.

What makes the D-shrink shrinking is the stipulation that there is a weighting function ϕ on strings over the alphabet $Q \cup \Gamma$ such that, for $\delta(q, a, b) = (q, w_1, w_2)$, the condition $\phi(w_1qw_2) < \phi(aqb)$ holds.

As mentioned in Section 1, a language is a GrCSL if and only if it is the language of a nondeterministic shrinking two-pushdown automaton ([13], Section 3). If a GrCSL is accepted by a D-shrink (i.e., a deterministic automaton of the variety) then the language is said to be a *DGrCSL (deterministic growing context-sensitive language)*. I shall argue in the remainder of this paper that the family of DGrCSL's is an important family of languages, perhaps more important than the larger family of all GrCSL's.

The language $\{ba^{2^n} | n \geq 1\}$ is a DGrCSL. One can design a D-shrink for it based on the grammar from Section 1:

$$S \to SK|baa$$

$$baaK \to baaaa$$

$$aK \to Kaa$$

Any word in the language of this grammar will be processed by the D-shrink according to a rightmost derivation in the grammar, which means that the word is processed from left to right. For example, the rightmost derivation of the word ba^{32} has the line ba^3KaaKK, to which the rule $aK \to Kaa$ is applied to the rightmost aK, resulting in the line $ba^3KaKaaK$. The automaton does things in reverse of the order in the derivation; for that step it might have Bba^3Ka on the left tape and KB on the right tape, and might be in a state showing that Kaa is between the ba^3Ka and the K. It then would push aK onto the left tape, and would go into a state showing that the null word is between the Bba^3KaaK on the left tape and the KB on the right tape.

We shall not verify that this automaton can be made deterministic, and therefore that the language $\{ba^{2^n} | n \geq 1\}$ is a DGrCSL. The language

$\{da^nb^nc^n|n \geq 1\}$ is also a DGrCSL; details beyond the discussion of this language in Section 1 are omitted.

It is not difficult to prove that the family of DGrCSL's is closed under complementation. This observation enables us to prove the existence of CFL's that are not DGrCSL's. Such a language is $\{a,b\}^* - \{ww|w \in \{a,b\}^*\}$. If this CFL were a DGrCSL then its complement $\{ww|w \in \{a,b\}^*\}$ would also be a DGrCSL, and hence would be a GrCSL, which (as mentioned in Section 1) it is not.

And so, although all null-word-free CFL's are GrCSL's, they are not all DGrCSL's. This may be an unpleasantness that might dissuade some theoreticians from accepting the family of GrCSL's as a member of the Chomsky hierarchy. As it now stands each family in the hierarchy is a subclass of the deterministic subclass of the family at the next level. (Incidentally, as noted in [12] and [13], all null-word-free deterministic CFL's are DGrCSL's.)

Whether or not it deserves a place in the Chomsky hierarchy, the family of GrCSL's is an important family. Indeed there is reason to regard the subclass of DGrCSL's as being more important than the larger family of GrCSL's. As will be shown in the next section, the DGrCSL's can be characterized in terms of confluent rewriting systems, which gives them perhaps even more significance.

(Before going on to Section 3, let us pause to observe that the family of DGrCSL's is not closed under union or intersection. The simple argument for intersection [14] is as follows: It is easy to see that the Gladkij language $\{wcw^Rcw|w \in \{a,b\}^*\}$ is equal to the intersection of two deterministic CFL's, which are therefore both DGrCSL's; but the Gladkij language itself is not a DGrCSL. That this family is also not closed under union follows by the DeMorgan law, since the family is closed under complementation. Other such results can be found in Section 5 of [14].)

3 Confluent string rewriting systems

Perhaps the greatest selling point for the family of GrCSL's is its link with the theory of rewriting systems as it has been developing since 1970. More specifically, the selling point is for the family of DGrCSL's, which turns out to be identical to a family of languages definable in a certain way by confluent string rewriting systems, as discovered recently by Niemann and Otto [14].

Briefly, a *string rewriting system* is a semi-Thue system. We focus on systems in which the application of a rule to a word results in a simplification of the word: for example, it may be that $|\beta| < |\alpha|$ holds for each rule $\alpha \to \beta$. Such systems are often called *reduction systems,* since their purpose is to take a long word and gain some sort of understanding by reducing it to a shorter word. The length requirement is not a strict requirement, but one necessary property of a reduction system is that there be no infinite derivations. Consequently every word can be reduced to an irreducible word (the Noetherian property).

A further property that is desirable for a reduction system is that no word can be reduced to two distinct irreducible words (the confluence property). In reducing a word according to a confluent reduction system, it is sometimes possible to start reducing in two distinct ways at the same point in the word. But then the two reduction sequences must eventually come together. (For a good exposition of string rewriting systems, see the first two chapters of [9].)

Some languages can be defined by confluent reduction systems; if the alphabet of the system is the same as the alphabet of the language then the language is a *congruential language,* i.e., the union of some of the congruence classes of a congruence relation over Σ^*. Unfortunately, most interesting formal languages are not congruential.

However, if we allow ourselves to supplement the alphabet of the reduction system to include some control characters along with the alphabet of the language, we get something that is more fruitful. The paper [7] investigated the question of which formal languages could be defined in this way. The results and conceptual development of that paper have recently been surpassed in a remarkable way by Niemann and Otto [14], whom the present exposition will follow.

Another desirable property that our reduction systems must have is that they be weight reducing in the sense defined in Section 1; that is to say, there is a weight function ϕ such that, for every rule $\alpha \to \beta$, $\phi(\beta) < \phi(\alpha)$. A reduction system has the *generalized Church-Rosser property* if it is confluent and weight reducing. (It has the *strict Church-Rosser property* if it is confluent and length-reducing. Except for a few isolated remarks, we shall generally ignore the strict property for the remainder of this paper in favor of the more general property.)

A language $L \subseteq \Sigma^*$ is a *GenCRL (generalized Church-Rosser language)* if there exists a reduction system S with the generalized Church-Rosser property satisfying the following conditions:

(1) The alphabet Γ of S contains Σ as a proper subset;
(2) There are Y, t_1 and t_2, where $t_1, t_2 \in (\Gamma - \Sigma)^*$, $Y \in \Gamma - \Sigma$ and Y is irreducible in S, such that, for all $w \in \Sigma^*$, $w \in L$ if and only if $t_1 w t_2 \to^* Y$ (viz., Y is derivable from $t_1 w t_2$ in S)

Notice that if we have a language L that is of interest to us then such a system S would be a nice thing to have for testing membership in L. Given any $w \in \Sigma^*$, we would form the word $t_1 w t_2$ and reduce it modulo S. If the reduced word is Y then $w \in L$; if not then it is not. Since every rule of S is weight-reducing, the length of the reduction of $t_1 w t_2$ is linear in $|\phi(w)|$, and hence linear in $|w|$. Each step of the reduction is rather easy; we simply scan the word that we have for a subword that is the left side of a rule. (Things can be done so that the total amount of time spent in scanning during the entire reduction is insignificant.) When we find such a subword we reduce the word accordingly. If we find that there is no such subword and the word at that point is not simply Y then we know that the original word w is not in L.

As proved in [13] and [14], a language L is a GenCRL if and only if it is a DGrCSL. In effect, the D-shrink is a suitable mechanism for reduction of the word in the reduction system; in fact, precisely suitable. This automaton is similar to the automaton conceived by Ron Book [4] to reduce a word according to a reduction system with the Church-Rosser property.

In [14] it is demonstrated that every GenCRL is also a CRL; in other words the reduction system can be modified so as to allow the length function as the weighting function (i.e., for each $x \in \Gamma$, $\phi(x) = 1$). In the same paper, it is also demonstrated that the reduction system can be modified so that, for every $w \in \Sigma^*$, t_1wt_2 reduces either to Y (indicating "yes," that $w \in L$), or to N (indicating "no," that it is not); both $N$and Y are in $\Gamma - \Sigma$. These results settled questions left open in [7]. Furthermore, they make the characterization of DGrCSL's in terms of rewriting systems even more significant than they appear at first. They also strengthen our feeling that the DGrCSL's constitute an important family of languages.

An interesting open question concerns the language $\{ww^R | w \in \{a,b\}^*\}$, which is clearly a GrCSL, since it is context-free. I conjecture, however, that it is not a DGrCSL. In [7] there is a plausibility argument that it is not a CRL.

In conclusion, I suspect there will probably be few theoreticians who will press for any modification of the Chomsky hierarchy. Nevertheless, I hope many will come to realize that both the family of GrCSL's and the family of DGrCSL's will play important roles in the future of computer science. I am especially convinced of the importance of the DGrCSL's, since they have such a solid link with the contemporary theory of rewriting systems, and, in particular, with string rewriting systems having the confluence property.

References

1. Gladkij, A.W., "On the complexity of derivations in context-sensitive grammars," *Algebri i Logika Sem.*, vol. 3, pp. 29–44, 1964. In Russian.
2. Book, R.V. *Grammars with time functions,* Dissertation, Harvard University, 1969.
3. Salomaa, A., *Theory of Automata,* Pergamon Press, Oxford, England, 1969.
4. Book, R.V., "Confluent and other types of Thue systems," *J. ACM,* vol. 29, pp. 171–182, 1982.
5. Dahlhaus, E. and M.K. Warmuth, "Membership for growing context-sensitive grammars is polynomial," *J. Computer and System Sciences,* vol. 33, pp. 456–472, 1986.
6. Lautemann, C. "One pushdown and a small tape," in *Dirk Siefkes zum 50. Geburtstag* (K.W. Wagner, ed.), pp. 42–47, Technische Universität Berlin and Universität Augsburg, 1988.

7. McNaughton, R., P. Narendran and F. Otto, "Church-Rosser Thue systems and formal languages," *J. ACM,* vol. 35, pp. 324–344, 1988.
8. Buntrock, G. and Loryś, K., "On growing context-sensitive languages," *Proc. 19th ICALP, Lecture Notes in Computer Science* (W. Kuich, ed.), vol. 623, Springer-Verlag, pp. 77–88, 1992.
9. Book, R.V. and F. Otto, *String-rewriting systems,* Texts and Monograps in Computer Science, Springer-Verlag, 1993.
10. Buntrock, G., "Growing context-sensitive languages and automata," Preprint-Reihe, Nr. 69, Inst. für Informatik, Universität Würzburg, 1993.
11. Buntrock, G. and Loryś, K., "The variable membership problem: succinctness versus complexity," *Proc. 11th STACS, Lecture Notes in Computer Science* (P. Enjalbert, E.W. Mayr and K.W. Wagner, eds.), vol. 775, Springer-Verlag, pp. 595–606, 1994.
12. Buntrock, G., *Wachsend kontextsensitive Sprachen,* Habilitationsschrift, Fakultät für Mathematik und Informatik, Universität Würzburg. 1996.
13. Buntrock, G. and F. Otto, "Growing context-sensitive languages and Church-Rosser languages," *Inf. and Computation,* vol. 141, pp. 1–36, 1998.
14. Niemann, G. and F. Otto, "The Church-Rosser languages are the deterministic variants of the growing context-sensitive languages," *Proc. Foundations of Software Science and Computation Structures, Lecture Notes in Computer Science (M. Nivat, ed.),* vol. 1378, Springer-Verlag, pp. 243–257, 1998.

Word Length Controlled DT0L Systems and Slender Languages

Taishin Yasunobu Nishida

Summary. We introduce a new controlled DT0L system, called a word length controlled DT0L system, or a wlcDT0L system for short. A wlcDT0L system is a DT0L system with a control function which maps from the set of nonnegative integers to the set of tables. A wlcDT0L system derives exactly one word from a given word by iterating the table which is the value of the control function of the length of the given word. Thus a wlcDT0L system generates a sequence of words which starts from the axiom. We prove that every wlcPDT0L system generates a slender language. We also prove that there is a wlcDT0L language which is not slender. Since a slender language is applicable to cryptography, the family of wlcPDT0L languages may be useful for cryptography.

1 Introduction

The notion of slender languages and its applicability to cryptography is first introduced in [1]. A language L is slender if there exists a constant k such that L has at most k words of the same length. A slender language may be used as a key generation mechanism of a cryptosystem. Although the paper [1] mentions the Richelieu system, we prefer a system like the one time pad system since the former makes much longer ciphertext than the latter. As for the details of cryptography, see, for example, [3].

The cryptosystem uses slender language is as follows. The sender and receiver have the same slender language L. The sender selects a key from L which has the same length with the plaintext and enciphers the plaintext by adding letter to letter with the key. The receiver first selects candidates of keys which are included in L and have the same length with the ciphertext. Then the plaintext is recovered by subtracting one of the keys from the ciphertext. We note that the addition and the subtraction are performed under modulo n where n is the cardinality of the alphabet of plaintexts.

The above cryptosystem will be strong and easy to handle if there is a family of slender languages which seem as if sets of random sequences and are generated in polynomial time. The necessary and sufficient conditions for a context-free language to be slender are known [2,7,8]. But context-free slender languages have simple forms and are easily inferred [12]. For the 0L languages, a subfamily in which the slenderness problem is decidable is known [4–6]. The known slender 0L languages also have simple repetitive forms. So the search for language family which generates slender languages that are suitable to cryptography did not succeed.

In this paper we introduce new controlled DT0L systems, called word length controlled DT0L systems, or wlcDT0L systems for short. A wlcDT0L system is a DT0L system added a control function which maps from the set of nonnegative integers to the set of tables. A wlcDT0L system selects the table or morphism to be iterated current word by the value of the control function of the length of current word. We show that every wlcPDT0L language is slender (Theorem 1). We give an example of wlcPDT0L system which generates a very complex language (Example 1). So the family of wlcPDT0L languages may be useful for cryptography. We consider other properties of wlcDT0L systems, including relationship between the family of wlcPDT0L languages and the family of DT0L languages and the finiteness problem for wlcDT0L languages.

2 Preliminaries

Let Σ be a finite alphabet. The element of Σ is called a letter. The set of all finite words over Σ including the empty word λ is denoted by Σ^*. For a word $w \in \Sigma^*$, the length of w is denoted by $|w|$. Let a be a letter in Σ. We denote by $|w|_a$ the number of occurrences of a in w. For a subset $P \subseteq \Sigma$, $|w|_P$ is the number of occurrences of letters in P, i.e.,

$$|w|_P = \sum_{a \in P} |w|_a.$$

Let S be an arbitrary set. The cardinality of S is denoted by card(S).

We denote by $\mathbb{N}$ the set of nonnegative integers and $\mathbb{N}_+$ the set of positive integers.

Let Σ and Γ be finite alphabets. A mapping h from Σ^* to Γ^* is said to be a morphism if h satisfies

$$h(uv) = h(u)h(v)$$

for every $u, v \in \Sigma^*$. A morphism from Σ^* to Σ^* is called a morphism over Σ. A morphism h is said to be λ-free if for every $a \in \Sigma$, $h(a) \neq \lambda$. Let h be a morphism over Σ. For every $n \in \mathbb{N}$ and $w \in \Sigma^*$, h^n is defined by

$$\begin{aligned} h^0(w) &= w \quad \text{and} \\ h^n(w) &= h(h^{n-1}(w)) \quad \text{for } n > 0. \end{aligned}$$

A triplet $G = \langle \Sigma, h, w \rangle$ is said to be a D0L system if Σ is a finite alphabet, h is a morphism over Σ, and $w \in \Sigma^*$. A D0L system G generates a sequence of words (w_i) where $w_i = h^i(w)$. The language generated by G, denoted by $L(G)$, is given by $L(G) = \{h^i(w) \mid i \geq 0\}$. A D0L system is called a PD0L system if h is λ-free.

A triplet $G = \langle \Sigma, \Pi, w \rangle$ is said to be a DT0L system if Σ is a finite alphabet, Π is a finite set of morphisms over Σ, and $w \in \Sigma^*$. The language generated by G, denoted by $L(G)$, is given by

$$L(G) = \{u \mid u = w \text{ or } u = h_1 \cdots h_k(w) \text{ where } h_1, \ldots, h_k \in \Pi\}.$$

A DT0L system is called a PDT0L system if every morphism in Π is λ-free.

A language L over Σ is called slender if there exists a constant $k \in \mathbb{N}_+$ such that

$$\mathrm{card}(\{w \in L \mid |w| = l\}) \leq k$$

for every $l \in \mathbb{N}$.

We assume the reader is familiar with the rudiments of the formal language theory and the theory of L systems, see, for example, [9,10].

3 Definition of word length controlled DT0L systems

We give the central notion of this paper.

Definition 1. A *word length controlled DT0L system*, or a *wlcDT0L system* for short, is a 4-tuple $\langle \Sigma, \Pi, w, f \rangle$ where Σ is a finite alphabet, Π is a set of morphisms over Σ called the set of tables, $w \in \Sigma^*$ is the axiom, and f is a partial recursive function from $\mathbb{N}$ to Π called the *control function.*

A derivation by a wlcDT0L system is defined as follows.

Definition 2. Let $G = \langle \Sigma, \Pi, w, f \rangle$ be a wlcDT0L system. Let x and y be words over Σ. Then G *directly derives* y from x if $y = f(|x|)(x)$. If $f(|x|)$ is not defined, then G derives nothing from x.

By Definition 2, a wlcDT0L system $G = \langle \Sigma, \Pi, w, f \rangle$ generates a sequence of words $w = w_0, w_1, \ldots, w_i, \ldots$ which is given by $w_{i+1} = f(|w_i|)(w_i)$ for $i \in \mathbb{N}$. The sequence (w_i) is called the sequence generated by G. The language generated by G, denoted by $L(G)$, is defined by

$$L(G) = \{u \mid u \text{ is in the sequence generated by } G\}.$$

A wlcDT0L system $G = \langle \Sigma, \Pi, w, f \rangle$ is said to be a wlcPDT0L system if every morphism $h \in \Pi$ is λ-free. A language generated by a wlcDT0L (resp. wlcPDT0L) system is called a wlcDT0L (resp. wlcPDT0L) language. We denote by $\mathcal{L}(X)$ the family of X languages, where X is DT0L, PDT0L, wlcDT0L, or wlcPDT0L.

Now we give an example of wlcPDT0L system.

Example 1. Let $G = \langle \{A, a, b\}, \{h_1, h_2\}, A, f \rangle$ be a wlcPDT0L system where

$$h_1(A) = aA,\ h_1(a) = a,\ h_1(b) = b,$$

$$h_2(A) = bA,\ h_2(a) = b,\ h_2(b) = a$$

and

$$f(n) = \begin{cases} h_1 & \text{if } n \text{ is a prime number} \\ h_2 & \text{otherwise} \end{cases} .$$

The first few words in the sequence generated by G is as follows:

$$w_0 = A,\ w_1 = h_2(A) = bA,\ w_2 = h_1(w_1) = baA,\ w_3 = h_1(w_2) = baaA,$$

$$w_4 = h_2(w_3) = abbbA,\ w_5 = h_1(w_4) = abbbaA,\ w_6 = h_2(w_5) = baaabbA,$$

$$w_7 = h_1(w_6) = baaabbaA, \ldots.$$

Since f is a total recursive function, $L(G)$ is infinite. We cannot characterize $L(G)$ because we do not have an entire characterization of prime numbers.

4 WlcDT0L languages and slender languages

In this section we prove that every wlcPDT0L language is slender and that there is a wlcDT0L language which is not slender.

First we establish a lemma concerning D0L sequence.

Lemma 1. *Let $G = \langle \Sigma, h, w \rangle$ be a D0L system with* card$(\Sigma) = k$. *Let (w_i) be the sequence generated by G. If $|w_n| = |w_{n+1}| = \cdots = |w_{n+k}|$ for some $n \geq 0$, then $|w_n| = |w_i|$ for every $i \geq n$.*

Proof. Assume contrary, that is, if there exists an integer $i > n$ such that $|w_i| \neq |w_n| = |w_{n+1}| = \cdots = |w_{i-1}|$, then by Theorem I.3.6 of [9] (w_j) is not a D0L sequence. □

Then we state the first main theorem of this section.

Theorem 1. *Every wlcPDT0L system generates a slender language.*

Proof. Let $G = \langle \Sigma, \Pi, w, f \rangle$ be a wlcPDT0L system. We first note that the sequence $w = w_0, w_1, \ldots, w_i, \ldots$ generated by G satisfies $|w_0| \leq |w_1| \leq \cdots \leq |w_i| \leq \cdots$. Since every finite language is slender, we assume $L(G)$ is infinite. If $L(G)$ is not slender, then for every $k \in \mathbb{N}_+$ there exists $i \geq 0$ such that $|w_i| = |w_{i+1}| = \cdots = |w_{i+k}|$. Because $w_i, \ldots, w_{i+k}$ have the same length, there is a morphism $h \in \Pi$ such that $w_{j+1} = h(w_j)$ for $j = i, \ldots, i+k$. In other words, the sequence $(w_i, \ldots, w_{i+k}, w_{i+k+1})$ is a PD0L sequence generated by $\langle \Sigma, h, w_i \rangle$. But by Lemma 1, this is impossible. Thus $L(G)$ is slender. □

The propagating restriction (that is, the assumption of wlc*P*DT0L system) of Theorem 1 is necessary. Indeed we have the next theorem.

Theorem 2. *There is a wlcDT0L language which is not slender.*

In order to prove this theorem, we consider a wlcDT0L system $G_0 = \langle\{A, B, a\}, \{h_1, h_2\}, A, f\rangle$ where

$$h_1(A) = aA,\ h_1(B) = aB,\ h_1(a) = a,$$

$$h_2(A) = AB,\ h_2(B) = B,\ h_2(a) = \lambda,$$

and

$$f(x) = \begin{cases} h_1 & \text{if } x \text{ is not a perfect square number} \\ h_2 & \text{if } x \text{ is a perfect square number} \end{cases}.$$

Let $w_0 = A, w_1, \ldots, w_i, \ldots$ be the sequence generated by G_0. Then for every $i \geq 0$ the following properties hold:

1. $|w_i|_A = 1$.
2. $$|w_{i+1}|_{\{A,B\}} = \begin{cases} |w_i|_{\{A,B\}} + 1 & \text{if } |w_i| \text{ is a perfect square number} \\ |w_i|_{\{A,B\}} & \text{otherwise} \end{cases}.$$
3. $$|w_{i+1}| = \begin{cases} |w_{i+1}|_{\{A,B\}} & \text{if } |w_i| \text{ is a perfect square number} \\ |w_i| + |w_i|_{\{A,B\}} & \text{otherwise} \end{cases}.$$

Now we establish three lemmas.

Lemma 2. *For every $i \geq 0$, there exists an integer m such that $|w_i| = m|w_i|_{\{A,B\}}$ and $1 \leq m \leq |w_i|_{\{A,B\}}$.*

Proof. We prove this lemma by induction on i.

If $i = 0$, then the lemma holds with $m = 1$.

Assume that the lemma holds for every nonnegative integer less than $i+1$, i.e., $|w_i| = m|w_i|_{\{A,B\}}$ for some m. If $|w_i|$ is a perfect square number, then $|w_{i+1}| = |w_{i+1}|_{\{A,B\}}$. Hence the lemma holds with $m = 1$. If $|w_i|$ is not a perfect square number, then $|w_{i+1}| = |w_i| + |w_i|_{\{A,B\}} = (m+1)|w_i|_{\{A,B\}}$, in which the second equality is due to the inductive hypothesis. Moreover we have that $|w_i|_{\{A,B\}} = |w_{i+1}|_{\{A,B\}}$ and $m < |w_i|_{\{A,B\}}$. Therefore $|w_{i+1}| = (m+1)|w_{i+1}|_{\{A,B\}}$ and $m + 1 \leq |w_{i+1}|_{\{A,B\}}$. □

Lemma 3. *For every $k > 0$, G_0 generates a word w such that $|w|_{\{A,B\}} = k$.*

Proof. We prove this lemma by induction on k.

For $k = 1$, the axiom A satisfies the result.

Assume that the lemma holds for every positive integer less than $k+1$, i.e., $|w|_{\{A,B\}} = k$ for some $w \in L(G_0)$. If $|w|$ is a perfect square number, then the word $w' = h_2(w)$ satisfies $|w'|_{\{A,B\}} = |w|_{\{A,B\}} + 1 = k + 1$. If $|w|$ is not a perfect square number, then there exists a sequence of $l + 1$ words $w = w_0, w_1, \ldots, w_l$ such that $w_i = h_1(w_{i-1})$ and $|w_i|$ is not perfect square number for $i = 1, 2, \ldots, l$. Then by Lemma 2, there is a positive integer m such that $|w| = m|w|_{\{A,B\}}$. By the property 2, we have

$$|w_i| = |w| + i|w|_{\{A,B\}} \\ = (m+i)|w|_{\{A,B\}}.$$

Because l is the largest number which is derived by h_1 from w and satisfies that $|w_l|$ is not a perfect square number, we have that $m + l + 1 = |w|_{\{A,B\}}$ and $|w_{l+1}| = (m + l + 1)|w|_{\{A,B\}} = (m + l + 1)^2$. The word w_{l+2} which is derived by h_2 from w_{l+1}, i.e., $w_{l+2} = h_2(w_{l+1})$, satisfies

$$|w_{l+2}|_{\{A,B\}} = |w_{l+1}|_{\{A,B\}} + 1 = |w|_{\{A,B\}} + 1 = k + 1$$

by property 2. Then the lemma holds for $k + 1$. □

Let $p_1, p_2, \ldots, p_k$ be arbitrary prime numbers such that $p_1 < p_2 < \cdots < p_k$. Let P be the set of words which is given by $P = \{w \in L(G_0) \mid |w| = p_1^2 p_2^2 \cdots p_k^2\}$. We denote by $|P|_{\{A,B\}}$ the set of integers $\{|w|_{\{A,B\}} \mid w \in P\}$.

Lemma 4. *The set* $|P|_{\{A,B\}}$ *satisfies*

$$|P|_{\{A,B\}} \supseteq \{p_1^{i_1} p_2^{i_2} \cdots p_k^{i_k} \mid (i_1, i_2, \ldots, i_k) \in \{1,2\}^k \text{ such that not all } i_j = 2\}.$$

Proof. By Lemma 3, for every $(i_1, i_2, \ldots, i_k) \in \{1, 2\}^k$ such that not all $i_j = 2$, there exists a word $u \in L(G_0)$ which satisfies

$$|u|_{\{A,B\}} = p_1^{i_1} p_2^{i_2} \cdots p_k^{i_k}.$$

We note that $|u|_{\{A,B\}}$ is not a perfect square number. Let $i'_j = 2 - i_j$ for $j = 1, 2, \ldots, k$. Then there exists j such that $i'_j = 1$ because not all $i_j = 2$. We have a sequence of words $u = u_0, u_1, \ldots, u_l$ such that $u_{i+1} = h_1(u_i)$, $|u_i|$ is not a perfect square number for $i = 0, 1, \ldots, l-1$, and $m+l = p_1^{i'_1} p_2^{i'_2} \cdots p_k^{i'_k}$ where $|u| = m|u|_{\{A,B\}}$. Therefore we have $|u_l| = (m + l)|u|_{\{A,B\}} = p_1^2 p_2^2 \cdots p_k^2$, i.e., $u_l \in P$. Since $|u|_{\{A,B\}} = |u_1|_{\{A,B\}} = \cdots = |u_l|_{\{A,B\}}$, we have $p_1^{i_1} p_2^{i_2} \cdots p_k^{i_k} \in |P|_{\{A,B\}}$. □

Proof of Theorem 2. For two words $w_1, w_2 \in P$, if $|w_1|_{\{A,B\}} \neq |w_2|_{\{A,B\}}$, then we have $w_1 \neq w_2$. Now by Lemma 4, $\text{card}(P) \geq \text{card}(|P|_{\{A,B\}}) \geq 2^k - 1$. This means that $L(G_0)$ is not slender. □

5 Other properties of wlcDT0L systems

This section is devoted to the relationships among $\mathcal{L}$(DT0L), $\mathcal{L}$(PDT0L), and $\mathcal{L}$(wlcPDT0L) and to the finiteness problem of wlcDT0L languages. First we consider a wlcPDT0L system which generates a language that is not DT0L language.

Example 2. Let $G = \langle\{a\},\{h_1,h_2\},a^2,f\rangle$ be a wlcPDT0L system where $h_1(a) = a^2$, $h_2(a) = a^3$, and

$$f(x) = \begin{cases} h_1 & \text{if } x = 2^n 3^n \text{ for every } n \geq 1 \\ h_2 & \text{otherwise} \end{cases}.$$

Then G generates the wlcPDT0L sequence

$$a^2,\ h_2(a^2) = a^{2\cdot 3},\ h_1(a^{2\cdot 3}) = a^{4\cdot 3}, \ldots.$$

Clearly $L(G) = \{a^{2^{n+1}3^n} \mid n \geq 0\} \cup \{a^{2^n 3^n} \mid n \geq 1\}$. We denote $L(G)$ by L_1.

We show that L_1 is not a DT0L language.

Lemma 5. *The language L_1 is not a DT0L language.*

Proof. If a DT0L system $G = \langle\{a\},\{h_1,h_2,\ldots,h_k\},w\rangle$ generates L_1, then $w = a^2$ and for every $i = 1,2,\ldots,k$, $h_i(a) \neq \lambda$ because $\lambda \notin L_1$. Since a^6 is the second shortest word in L_1, $h_j(a) = a^3$ for some j. Then G derives $a^{2\cdot 3^i}$ for every $i \geq 1$. This is a contradiction. □

Then we show that there is a PDT0L language which is not slender.

Lemma 6. *There is a PDT0L language which is not slender.*

Proof. Let $G = \langle\{a,b,c\},\{h_1,h_2\},a\rangle$ be a PDT0L system where

$$h_1(a) = ab,\ h_1(b) = b,\ h_1(c) = c,$$

$$h_2(a) = ac,\ h_2(b) = b,\ h_2(c) = c.$$

Then it is obvious that $L(G) = a\{b,c\}^*$ is not slender. □

The above lemmas lead the next theorem.

Theorem 3. *We have the following relations: $\mathcal{L}$(wlcPDT0L) and $\mathcal{L}$(PDT0L) are incomparable and $\mathcal{L}$(wlcPDT0L) and $\mathcal{L}$(DT0L) are incomparable.*

Proof. By Lemma 5, $\mathcal{L}$(DT0L) cannot include $\mathcal{L}$(wlcPDT0L) and by Lemma 6 and Theorem 1, $\mathcal{L}$(wlcPDT0L) cannot include $\mathcal{L}$(PDT0L). Then the results follow immediately. □

Now we discuss the finiteness problem for wlcDT0L languages. First we note that $\mathcal{L}$(wlcDT0L) is not include the family of all finite languages.

Proposition 1. *There is a finite language which is not a wlcDT0L language.*

Proof. We will show that the language $L_2 = \{a^2, a^3\}$ is not generated by any wlcDT0L system. If $G = \langle\{a\},\Pi,w,f\rangle$ generates L_2, then for every $h \in \Pi$ such that h is eventually used in the derivation, h is λ-free because $\lambda \notin L_2$. Hence a^2 must be the axiom. But no morphism over $\{a\}$ can obtain a^3 from a^2. Thus L_2 is not a wlcDT0L language. □

Then we show that the finiteness problem for wlcDT0L languages is undecidable.

Theorem 4. *Finiteness problem for wlcPDT0L languages is undecidable.*

Proof. For every positive integer k, we construct the wlcPDT0L system $G_k = \langle \{a, b\}, \{h_1, h_2\}, a, f_k \rangle$ where

$$h_1(a) = b, \ h_1(b) = b,$$

$$h_2(a) = ab, \ h_2(b) = b,$$

and

$$f_k(x) = \begin{cases} h_1 & \text{if } x = k \text{ and } TM_k(k) \text{ is not defined} \\ h_2 & \text{otherwise} \end{cases},$$

in which TM_k is the one argument partial recursive function computed by the k-th Turing machine under some numbering of Turing machines. Then f_k is defined and has the value h_2 unless $x = k$ and $TM_k(k)$ is not defined. Therefore $L(G_k)$ is finite if and only if $TM_k(k)$ is not defined. This implies that an algorithm which solves the finiteness problem for wlcPDT0L languages leads that the set $\{k \in \mathbb{N}_+ \mid TM_k(k) \text{ is not defined}\}$ is recursive. This is a contradiction (see, for example, [11]). □

Theorem 4 is rather obvious because the control functions may be partial recursive functions. If the control functions are restricted to total recursive functions, then it is interesting whether or not the finiteness problem for wlcDT0L languages is still undecidable.

References

[1] M. Andraşiu, G. Păun, J. Dassow, and A. Salomaa (1990) Language-theoretic problems arising from Richelieu cryptosystems, *Theoret. Comput. Sci.* **116**, 339–357.

[2] L. Ilie (1994) On a conjecture about slender context-free languages, *Theoret. Comput. Sci.* **132**, 427–434.

[3] A. G. Konheim (1981) *Cryptography: A Primer*, John Wiley & Sons, New York.

[4] T. Y. Nishida and A. Salomaa (1996) Slender 0L languages, *Theoret. Comput. Sci.* **158**, 161–176.

[5] T. Y. Nishida and A. Salomaa, A note on slender 0L languages, *Theoret. Comput. Sci.* to appear.

[6] G. Păun and A. Salomaa (1992) Decision problems concerning the thinness of D0L languages, *Bull. EATCS* **46**, 171–181.

[7] G. Păun and A. Salomaa (1995) Thin and slender languages, *Discrete Applied Mathematics* **61**, 257–270.

[8] D. Raz (1995) On slender context-free languages, in: *STACS 95*, Lecture Notes in Computer Science, Vol. 900, Springer, Berlin, 445–454.

[9] G. Rozenberg and A. Salomaa (1980) *The Mathematical Theory of L Systems*, Academic Press, New York.

[10] A. Salomaa (1973) *Formal Languages*, Academic Press, New York.
[11] A. Salomaa (1985) *Computation and Automata*, Cambridge University Press, Cambridge.
[12] Y. Takada and T. Y. Nishida (1996) A note on grammatical inference of slender context-free languages, in: L. Miclet and C. de la Higuera, ed., *Grammatical Inference: Learning Syntax from Sentences*, Lecture Notes in Computer Science (subseries Artificial Intelligence), Vol. 1147, Springer, Berlin, 117–125.

Part V

Algorithms and Complexity

Program-Size Complexity of Initial Segments and Domination Reducibility*

Cristian S. Calude and Richard J. Coles

Summary. A theorem of Solovay states that there is a noncomputable Δ_2^0 real $x = (x_n)_n$ such that $H(x_n) \leqslant H(n) + O(1)$. We improve this result by showing there is a noncomputable c.e. real x with the same property. This answers a question concerning the relationship between the domination relation and program-size complexity of initial segments of reals (informally, a real x dominates a real y if from a good approximation of x from below one can compute a good approximation of y from below). Solovay proved that if x and y are two c.e. reals and y dominates x, then $H(x_n) \leqslant H(y_n) + O(1)$. The result above shows that the converse is false, namely there are c.e. reals x and y such that $H(x_n) \leqslant H(y_n) + O(1)$ and y does not dominate x.

1 Introduction

Solovay [10] introduced the domination relation which plays an important role in defining the so-called Ω–like reals. The class of Ω–like reals coincides with the class of Chaitin Ω reals (halting probabilities of universal self-delimiting Turing machines [5–7,2,9]) (cf. [4]) and the class of c.e. random reals (cf. [11]). Solovay proved that if x and y are two c.e. reals and y dominates x, then $H(x_n) \leqslant H(y_n) + O(1)$. In this paper we prove that the converse implication is false, namely there are c.e. reals x and y such that $H(x_n) \leqslant H(y_n) + O(1)$ and y does not dominate x. We do this by constructing a noncomputable c.e. real x such that $H(x_n) \leqslant H(n) + O(1)$. In Section 2 we introduce our notation and basic concepts. Sections 3, 4 and 5 deal with constructing the c.e. real mentioned above. The methods of these sections generalise notions first appearing in Solovay [10]. In Section 6 we apply our result to the question of domination.

2 Preliminaries

Suppose $a, b \in \{0,1\}^*$, the set of binary strings. The concatenation of a, b is denoted by $a^\frown b$. Let $|a|$ denote the length of a. For $j \geqslant 0$, we write $a(j) = k$

* Calude was partially supported by AURC A18/XXXXX/62090/F3414056, 1996. Coles was supported by a UARC Post-Doctoral Fellowship.

iff the jth bit of a is k. We let $\prec$ denote the quasi-lexicographical ordering of finite binary strings. We write *string*(n) to denote the nth string with respect to $\prec$. By $\min_{\prec}$ we denote the minimum operation taken according to $\prec$.

We fix a computable bijective function $\langle\cdot,\cdot\rangle$ from $\mathbb{N}\times\mathbb{N}$ to $\mathbb{N}$. We write $\log n$ to denote $\log_2 n$. For a function $f:\{0,1\}^*\to\mathbb{N}$, we define $O(f)=\{g:\{0,1\}^*\to\mathbb{N}\mid \exists c\in\mathbb{N}\forall a\in\{0,1\}^*\,(g(a)\leqslant c\cdot f(a))\}$. Suppose $h_0,h_1:\{0,1\}^*\to\mathbb{N}$. We write $O(1)$ for $O(f)$ when f is the constant function $f(a)=1$ for all $a\in\{0,1\}^*$. We write $h_0\leqslant h_1+O(f)$ if there is a function $g\in O(f)$ such that $h_0(a)\leqslant h_1(a)+g(a)$ for all $a\in\{0,1\}^*$.

We will look at real numbers in the interval $[0,1]$ through their binary expansions, i.e., in terms of functions $n\mapsto x_n$ (from $\mathbb{N}$ into $\{0,1\}$). We write $(x_n)_n$ for the sequence $n\mapsto x_n$. A real is computable if the function $n\mapsto x_n$ is computable. A real $x=(x_n)_n$ is computable enumerable (c.e.) if it is the limit of a computable, increasing, converging sequence of rationals. Equivalently, $\alpha=0.\chi_A$ is a c.e. real if A has a computable approximation $\{A[s]\}_{s\geqslant 0}$ such that whenever $i\in A[s]$ and $i\notin A[s+1]$, then there is some $j<i$ such that $j\notin A[s]$ and $j\in A[s+1]$; χ_A is the characteristic function of A. Following Solovay [10] and Chaitin [5] we say that a real x *dominates* a real y if from a good approximation of x from below one can compute a good approximation of y from below (see Section 6 for a formal definition).

Let ϕ_e be a standard list of all partial computable functions from $\mathbb{N}$ into $\mathbb{N}$. In case $\phi_e(x)$ halts (and produces y) we write $\phi_e(x)\!\downarrow$ ($\phi_e(x)\!\downarrow=y$); otherwise, $\phi_e(x)\!\uparrow$. By $\phi_e(x)[t]$ we denote the time relativised version of $\phi_e(x)$, i.e., $\phi_e(x)[t]=\phi_e(x)$ in case $\phi_e(x)$ halts in time t. The binary predicate $\phi_e(x)[t]\!\downarrow$ is primitive recursive. We will adopt the following *convention*: if $\phi_e(x)[t]\!\downarrow$, then $\phi_e(x)\leqslant t$.[1] By $\operatorname{dom}\phi_e$ we denote the domain of the partial function ϕ_e.

A self-delimiting computer is a partial computable function C from $\{0,1\}^*\times\{0,1\}^*$ with values in $\{0,1\}^*$ such that for every $y\in\{0,1\}^*$ the set $\{x\mid C(x,y)\!\downarrow\}$ is prefix-free. Here C stands for the interpreter, x for the program, and y for the input data.

The Invariance Theorem ([5,7,2,10]) states the existence of a universal computer U with the property that for every computer C there is a constant d (depending upon U and C) such that if $C(x,y)=z$, then $U(x',y)=z$, for some program x' with the length $|x'|\leqslant|x|+d$. Note that U does not need more than a primitive recursive extra time to simulate C, i.e., there is a primitive recursive function h such that if $C(x,y)[t]\!\downarrow=z$, then $U(x',y)[h(t)]\!\downarrow=z$. In what follows we will fix a universal computer U. Let a^* be the quasi-lexicographical least p such that $U(p,\lambda)=a$.

---

[1] Sometimes we will be concerned with constructions occurring over ω-many stages and thus often append $[t]$ to parameters to denote the value of the parameter at the end of stage t.

Define the following program-size complexities ([6]):

$$H(a) = \min\{|p| \mid U(p, \lambda) = a\},\ H(a, b) = H(\langle a, b\rangle),$$

$$H(a/b) = \min\{|p| \mid U(p, b^*) = a\},\ \tilde{H}(a/b) = \min\{|p| \mid U(p, b) = a\}.$$

Note that $\tilde{H}(a/b^*) = H(a/b)$.

In expressions relating to strings, such as $U(p, \lambda) = n$, we are identifying n with the binary string 1^n of length n. We also write $H(n)$ for $H(1^n)$. In fact notice that $|H(n) - H(string(n))| = O(1)$. For integers n, we write n^* for $\min_{\prec}\{p \mid U(p, \lambda) = n\}$.

We continue by defining some useful functions. For all $j, n, t \in \mathbb{N}$, define

$$H(n)[t] = \min\{|p| \mid U(p, \lambda)[t] = n \,\&\, |p| \leqslant t\},$$

if such a p exists, $H(n)[t] = 0$, otherwise. Further let,

$$\alpha(n)[t] = \min\{H(j)[t] \mid j \geqslant n\}, \alpha(n) = \min\{H(j) \mid j \geqslant n\}.$$

It is seen that $H(n)[t]$ and $\alpha(n)[t]$ are primitive recursive functions (reason: for the computation we only need the set $\{p \mid U(p, \lambda)[t], |p| \leqslant t\}$), decreasing in t, and $H(n) = \lim_t H(n)[t]$, $\alpha(n) = \lim_t \alpha(n)[t]$.

Assume that D is an oracle and consider the relativised computation U^D. Then the relativised program-size complexities are defined in the obvious way, for example,

$$H^D(a) = \min\{|p| | U^D(p, \lambda) = a\}.$$

If instead of a self-delimiting universal computer we work with a universal partial computable function V, then the induced complexities will be denoted by $K(a), K(a/b), \tilde{K}(a/b)$. We now summarise the known results relating the complexity of initial segments to the computability of a real. Let $x = (x_n)_n$ be a binary sequence.

Theorem 1 (Loveland [8]) *x is computable iff $\tilde{K}(x_n/n) = O(1)$.*

Corollary 2 *x is computable iff $\tilde{H}(x_n/n) = O(1)$.*

Theorem 3 (Chaitin [6]) *x is computable iff $K(x_n) \leqslant K(n) + O(1)$.*

Theorem 4 (Chaitin [6]) *If $H(x_n) \leqslant H(n) + O(1)$, then $x \in \Delta_2^0$.*

Proof. Start by noting that

$$H(x_n) = H(n, x_n) + O(1) = H(x_n/n) + H(n) + O(1).$$

Now if x satisfies the hypothesis of the theorem, then we have

$$H(x_n/n) + H(n) + O(1) \leqslant H(n) + O(1).$$

Therefore $H(x_n/n) = O(1)$, and so $\tilde{H}(x_n/n^*) = O(1)$.

Relativising to the oracle $D = \{p \mid U(p, \lambda)\downarrow\}$, we have $\tilde{H}^D(n^*/n) = O(1)$, since the mapping $n \mapsto n^*$ is Turing reducible to D. Consequently,

$$\tilde{H}^D(x_n/n) \leqslant \tilde{H}^D(x_n/n^*) + \tilde{H}^D(n^*/n) = O(1),$$

therefore $\tilde{H}^D(x_n/n) \leqslant O(1)$. So by the relativised version of Corollary 2 we see that x is Δ_2^0. □

Theorem 5 (Solovay [10]) *There is a noncomputable real x such that*

$$H(x_n) \leqslant H(n) + O(1). \qquad (\natural)$$

In section 5 we show the following stronger result:

Theorem 6 *There is a noncomputable c.e. real x such that for all $n \in \mathbb{N}$,*

$$H(x_n) \leqslant H(n) + O(1).$$

In Section 6 we make an application of this result to the domination degrees of c.e. reals. First we consider how to obtain a real with the desired initial segment complexity.

3 Achieving $H(x_n) \leqslant H(n) + O(1)$

Suppose $(t_i)_i$ is a computable increasing sequence of natural numbers. Define the total computable function σ as follows:

$$\sigma(i) = \max\{j \leqslant i \mid H(j)[t_i] = H(j)[t_{i+1}]\}.$$

Notice that the graph of $\sigma(i)$ is primitive recursive. and that $\sigma(i) \leqslant i$.

Let $(t_i)_i$ be a computable sequence of times and $\{p_i \mid i \geqslant 0\}$ a c.e. set of programs such that $i < t_i < t_{i+1}$ and $U(p_i, \lambda)[t_i] = i$. For $i \in \mathbb{N}$ define

$$A_i = \{p \mid U(p, \lambda)[t_i]\downarrow \;\&\; |U(p, \lambda)| \leqslant i\}.$$

Notice that for every $i \in \mathbb{N}$, A_i is computable, $A_i \subset A_{i+1}$, and for every $n \in \mathbb{N}$, $n^* \in A_{i+1} \setminus A_i$ for some i.

Proposition 7 *If $n^* \in A_{i+1} \setminus A_i$, then* a) *$n^* \in A_j$ for all $j \geqslant i+1$,* b) *$H(n)[t] = |n^*|$ for all $t \geqslant t_{i+1}$,* c) *$\sigma(i') \geqslant n$ for all $i' \geqslant i+1$.*

Definition 8 *Let $f : \mathbb{N} \to \mathbb{N}$ be a total computable increasing function. We say that a partial computable function $\psi : \{0,1\}^* \to \{0,1\}^*$ is quasi-universal with respect to f if* $\operatorname{dom}\psi = \operatorname{dom}U$, *and for all $m, n \in \mathbb{N}$, $m < n$ implies that $\psi((m + f(m))^*) \preccurlyeq \psi((n + f(n))^*)$.*

Lemma 9 *If ψ is quasi-universal with respect to f then the real $x = (x_n)_n$ defined by $x_n = \psi((n + f(n))^*)$ satisfies the relation $H(x_n) \leqslant H(n) + O(1)$.*

Proof. We have $H(x_n) \leqslant H(x_n/n + f(n)) + H(n + f(n)) + O(1)$. However $H(n + f(n)) \leqslant H(n) + O(1)$ since f is computable, and

$$H(x_n/n + f(n)) = \tilde{H}(x_n/(n + f(n))^*) = O(1),$$

since $x_n = \psi((n + f(n))^*)$ and ψ is partial computable with $(n + f(n))^*$ in $\operatorname{dom}\psi$. Therefore $H(x_n) \leqslant H(n) + O(1)$ as required. □

Let $f : \mathbb{N} \to \mathbb{N}$ be a total computable increasing function. Now suppose $(z[i])_i$ is a computable sequence of finite binary strings such that $|z[i]| \geqslant i$ and

$$(n + f(n))^* \in A_{i+1} \setminus A_i \implies \forall i' > i, \forall j \leqslant n\,(z[i+1](j) = z[i'](j)) \qquad (\dagger)$$

For $p \in \{0,1\}^*$ define $\psi(p)$ as follows. If $p \in A_{i+1} \setminus A_i$ and $|U(p,\lambda)| = n+f(n)$ for some $n \leqslant i$ then let $\psi(p)$ be the initial segment of $z[i+1]$ of length n. Otherwise let $\psi(p)$ be the initial segment of $z[i+1]$ of length $|U(p,\lambda)|$. Note that ψ is partial computable.

Lemma 10 *ψ is quasi-universal with respect to f.*

Proof. Clearly ψ has the same domain as U. So suppose $m, n \in \mathbb{N}$ with $m < n$. Then $\psi((n+f(n))^*)$ is the initial segment of $z[i+1]$ of length n, where $i+1$ is such that $(n + f(n))^* \in A_{i+1} \setminus A_i$. But by (†), $z[i+1](j) = z[i'](j)$ for all $i' > i$ and all $j \leqslant n$. The same is true for m in place of n. Hence we must have $\psi((m + f(m))^*) \preccurlyeq \psi((n + f(n))^*)$. □

So if we can construct a c.e. real z satisfying (†) for some suitable function f, then $x = (x_n)_n = z$ defined by $x_n = \psi((n + f(n))^*)$ will prove Theorem 6.

4 Two Useful Results

Before proving Theorem 6 we need two useful theorems, namely Theorems 11 and 12 below due to Solovay [10]. Let B be a total increasing computable function which is not primitive recursive, but has a primitive recursive graph. For example take B to be Ackermann's diagonal function. Then in fact, B dominates any primitive recursive function, that is, for every primitive recursive function g, $B(n) > g(n)$, for almost all natural n; see [1].

Theorem 11 *There is a computable sequence of times $(t_n)_n$, such that $t_n < t_{n+1}$ for all $n \in \mathbb{N}$, and for all $n > 0$,*

(1) there is a natural number s_n such that $B(t_{n-1}) \leqslant s_n < B(s_n) = t_n$,

(2) for all programs p with $|p| \leqslant B(t_{n-1})$, $U(p,\lambda)\downarrow$ implies that either $U(p,\lambda)[s_n]\downarrow$ or $U(p,\lambda)[B(s_n)]\uparrow$,

(3) the predicate $t_n = j$ is primitive recursive, but the function $n \mapsto t_n$ is not primitive recursive.

Proof. Let $t_0 = 0$ and assume that $s_1, \ldots, s_{n-1}, t_1, \ldots, t_{n-1}$ have been defined. Let Q be the set of all programs of length less than or equal to $B(t_{n-1})$. Let $r_1 = t_{n-1}$ and define the following three sets: $X_1 = \{p \in Q \mid U(p,\lambda)[r_1]\downarrow\}$, $Y_1 = Q \setminus X_1$, $Z_1 = \{p \in Y_1 \mid U(p,\lambda)[B(r_1)]\downarrow\}$.

If $Z_1 = \emptyset$, then let $s_n = r_1$ and $t_n = B(s_n)$. Otherwise $Z_1 \neq \emptyset$, in which case set $r_2 = B(r_1)$ and define $X_2 = X_1 \cup Z_1, Y_2 = Q \setminus X_2, Z_2 = \{p \in Y_2 \mid U(p,\lambda)[B(r_2)]\downarrow\}$.

If $Z_2 = \emptyset$, then let $s_n = r_2$ and $t_n = B(s_n)$. Otherwise continue this process until reaching a step i with $Z_i = \emptyset$ and hence let $s_n = r_i$ and $t_n = B(s_n)$. Such an i must be reached since $X_1 \subset X_2 \subset \ldots \subseteq Q \supseteq Y_1 \supset Y_2 \supset \ldots$, Q is finite and $Y_i \supset Z_i$.

It follows that (1) and (2) are satisfied. Part (3) follows from properties of Ackermann's diagonal function B. □

We use the sequence of times $(t_n)_n$ constructed above in the following theorem, which plays a crucial role in the priority argument in Section 5. In particular, the function σ we work with is defined with respect to this sequence of times.

Theorem 12 *There is a computable sequence of times $(t_i)_{i \geqslant 0}$, $t_i < t_{i+1}$ and a c.e. set $\{p_i\}_{i \geqslant 0}$ of programs such that $U(p_i,\lambda)[t_i] = i$, for which the following condition holds true: if $g : \mathbb{N} \to \mathbb{N}$ is a total computable function, then for infinitely many natural numbers i, $g(\sigma(i)) < i$.*

Proof. Take the sequence of times $(t_i)_i$ constructed in Theorem 11. Suppose for a contradiction that $g(\sigma(n)) \geqslant n$ for almost all n. We may assume that g is increasing, and in fact that it dominates all primitive recursive functions.

Let $G(n) = \min\{m \mid g(m+1) \geqslant n\}$. Then our assumption on g implies that $G(n) \leqslant \sigma(n)$ for almost all n. Note that $G(n) \leqslant t_n$ under this assumption.

The intuition for obtaining the contradiction is as follows. The function $\sigma(i)$, roughly speaking, is the largest j such that the approximation to $H(j)$ does not change between time steps t_i and t_{i+1}. Therefore to show $G(n) > \sigma(n)$ we would like to construct short programs (of length $\leqslant B(t_n)$) for $j \in [G(n), n]$ which halt by time t_{n+1}. We will be able to do this if

(i) n has a short program that converges in approximately time t_n,

(ii) $\sum_{G(n) \leqslant j \leqslant n} 2^{-H(j)[t_n]}$ is small,

because then we can use the Kraft-Chaitin Theorem to construct new codes (programs) for all $j \in [G(n), n]$. The next two lemmata perform this task.

Lemma 13 *There is a primitive recursive function h such that for almost all k, there is a number n_k such that $\alpha(G(n_k))[t_{n_k}] \geqslant k$, and a program p_{n_k} such that $U(p_{n_k}, \lambda)[h(t_{n_k})]\downarrow = n_k$, and $|p_{n_k}| = O(\log k)$.*

Proof. Fix $k \in \mathbb{N}$. If n is sufficiently large then $\alpha(G(n)) \geqslant k$ because g is unbounded, and so $\alpha(G(n))[t_n] \geqslant k$. Let

$$n_k = \min\{m \mid m > k \text{ and } \alpha(G(m))[t_m] \geqslant k\}.$$

For large n,

$$G(n) = \min\{m \mid m \leqslant \sigma(n) \text{ and } g(m+1) \geqslant n\},$$

so G is primitive recursive. Further, the predicate, $\alpha(n)[t] \geqslant k$ is also primitive recursive as it can be expressed by the formula

$$(\forall n \leqslant j \leqslant t, |p| \leqslant t(U(p,\lambda)[t] = j \implies |p| \geqslant k)) \text{ or } (n > t).$$

So the function $k \mapsto n_k$ has a primitive recursive graph: $U(p,\lambda)[t_m] = j$ iff $U(p,\lambda)[s] = j$, for some $s \leqslant t_m$. Let $h_1(k,s)$ be the primitive recursive function evaluating the running time of the predicate $n_k = s$.

Now consider the self-delimiting computer $C(1^{\log k}0, \lambda) = n_k$. Use the Invariance Theorem to show that for every natural k there is a program, p_{n_k} say, such that $|p_{n_k}| = O(\log k)$ such that $U(p_{n_k}, \lambda) = n_k$. The time necessary to compute p_{n_k} from k is the sum between the time to get n_k from k (a primitive recursive function in n_k, $h_1(k, n_k)$) and the time to simulate C on U, which is primitive recursive in the running time of C, i.e., $h(h_1(k,n_k))$. □

Lemma 14 *For sufficiently large k,*

$$\sum_{G(n_k)\leqslant j \leqslant n_k} 2^{-H(j)[t_{n_k}]} \leqslant 2^{-k/2}.$$

Proof. Again, let k be sufficiently large so that $\alpha(G(n_k)) \geqslant k$. Suppose for a contradiction that

$$\sum_{G(n_k)\leqslant j \leqslant n_k} 2^{-H(j)[t_{n_k}]} > 2^{-k/2}.$$

The number $\sum_{G(n_k)\leqslant j \leqslant n_k} 2^{-H(j)[t_{n_k}]}$ can be computed in a primitive recursive way and is less than 1. Computing its most significant $k/2$ digits we get a $j \in [G(n_k), n_k]$ and a program p_j such that $|p_j| \leqslant k/2 + O(1)$, $U(p_j, \lambda) = j$, and $U(p_j,\lambda)[z]\downarrow$ for some z. Consequently, $H(j) \leqslant k/2 + O(1)$, so

$$\alpha(G(n_k)) = \min\{H(j) \mid j \geqslant G(n_k)\} \leqslant k/2 + O(1).$$

For sufficiently large k, this contradicts $\alpha(G(n_k)) \geqslant k$, hence the required inequality has been demonstrated. □

We now continue with the proof of Theorem 12. Consider the following self-delimiting computer:

(1) input $string(k)^\frown y$,
(2) compute n_k using the procedure in Lemma 13,
(3) compute t_{n_k} and $G(n_k)$,
(4) now use the Kraft-Chaitin Theorem[2] to construct a prefix-free set $E = \{e_j \mid j \in [G(n_k), n_k]\}$ such that $|e_j| \leqslant H(j)[t_{n_k}] - k/2$, which is possible by Lemma 14,
(5) output the code e_y of y if $y \in [G(n_k), n_k]$.

The time of this procedure is $h_0(t_{n_k})$ for some primitive recursive function h_0, and so $t_{n_k+1} > B(t_{n_k}) \geqslant h_0(t_{n_k})$. Hence for $j \in [G(n_k), n_k]$,

$$H(j)[t_{n_k+1}] \leqslant H(j)[t_{n_k}] - k/2 + O(\log k).$$

The $O(\log k)$ term is derived from the length of the program in Lemma 13 for computing n_k from k. Hence for sufficiently large k, $\sigma(n_k) \leqslant G(n_k)$, contradicting our assumption that $G(n) < \sigma(n)$ for almost all n. □

5 The Priority Argument

We now return to the proof of Theorem 6. In this section we construct a noncomputable c.e. real that is consistent with the ideas of Section 3.[3]

Proof. We want to construct a c.e. real z such that the following requirements are satisfied for all $e > 0$:

$\mathrm{R}_e : \exists m(\phi_e(m) \neq z(m))$.

We also want to ensure (†) is met so that the initial segments of z have the desired complexity. To understand how this is achieved, begin by thinking of the function f in (†) as the constant function 0. The idea is that once n^* is in A_{i+1} then $\sigma(s+1) \geqslant n$ for all $s > i$. By using the function σ to provide witnesses for diagonalisation, we can ensure that the initial segment of the approximation to z of length n does not change after stage i because all future witnesses are greater than or equal to n. This method is essentially the one used by Solovay [10] to obtain a Δ_2^0 real x. We modify this to obtain a c.e. real z as follows.

For a single requirement R_1 in isolation, the naive approach is to wait for a stage $s+1$ such that $\phi_1(\sigma(s+1))\downarrow$ in less than $s+1$ steps. Theorem

[2] Given a recursive list of "requirements" $\langle n_i, s_i \rangle$ $(i \geq 0, s_i \in \Sigma^*, n_i \geqslant 0)$ such that $\sum_i 2^{-n_i} \leq 1$, we can effectively construct a self-delimiting computer C and a recursive one-to-one enumeration $x_0, x_1, x_2, \ldots$ of words x_i of length n_i such that $C(x_i, \lambda) = s_i$ for all i and $C(x, \lambda) = \infty$ if $x \notin \{x_i \mid i \geqslant 0\}$; see [2].

[3] A full proof of Solovay's Theorem 5 can be found in Calude and Coles [3] and may help the reader to understand the c.e. case below.

12 tells us that there are infinitely many such stages if ϕ_e is total. Now set $z(\sigma(s+1)) \neq \phi_1(\sigma(s+1))$.

For two requirements, R_1 of higher priority than R_2, we may diagonalise against ϕ_2 on argument $\sigma(s+1)$ only to find at a stage $s'+1 > s+1$ that $\sigma(s'+1) = \sigma(s+1)$ and we would like to diagonalise against ϕ_1 on argument $\sigma(s'+1)$. If $z[s+1](\sigma(s+1))$ was set to 1 and we want to set $z[s'+1](\sigma(s'+1))$ to 0 then we risk not constructing z to be a c.e. real. To set $z[s'+1](\sigma(s+1)) = 0$ we must find some $c < \sigma(s'+1)$ such that $z[s'](c) = 0$. Then we can define $z[s'+1](c) = 1$ allowing us to set $z[s'+1](\sigma(s'+1)) = 0$ so that z is c.e. Thus before using $\sigma(s+1)$ as a witness for R_2 we must make sure there is some $c < \sigma(s+1)$ which is available for correction if it turns out R_1 would like to use $\sigma(s+1)$ as a witness at some later stage. Therefore in the construction, R_2 may only diagonalise with $\sigma(s+1)$ if $\sigma(s+1) > 2$.

For an arbitrary requirement R_e, this generalises to R_e using witnesses $\sigma(s+1)$ only if $\sigma(s+1) > e+w(e)[s+1]$, where $w(e)[s+1]$ is the largest witness seen so far for requirements $R_{e'}$ such that $e' \leqslant e$. This ensures enough space is left for correction in case higher priority requirements require witnesses used by lower priority requirements.

If a higher priority requirement wants to use a witness $\sigma(s+1)$ that has already been used by a lower priority requirement, but $z[s](\sigma(s+1)) = 0$, then of course the higher priority requirement could set $z[s+1](\sigma(s+1)) = 1$ without using a correcting bit c and still keep z c.e., but to simplify the presentation we always use a correcting bit c whenever a requirement receives attention. Consequently, we insist witnesses used for R_e are actually greater than $2e+w(e)[s+1]$ so enough space is available for correction. This approach means we can reset $z[s](j) = 0$ for all $j > c$ such that $j \neq \sigma(s+1)$ whenever we act for some requirement at stage $s+1$.

The need to leave space for correcting bits to make z a c.e. real is the reason for the computable function f mentioned in (†). The particular f we use is described after the construction of z. The idea is that we have to look beyond n to find out when an initial segment of z of length n has settled down. That is, once $(n+f(n))^* \in A_{i+1} \setminus A_i$, then $\sigma(i') \geqslant n + f(n)$ for all $i' > i$. Therefore if $f(n)$ is large enough, all changes of bits of z will be above n.

The Construction.

Stage s = 0: Let $z[0] = \lambda$.
Stage s + 1: We have already defined $z[s]$.

We say that R_e requires attention at stage $s+1$ if

(i) $e < s+1$,
(ii) $\phi_e(\sigma(s+1))\downarrow$ in at most $s+1$ steps,
(iii) $\sigma(s+1)$ is not a witness for any requirement $R_{e'}$ with $e' < e$,
(iv) R_e has no witness at stage $s+1$ and $\sigma(s+1)$ was not previously a witness of R_e,

(v) $\sigma(s+1) \geqslant 2e + w(e)[s+1]$, where $w(e)[s+1]$ is the largest witness used so far for requirements $\mathrm{R}_{e'}$ with $e' \leqslant e$, if it exists, and $w(e)[s+1] = 0$ otherwise.

We act for the least e such that R_e requires attention, e_0 say. We say that R_{e_0} receives attention at stage $s+1$ and is satisfied.

Let c be the largest natural number such that $w(e)[s+1] < c < \sigma(s+1)$, $z[s](c) = 0$ and c is not a witness for a requirement $\mathrm{R}_{e'}$ for all $e' < e$. (The existence of such a c is proven below in Lemma 15.) Define

$$z[s+1](c) = 1,$$

$$z[s+1](\sigma(s+1)) \neq \phi_{e_0}(\sigma(s+1))\downarrow,$$

$$z[s+1](j) = 0 \text{ for all } j > c \text{ such that } j \neq \sigma(s+1),$$

$$z[s+1](j) = z[s](j) \text{ for all } j < c.$$

Finally cancel all witnesses for requirements $\mathrm{R}_{e'}$ for all $e' > e_0$. Declare $\mathrm{R}_{e'}$ unsatisfied for all $e' > e_0$. This completes the construction.

The Verification.

Lemma 15 *If* R_e *receives attention at stage* $s+1$ *then there is some* c *with* $w(e)[s+1] < c < \sigma(s+1)$ *such that* $z[s](c) = 0$ *and* c *is not a witness for requirements* $\mathrm{R}_{e'}$ *with* $e' < e$.

Proof. Suppose not for a contradiction and let $s+1$ be a least counterexample. Suppose $w(e)[s+1]$ is chosen as a witness for $\mathrm{R}_{e'}$ where $e' \leqslant e$ at a stage $s'+1 < s+1$. Let s_0+1 be the largest stage $< s+1$ for which some requirement $\mathrm{R}_{e'}$ with $e' < e$ received attention. Then there is a $c < \sigma(s_0+1)$ such that $z[s_0](c) = 0$ and c is not the witness for a requirement $\mathrm{R}_{e''}$ for all $e'' < e'$ at stage s_0+1. Then $z[s_0+1](j) = 0$ for all $j > w(e)[s_0+1]$.

We now consider what can happen between stages s_0+1 and $s+1$. As no requirement of higher priority than R_e receives attention between stages s_0+1 and $s+1$, the only requirements that can change the value of bits in the approximation to z of less than $\sigma(s+1)$ are those of lower priority than R_e. But by construction a lower priority requirement R_i, $i > e$, leaves at least $2i$ bits between any witness it may use and $w(e)[s+1]$. Hence, there is a c such that $w(e)[s+1] < c < \sigma(s+1)$ and $z[s](c) = 0$ and c is not the witness for a requirement $\mathrm{R}_{e'}$, $e' < e$. (Note that $\sigma(s+1)$ is strictly greater than $w(e)[s+1]$ as by construction, $\sigma(s+1) \geqslant w(e)[s+1] + 2e$.) □

Lemma 16 *For all* $e \in \mathbb{N}$, R_e *has at most a finite number of witnesses.*

Proof. An easy induction on e shows that R_e can only require attention finitely often. □

Lemma 17 $\lim_s z[s](i)$ *exists for all* $i \in \mathbb{N}$ *and* z *is a c.e. real.*

Proof. This follows from Lemmas 15 and 16, and by construction. □

Lemma 18 *If ϕ_e is total then* R_e *has a final witness.*

Proof. Suppose R_e does not get a final witness. Let s_0 be the least stage $> e$ such that for all $e' < e$, $\mathrm{R}_{e'}$ does not require attention at all stages $s \geqslant s_0$. Then clearly (i) holds for all stages $s \geqslant s_0$. Choose $s_1 > s_0$ so that (iii) and (v) hold for all stages $s \geqslant s_1$. This is possible since $\lim_s \sigma(s) = \infty$. Now let $g(m)$ be the number of steps it takes for $\phi_e(m)$ to converge. Then (ii) holds at infinitely many stages since there are infinitely many i such that $g(\sigma(i)) < i$ by Theorem 12. Hence (iv) must fail, for otherwise R_e would receive attention at some stage $s_2 \geqslant s_1$ and get a final witness. Therefore R_e has a witness at a stage after s_1 which causes (iv) to fail, but then this witness is permanent, contradicting the assumption. Hence R_e has a final witness. □

Lemma 19 *If m is the final witness of* R_e, *then $\phi_e(m) \neq z(m)$.*

Proof. By construction, if m is the final witness of R_e then $z(m) \neq \phi_e(m)$. □

It remains to show that the initial segments of z of length n have complexity $\leqslant H(n) + O(1)$.

We inductively define the computable function $f : \mathbb{N} \to \mathbb{N}$ as follows: Run the construction above looking for a stage $s+1$ such that:

(1) $\sigma(s+1)$ is used as a witness for some requirement R_e,
(2) $\sigma(s+1) - n > f(m)$ for all $m < n$,
(3) $\sigma(s+1) > n + 2e$,
(4) Upon finding such a $\sigma(s+1)$, set $f(n) = \sigma(s+1) - n$.

We remark that (3) is possible by waiting for a stage $s+1$ where $w(e)[s+1] > n$. Then f is a total computable and increasing function and $(z[i])_i$ is a computable sequence. Define $\psi : \{0,1\}^* \to \{0,1\}^*$ as in Section 3, namely: for $p \in \{0,1\}^*$, wait for $p \in A_{i+1} \setminus A_i$. Then $U(p,\lambda)\downarrow$ and $|U(p,\lambda)| \leqslant i$ by definition of A_{i+1}. If $|U(p,\lambda)| = f(n) + n$ for some $n \leqslant i$, then let $F(p)$ be the initial segment of $z[i+1]$ of length n. Otherwise let $F(p)$ be the initial segment of $z[i+1]$ of length $|U(p,\lambda)|$.

If we can show that f and $(z[i])_i$ meet (†) then by Lemma 10 ψ is quasi-universal with respect to f. Furthermore, Lemma 9 proves that defining x by $x_n = \psi(n + f(n))^*)$ satisfies Theorem 6.

Lemma 20 *Suppose that $(n+f(n))^*$ is in $A_{i+1} \setminus A_i$. Then for all $j \leqslant n$ and $i' > i$, $z[i+1](j) = z[i'](j)$.*

Proof. First notice that since $n + f(n)$ was a witness for R_e, then all lower priority requirements ($\mathrm{R}_{e'}$ for $e' > e$) use witnesses greater than $2e' + \sigma(s+1)$ by construction. Hence if action is taken for a requirement $\mathrm{R}_{e'}$ with $e' > e$ at a stage $s'+1 > s+1$ then there is always a c such that $\sigma(s+1) < c < \sigma(s'+1)$ that can be used for correction by $\mathrm{R}_{e'}$.

Thus we only need to consider requirements $\mathrm{R}_{e'}$ for $e' \leqslant e$. In order that $z[i+1](j) \neq z(j)$ for some $j \leqslant n$, some witness $y \geqslant f(n) + n$ must be used by a requirement $\mathrm{R}_{e'}$ for some $e' < e$ at a later stage, $s_1 > i$ say.

Now since $(n + f(n))^* \in A_{i+1} \setminus A_i$, $\sigma(s') \geqslant n + f(n)$ for all $s' \geqslant i+1$. At stage $s+1$ $\sigma(s+1)$ was used as a witness and hence $\sigma(s+1) > 2e + w(e)[s+1]$.

Let $s_0 + 1$ be the largest stage $< s+1$ such that some requirement $\mathrm{R}_{e'}$ with $e' \leqslant e$ received attention. Then $z[s_0+1](j) = 0$ for all $j > w(e)[s+1]$ by construction and Lemma 15.

There are $e - 1$ requirements of higher priority than e, and each may want to use $\sigma(s+1)$ as a witness at some stage after stage $s+1$. However $\sigma(s+1) > n + 2e$, and so if such action is taken at a stage $s'+1 > s+1$ for a requirement $\mathrm{R}_{e'}$ where $e' < e$, then there is a c such that $n < c < \sigma(s+1)$, $z[s'](c) = 0$ and c is a not a witness for a requirement $\mathrm{R}_{e''}$ for $e'' < e'$.

So (†) is satisfied. Hence $n < m$ implies that $F(n^*) \preccurlyeq F(m^*)$. and $x_n = F(n^*)$. □

This completes the proof of Theorem 6. □

6 Domination Degrees and Program-size Complexity

A real x is said to *dominate* the real y if there is a partial computable function f from rationals to rationals and a constant $c > 0$ with the property that if p is a rational number less than x, then $f(p)$ is (defined and) less than y, and it satisfies the inequality $c(x - p) \geqslant y - f(p)$. In this case we write $x \geqslant_{dom} y$. A c.e. real is universal if it dominates every c.e. real; such a real is called Solovay-Chaitin Ω-like [10,5]. The class of Chaitin Ω reals [5,7,9] coincides with the class of Solovay-Chaitin Ω-like reals, see Calude, Hertling, Khoussainov, Wang [4], which equals the class of c.e. random reals, Slaman [11]. For the theory of randomness see Chaitin [7] and Calude [2].

Does domination reducibility capture randomness of c.e. reals? On the one hand we have:

Theorem 21 (Solovay [10]) *Let x and y be two c.e. reals. Then $x \leqslant_{dom} y$ implies $H(x_n) \leqslant H(y_n) + O(1)$.*

So, the natural question is whether the converse implication is true. We can now show that the answer is negative.

Theorem 22 *There exist c.e. reals x and y such that $y \not\geqslant_{dom} x$ and $H(x_n) \leqslant H(y_n) + O(1)$.*

Proof. Following Theorem 6 let x be a noncomputable c.e. real such that $H(x_n) \leqslant H(n) + O(1)$. Let y be any computable real. Then $x \geqslant_{dom} y$ but $y \not\geqslant_{dom} x$.

Now since y is computable, $H(y_n) = H(n) + O(1)$. Furthermore, since $x \geqslant_{dom} y$ we have $H(y_n) \leqslant H(x_n) + O(1)$ by Theorem 21. Suppose that there is no constant c such that $H(x_n) \leqslant H(y_n) + c$. Then for every constant c there are infinitely many n such that $H(x_n) > H(y_n) + c$. Consequently for every constant c there are infinitely many n such that $H(x_n) > H(n) + c$, a contradiction. Therefore $H(x_n) \leqslant H(y_n) + O(1)$. □

References

1. C. Calude. *Theories of Computational Complexity*, North-Holland, Amsterdam, 1988.
2. C. Calude. *Information and Randomness. An Algorithmic Perspective,* Monographs in Theoretical Computer Science - An EACTS Series, Springer-Verlag, Berlin, 1994.
3. C. S. Calude and R. J. Coles. On a Theorem of Solovay, *CDMTCS Research Report*, 094 (1999), 14pp.
4. C. S. Calude, P. Hertling, B. Khoussainov, Y. Wang. Recursively enumerable reals and Chaitin Ω numbers, in M. Morvan, C. Meinel, D. Krob (eds.). *Proceedings of STACS'98, Paris, 1998*, Springer-Verlag, Berlin, 1998, 596–606.
5. G. J. Chaitin. Algorithmic information theory, *IBM J. Res. Develop.* 21 (1977), 350–359, 496.
6. G. J. Chaitin. Information-theoretic characterizations of recursive infinite strings, *Theoretical Computer Science,* 2 (1976), 45–48.
7. G. J. Chaitin. *The Limits of Mathematics*, Springer-Verlag, Singapore, 1997.
8. D. W. Loveland. A variant of the Kolmogorov concept of complexity, *Information and Control,* 15 (1969), 510–526.
9. G. Rozenberg, A. Salomaa. *Cornerstones of Undecidability*, Prentice Hall, Englewood Cliffs, 1994.
10. R. M. Solovay. *Draft of a paper (or series of papers) on Chaitin's work ... done for the most part during the period of Sept. to Dec. 1974*, unpublished manuscript, IBM Thomas J. Watson Research Center, Yorktown Heights, New York, May 1975, 215 pp.
11. T. A. Slaman. *Random Implies Ω-Like,* manuscript, 14 December 1998, 2 pp.

Stability of Approximation Algorithms and the Knapsack Problem

Juraj Hromkovič

Summary. The investigation of the possibility or the impossibility to efficiently compute approximations of hard optimization problems becomes one of the central and most fruitful areas of current algorithm and complexity theory. Currently, optimization problems are considered to be tractable if there exist randomized polynomial-time approximation algorithms that solve them with a reasonable approximation ratio. Our opinion is that this definition of tractable problems is still too hard. This is because one usually considers the worst-case complexity and what is really important is the complexity of algorithms on "natural" problem instances (real data coming from the practice). Nobody exactly knows what "natural" data are and how to mathematically specify them. But, what one can do is to try to separate the easy problem instances from the hard ones. The aim of this paper is to develop an approach going in this direction.

More precisely, a concept for measuring the stability of approximation algorithms is presented. This concept is of theoretical and practical interest because it can be helpful to determine the border between easy problem instances and hard problem instances of complex optimization problems that do not admit polynomial-time approximation algorithms. We illustrate the usefulness of our approach in an exemplary study of the knapsack problem.

1 Introduction

Immediately after introducing NP-hardness (completeness) [5] as a concept for proving intractability of computing problems [13], the following question has been posed: If an optimization problem does not admit an efficiently computable optimal solution, is there a possibility to efficiently compute at least an approximation of the optimal solution? Several researchers [4], [10], [11], [12], [14], [19], [20] provided already in the middle of the seventies a positive answer for some optimization problems. It may seem to be a fascinating effect if one jumps from the exponential complexity (a huge inevitable amount of physical work) to the polynomial complexity (tractable amount of physical work) due to a small change in the requirement – instead of an exact optimal solution one forces a solution whose quality differs from the quality of an optimal solution at most by $\varepsilon \cdot 100$ % for some ε. This effect is very strong, especially, if one considers problems for which this approximation concept works for any small ε (see the concept of approximation schemes in [10], [16], [18], [3]).

There is also another possibility to jump from NP to P. Namely, to consider the subset of inputs with a special, nice property instead of the whole set of inputs for which the problem is well-defined [13]. A nice example is the Travelling Salesman Problem (TSP). TSP is not only NP-complete, but also the search of an approximation solution for TSP is NP-complete for every ε.[1] But if one considers TSP for inputs satisfying the triangle inequality (so called Δ-TSP), one can even design a polynomial-time approximation algorithm [4] with the quality $\varepsilon = \frac{1}{2}$. The situation is still more interesting, if one considers the Euclidean TSP, where the distances between the nodes correspond to the distances in the Euclidean metrics. The Euclidean TSP is NP-complete [17], but for every small $\varepsilon > 0$ one can design a polynomial-time ε-approximation algorithm [1], [2], [15], whose randomized version has an almost linear time complexity.[2]

The fascinating observations of huge quantitive changes mentioned above led to the development of the concept of "stability" of approximation algorithms in [9]. In this paper we give the formal description of this concept. Then, we apply it for the knapsack problem in order to illustrate its usefulness.

Informally, one can introduce the measurement of the stability of approximation algorithms as follows: Let us consider the following scenario. One has an optimization problem P for two sets of inputs L_1 and L_2, $L_1 \subsetneq L_2$. For L_1 there exists an polynomial-time ε-approximation algorithm A, but for L_2 there is no polynomial-time δ-approximation algorithm for any $\delta > 0$ (if NP is not equal to P). We pose the following question: Is the algorithm A really useful for inputs from L_1 only? Let us consider a distance measure M in L_2 determining the distance between L_1 and any given input $x \in L_2 - L_1$. Now, one can consider an input $x \in L_2 - L_1$, for which there $distance_M(x, L_1) \leq k$ for some positive real k. One can look for how "good" the algorithm A is for the input $x \in L_2 - L_1$. If for every $k > 0$ and every x with the distance at most k to L_1, A computes an $\delta_{\varepsilon,k}$ approximation of an optimal solution for x ($\delta_{\varepsilon,k}$ is considered to be a constant depending on k and ε only), then one can say that A is "(approximation) stable" according to the distance measure M. Obviously, such a concept enables to show positive results extending the applicability of known approximation algorithms. On the other hand it can help to show the boundaries of the use of approximation algorithms and possibly even a new kind of hardness of optimization problems.

---

[1] In fact Håstad [7] even proved that there is no polynomial-time $O(n^{1-\delta})$-approximation algorithm for TSP for any $\delta > 0$ (n is the input size) if NP is not a subset of ZPP.

[2] Obviously, we know a lot if similar examples where with restricting the set of inputs one crosses the border between decidability and undecidability (Post Correspondence Problem) or the border between P and NP (SAT and 2-SAT, or vertex cover problem).

This paper is organized as follows: In Section 2 we present the concept of approximation stability. In order to do this, we give a new definition of optimization problems that enables to investigate some new aspects of optimization problems. In Section 3 we investigate the stability of the well-known PTAS [19] for the simple knapsack problem according to a reasonable distance measure. We do not prove any new result for the knapsack problem because we have already efficient FPTAS for it [10]. But the aim is to present an example that exemplary shows the usefulness of our concept. In the last Section 4 we discuss the possibilities of the use of the presented concept in the investigation in the area of combinatorial optimization.

2 Definition of the Stability of Approximation Algorithms

We assume that the reader is familiar with the basic concepts and notions of algorithmics and complexity theory as presented in standard textbooks like [3], [6], [8], [18], [21]. Next, we present the new definition of the notion of an optimization problem [9]. The reason to do this is to obtain the possibility to study the influence of the input sets on the hardness of the problem considered. Let $\mathbb{N} = \{0, 1, 2, ...\}$ be the set of nonnegative integers, and let $\mathbb{R}^+$ be the set of positive reals.

Definition 1. An **optimization problem** U is an 7-tuple $U = (\Sigma_I, \Sigma_O, L, L_I, \mathcal{M}, cost, goal)$, where

1. Σ_I is an alphabet called **input alphabet**,
2. Σ_O is an alphabet called **output alphabet**,
3. $L \subseteq \Sigma_I^*$ is a language over Σ_I called the **language of consistent inputs**,
4. $L_I \subseteq L$ is a language over Σ_I called the **language of actual inputs**,
5. $\mathcal{M}$ is a function from L to $2^{\Sigma_O^*}$, where, for every $x \in L$, $\mathcal{M}(x)$ is called the **set of feasible solutions** for the input x,
6. *cost* is a function, called **cost function**, from $\bigcup_{x \in L} \mathcal{M}(x)$ to $\mathbb{R}^+$,
7. $goal \in \{minimum, maximum\}$.

For every $x \in L$, we define

$$\boldsymbol{Output_U(x)} = \{y \in \mathcal{M}(x) | cost(y) = goal\{cost(z) | z \in \mathcal{M}(x)\}\},$$

and

$$\boldsymbol{Opt_U(x)} = cost(y)$$

for some $y \in Output_U(x)$.

Clearly, the meaning for Σ_I, Σ_O, $\mathcal{M}$, *cost* and *goal* is the usual one. L may be considered as a set of consistent inputs, i.e., the inputs for which the optimization problem is consistently defined. L_I is the set of inputs considered and only these inputs are taken into account when one determines the complexity of the optimization problem U. This kind of definition is useful for considering the complexity of optimization problems parameterized according to their languages of actual inputs. In what follows $\boldsymbol{Language(U)}$ denotes the language L_I of actual inputs of U.

To illustrate Definition 1 consider the Knapsack Problem (KP). In what follows $\boldsymbol{bin(u)}$ denotes the integer binary coded by the string $u \in \{0,1\}^*$. The input of KP consists of $2n+1$ integers $w_1, w_2, ..., w_n, b, c_1, c_2, ..., c_n, n \in \mathbb{N}$. So, one can consider $\Sigma_I = \{0,1,\#\}$ with the binary coding of integers and # for ",". The output is a vector $x \in \{0,1\}^n$, and so we set $\Sigma_O = \{0,1\}$. $L = \{0,1\}^* \cdot \bigcup_{i=0}^{\infty} (\#\{0,1\}^*)^{2i}$. We speak about the Simple Knapsack Problem (SKP) if $w_i = c_i$ for every $i = 1, ..., n$. So, we can consider $L_I = \{z_1\#z_2\#...\#z_n\#b\#z_1\#z_2\#...\#z_n \mid b,\ z_i \in \{0,1\}^* \text{ for } i = 1, ..., n,\ n \in \mathbb{N}\}$ as a subset of L. $\mathcal{M}$ assigns to each $I = y_1\#...\#y_n\#b\#u_1\#...\#u_n$ the set of words $\mathcal{M}(I) = \{x = x_1x_2...x_n \in \{0,1\}^n \mid \sum_{i=1}^{n} x_i \cdot bin(y_i) \leq bin(b)\}$. For every $x = x_1...x_n \in \mathcal{M}(I)$, $cost(x) = \sum_{i=1}^{n} x_i \cdot bin(u_i)$. The goal is maximum. So, $KP = (\Sigma_I, \Sigma_O, L, L, \mathcal{M}, cost, goal)$, and $SKP = (\Sigma_I, \Sigma_O, L, L_I, \mathcal{M}, cost, goal)$.

Definition 2. Let $U = (\Sigma_I, \Sigma_O, L, L_I, \mathcal{M}, cost, goal)$ be an optimization problem. We say that an algorithm A is a **consistent algorithm for U** if, for every input $x \in L_I$, A computes an output $A(x) \in \mathcal{M}(x)$. We say that **A solves U** if, for every $x \in L_I$, A computes an output $A(x)$ from $Output_U(x)$. The time complexity of A is defined as the function

$$Time_A(n) = \max\{Time_A(x) | x \in L_I \cap \Sigma_I^n\}$$

from $\mathbb{N}$ to $\mathbb{N}$, where $Time_A(x)$ is the length of the computation of A on x.

Next, we give the definitions of standard notions in the area of approximation algorithms.

Definition 3. Let $U = (\Sigma_I, \Sigma_O, L, L_I, \mathcal{M}, cost, goal)$ be an optimization problem, and let A be a consistent algorithm for U. For every $x \in L_I$, the **relative error $\varepsilon_A(x)$ is defined as**

$$\boldsymbol{\varepsilon_A(x)} = \frac{|cost(A(x)) - Opt_U(x)|}{Opt_U(x)}.$$

For any $n \in \mathbb{N}$, we define **the relative error of A**

$$\boldsymbol{\varepsilon_A(n)} = \max\{\varepsilon_A(x) | x \in L_I \cap \Sigma_I^n\}.$$

For any positive real δ, we say that A is an **δ-approximation algorithm for U** if $\varepsilon_A(x) \leq \delta$ for every $x \in L_I$.

For every function $f : \mathbb{N} \to \mathbb{R}$, we say that A is a **$f(n)$-approximation algorithm for U** if $\varepsilon_A(n) \leq f(n)$ for every $n \in \mathbb{N}$.

The best what can happen for a hard optimization problem U is that one has a polynomial-time ε-approximation algorithm for U for any $\varepsilon > 0$. In that case we call this collection of approximation algorithms a **polynomial-time approximation scheme (PTAS) for U**. A nicer definition of a PTAS than the above one is the following one. An algorithm B is a PTAS for an optimization problem U if B computes an output $B(x, \varepsilon) \in \mathcal{M}$ for every input $(x, \varepsilon) \in Language(U) \times \mathbb{R}^+$ with

$$\frac{|cost(B(x,\varepsilon)) - Opt(x)|}{Opt(x)} \leq \varepsilon$$

in time polynomial according to $|x|$.

Now, we define the complexity classes of optimization problems in the usual way.

Definition 4.

$$\begin{aligned} \boldsymbol{NPO} = \{U = (\Sigma_I, \Sigma_O, L, L_I, \mathcal{M}, cost, goal) \,|\, &U \text{ is an optimization} \\ &\text{problem}, L, L_I \in P;\ \text{for every } x \in L, \mathcal{M}(x) \in P; cost \\ &\text{is computable in polynomial time}\}, \end{aligned}$$

For every optimization problem $U = (\Sigma_I,\ \Sigma_O,\ L,\ L_I,\ \mathcal{M},\ cost,\ goal)$, the **underlying language of U** is

$$Under_U = \{(w,k) \,|\, w \in L_I, k \in \mathbb{N} - \{0\}, Opt_U(w) \leq k\}$$

if $goal = maximum$. Analogously, if $goal = minimum$

$$Under_U = \{(w,r) \,|\, w \in L_I,\ r \in \mathbb{N} - \{0\},\ Opt_U(w) \geq r\}.$$

$$\begin{aligned} \boldsymbol{PO} &= \{U \in NPO \,|\, Under_U \in P\} \\ \boldsymbol{APX} &= \{U \in NPO \,|\, \text{ there exists an } \varepsilon\text{-approximation algorithm} \\ &\qquad \text{for } U \text{ for some } \varepsilon \in \mathbb{R}^+\}. \end{aligned}$$

In order to define the notion of stability of approximation algorithms we need to consider something like a distance between a language L and a word outside L.

Definition 5. Let $U = (\Sigma_I,\ \Sigma_O,\ L,\ L_I,\ \mathcal{M},\ cost,\ goal)$ and $\overline{U} = (\Sigma_I,\ \Sigma_O,\ L,\ L,\ \mathcal{M},\ cost,\ goal)$ be two optimization problems with $L_I \subsetneq L$. A **distance function for U according to L_I** is any function $h_L : L \to \mathbb{R}^+$ satisfying the property

$$h_L(x) = 0 \text{ for every } x \in L_I.$$

We define, for any $r \in \mathbb{R}^+$,

$$\boldsymbol{Round_{r,h}(L_I)} = \{w \in L \,|\, h(w) \leq r\} \subseteq L.$$

Let A be a consistent algorithm for $\overline{U}$, and let A be an ε-approximation algorithm for U for some $\varepsilon \in \mathbb{R}^+$. Let p be a positive real. We say that A is **p-stable according to h** if, for every real $0 \leq r \leq p$, there exists a $\delta_{r,\varepsilon} \in \mathbb{R}^+$ such that A is an $\delta_{r,\varepsilon}$-approximation algorithm for $U_r = (\Sigma_I, \Sigma_O, L, Round_{r,h}(L_I), \mathcal{M}, cost, goal)$.[3]

A is **stable according to h** if A is p-stable according to h for every $p \in \mathbb{R}^+$. We say that A is **unstable according to h** if A is not p-stable for any $p \in \mathbb{R}^+$.

For every positive integer r, and every function $f_r : \mathbb{N} \to \mathbb{R}^+$ we say that A is **$(r, f(n))$-quasistable according to h** if A is an $f_r(n)$-approximation algorithm for $U_r = (\Sigma_I, \Sigma_O, L, Round_{r,h}(L_I), \mathcal{M}, cost, goal)$.

Note, that considering PTASs the situation may be a little bit more complicated. Let us consider a PTAS A as a collection of plynomial-time ε-approximation algorithms A_ε for every $\varepsilon > 0$. If A_ε is stable according to a distance measure M for every $\varepsilon > 0$, then it does not immediately imply that A is a PTAS for inputs from $\bigcup_{r\in\mathbb{N}} Round_{r,M}(L_I)$. This happens if, for instance, $\delta_{r,\varepsilon} = r + \varepsilon$. Then A_ε is a $(r+\varepsilon)$-approximation algorithm for $Round_{r,M}(L_I)$, but we have obtained no PTAS for $Round_{r,M}(L_I)$. Obviously, if $\delta_{r,\varepsilon} = \varepsilon \cdot r$, then A is a PTAS for $Round_{r,M}(L_I)$. This is the reason to define the stability of PTASs as follows.

Definition 6. Let U, $\overline{U}$, U_r and h have the same meaning as in Definition 5. Let A as collection of algorithms $\{A_\varepsilon\}_{\varepsilon>0}$ be a PTAS for U. If, for every $r > 0$, A_ε is a $\delta_{r,\varepsilon}$-approximative algorithm for U_r, we say that A is *stable according to h.*

If $\delta_{r,\varepsilon} \leq f(\varepsilon) \cdot g(r)$, where

1. f and g are some functions from $\mathbb{R}$ to $\mathbb{R}^+$, and
2. $\lim_{\varepsilon\to 0} f(\varepsilon) = 0$,

then we say that the PTAS A is *super-stable according to h.*

One may see that the notions of stability can be useful for answering the question how broadly a given approximation algorithm is applicable. So, if one is interested in positive results then one is looking for a suitable distance measure that enables to use the algorithm outside the originally considered set of inputs. In this way one can search for the border of the applicability of the given algorithm. If one is interested in negative results then one can try to show that for any reasonable distance measure the considered algorithm cannot be extended to work for a much larger set of inputs than the original one. In this way one can search for fine boundaries between polynomial approximability and polynomial non-approximability. A more involved discussion about the applicability of our concept is presented in the last section of this paper.

[3] Note, that $\delta_{r,\varepsilon}$ is a constant depending on r and ε only.

3 An Example – Stability of Approximation and the Knapsack Problem

In this section we consider the stability of the well known PTAS [19] for the simple knapsack problem (SKP). This PTAS is a combination of (compromise between) the total search and the greedy method. We show that this PTAS is stable according to a reasonable distance measure. This leads to an approximation algorithm for some extension of the simple SKP, but unfortunately not to a PTAS for this extension. This is the motivation for modifying this PTAS in order to obtain a PTAS for the extension according to the distance measure considered.

Note, that we do not develop any new algorithm here [19] and that one knows already a FPTAS for the general knapsack problem [10]. The aim of this section is only to illustrate how the study of the stability of an approximation algorithm (or a PTAS) may lead to the development of a modified version of this algorithm (PTAS) that can be applied for inputs for which the original algorithm cannot be successfuly applied.

First, we review the PTAS for the simple knapsack problem.

PTAS SKP

Input: Positive integers $w_1, w_2, ..., w_n, b$ for some $n \in \mathbb{N}$ and some positive real number $1 > \varepsilon > 0$.

Step 1: Sort $w_1, w_2, ..., w_n$. For simplicity we may assume $w_1 \geq w_2 \geq ... \geq w_n$.

Step 2: Set $k = \lceil 1/\varepsilon \rceil$.

Step 3: For every set $T = \{i_1, i_2, ..., i_l\} \subseteq \{1, 2, ..., n\}$ with $|T| = l \leq k$ and $\sum_{i \in T} w_i \leq b$, extend T to T' by using the greedy method and values $w_{i_l+1}, w_{i_l+2}, ..., w_n$. (The greedy method stops if $\sum_{i \in T'} w_i \leq b$ and $w_j > b - \sum_{i \in T'} w_i$ for all $j \notin T'$, $j \geq i_l$.)

Output: The best set T' constructed in Step 3.

For every given ε, we denote the above algorithm by SKP_ε. It is known that SKP_ε is an ε-approximation algorithm for SKP. Observe that it is consistent for KP. Now, we consider the following distance function $DIST$ for any input w_1, w_2, ..., w_n, b, c_1, ..., c_n of KP:

$$DIST(w_1, ..., w_n, b, c_1, ..., c_n) = \max\left\{ \max\left\{ \frac{c_i - w_i}{w_i} \,\middle|\, c_i \geq w_i,\ i \in \{1, ..., n\} \right\},\ \max\left\{ \frac{w_i - c_i}{c_i} \,\middle|\, w_i \geq c_i,\ i \in \{1, ..., n\} \right\} \right\}.$$

Let $KP_\delta = (\Sigma_I, \Sigma_O, L, Round_{\delta,DIST}(L_I), \mathcal{M}, cost, maximum)$ for any δ. Now, we show that PTAS SKP is stable according to $DIST$ but this result does not imply the existence of a PTAS for KP_δ for any $\delta > 0$.

Lemma 1. *For every $\varepsilon > 0$, $\delta > 0$, the algorithm* SKP$_\varepsilon$ *is an* $(\varepsilon + \delta(2+\delta) \cdot (1+\varepsilon))$*-approximation algorithm for* KP_δ.

Proof. Let $w_1 \geq w_2 \geq ... \geq w_n$ for an input $I = w_1, ..., w_n, b, c_1, ..., c_n$, and let $k = \lceil 1/\varepsilon \rceil$. Let $U = \{i_1, i_2, ..., i_l\} \subseteq \{1, 2, ..., n\}$ be an optimal solution for I. If $l \leq k$ then SKP$_\varepsilon$ outputs an optimal solution with $cost(U)$ because SKP$_\varepsilon$ has considered U as a candidate for the output in Step 3.

Consider the case $l > k$. SKP$_\varepsilon$ has considered the greedy extension of $T = \{i_1, i_2, ..., i_k\}$ in Step 2. Let $T' = \{i_1, i_2, ..., i_k, j_{k+1}, ..., j_{k+r}\}$ be the greedy extension of T. Obviously, it is sufficient to show that the difference $cost(U) - cost(T')$ is small relative to $cost(U)$, because the cost of the output of SKP$_\varepsilon$ is at least $cost(T')$. We distinguish the following two possibilities:

1. Let $\sum_{i \in U} w_i - \sum_{j \in T'} w_j \leq 0$. Obviously, for every i, $(1+\delta)^{-1} \leq \frac{c_i}{w_i} \leq 1 + \delta$. So, $cost(U) = \sum_{i \in U} c_i \leq (1+\delta) \cdot \sum_{i \in U} w_i$ and $cost(T') = \sum_{j \in T'} c_j \geq (1+\delta)^{-1} \cdot \sum_{j \in T'} w_j$. In this way we obtain

$$
\begin{aligned}
cost(U) - cost(T') &\leq (1+\delta) \cdot \sum_{i \in U} w_i - (1+\delta)^{-1} \cdot \sum_{j \in T'} w_j \\
&\leq (1+\delta) \cdot \sum_{i \in U} w_i - (1+\delta)^{-1} \cdot \sum_{i \in U} w_i \\
&= \frac{\delta \cdot (2+\delta)}{1+\delta} \cdot \sum_{i \in U} w_i \\
&\leq \frac{\delta \cdot (2+\delta)}{1+\delta} \cdot \sum_{i \in U} (1+\delta) \cdot c_i \\
&= \delta \cdot (2+\delta) \cdot \sum_{i \in U} c_i \\
&= \delta \cdot (2+\delta) \cdot cost(U) \,.
\end{aligned}
$$

   Finally,

$$
\frac{cost(U) - cost(T')}{cost(U)} \leq \frac{\delta \cdot (2+\delta) \cdot cost(U)}{cost(U)} = \delta \cdot (2+\delta).
$$

2. Let $d = \sum_{i \in U} w_i - \sum_{j \in T'} w_j > 0$. Let c be the cost of the first part of U with the weight $\sum_{j \in T'} w_j$. Then in the same way as in 1. one can establish $\frac{c - cost(T')}{c} \leq \delta \cdot (2+\delta)$.
   It remains to bound $cost(U) - c$, i.e. the cost of the last part of U with the weight d. Obviously, $d \leq b - \sum_{j \in T'} w_j \leq w_{i_r}$ for some $r > k$, $i_r \in U$ (if not, then SKP$_\varepsilon$ would add r to T' in the greedy procedure). Since $w_{i_1} \geq w_{i_2} \geq ... \geq w_{i_l}$,

$$
d \leq w_{i_r} \leq \frac{w_{i_1} + w_{i_2} + ... + w_{i_r}}{r} \leq \frac{\sum_{i \in U} w_i}{k+1} \leq \varepsilon \cdot \sum_{i \in U} w_i .
$$

Since $cost(U) \le c + d \cdot (1+\delta)$ we obtain

$$\begin{aligned}\frac{cost(U) - cost(T')}{cost(U)} &\le \frac{c + d \cdot (1+\delta) - cost(T')}{cost(U)} \\ &\le \frac{c - cost(T')}{cost(U)} + \frac{(1+\delta) \cdot \varepsilon \cdot \sum_{i \in U} w_i}{cost(U)} \\ &\le \delta \cdot (2+\delta) + (1+\delta) \cdot \varepsilon \cdot (1+\delta) \\ &= 2\delta + \delta^2 + \varepsilon \cdot (1+\delta)^2 \\ &= \varepsilon + \delta \cdot (2+\delta) \cdot (1+\varepsilon)\,.\end{aligned}$$

□

Corollary 1. *The PTAS* SKP *is stable according to DIST, but not super-stable according to DIST.*

Proof. The first assertion directly follows from 1. To see that SKP is not super-stable according to $DIST$ it is sufficient to consider the following input:[4]

$$w_1, w_2, ..., w_m, u_1, u_2, ..., u_m, b, c_1, c_2, ..., c_{2m}\,,$$

where $w_1 = w_2 = ... = w_m$, $u_1 = u_2 = ...u_m$, $w_1 = u_1 + 1$, $b = \sum_{i=1}^{m} w_i$, $c_1 = c_2 = ... = c_m = (1-\delta)w_1$ and $c_{m+1} = c_{m+2} = ... = c_{2m} = (1+\delta)u_1$.

□

We see that the PTAS SKP is stable according to $DIST$, but this does not suffice to get a PTAS for KP_δ for any $\delta > 0$. This is because in the approximation ratio we have the additive factor $\delta \cdot (2+\delta)$ that is independent on ε. In what follows we change the PTAS SKP a little bit in such a way that we obtain a PTAS for every KP_δ, $\delta > 0$.

PTAS MOD-SKP

Input: Positive integers $w_1, w_2, ..., w_n, b, c_1, ..., c_n$ for some $n \in \mathbb{N}$ and some positive real number ε, $1 > \varepsilon > 0$.
Step 1: Sort $\frac{c_1}{w_1}, \frac{c_2}{w_2}, ..., \frac{c_n}{w_n}$. For simplicity we may assume $\frac{c_i}{w_i} \ge \frac{c_{i+1}}{w_{i+1}}$ for $i = 1, ..., n-1$.
Step 2: Set $k = \lceil 1/\varepsilon \rceil$.
Step 3: The same as Step 3 of PTAS SKP, but the greedy procedure follows the ordering of the w_i's of Step 1.
Output: The best T' constructed in Step 3.

Let MOD-SKP$_\varepsilon$ denote the algorithm given by PTAS MOD-SKP for a fixed $\varepsilon > 0$.

---

[4] Note, that m should be chosen to be essentially larger than ε^{-1} for a given ε.

Lemma 2. *For every ε, $1 > \varepsilon > 0$ and every $\delta \geq 0$,* MOD-SKP$_\varepsilon$ *is an* $\varepsilon \cdot (1+\delta)^2$*-approximation algorithm for* SK_δ.

Proof. Let $U = \{i_1, i_2, ..., i_l\} \subseteq \{1, 2, ..., n\}$, where $w_{i_1} \leq w_{i_2} \leq ... \leq w_{i_l}$, be an optimal solution for the input $I = w_1, ..., w_n, b, c_1, ..., c_n$.

If $l \leq k$ then MOD-SKP$_\varepsilon$ provides an optimal solution.

If $l > k$, then we consider a $T' = \{i_1, i_2, ..., i_k, j_{k+1}, ..., j_{k+r}\}$ as a greedy extension of $T = \{i_1, i_2, ..., i_k\}$. Again, we distinguish two possibilities:

1. Let $\sum_{i \in U} w_i - \sum_{j \in T'} w_j < 0$. Now, we show that this contradicts the optimality of U. Both, $cost(U)$ and $cost(T')$ contain $\sum_{s=1}^{k} c_{i_s}$. For the rest T' contains the best choice of w_i's according to the cost of one weight unit. The choice of U per one weight unit cannot be better. So, $cost(U) < cost(T')$.
2. Let $d = \sum_{i \in U} w_i - \sum_{j \in T'} w_j \geq 0$. Because of the optimal choice of T' according to the cost per one weight unit, the cost c of the first part of U with the weight $\sum_{j \in T'} w_j$ is at most $cost(T')$.
   Since U and T' contain the same k indices $i_1, i_2, ..., i_k$ and $w_{i_1}, ..., w_{i_k}$ are the largest weights in both U and T', the same same consideration as in the proof of Lemma 1 yields $d \leq \varepsilon \cdot \sum_{i \in U} w_i$, and $cost(U) \leq c + d \cdot (1+\delta)$. Thus,

$$\frac{cost(U) - cost(T')}{cost(U)} \leq \frac{c + d \cdot (1+\delta) - cost(T')}{cost(U)}$$

$$\leq \frac{d \cdot (1+\delta)}{cost(U)} \leq \varepsilon \cdot (1+\delta) \cdot \frac{\sum_{i \in U} w_i}{cost(U)} = \varepsilon \cdot (1+\delta)^2$$

□

We observe that the collection of MOD-SKP$_\varepsilon$ algorithms is a PTAS for every KP_δ with a constant $\delta \geq 0$ (independent of the size $2n+1$ of the input).

Corollary 2. MOD-SKP *is super-stable according to DIST.*

4 Conclusion and Discussion

In the previous sections we have introduced the concept of stability of approximations. Here we discuss the potential applicability and usefulness of this concept.

Using this concept, one can establish positive results of the following types:

1. An approximation algorithm or a PTAS can be successfully used for a larger set of inputs than the set usually considered (see Lemma 1).

2. We are not able to successfully apply a given approximation algorithm A (a PTAS) for additional inputs, but one can simply modify A to get a new approximation algorithm (a new PTAS) working for a larger set of inputs than the set of inputs of A (see Lemma 2).
3. To learn that an approximation algorithm is unstable for a distance measure could lead to the development of completely new approximation algorithms that would be stable according to the considered distance measure.

The following types of negative results may be achieved:

4. The fact that an approximation algorithm is unstable according to all "reasonable" distance measures and so that its use is really restricted to the original input set.
5. Let $Q = (\Sigma_I, \Sigma_O, L, L_I, \mathcal{M}, cost, goal) \in NPO$ be well approximable. If, for a distance measure D and a constant r, one proves the nonexistence of any approximation algorithm for $Q_{r,D} = (\Sigma_I, \Sigma_O, L, Round_{r,D}(L_I), \mathcal{M}, cost, goal)$, then this means that the problem Q is "unstable" according to D.

Thus, using the notion of stability one can search for a spectrum of the hardness of a problem according to the set of inputs. For instance, considering a hard problem like TSP one could get an infinite sequence of input languages $L_0, L_1, L_2, ...$ given by some distance measure, where $\varepsilon_r(n)$ is the best achievable relative error for the language L_r. Results of this kind can essentially contribute to the study of the nature of hardness of specific computing problems, and to a sharp determination of the border between tractable and intractable optimization problems.

References

1. S. Arora: Polynomial time approximation schemes for Euclidean TSP and other geometric problems. In: *Proc. 37th IEEE FOCS*, IEEE 1996, pp. 2–11.
2. S. Arora: Nearly linear time approximation schemes for Euclidean TSP and other geometric problems. In: *Proc. 38th IEEE FOCS*, IEEE 1997, pp. 554–563.
3. D. P. Bovet, C. Crescenzi: *Introduction to the Theory of Complexity*, Prentice-Hall 1993.
4. N. Christofides: Worst-case analysis of a new heuristic for the travelling salesman problem. Technical Report 388, Graduate School of Industrial Administration, Carnegie-Mellon University, Pittsbourgh, 1976.
5. S. A. Cook: The complexity of theorem proving procedures. In: Proc *3rd ACM STOC*, ACM 1971, pp. 151–158.
6. M. R. Garey, D. S. Johnson: *Computers and Intractability. A Guide to the Theory on NP-Completeness*. W. H. Freeman and Company, 1979.
7. J. Håstad: Clique is hard to approximate within $n^{1-\varepsilon}$. In: Proc. *37th IEEE FOCS*, IEEE 1996, pp. 627–636.

8. D. S. Hochbaum (Ed.): *Approximation Algorithms for NP-hard Problems.* PWS Publishing Company 1996.
9. J. Hromkovič: Towards the notion of stability of approximation algorithms for hard optimization problems, Dept. of Computer Science I, RWTH Aachen, May 1998.
10. O. H. Ibarra, C. E. Kim: Fast approximation algorithms for the knapsack and sum of subsets problem. *J. of the ACM* 22 (1975), pp. 463–468.
11. D. S. Johnson: Approximation algorithms for combinatorial problems. In: Proc. 5th *ACM STOC*, ACM 1973, pp. 38–49.
12. D. S. Johnson: Approximation algorithms for combinatorial problems *JCSS* 9 (1974), pp. 256–278.
13. R. M. Karp: Reducibility among combinatorial problems. In: R. E. Miller, J.W. Thatcher (eds.): *Complexity of Computer Computations*, Plenum Press 1972, pp. 85–103.
14. L.Lovász: On the ratio of the optimal integral and functional covers. *Discrete Mathematics* 13 (1975), pp. 383–390.
15. I. S. B. Mitchell: Guillotine subdivisions approximate polygonal subdivisions: Part II – a simple polynomial-time approximation scheme for geometric k-MST, TSP and related problems. Technical Report, Dept. of Applied Mathematics and Statistics, Stony Brook 1996.
16. E. W. Mayr, H. J. Prömel, A. Steger (Eds.): *Lecture on Proof Verification and Approximation Algorithms. Lecture Notes in Computer Science* 1967, Springer 1998.
17. Ch. Papadimitriou: The Euclidean travelling salesman problem is NP-complete. *Theoretical Computer Science* 4 (1977), pp. 237–244.
18. Ch. Papadimitriou: *Computational Complexity*, Addison-Wesley 1994.
19. S. Sahni: Approximate algorithms for the 0/1 knapsack problem. *J. ACM* 22 (1975), pp. 115–124.
20. S. Sahni, T. Gonzales: P-complete problems and approximate solutions. Comput. Sci. Techn. Rep. 74-5, University of Minnesota, Minneapolis, Minn., 1974.
21. I. Wegener: *Theoretische Informatik: eine algorithmenorientierte Einführung.* B.G. Teubner 1993.

Some Examples of Average-case Analysis by the Incompressibility Method

Tao Jiang, Ming Li, and Paul Vitányi

Summary. The incompressibility method is an elementary yet powerful proof technique. It has been used successfully in many areas, including average-case analysis of algorithms [14]. In this expository paper, we include several new simple average-case analyses to further demonstrate the utility and elegance of the method.

1 Introduction

The incompressibility of individual random objects yields a simple but powerful proof technique, namely *the incompressibility method.* This method is a versatile tool that can be used to prove lower bounds on computational problems, to obtain combinatorial properties of concrete objects, and to analyze the average-case complexity of algorithms. Since the early 1980's, the incompressibility method has been successfully used to solve many well-known questions that had been open for a long time and to supply new simplified proofs for known results. A comprehensive survey can be found in [14].

In this short expository paper, we use four simple examples of diverse topics to further demonstrate how easy the incompressibility method can be used to obtain (upper and lower) bounds which are useful in the domain of average-case analysis. The topics covered in this paper are well-known ones such as sorting, matrix multiplication, longest common subsequences, and majority finding. The proofs that we choose to include are not difficult ones and all the results are known before. However, our new proofs are much simpler than the old ones and are easy to understand. More such new proofs are contained in [5,13].

To make the paper self-contained, we give an overview of Kolmogorov complexity and the incompressibility method in the next section. We then consider the four diverse problems, namely sorting, boolean matrix multiplication, longest common subsequences, and majority finding, in four separate sections.

2 Kolmogorov Complexity and the Incompressibility Method

We use the following notation. Let x be a finite binary string. Then $l(x)$ denotes the *length* (number of bits) of x. In particular, $l(\epsilon) = 0$ where ϵ denotes the *empty word.*

We can map $\{0,1\}^*$ one-to-one onto the natural numbers by associating each string with its index in the length-increasing lexicographical ordering

$$(\epsilon, 0), (0, 1), (1, 2), (00, 3), (01, 4), (10, 5), (11, 6), \ldots . \tag{1}$$

This way we have a binary representation for the natural numbers that is different from the standard binary representation. It is convenient not to distinguish between the first and second element of the same pair, and call them "string" or "number" arbitrarily. As an example, we have $l(7) = 00$. Let $x, y, \in \mathcal{N}$, where $\mathcal{N}$ denotes the natural numbers. Let $T_0, T_1, \ldots$ be a standard enumeration of all Turing machines. Let $\langle \cdot, \cdot \rangle$ be a standard one-one mapping from $\mathcal{N} \times \mathcal{N}$ to $\mathcal{N}$, for technical reasons chosen so that $l(\langle x, y \rangle) = l(y) + O(l(x))$.

Informally, the Kolmogorov complexity, [15], of x is the length of the *shortest* effective description of x. That is, the *Kolmogorov complexity* $C(x)$ of a finite string x is simply the length of the shortest program, say in FORTRAN (or in Turing machine codes) encoded in binary, which prints x without any input. A similar definition holds conditionally, in the sense that $C(x|y)$ is the length of the shortest binary program which computes x on input y. Kolmogorov complexity is absolute in the sense of being independent of the programming language, up to a fixed additional constant term which depends on the programming language but not on x. We now fix one canonical programming language once and for all as reference and thereby $C()$. For the theory and applications, as well as history, see [14]. A formal definition is as follows:

Definition 1. Let U be an appropriate universal Turing machine such that

$$U(\langle\langle i, p\rangle, y\rangle) = T_i(\langle p, y\rangle)$$

for all i and $\langle p, y\rangle$. The *conditional Kolmogorov complexity* of x given y is

$$C(x|y) = \min_{p \in \{0,1\}^*} \{l(p) : U(\langle p, y\rangle) = x\}.$$

The unconditional Kolmogorov complexity of x is defined as $C(x) := C(x|\epsilon)$.

By a simple counting argument one can show that whereas some strings can be enormously compressed, the majority of strings can hardly be compressed at all. For each n there are 2^n binary strings of length n, but only $\sum_{i=0}^{n-1} 2^i = 2^n - 1$ possible shorter descriptions. Therefore, there is at least one binary string x of length n such that $C(x) \geq n$. We call such strings *incompressible*.

Definition 2. For each constant c we say a string x is *c-incompressible* if $C(x) \geq l(x) - c$.

Strings that are incompressible (say, c-incompressible with small c) are patternless, since a pattern could be used to reduce the description length. Intuitively, we think of such patternless sequences as being random, and we use "random sequence" synonymously with "incompressible sequence." It is possible to give a rigorous formalization of the intuitive notion of a random sequence as a sequence that passes all effective tests for randomness, see for example [14].

How many strings of length n are c-incompressible? By the same counting argument we find that the number of strings of length n that are c-incompressible is at least $2^n - 2^{n-c} + 1$. Hence there is at least one 0-incompressible string of length n, at least one-half of all strings of length n are 1-incompressible, at least three-fourths of all strings of length n are 2-incompressible, ..., and at least the $(1 - 1/2^c)$th part of all 2^n strings of length n are c-incompressible. This means that for each constant $c \geq 1$ the majority of all strings of length n (with $n > c$) is c-incompressible. We generalize this to the following simple but extremely useful *Incompressibility Lemma.*

Lemma 1. *Let c be a positive integer. For each fixed y, every set A of cardinality m has at least $m(1-2^{-c})+1$ elements x with $C(x|A,y) \geq \lfloor \log m \rfloor - c$.*

Proof. By simple counting. □

Definition 3. A *prefix set*, or prefix-free code, or prefix code, is a set of strings such that no member is a prefix of any other member. A prefix set which is the domain of a partial recursive function (set of halting programs for a Turing machine) is a special type of prefix code called a *self-delimiting* code because there is an effective procedure which reading left-to-right determines where a code word ends without reading past the last symbol. A one-to-one function with a range that is a self-delimiting code will also be called a self-delimiting code.

A simple self-delimiting code we use throughout is obtained by reserving one symbol, say 0, as a stop sign and encoding a natural number x as 1^x0. We can prefix an object with its length and iterate this idea to obtain ever shorter codes:

$$E_i(x) = \begin{cases} 1^x0 & \text{for } i = 0, \\ E_{i-1}(l(x))x & \text{for } i > 0. \end{cases} \tag{2}$$

Thus, $E_1(x) = 1^{l(x)}0x$ and has length $l(E_1(x)) = 2l(x)+1$; $E_2(x) = E_1(l(x))x$ $= 1^{l(l(x))}0x$ and has length $l(E_2(x)) = l(x)+2l(l(x))+1$. We have for example

$$l(E_3(x)) \leq l(x) + \log l(x) + 2 \log \log l(x) + 1.$$

Define the pairing function

$$\langle x, y \rangle = E_2(x)y \tag{3}$$

with inverses $\langle \cdot \rangle_1, \langle \cdot \rangle_2$. This can be iterated to $\langle \langle \cdot, \cdot \rangle, \cdot \rangle$.

The Incompressibility Method. In a typical proof using the incompressibility method, one first chooses an individually random object from the class under discussion. This object is effectively incompressible. The argument invariably says that if a desired property does not hold, then the object can be compressed. This yields the required contradiction. Then, since most objects are random, the desired property usually holds on the average.

3 Lower Bound for Sorting

We begin this paper with a very simple incompressibility proof for a well-known lower bound on comparison based sorting.

Theorem 1. *Any comparison based sorting algorithm requires* $\log n!$ *comparisons to sort an array of* n *elements.*

Proof. Let A be any comparison based sorting algorithm. Consider permutation I of $\{1, \ldots, n\}$ such that

$$C(I|A, P) \geq \log n!$$

where P is a fixed program to be defined. Suppose A sorts I in m comparisons. We can describe I by recording the binary outcomes of the m comparisons, which requires a total of m bits. Let P be such a program converting m to I (given A). Thus,

$$m \geq C(I|A, P) \geq \log n!$$

Hence, $m \geq \log n!$. □

The above proof in fact also implies a lower bound of $\log n! - 2\log n$ on the average number of comparisons required for sorting.

Corollary 1. *Any comparison based sorting algorithm requires* $\log n! - 2\log n$ *comparisons to sort an array of* n *elements, on the average.*

Proof. In the above proof, let I have Kolmogorov complexity:

$$C(I|A, P) \geq \log n! - \log n$$

Then we obtain that the (arbitrary) algorithm A requires

$$m \geq \log n! - \log n$$

comparisons on the permutation I. It follows from the Incompressibility Lemma that on the average, A requires at least

$$\frac{n-1}{n} \cdot (\log n! - \log n) + \frac{1}{n} \cdot (n-1) \geq \log n! - 2\log n$$

comparisons. □

4 Average Time for Boolean Matrix Multiplication

We begin with a simple (almost trivial) illustration of average-case analysis using the incompressibility method. Consider the well-known problem of multiplying two $n \times n$ boolean matrices $A = (a_{i,j})$ and $B = (b_{i,j})$. Efficient algorithms for this problem have always been a very popular topic in the theoretical computer science literature due to the wide range of applications of boolean matrix multiplication. The best worst-case time complexity obtained so far is $O(n^{2.376})$ due to Coppersmith and Winograd [7]. In 1973, O'Neil and O'Neil devised a simple algorithm described below which runs in $O(n^3)$ time in the worst case but achieves an average time complexity of $O(n^2)$ [16].

Algorithm QuickMultiply(A, B)

1. Let $C = (c_{i,j})$ denote the result of multiplying A and B.
2. For $i := 1$ to n do
3. Let $j_1 < \cdots < j_m$ be the indices such that $a_{i,j_k} = 1$, $1 \le k \le m$.
4. For $j := 1$ to n do
5. Search the list $b_{j_1,j}, \ldots, b_{j_m,j}$ sequentially for a bit 1.
6. Set $c_{i,j} = 1$ if a bit 1 is found, or $c_{i,j} = 0$ otherwise.

An analysis of the average-case time complexity of QuickMultiply is given in [16] using simple probabilitistic arguments. Here we give an analysis using the incompressibility method, to illustrate some basic ideas.

Theorem 2. *Suppose that the elements of A and B are drawn uniformly and independently. Algorithm QuickMultiply runs in $O(n^2)$ time on the average.*

Proof. Let n be a sufficiently large integer. Observe that the average time of QuickMultiply is trivially bounded between $O(n^2)$ and $O(n^3)$. By the Incompressibility Lemma, out of the 2^{2n^2} pairs of $n \times n$ boolean matrices, at least $(n-1)2^{2n^2}/n$ of them are $\log n$-incompressible. Hence, it suffices to consider $\log n$-incompressible boolean matrices.

Take a $\log n$-incompressible binary string x of length $2n^2$, and form two $n \times n$ boolean matrices A and B straightforwardly so that the first half of x corresponds to the row-major listing of the elements of A and the second half of x corresponds to the row-major listing of the elements of B. We show that QuickMultiply spends $O(n^2)$ time on A and B.

Consider an arbitrary i, where $1 \le i \le n$. It suffices to show that the n sequential searches done in Steps 4 – 6 of QuickMultiply take a total of $O(n)$ time. By the statistical results on various blocks in incompressible strings given in Section 2.6 of [14], we know that at least $n/2 - O(\sqrt{n \log n})$ of these searches find a 1 in the first step, at least $n/4 - O(\sqrt{n \log n})$ searches find a 1 in two steps, at least $n/8 - O(\sqrt{n \log n})$ searches find a 1 in three steps, and so on. Moreover, we claim that none of these searches take more than $4 \log n$ steps. To see this, suppose that for some j, $1 \le j \le n$, $b_{j_1,j} = \cdots = b_{j_{4 \log n},j} = 0$. Then we can encode x by listing the following items in a self-delimiting manner:

1. A description of the above discussion.
2. The value of i.
3. The value of j.
4. All bits of x except the bits $b_{j_1,j}, \ldots, b_{j_{4\log n},j}$.

This encoding takes at most

$$O(1) + 2\log n + 2n^2 - 4\log n + O(\log\log n) < 2n^2 - \log n$$

bits for sufficiently large n, which contradicts the assumption that x is $\log n$-incompressible.

Hence, the n searches take at most a total of

$$\begin{aligned}
&(\sum_{k=1}^{\log n}(n/2^k - O(\sqrt{n\log n})) \cdot k) + (\log n) \cdot O(\sqrt{n\log n}) \cdot (4\log n) \\
&< (\sum_{k=1}^{\log n} kn/2^k + O(\log^2 n\sqrt{n\log n}) \\
&= O(n) + O(\log^2 n\sqrt{n\log n}) \\
&= O(n)
\end{aligned}$$

steps. This completes the proof. □

5 Expected Length of a Longest Common Subsequence

For two sequences (*i.e.* strings) $s = s_1 \ldots s_m$ and $t = t_1 \ldots t_n$, we say that s is a *subsequence* of t if for some $i_1 < \ldots < i_m$, $s_j = t_{i_j}$. A *longest common subsequence* (LCS) of sequences s and t is a longest possible sequence u that is a subsequence of both s and t. For simplicity, we will only consider binary sequences over the alphabet $\Sigma = \{0, 1\}$.

Let n be an arbitrary positive integer and consider two random strings s and t that are drawn independently from the uniformly distributed space of all binary string of length n. We are interested in the expected length of an LCS of s and t. Tight bounds on the expected LCS length for two random sequences is a well-known open question in statistics [17,19]. After a series of papers, the best result to date is that the length is between $0.762n$ and $0.838n$ [6,8–10]. The proofs are based on intricate probablistic and counting arguments. Here we give simple proofs of some nontrivial upper and lower bounds using the incompressibility method.

Theorem 3. *The expected length of an LCS of two random sequences of length n is at most* $0.867n + o(n)$.

Proof. Let n be a sufficiently large integer. Observe that the expected length of an LCS of two random sequences of length n is trivially bounded between

$n/2$ and n. By the Incompressibility Lemma, out of the 2^{2n} pairs of binary sequences of length n, at least $(n-1)2^{2n}/n$ of them are $\log n$-incompressible. Hence, it suffices to consider $\log n$-incompressible sequences.

Take a $\log n$-incompressible string x of length $2n$, and let s and t be the first and second halves of x respectively. Suppose that string u is an LCS of s and t. In order to relate the Kolmogorov complexity of s and t to the length of u, we re-encode the strings s and t using the string u as follows. (The idea was first introduced in [12].)

We first describe how to re-encode s. Let the LCS $u = u_1 u_2 \cdots u_m$, where $m = l(u)$. We align the bits of u with the corresponding bits of s greedily from left to right, and rewrite s as follows:

$$s = \alpha_1 u_1 \alpha_2 u_2 \cdots \alpha_m u_m s'.$$

Here α_1 is the longest prefix of s containing no u_1, α_2 is the longest substring of s following the bit u_1 containing no u_2, and so on, and s' is the remaining part of s after the bit u_m. Thus α_i does not contain bit u_i, for $i = 1, \ldots, m$. In other words, each α_i is a unary string consisting of the bit complementary to u_i. We re-encode s as string:

$$s(u) = 0^{l(\alpha_1)} 1 0^{l(\alpha_2)} 1 \cdots 0^{l(\alpha_m)} 1 s'.$$

Clearly, given u we can uniquely decode the encoding $s(u)$ to obtain s.

Similarly, the string t can be rewritten as

$$t = \beta_1 u_1 \beta_2 u_2 \cdots \beta_m u_m t',$$

where each β_i is a unary string consisting of the bit complementary to u_i, and we re-encode t as string:

$$t(u) = 0^{l(\beta_1)} 1 0^{l(\beta_2)} 1 \cdots 0^{l(\beta_m)} 1 t'.$$

Hence, the string x can be described by the following information in the self-delimiting form:

1. A description of the above discussion.
2. The LCS u.
3. The new encodings $s(u)$ and $t(u)$ of s and t.

Now we estimate the Kolmogorov complexity of the above description of x. Items 1 and 2 take $m + O(1)$ bits. Since $s(u)$ contains at least m 1's, it is easy to see by simple counting and Stirling approximation (see *e.g.* [14]) that

$$\begin{aligned} C(s(u)) &\leq \log \sum_{i=m}^{n} \binom{n}{i} + O(1) \\ &\leq \log \left[\frac{n}{2} \binom{n}{m} \right] + O(1) \\ &\leq \log n + \log \binom{n}{m} + O(1) \\ &\leq 2 \log n + n \log n - m \log m - (n-m) \log(n-m) + O(1) \end{aligned}$$

The second step in the above derivation follows from the trivial fact that $m \geq n/2$. Similarly, we have

$$C(t(u)) \leq 2\log n + n\log n - m\log m - (n-m)\log(n-m) + O(1)$$

Hence, the above description requires a total size of

$$O(\log n) + m + 2n\log n - 2m\log m - 2(n-m)\log(n-m).$$

Let $p = n/m$. Since $C(x) \geq 2n - \log n$, we have

$$\begin{aligned} 2n - \log n &\leq O(\log n) + m + 2n\log n - 2m\log m - 2(n-m)\log(n-m) \\ &= O(\log n) + pn - 2np\log p - 2n(1-p)\log(1-p) \end{aligned}$$

Dividing both sides of the inequality by n, we obtain

$$2 \leq o(1) + p - 2p\log p - 2(1-p)\log(1-p)$$

Solving the inequality numerically we get $p \leq 0.867 - o(1)$. □

Next we prove that the expected length of an LCS of two random sequences of length n is at least $0.66666n - O(\sqrt{n\log n})$. To prove the lower bound, we will need the following greedy algorithm for computing common subsequences (not necessarily the longest ones).

Algorithm Zero-Major$(s = s_1 \cdots s_n, t = t_1 \cdots t_n)$

1. Let $u := \epsilon$ be the empty string.
2. Set $i := 1$ and $j := 1$;
3. **Repeat** steps 4–6 until $i > n$ or $j > n$:
4.   **If** $s_i = t_j$ **then begin** append bit s_i to string u; $i := i + 1$; $j := j + 1$ **end**
5.   **Elseif** $s_i = 0$ **then** $j := j + 1$
6.   **Else** $i := i + 1$
7. Return string u.

Theorem 4. *The expected length of an LCS of two random sequences of length n is at least* $0.66666n - O(\sqrt{n\log n})$.

Proof. Again, let n be a sufficiently large integer, and take a $\log n$-incompressible string x of length $2n$. Let s and t be the first and second halves of x respectively. It suffices to show that the above algorithm Zero-Major produces a common subsequence u of length at least $0.66666n - O(\sqrt{n\log n})$ for strings s and t.

The idea is to encode s and t (and thus x) using information from the computation of Zero-Major on strings s and t. We consider the comparisons made by Zero-Major in the order that they were made, and create a pair of strings y and z as follows. For each comparison (s_i, t_j) of two complementary bits, we simply append a 1 to y. For each comparison (s_i, t_j) of two identical bits, append a bit 0 to the string y. Furthermore, if this comparison of identical bits is preceded by a comparison $(s_{i'}, t_{j'})$ of two complementary bits, we then append a bit 0 to the string z if $i' = i - 1$ and a bit 1 if $j' = j - 1$. When one string (s or t) is exhausted by the comparisons, we append the remaining part (call this w) of the other string to z.

As an example of the encoding, consider strings s = 1001101 and t = 0110100. Algorithm Zero-Major produces a common subsequence 0010. The following figure depicts the comparisons made by Zero-Major, where a "*" indicates a mismatch and a "|" indicates a match.

```
s =                  10  01101
comparisons          *|**||*|*
t =                   01101 0 0
```

Following the above encoding scheme, we obtain y = 101100101 and z = 01100.

It is easy to see that the strings y and z uniquely encode s and t and, moreover, $l(y)+l(z) = 2n$. Since $C(yz) \geq C(x) - 2\log n \geq 2n - 3\log n - O(1)$, and $C(z) \leq l(z) + O(1)$, we have

$$C(y) \geq l(y) - 3\log n - O(1)$$

Similarly, we can obtain

$$C(z) \geq l(z) - 3\log n - O(1)$$

and

$$C(w) \geq l(w) - 3\log n - O(1)$$

where w is the string appended to z at the end of the above encoding.

Now let us estimate the length of the common subsequence u produced by Zero-Major on strings s and t. Let $\#zeroes(s)$ and $\#zeroes(t)$ be the number of 0's contained in s and t respectively. Clearly, u contains $\min\{\#zeroes(s), \#zeroes(t)\}$ 0's. From [14] (page 159), since both s and t are $\log n$-incompressible, we know

$$n/2 - O(\sqrt{n\log n}) \leq \#zeroes(s) \leq n/2 + O(\sqrt{n\log n})$$

$$n/2 - O(\sqrt{n\log n}) \leq \#zeroes(t) \leq n/2 + O(\sqrt{n\log n})$$

Hence, the string w has at most $O(\sqrt{n\log n})$ 0's. Combining with the fact that $C(w) \geq l(w) - 3\log n - O(1)$ and the above mentioned result in [14], we claim

$$l(w) \leq O(\sqrt{n\log n}).$$

Since $l(z) - l(w) = l(u)$, we have a lower bound on $l(u)$:

$$l(u) \geq l(z) - O(\sqrt{n\log n}).$$

On the other hand, since every bit 0 in the string y corresponds to a unique bit in the common subsequence u, we have $l(u) \geq \#zeroes(y)$. Since $C(y) \geq l(y) - 2\log n - O(1)$,

$$l(u) \geq \#zeroes(y) \geq l(y)/2 - O(\sqrt{n \log n}).$$

Hence,

$$3l(u) \geq l(y) + l(z) - O(\sqrt{n \log n}) \geq 2n - O(\sqrt{n \log n}).$$

That is,

$$l(u) \geq 2n/3 - O(\sqrt{n \log n}) \approx 0.66666n - O(\sqrt{n \log n}).$$

□

Our above upper and lower bounds are not as tight as the ones in [6,8–10]. Recently, Baeza-Yates and Navarro improved our analysis and obtained a slightly better upper of 0.860 [4]. It will be interesting to know if stronger bounds can be obtained using the incompressibility method by more clever encoding schemes.

6 Average Complexity of Finding the Majority

Let $x = x_1 \cdots x_n$ be a binary string. The *majority bit* (or simply, the *majority*) of x is the bit (0 or 1) that appears more than $\lfloor n/2 \rfloor$ times in x. The majority problem is that, given a binary string x, determine the majority of x. When x has no majority, we must report so.

The time complexity for finding the majority has been well studied in the literature (see, *e.g.* [1–3,11,18]). It is known that, in the worst case, $n - \nu(n)$ bit comparisons are necessary and sufficient [2,18], where $\nu(n)$ is the number of occurrences of bit 1 in the binary representation of number n. Recently, Alonso, Reingold and Schott [3] studied the average complexity of finding the majority assuming the uniform probability distribution model. Using quite sophisticated arguments based on decision trees, they showed that on the average finding the majority requires at most $2n/3 - \sqrt{8n/9\pi} + O(\log n)$ comparisons and at least $2n/3 - \sqrt{8n/9\pi} + \Theta(1)$ comparisons.

In this section, we consider the average complexity of finding the majority and prove an upper bound tight up to the first major term, using simple incompressibility arguments.

The following standard tournament algorithm is needed.

Algorithm Tournament($x = x_1 \cdots x_n$)

1. If $n = 1$ then return x_1 as the majority.
2. Elseif $n = 2$ then

3. If $x_1 = x_2$ then return x_1 as the majority.
4. Else return "no majority".
5. Elseif $n = 3$ then
6. If $x_1 = x_2$ then return x_1 as the majority.
7. Else return x_3 as the majority.
8. Let $y = \epsilon$.
9. For $i := 1$ to $\lfloor n/2 \rfloor$ do
10. If $x_{2i-1} = x_{2i}$ then append the bit x_{2i} to y.
11. If n is odd and $\lfloor n/2 \rfloor$ is even then append the bit x_n to y.
12. Call Tournament(y).

Theorem 5. *On the average, algorithm Tournament requires at most* $2n/3 + O(\sqrt{n})$ *comparisons.*

Proof. Let n be a sufficiently large number. Again, since algorithm Tournament makes at most n comparisons on any string of length n, by the Incompressibility Lemma, it suffices to consider running time of Tournament on δ-incompressible strings, where $\delta \leq \log n$. Consider an arbitrary $\delta \leq \log n$ and let $x = x_1 \cdots x_n$ be a fixed δ-incompressible binary string. For any integer $m \leq n$, let $\sigma(m)$ denote the maximum number of comparisons required by algorithm Tournament on any δ-incompressible string of length m.

We know from [14] that among the $\lfloor n/2 \rfloor$ bit pairs $(x_1, x_2), \ldots,$ $(x_{2\lfloor n/2 \rfloor - 1}, x_{2\lfloor n/2 \rfloor})$ that are compared in step 10 of Tournament, there are at least $n/4 - O(\sqrt{n\delta})$ pairs consisting of complementary bits. Clearly, the new string y obtained at the end of step 11 should satisfy

$$C(y) \geq l(y) - \delta - O(1)$$

Hence, we have the following recurrence relation for $\sigma(m)$:

$$\sigma(m) \leq \lfloor m/2 \rfloor + \sigma(m/4 + O(\sqrt{m\delta}))$$

By straightforward expansion, we obtain that

$$\begin{aligned}
\sigma(n) &\leq \lfloor n/2 \rfloor + \sigma(n/4 + O(\sqrt{n\delta})) \\
&\leq n/2 + \sigma(n/4 + O(\sqrt{n\delta})) \\
&\leq n/2 + (n/8 + O(\sqrt{n\delta})/2) + \sigma(n/16 + O(\sqrt{n\delta})/4 + O(\sqrt{(n\delta)/4})) \\
&= n/2 + (n/8 + O(\sqrt{n\delta})/2) + \sigma(n/16 + (3/4) \cdot O(\sqrt{n\delta})) \\
&\leq \cdots \\
&\leq 2n/3 + O(\sqrt{n\delta})
\end{aligned}$$

Using the Incompressibility Lemma, we can calculate the average complexity of algorithm Tournament as:

$$\sum_{\delta=1}^{\log n} \frac{1}{2^\delta}(2n/3 + O(\sqrt{n\delta})) + \frac{1}{n} n = 2n/3 + O(\sqrt{n})$$

□

References

1. L. Alexanderson, L.F. Klosinski and L.C. Larson, *The William Lowell Putnam Mathematical Competition, Problems and Solutions: 1965-1984*, Mathematical Association of America, Washington, DC, 1985.
2. L. Alonso, E. Reingold and R. Schott, Determining the majority, *Information Processing Letters* 47, 1993, pp. 253-255.
3. L. Alonso, E. Reingold and R. Schott, The average-case complexity of determining the majority, *SIAM Journal on Computing* 26-1, 1997, pp. 1-14.
4. R. Baeza-Yates and G. Navarro, Bounding the expected length of longest common subsequences and forests, *Manuscript*, 1997.
5. H. Buhrman, T. Jiang, M. Li, and P.M.B. Vitányi, New applications of the incompressibility method: Part II, accepted to *Theoret. Comput. Sci.* (Special Issue for "Int'l Conf. Theoret. Comput. Sci.", Hong Kong, April, 1998).
6. V. Chvátal and D. Sankoff. Longest common subsequences of two random sequences. *J. Appl. Probab.* 12, 1975, 306-315.
7. D. Coppersmith and S. Winograd. Matrix multiplication via arithmetic progressions. *Proc. of 19th ACM Symp. on Theory of Computing*, 1987, pp. 1-6.
8. V. Dančík and M. Paterson, Upper bounds for the expected length of a longest common subsequence of two binary sequences, *Proc. 11th Annual Symposium on Theoretical Aspects of Computer Science*, LNCS 775, Springer, pp. 669-678, Caen, France, 1994.
9. J.G. Deken, Some limit results for longest common subsequences, *Discrete Mathematics* 26, 1979, pp. 17-31.
10. J.G. Deken, Probabilistic behavior of longest-common-subsequence length, *Time Warps, String Edits, and Macromolecules: The Theory and Practice of Sequence Comparison.* (D. Sankoff and J. Kruskall, Eds.) , Addison-Wesley, Reading, MA., 1983, pp. 359-362.
11. D.H. Greene and D.E. Knuth, *Mathematics for the Analysis of Algorithms*, 3rd ed., Birkhäuser, Boston, MA, 1990.
12. T. Jiang and M. Li, On the approximation of shortest common supersequences and longest common subsequences, *SIAM Journal on Computing* 24-5, 1122-1139, 1995.
13. T. Jiang, M. Li, and P.M.B. Vitányi, New applications of the incompressibility method, accepted to *The Computer Journal*, 1998.
14. M. Li and P.M.B. Vitányi, *An Introduction to Kolmogorov Complexity and its Applications*, Springer-Verlag, New York, 2nd Edition, 1997.
15. A.N. Kolmogorov, Three approaches to the quantitative definition of information. *Problems Inform. Transmission*, 1(1):1-7, 1965.
16. P. O'Neil and E. O'Neil. A fast expected time algorithm for boolean matrix multiplication and transitive closure. *Information and Control* 22, 1973, pp. 132-138.
17. M. Paterson and V. Dančík, Longest common subsequences, *Proc. 19th International Symposium on Mathematical Foundations of Computer Science*, LNCS 841, Springer, pp. 127-142, Kosice, Slovakia, 1994.
18. M.E. Saks and M. Werman, On computing majority by comparisons, *Combinatorica* 11, 1991, pp. 383-387.
19. D. Sankoff and J. Kruskall (Eds.) *Time Warps, String Edits, and Macromolecules: The Theory and Practice of Sequence Comparison.* Addison-Wesley, Reading, MA., 1983.

Complexity of Language Recognition Problems for Compressed Words

Wojciech Plandowski * and Wojciech Rytter **

Summary. The **compressed recognition** problem consists in checking if an input word w is in a given language L, when we are only given a compressed representation of w. We present several new results related to language recognition problems for compressed texts. These problems are solvable in polynomial time for uncompressed words and some of them become NP-hard for compressed words.

Two types of compression are considered: Lempel-Ziv compression and compression in terms of straight-line programs (or sequences of recurrences, or context-free grammars generating single texts). These compressions are polynomially related and most of our results apply to both of them.

Denote by $LZ(w)$ ($SLP(w)$) the version of a string w produced by *Lempel-Ziv encoding* (*straight-line program*). The complexity of the following problem is considered:
given a compressed version ($LZ(w)$ or $SLP(w)$) of the input word w, test the membership $w \in L$, where L is a formal language.
The complexity depends on the type and description of the language L. Surprisingly the proofs of NP-hardness are in this area usually easier than the proofs that a problem in in NP.

In particular the membership problem is in polynomial-time for regular expressions. However it is NP-hard for semi-extended regular expressions and for (linear) context-free languages, and we don't know if it is in NP in these cases. The membership problem is NP-complete for unary regular expressions with compressed constants.

The membership problem is in $DSPACE(n^2)$ for general context-free sets L and in $NSPACE(n)$ for linear languages. We show that for unary languages compressed recognition of context-free languages is NP-complete.

We also briefly discuss some known results related to the membership problem for the string-matching languages and for languages related to string-matching: square-free words, squares, palindromes, and primitive words.

* Instytut Informatyki, Uniwersytet Warszawski, Banacha 2, 02–097 Warszawa, Poland. Email:`wojtekpl@mimuw.edu.pl`.

** Instytut Informatyki, Uniwersytet Warszawski, Banacha 2, 02–097 Warszawa, Poland, and Department of Computer Science, University of Liverpool. Supported partially by the grant KBN 8T11C03915. Email:`rytter@mimuw.edu.pl`

1 Introduction

One of the basic language recognition problems is the *string-matching* problem:

check if $P\#T \in L_{string-match}$,

where

$L_{string-match} = \{x\#y : x \text{ is a subword of } y, \text{ and } x,\ y \text{ do not contain } \#\}$.

The ***compressed string-matching*** problem (when the input is given in the compressed form) has been investigated in [1], [2], [3], [12], [13], [10]. The ***fully compressed*** matching occurs when both P and T are given in compressed forms.

However other typical membership problems have not been considered, e.g. recognition of context-free languages.

In this paper we discuss the complexity of several language recognition problems for input words given in a compressed form. We also briefly discuss the complexity of the compressed string-matching.

1.1 Lempel-Ziv encodings

The LZ compression (see [15]) gives a very natural way of representing a string and it is a practically successful method of text compression. We consider the same version of the LZ algorithm as in [11] (this is called $LZ1$ in [11]). Intuitively, LZ algorithm compresses the text because it is able to discover some repeated subwords. We consider here the version of LZ algorithm without *self-referencing* but our algorithms can be extended to the general self-referential case. Assume that Σ is an underlying alphabet and let w be a string over Σ. The factorization of w is given by a decomposition:

$$w = c_1 f_1 c_2 \ldots f_k c_{k+1},$$

where $c_1 = w[1]$ and for each $1 \le i \le k$ $c_i \in \Sigma$ and f_i is the longest prefix of $f_i c_{i+1} \ldots f_k c_{k+1}$ which appears in $c_1 f_1 c_2 \ldots f_{i-1} c_i$.

We can identify each f_i with an interval $[p, q]$, such that $f_i = w[p..q]$ and $q \le |c_1 f_1 c_2 \ldots f_{i-1} c_{i-1}|$. If we drop the assumption related to the last inequality then it occurs a *self-referencing* (f_i is the longest prefix which appears before but not necessarily terminates at a current position). We assume that this is not the case.

Example.
The factorization of a word *aababbabbaababbabba#* is given by:

$$c_1\ f_1\ c_2\ f_2\ c_3\ f_3\ c_4\ f_4\ c_5 = a\ a\ b\ ab\ b\ abb\ a\ ababbabba\ \#.$$

After identifying each subword f_i with its corresponding interval we obtain the LZ encoding of the string. Hence

$$LZ(aababbabbababbabb\#) = a[1,1]b[1,2]b[4,6]a[2,10]\#.$$

1.2 Straight-line programs encodings

In [17] and [10] the compressed strings were considered in terms of straight-line programs (context-free grammars generating single texts). A *straight-line program* (SLP for short) $\mathcal{R}$ is a sequence of assignment statements:

$$X_1 = expr_1;\ X_2 = expr_2; \ldots ;\ X_n = expr_n,$$

where X_i are variables and $expr_i$ are expressions of the form:

- $expr_i$ is a symbol of a given alphabet Σ, or
- $expr_i = X_j \cdot X_k$, for some $j, k < i$, where $\cdot$ denotes the concatenation of X_j and X_k.

Example.
The following program $\mathcal{T}$ describes the 8*th Fibonacci word*, see [16].

$$X_1 = \mathsf{b};\ X_2 = \mathsf{a};\ X_3 = X_2 \cdot X_1;\ X_4 = X_3 \cdot X_2;$$
$$X_5 = X_4 \cdot X_3;\ X_6 = X_5 \cdot X_4;\ X_7 = X_6 \cdot X_5;\ X_8 = X_7 \cdot X_6,$$

Both types of encodings are polynomially related, so from the point of view of polynomial-time computability we can use term *compressed* to mean LZ or SLP compression.

Theorem 1.
(a). *Let $n = |LZ(w)|$. Then we can construct a context-free grammar G of size $O(n^2 \log n)$ which generates w and which is in the Chomsky normal form. Moreover the height of the derivation tree of w with respect to G is $O(n \cdot \log\log n)$.*
(b). *Assume w is generated by a straight-line program of size n then $|LZ(w)| = O(n)$.*

2 Compressed recognition problems for regular expressions

We consider three classes of regular expressions as descriptions of regular languages:

1. (standard) regular expressions (using uncompressed constants and operations $\cup$, *, $\cdot$);
2. regular expressions with compressed constants (constants given in compressed forms);
3. semi-extended regular expressions (using additionally the operator $\cap$ and only uncompressed constants)

The size of the expression will be a part of the input size.

Theorem 2.
(a) *We can decide for SLP-compressed words the membership in a language described by given regular expression W in $O(n \cdot m^3)$ time, where $m = |W|$.*
(b) *We can decide for SLP-compressed words the membership in a language described by given determiniatic automaton M in $O(n \cdot m)$ time, where m is the number of states of M.*

Proof.
(a) Construct a nondeterministic finite automaton M accepting the language described by W with $O(m)$ states. For each variable X denote by $TransTable_X$ a relation on states of M such that (p, q) are in this relation iff there is a path from the state p to q when reading the input word generated by X.

If we have an equation $X = Y \cdot Z$ then
$$TransTable_X = TransTable_Y \otimes TransTable_Z,$$
where $\otimes$ is the boolean matrix multiplication.

In this way we compute bottom-up the transition table for X_n. The most costly operation is the multiplicaton $\otimes$ of boolean matrices, which can be done in $O(m^3)$ time.

(b) In this case the table $TransTable_X$ becomes a function, which can be represented by a vector. Composition of such functions can be done in $O(m)$ time, so the total cost is $O(n \cdot m)$. This completes the proof.

We use the following problem to show NP-hardness of several compressed recognition problems.

SUBSET SUM problem:
Input instance: Finite set $A = \{a_1, a_2, \ldots, a_n\}$ of integers and an integer K. The size of the input is the number of bits needed for the binary representation of numbers in A and K.
Question: Is there a subset $A' \subseteq A$ such that the sum of the elements in A' is exactly K?

Lemma 1. *The problem SUBSET SUM is NP-complete.*

Proof. see [9], and [8], pp. 223.

Theorem 3.
The problem of checking membership of a compressed unary word in a language described by a star-free *regular expression with compressed constants is NP-complete.*

Proof.
The proof of NP-hardness is a reduction from the SUBSET SUM problem. We can construct easily a straight-line program such that $value(X_i) = d^{a_i}$ and $w = d^K$. Then the SUBSET SUM problem is reduced to the membership:

$$w \in (value(X_1) \cup \varepsilon) \cdot (value(X_2) \cup \varepsilon) \cdots (value(X_n) \cup \varepsilon).$$

The empty string ε can be easily eliminated in the following way. We replace each ε by a single symbol d and each number a_i by $(a_i + 1)$. Then we check whether d^{n+K} is generated by the obtained expression.

The problem is in NP since expressions are *star-free*. We can construct an equivalent nondeterministic finite automaton A and guess an accepting path. All paths are of polynomial length due to *star-free* condition. We can check in polynomial time if concatenation of constants on the path equals an input text P. This completes the proof.

It is not obvious if the previously considered problem is in NP for regular expressions containg the operation *, in this case there is no polynomial bound on the length of accepting paths of A. There is a simple argument in case of unary languages.

In the next theorem we show an interesting application of the Euler path technique to a unary language recognition.

Theorem 4. *[unary expressions]*
The problem of checking membership of a compressed unary word in a language described by a given regular expression with compressed constants is in NP.

Proof.
Let A be a nondeterministic finite automaton accepting a given regular expression. We can assume that there is only one accepting state, the initial state has zero indegree and the accepting state has zero outdegree. Let S be a set of edges of A. Denote by $in(S,q)$ and $out(S,q)$ the number of edges in S entering (outgoing) a given state a of A. Denote by q_0, q_a the initial (accepting) state of A.

We say that a multiset S is *valid* for A iff it is a set of edges of an accepting path of A.

Claim. (Euler Path Property)
S is valid for A iff the following conditions are satisfied:
(a) $in(S,q) = out(S,q)$ for each noninitial and nonacepting state of A,
(b) $in(S,q_0) = out(S,q_a) = 0$ and $out(S,q_0) = in(S,q_a) = 1$.
(c) the graph which is formed from the edges in S is connected,

Proof. (of the claim)
Obviously each accepting path corresponds to a valid multiset S. On the other hand if S is valid and if we continue arbitrarily a path starting at q_0 and using each time a single edge from S (deleting it after usage) then the only possibility is to finish in q_a. In this way an accepting path corresponding to S is constructed in a similar way as an Euler path in a graph in the proof of Euler theorem.

We can guess a multiset S since number of edges of A is linear. Knowing the multiplicity of each edge we can now deterministically check if the total length of all edges in the multiset S (counting with multiplicity) equals the length of the input word. We have numbers with polynomial number of bits, hence all computations are in polynomial time.

Theorem 5.
The problem of checking membership of a compressed word in a language described by a semi-extended regular expression is NP-hard.

Proof.
We use the following number-theoretic fact.

Chinese Remainder Theorem.

Let $p_1, p_2, \ldots, p_k$ be relatively prime integers. Then for each sequence $r_1, r_2, \ldots, r_k$ there is exactly one integer $0 \leq a < p_1 p_2 \ldots p_k$ such that $a \bmod p_i = r_i$ for each $1 \leq i \leq k$.

We use Thorem 3. Assume the total number of bits representing compressed unary constants and size of the input in this theorem is m. Then we replace the compressed strings d^{a_i} by $(d^{p_1})^* d^{r_1} \cap (d^{p_2})^* d^{r_2} \cap ...(d^{p_k})^* d^{r_k}$, where p_i are all prime numbers smaller than m and $r_j = a_i \bmod p_j$.

3 Compressed recognition problems for context-free languages

Theorem 6.
The problem of checking membership of a compressed word in a given linear cfl L is NP-hard, even if L is given by a context-free grammar of a constant size.

Proof.
Take an instance of the subset-sum problem with the set $A = \{a_1, a_2, \ldots, a_n\}$ of integers and an integer K. Define the following language:

$$L = \{d^R \$ d^{v_1} \# d^{v_2} \# \ldots \# d^{v_t} \ : \ t \geq 1 \text{ and there is a subset } A' \subseteq \{v_1, \ldots, v_t\} \text{ such that } \textstyle\sum_{u \in A'} u \ = \ R\} \ ?$$

L is obviously a linear context-free language generated by a linear context-free grammar of a constant size. We can reduce an instance of the subset sum-problem to the memebership problem:

$$d^K \$ d^{a_1} \# d^{a_2} \# \ldots \# d^{a_n} \ \in \ L.$$

Theorem 7.
(a) *The problem of checking membership of a compressed word in a given linear cfl is in $NSPACE(n)$.*
(b) *The problem of checking membership of a compressed word in a given cfl is in $DSPACE(n^2)$.*

Proof.
We can easily compute in linear space a symbol on a given position in a compressed input word. Now we can use a space-efficient algorithm for the recognition of context-free languages. It is known that linear languages can be recognized in $O(logN)$ nondeterministic space and general cfls can be done in $O(log^2 N)$ deterministic space, where N is the size of the uncompressed input word. In our case $N = O(2^n)$, this gives required $NSPACE(n)$ and $DSPACE(n^2)$ complexities.

Theorem 8.
The problem of checking membership of a compressed unary word in a given cfl is NP-complete.

Proof.
(1) The proof of NP-hardness.
We use the construction from Theorem 3 and construct a grammar generating a language described by a *star-free* regular expression. Each compressed unary constant can be generated by a polynomial size grammar since there is grammar generating d^{2^k} with $O(k)$ productions.

(2) Containment in NP.
Assume the grammar G is in Chomsky normal form. Let M be a multiset of productions of the grammar G. We say that M is ***valid*** for G iff there is a derivation tree T of G such that M is the multiset of productions used in T.

Denote by $\#fathers_A(M)$ the number of occurrences of a nonterminal A as a root in a production in M, and by $\#sons_A(M)$ the number of occurrences of a nonterminal A as a son in a production in M. Let S be the starting nonterminal of G.

Claim.
A multiset M of productions is valid for G iff the following conditions are satisfied:
(a) $\#sons_A(M) = \#fathers_A(M)$ for each nonterminal $A \neq S$;
(b) $\#sons_S(M) = \#fathers_S(M) - 1$.
(c) each nonterminal occuring in productions in M is reachable from S using the productions in M.

Proof. (of the claim)
If T is a derivation tree then the multiset of productions in T certainly satisfies the conditions. Conversely, assume M satisfies both conditions. Then we can construct the tree top-down starting with a production which has on the left side S and belongs to M. Whenever we have a leaf of a partially constructed tree which is a nonterminal we can expand it using a production from M. Each time we use a production from M we delete it from M. The conditions guarantee that we finish successfully.

We can guess a multiset M, check its validity and check if the number (counting multiplicites) of terminal productions (of the form $A \to d$) equals the length of the input unary word. This completes the proof.

Observe here that we cannot strengthen Theorem 8 as in Theorem 6 by bounding the size of the input grammar by a constant since for constant-size grammars and unary alphabets we can transform the grammar to a finite automaton of a constant size and obtain a deterministic polynomial time algorithm for the problem, due to Theorem 2.

4 Languages related to string-matching

The key role in string-matching problems is the size of the description of an exponential size set of positions. Small descriptions use the fact that exponential length arithmetic progression can be identified with three integers. A set of integers forming an arithmetic progression is called here *linear*. We say that a set of positive integers from $[1 \dots U]$ is *linearly-succinct* iff it can be decomposed in at most $\lfloor \log_2(U) \rfloor + 1$ linear sets. Denote $Periods(w) = \{p : p \text{ is a period of } w\}$. The following lemma was shown in [10].

Lemma 2 (linearly-succinct sets lemma).
The set Periods(w) is linearly-succinct. The set of occurrences of a word overlapping a fixed position is linear.

Surprisingly the set of occurrences of a fixed pattern in a compressed text does not necessarily have a linearly succinct representation, which shows nontriviality of the compressed string-matching.

Example 1. The set of all occurrences of a given pattern in a "well" compressible text T is not necessarily linearly-succinct, even if the pattern is a single letter. Consider the recurrently defined sequence of words $\{a_i\}_{i \geq 0}$ which is defined in the following way:

$$a_0 = a \quad a_i = a_{i-1} b^{|a_{i-1}|} a_{i-1} \text{ for } i \geq 1.$$

Let S_i be the set of positions of occurrences of the letter a in the word a_i. Clearly, $|S_i| = 2^i$. It is not difficult to prove that each arithmetic sequence in S_i has length at most 2. This means that each decomposition of the set S_i into arithmetic sequences contains at least $|S_i|/2 = 2^{i-1}$ sequences and the set S_i is not linearly-succinct. Note, that the words a_i are "well" compressible since $|LZ(a_i)| \leq 4i + 1$ and $|a_i| = 3^i$.

The definition of words a_i resembles the definition of Cantor's set which is the set of points in interval $[0, 1]$ remaining after applying the following procedure. Divide all existing intervals into three equal parts and remove the middle one. In the definition of a_i we divide a_i into three equal parts and the middle part is filled by letters b. We do the same recurently with remaining parts of a_i. What remains after applying this procedure is replaced by a i.e. the set of positions of a corresponds to Cantor's set. Indeed, those two sets satisfy similar dependence. Cantor's set consists of points ternary expansion of which consists of zeroes and twos. Similarly positions of letters a in a_i consists of zeroes and twos in ternary expansions under the assumption that the position of the first letter is zero not one.

In the example occurrences of a pattern are well compressible by the word a_i itself if we assume that the symbol a represents a starting position of an occurrence of the pattern a in a_i and the symbol b represents a position which is not a starting position of an occurrence of the pattern a in a_i. It appears that this is always the case. Let p be a pattern and t be a compressed text. Let $occ(p,t)$ be a word of length $|t|$ over the alphabet $\{0,1\}$ such that $occ[i] = 1$ iff i is an ending position of an occurrence of p in t.

Theorem 9. *There is SLP for $occ(p,t)$ which is of polynomial size with respect to SLP for t and to $|p|$.*

Proof. First we construct a deterministic automaton $\mathcal{A}$ accepting words with suffix p [7]. The automaton has $|p|$ states. Let $\mathcal{P}$ be SLP for t. We construct a SLP $\mathcal{R}$ for the path in $\mathcal{A}$ labeled t which starts in the initial state q_0 of $\mathcal{A}$. The constants in $\mathcal{R}$ are edges of $\mathcal{A}$. The variables in $\mathcal{R}$ are triples (q, X, q') where X is a variable from SLP for t and q' is the state reached from q after reading the word corresponding to X. The variable (q, X, q') generates the path in $\mathcal{A}$ labeled the word corresponding to X which goes from q to q'. The construction of $\mathcal{R}$ is as follows:

- $(q, X, q') = (q, a, q')$ iff $X = a$ in SLP for t and (q, a, q') is an edge in $\mathcal{A}$,
- $(q, V, q') = (q, X, q'')(q'', Y, q')$ iff $V = XY$ in SLP for t and q' is reachable by S starting from q and q'' is reachable by X starting from q.

The variable of the form (q_0, S, q') generates the path for t if S generates t in SLP for t. To obtain SLP for $occ(p,t)$ we replace each edge (q, a, q') in $\mathcal{R}$ by 1 if q' is accepting state in $\mathcal{A}$ and by 0 otherwise.

Let U denote the size of uncompressed text.

Theorem 10. *[13]*
The Fully Compressed Matching Problem can be solved in $O((n \log n)^5 \log^4 U)$ time.
We can compute within the same complexity the period of the compressed text and check if it is primitive.

Algorithms for compressed palindromes and squares use ideas from [4]: palindromes are searched using *periodicities* implied by sequences of many palindromes which are *close to each other* and searching of squares is reduced to multiple application of pattern-matching.

Theorem 11. *[13]*
We can check if a text is square-free in $O((n \log n)^6 \log^5 U)$ *time.*

Theorem 12. *[13]*
The compressed representation of all palindromes in the compressed text can be computed in $O((n \log n)^5 \log^4 U)$ *time.*

5 Open problems

There are several interesting questions remaining:

1. The complexity of our algorithm for compressed recognition problem for regular expressions works in $O(nm^3)$ time while our algorithm for the same problem for deterministic automata works in $O(nm)$ time. Does there exist an algorithm for the former problem which is of better complexity than $O(nm^3)$?
2. Is the problem of compressed recognition for semi-extended regular expressions in NP ? We conjecture it is.
3. Is the problem of compressed recognition for regular expressions with compressed constants in NP ? We conjecture it is.
4. Is the problem of compressed recognition for regular expressions with compressed nonperiodic constants in P ? We conjecture it is.
5. Is the problem of compressed recognition for context-free languages in NP ? We conjecture that this problem is *P-SPACE*-hard.
6. Does the set of all occurrences of the letter a in the n-th Fibonacci word consist of a polynomial number of arithmetic progressions ?
7. Does there exist SLP for $occ(p,t)$ which is of polynomial size with respect to a SLP for t and SLP for p?

References

1. A. Amir and G. Benson, Efficient two dimensional compressed matching, *Proc. of 2nd IEEE Data Compression Conference*, pp. 279–288, March 1992.
2. A. Amir, G. Benson, and M. Farach, Let sleeping files lie: Pattern matching in Z-compressed files, *Proc. of 5th Annual ACM-SIAM Symposium on Discrete Algorithms*, January 1994.
3. A. Amir, G. Benson, and M. Farach, Optimal two-dimensional compressed matching, *Proc. of 21st International Colloquium on Automata, Languages, and Programming (ICALP'94)*, LNCS 820, Springer-Verlag, 1994, pp. 215–226.
4. A. Apostolico, D. Breslauer, Z. Galil, Optimal Parallel Algorithms for Periods, Palindromes and Squares, *Proc. of 19th International Colloquium on Automata, Languages, and Programming (ICALP'92)*, LNCS 623, Springer-Verlag, 1992, pp. 296–307.
5. P. Berman, M. Karpiński, L. Larmore, W. Plandowski, and W. Rytter, The complexity of pattern matching of highly compressed two-dimensional texts, Proc. of *8th Annual Symposium on Combinatorial Pattern Matching*, CPM'97, July 1997.

6. B.S. Chlebus and L. Gąsieniec, Optimal pattern matching on meshes, *Proc. 11th Symposium on Theoretical Aspects of Computer Science*, 1994, pp. 213–224.
7. M. Crochemore, W. Rytter, *Text Algorithms*, Oxford University Press, 1994.
8. M.R. Garey and D.S. Johnson, Computers and Intractability: A Guide to the Theory of NP-Completeness, W.H. Freeman, 1979
9. R.M. Karp, Reducibility among combinatorial problems, in: Complexity of Computer Computations, Plenum Press, New York, 1972 (editors R.E. Miller and J.W. Thatcher)
10. M. Karpiński, W. Rytter, and A. Shinohara, A pattern matching for strings with short description, *Proc. 6th Combinatorial Pattern Matching*, June 1995.
11. M. Farach and M. Thorup, String Matching in Lempel-Ziv Compressed Strings, *Proc. 27th ACM Symposium on Theory of Computing*, pp. 703–713, 1994.
12. L. Gąsieniec, M. Karpiński, W. Plandowski, and W. Rytter, Randomised efficient algorithms for compressed strings: the finger-print approach, *Proc. 7th Combinatorial Pattern Matching*, LNCS 1075, Springer-Verlag, June 1996, pp. 39–49.
13. L. Gąsieniec, M. Karpiński, W. Plandowski, and W. Rytter, Efficient algorithms for Lempel-Ziv encoding, *Proc. 5th Scandinavian Workshop on Algorithms Theory*, LNCS 1097, Springer-Verlag, July 1996, pp. 392–403.
14. L. Gąsieniec, W. Rytter, Fully compressed pattern matching for LZW encoding, to appear at IEEE Data Compression Conference (1999)
15. A. Lempel and J. Ziv On the complexity of finite sequences, *IEEE Transactions on Information Theory*, 22:75–81, 1976.
16. M. Lothaire, *Combinatorics on Words*, Addison-Wesley, 1983.
17. M. Miyazaki , A. Shinohara, and M. Takeda, An improved pattern matching for strings in terms of straight-line programs, *8th Combinatorial Pattern Matching*, LNCS 1264, Springer-Verlag, July 1997, pp. 1–11.
18. W. Plandowski and W. Rytter, Application of Lempel-Ziv encodings to the solution of words equations, *Proc. of 25th International Colloquium on Automata, Languages, and Programming (ICALP'98)*, LNCS 1443, Springer-Verlag, 1998, pp. 731–742.
19. T.A. Welch, A technique for high performance data compression, *IEEE Transactions on Computers*, 17:8–19, 1984.
20. J. Ziv and A. Lempel, A universal algorithm for sequential data compression, *IEEE Transactions on Information Theory*, 23(3):337–343, 1977.

Algorithms on Continued Fractions *

Octavian Soldea and Azaria Paz

Summary. Some algorithms for performing arithmetical operations, on line, and fit for parallel and concurrent computation are described and investigated. The algorithms are based on the continued fractions representation of numbers and the continued fraction representation is generalized so as to allow rational quotients (instead of integer quotients). This generalization is intended to minimize the delay in the output stream of quotients provided by the units when a continuous stream of quotients is received by them at input.

Real numbers can be represented in many ways. The most common representations are the positional representations e.g. the decimal or binary representation where a real number α is represented in the form

$$\alpha = a_k a_{k-1} \dots a_0 . a_{-1} a_{-2} \dots$$

with the interpretation $\alpha = \sum_{i=-k}^{\infty} a_i \cdot 10^{-i}$ or $\alpha = \sum_{i=-k}^{\infty} a_i \cdot 2^{-i}$ etc. The positional representation is natural and handy and it enables straightforward and very simple algorithms for performing the basic arithmetic operations. Very efficient implementation of those algorithms have been designed and incorporated in the arithmetical units. On the other hand, for computers that are designed to perform parallel or concurrent on line computations the above mentioned algorithms may create a serious problem which can be explained as follows. Suppose that a computer uses several arithmetical units that operate in a concurrent way. Let U be such a unit, designed to compute $z = x \otimes y$, with x and y received at input from other units and z sent at output to still another unit. Assume that at time t the unit completed the computation of z for x and y received at input with k binary digits. If at time $t+1$ the unit receives an additional digit of x and of y - refining the accuracy of the input numbers then the unit may have to redo the whole computation of z since the last digits of x and y may induce a change in the first digit of z. The unit receiving z at input will have now to reset its whole computation

---

* This extended abstract is based on a thesis submitted to the Senate of the Technion, Israel Institute of Technology, in partial fulfillment of the requirements for the degree of Master of Science in Computer Science done by the first author under the supervision of the second author. A copy of the full thesis in the English language is available from octavian@cs.technion.ac.il

based on the new z which may differ in many digits from the old z. This problem can be resolved if numbers are represented in the form of continued fractions. In this form, a real number α is represented as $\alpha = [k_0, k_1, \ldots]$ where the k_i (called quotients) are defined as follows: Set $\alpha = \alpha_0$, and for $j \geq 0$ $\alpha_j = \lfloor \alpha_j \rfloor + \frac{1}{\alpha_{j+1}}$. Define $k_j = \lfloor \alpha_j \rfloor$. This representation of numbers is well known see e.g. [2] and has numerous applications in various branches of mathematics and in particular in number theory.

It has been known for quite a while that it is possible to perform the basic arithmetical operations $(+, -, \cdot, /)$ based on this representation but the corresponding algorithms are more complicated (see [1], [8], [13]). Straightforward and fast algorithms are also known for converting one form of representation to another (i.e. from positional representation to continued fractions representation and vice-versa). The continued fraction representation has also some nice properties with regard to closeness of the approximation achieved for a bounded number of quotients. See e.g. [3], [4], [6], [7], [9]. In addition, and this is the most important property for on line concurrent and parallel computations, additional new quotients for input values of x and y, refining the accuracy of the input, *do not* affect the quotients of the output of the performed operation done before the new quotients were received. The new quotients at input may only result in new quotients for the output thus refining it's accuracy in a way which fits the on line mode of computation. Several authors have tried to exploit this property in order to design arithmetic units fit for this mode of operation. See e.g. [1], [5]. We mention here several additional works that were done in the framework of intervals and, more generally, in the framework of integer algorithms and are strongly connected to the above works. See e.g. [10], [11], [12].

In this work we introduce a new algorithm for arithmetical operations, which differs from the algorithm suggested in [8]. Our algorithm seems to be simple and easy to implement and deals in a uniform way with all the basic operations $(+, -, \cdot, /)$ over positive and negative numbers. The algorithm can be extended to more complex operations such as square roots or exponentiation and to basic vector operations, too.

One of the problems encountered when continued fraction arithmetic is used has to do with waiting time. A unit receiving x and y at input, on line, may have to wait for several input steps before it can provide output. This problem does not exist with regular arithmetic where any additional input digit increasing the accuracy of the input induces an additional output digit increasing the accuracy of the output. We try to overcome this problem here by extending the continued fraction representation so as to allow rational quotients too (the quotients for regular continued fractions must be integers). This extension creates redundancy in the representation (a real number may have now several different representations) but the redundancy can be handled in a proper way and is not affecting the functionality of the units in a net of processors working concurrently or in parallel.

We shall use the following notations:

N	the set of natural numbers
Z	the set of integer numbers
Q	the set of rational numbers
R	the set of real numbers
A_+	the nonnegative numbers of A, where $A \subseteq R$
A^*	$A - \{0\}$, where $A \subseteq R$.

We refer to the continued fraction representation form of $x \in R$ as the expression

$$x = \sigma(x) \left(a_0 + \cfrac{1}{a_1 + \cfrac{1}{a_2 + \ddots}} \right)$$

where a_i are numbers for all $i \in N$ and $\sigma(x)$ is the sign of x. The numbers a_i are called quotients. We assume $a_0 \in Q_+$, $a_i \in Q_+^*$ for all $i \in N$, and take separately into account the sign $\sigma(x)$ of x. We use the notation $x = [\sigma(x), a_0, a_1, \ldots]$ for the continued fraction of x.

The values

$$C(x,k) = a_0 + \cfrac{1}{a_1 + \cfrac{1}{a_2 + \ddots + \cfrac{1}{a_k}}}$$

are called the convergents of x, where $k \in N$. Each convergent can be written in the form

$$C(x,k) = \frac{N(x,k)}{D(x,k)}$$

where $N(x,k), D(x,k) \in Q_+$ for any $k \in N$. In the following we exhibit several properties that hold for the values above defined. These properties represent, in the field of continued fractions with integer quotients, classic results that are reformulated in our new framework of continued fractions with rational quotients. These properties are used in the proofs of correctness of the algorithms that we describe.

Let

$$N(x,-1) = 1, D(x,-1) = 0, N(x,-2) = 0, D(x,-2) = 1,$$

and define

$$C(x,-2) = \frac{N(x,-2)}{D(x,-2)} = \frac{0}{1} = 0 \text{ and } C(x,-1) = \frac{N(x,-1)}{D(x,-1)} = \frac{1}{0} = \infty.$$

From the definitions of the convergents of the continued fractions it follows that the following properties hold: For any $k \geq 0$

$$\begin{cases} N(x,k) = a_k \cdot N(x,k-1) + N(x,k-2) \\ D(x,k) = a_k \cdot D(x,k-1) + D(x,k-2) \end{cases}.$$

(This theorem is known as the law of formation of the convergents.)

Lemma 1. *The slopes of the vectors $(D(x,k), N(x,k))$ with $k \geq -2$ approximate x alternately from below and above. If $C(x,k)$ converges when $k \to \infty$ then $\lim_{k\to\infty} C(x,k) = x$.*

Example
It is well known that

$$e = [2,1,2,1,1,4,1,1,\ldots,2\cdot k,1,1,\ldots] = 2.7182818\ldots.$$

The first 5 convergents of e are shown below

$$\begin{aligned}
&C(e,-2) = \tfrac{N(x,-2)}{D(x,-2)} = \tfrac{0}{1} = 0,\ C(e,-1) = \tfrac{N(x,-1)}{D(x,-1)} = \tfrac{1}{0} = \infty,\\
&C(e,0) = \tfrac{2}{1} = 2 = \tfrac{2\cdot1+0}{2\cdot0+1} = \tfrac{a_0\cdot N(x,-1)+N(x,-2)}{a_0\cdot D(x,-1)+D(x,-2)},\\
&C(e,1) = 2+\tfrac{1}{1} = \tfrac{3}{1} = 3 = \tfrac{1\cdot2+1}{1\cdot1+0} = \tfrac{a_1\cdot N(x,0)+N(x,-1)}{a_1\cdot D(x,0)+D(x,-1)}, \text{ and}\\
&C(e,2) = 2+\tfrac{1}{1+\frac{1}{2}} = 2+\tfrac{1}{\frac{3}{2}} = 2+\tfrac{2}{3} = \tfrac{8}{3} = \tfrac{2\cdot3+2}{2\cdot1+1} = \tfrac{a_2\cdot N(x,1)+N(x,0)}{a_2\cdot D(x,1)+D(x,0)}.
\end{aligned}$$

We design our algorithms based on matrix descriptions. We associate with each quotient a_i a matrix of the following two types

$$\begin{pmatrix} 1 & 0 \\ a_i & 1 \end{pmatrix} \text{ or } \begin{pmatrix} 1 & a_i \\ 0 & 1 \end{pmatrix}.$$

For any $i \geq 0$ the matrix associated with a_i differs in type from that associated to a_{i+1} and the matrix associated to a_0 is of type $\begin{pmatrix} 1 & a_0 \\ 0 & 1 \end{pmatrix}$.

It can be shown that for any $i \geq 0$ one of the following relations holds:

$$\begin{pmatrix} 1 & a_i \\ 0 & 1 \end{pmatrix} \cdot \ldots \cdot \begin{pmatrix} 1 & 0 \\ a_1 & 1 \end{pmatrix} \cdot \begin{pmatrix} 1 & a_0 \\ 0 & 1 \end{pmatrix} = \begin{pmatrix} D(x,i) & N(x,i) \\ D(x,i-1) & N(x,i-1) \end{pmatrix}$$

or

$$\begin{pmatrix} 1 & 0 \\ a_i & 1 \end{pmatrix} \cdot \ldots \cdot \begin{pmatrix} 1 & 0 \\ a_1 & 1 \end{pmatrix} \cdot \begin{pmatrix} 1 & a_0 \\ 0 & 1 \end{pmatrix} = \begin{pmatrix} D(x,i-1) & N(x,i-1) \\ D(x,i) & N(x,i) \end{pmatrix}.$$

These relations can be proved by induction.

Let $U = (u_{i,j})$, $V = (v_{i,j})$, and $W = (w_{i,j})$ be 2x2 matrices, where $i,j \in \{0,1\}$. Assume the elements of U and V are rational numbers and the elements of W are row vectors in the 2-dimensional vector space of Q^2.

Definition 1. Let W and U be matrices as above. We define the matrix $W \circ U$ whose entries are 2-vectors as follows: $W \circ U = (r_{i,j})$ where $r_{i,j} = w_{i,j} \cdot U$, $w_{i,j} \cdot U$ is the ordinary vector by matrix product, and $i, j \in \{0, 1\}$.

For example,

$$\begin{pmatrix} (1,2) & (1,3) \\ (2,2) & (2,3) \end{pmatrix} \circ \begin{pmatrix} 1 & -1 \\ 0 & 1 \end{pmatrix} = \begin{pmatrix} (1,1) & (1,2) \\ (2,0) & (2,1) \end{pmatrix}.$$

Notice that, as the elements of W are row vectors, the operation $U \circ W$ can not be defined in a similar way.

Definition 2. Let W and U be matrices as above. We define the matrix $W \circ \circ U$ whose entries are 2-vectors as follows: $W \circ \circ U = (s_{i,j})$ where

$$s_{i,j} = \sum_{k=0}^{1} w_{i,k} \cdot u_{k,j},$$

$w_{i,k} \cdot u_{k,j}$ is the scalar multiplication of the vector $w_{i,k}$ by the scalar $u_{k,j}$, and $i, j \in \{0, 1\}$. In a similar way, $U \circ \circ W = (t_{i,j})$ where

$$t_{i,j} = \sum_{k=0}^{1} u_{i,k} \cdot w_{k,j}$$

For example, in the first case

$$\begin{pmatrix} 1 & 2 \\ 0 & 1 \end{pmatrix} \circ \circ \begin{pmatrix} (1,1) & (1,2) \\ (2,0) & (2,1) \end{pmatrix} = \begin{pmatrix} (5,1) & (5,4) \\ (2,0) & (2,1) \end{pmatrix}$$

and in the second case

$$\begin{pmatrix} (4,1) & (1,4) \\ (2,0) & (1,1) \end{pmatrix} \circ \circ \begin{pmatrix} 1 & 0 \\ 2 & 1 \end{pmatrix} = \begin{pmatrix} (6,9) & (1,4) \\ (4,2) & (1,1) \end{pmatrix}.$$

Theorem 1. *With the notations above the following equalities hold:*

a) $(W \circ U) \circ V = W \circ (U \cdot V)$
b) $U \circ \circ (V \circ \circ W) = (U \cdot V) \circ \circ W$
c) $(W \circ \circ U) \circ \circ V = W \circ \circ (U \cdot V)$
d) $V \circ \circ (W \circ \circ U) = (V \circ \circ W) \circ \circ U$
e) $V \circ \circ (W \circ U) = (V \circ \circ W) \circ U$
f) $(W \circ U) \circ \circ V = (W \circ \circ V) \circ U,$

where $U \cdot V$ represents the ordinary matrix multiplication.

The goal of this work is to design algorithms, that are based on the representation of the numbers in continued fraction form, for arithmetical units. In general, in the literature, the existing models deal with continued fractions which have integer quotients only. This constraint yields delays in the computations. We relax this constraint, by allowing the quotients of the continued fractions, that the described unit deals with, to be rational numbers. This relaxation considerably reduces those delays.

On the other hand the extension to rational quotients may create redundancy in the representation, as mentioned before, but this redundancy can be handled in a proper way. In order to bound the size of the numbers involved we shall allow the quotients to have the form $\frac{1}{\alpha}$, where $\alpha \in N$ is bounded.

We give here an outline of the algorithm only and will only provide a hint for the proof of its correctness. The algorithm is explicated by an example.

We assume the input to consist of two real numbers

$$x = [\sigma(x), x_0, x_1, \ldots, x_k, \ldots] \text{ and } y = [\sigma(y), y_0, y_1, \ldots, y_k, \ldots]$$

fed, on line, to the arithmetical unit.

At output the arithmetical unit produces, on line, the result of the arithmetical operation

$$z = [\sigma(z), z_0, z_1, \ldots, z_k, \ldots] = x \otimes y$$

where $\otimes \in \{+, -, \cdot, /\}$. Let $U^{(t)}$ be the arithmetical unit at the tth iteration. The unit is represented by a 2x2 matrix whose elements are row vectors (u, v) with $u, v \in Q_+$. For example,

$$U^{(t)} = \begin{bmatrix} (a,b) & (c,d) \\ (e,f) & (g,h) \end{bmatrix}$$

represents the unit at iteration t with $a, b, c, d, e, f, g, h \in Q_+$ and $U_{0,0}^{(t)} = (a, b)$.

The algorithm outline is as follows:

Initialization
ready_for_output := false
while (condition)
    $U := U^{(t-1)}$
    Input step
    If *ready_for_output* = false then
        Try to format step
    If *ready_for_output* = true then
        Output step
    $U^{(t)} := U$.

The algorithm is designed in a way such that it has the same form for all 4 operations and the distinction between the operations is set at the initialization step. The iterations of the algorithm consist of 3 main operations. The first operation consists of reading the input, the second operation is the introduction of the information of the input in the buffers of the unit, and the third operation is the output. All these operations can be explicated in the following.

Example
Let

$$x = \left[+, 2, 1, \frac{1}{4}, 1 \ldots\right] = \frac{0}{1}, \frac{1}{0}, 2, \frac{3}{1}, \frac{\frac{11}{4}}{\frac{5}{4}}, \frac{\frac{23}{4}}{\frac{9}{4}}, \ldots$$

and

$$y = \left[+, 1, 1, \frac{1}{4}, 1 \ldots\right] = \frac{0}{1}, \frac{1}{0}, 1, \frac{2}{1}, \frac{\frac{3}{2}}{\frac{5}{4}}, \frac{\frac{7}{2}}{\frac{9}{4}}, \ldots.$$

We assume that all the quotients that are pulled in or out are greater than $\frac{1}{4}$.

If at the first iteration one knows the signs $\sigma(x)$, $\sigma(y)$, and the operation $\otimes = /$ then the formatting step is a relatively easy task. This is the case in this example. Special and more complicated steps are provided for the other cases in the full thesis, but will not be discussed here.

Iteration 0

Initialization

$$U'^{(0)} := \begin{bmatrix} (0,0) & (1,0) \\ (0,1) & (0,0) \end{bmatrix}$$

{ this initialization fits the operation $\otimes = /$ }

$phase := 0$
$output\_zero :=$ false
$ready\_for\_output :=$ false

Input step

Draw $\sigma(x) = +$
$phase := 1$ { signifies the fact that $\sigma(x)$ is known }
Draw $\sigma(y) = +$
$phase := 2$
{ signifies the fact that both $\sigma(x)$ and $\sigma(y)$ are known }

$$U^{(0)} := \begin{bmatrix} (0,0) & (1,0) \\ (0,1) & (0,0) \end{bmatrix} = U'^{(0)}$$

$\sigma(z) := +$
$ready\_for\_output :=$ true

Output step

Pull out $\sigma(z) := +$
phase := 3 { signifies the fact that the output sign is known }

Iteration 1

Input step

$x_0 = 2$ and $y_0 = 1$

Updates

$$U'^{(1)} := \begin{bmatrix} (0,2) & (1,0) \\ (0,1) & (0,0) \end{bmatrix} = \begin{pmatrix} 1 & 2 \\ 0 & 1 \end{pmatrix} \circ \circ U^{(0)}$$

{ The left $' \circ \circ'$ operation inserts the x-input into the $U^{(t)}$ unit for $t \geq 0$. }

$$U^{(1)} := \begin{bmatrix} (1,2) & (1,0) \\ (0,1) & (0,0) \end{bmatrix} = U'^{(1)} \circ \circ \begin{pmatrix} 1 & 0 \\ 1 & 1 \end{pmatrix}$$

{ Now both quotients of x and of y are incorporated in the unit. The y-input update was achieved by the right $' \circ \circ'$ operation. }

Output step

No output

Iteration 2

Input step

$x_1 = 1$ and $y_1 = 1$

Updates

$$U'^{(2)} := \begin{bmatrix} (1,2) & (1,0) \\ (1,3) & (1,0) \end{bmatrix} = \begin{pmatrix} 1 & 0 \\ 1 & 1 \end{pmatrix} \circ \circ U^{(1)} \text{ \{ see remark above \}}$$

$$U''^{(2)} := \begin{bmatrix} (1,2) & (2,2) \\ (1,3) & (2,3) \end{bmatrix} = U'^{(2)} \circ \circ \begin{pmatrix} 1 & 1 \\ 0 & 1 \end{pmatrix} \text{ \{ see remark above \}}$$

Output step

Pull out $\begin{pmatrix} 1 & 1 \\ 0 & 1 \end{pmatrix}$

{ first output quotient is determined, that is $z_0 = 1$ }

Updates

$$U^{(2)} := \begin{bmatrix} (1,1) & (2,0) \\ (1,2) & (2,1) \end{bmatrix} = U''^{(2)} \circ \begin{pmatrix} 1 & -1 \\ 0 & 1 \end{pmatrix}$$

{ the output (pull-out) was recorded in the unit via the right '$\circ$' operation }

The remarks given in the previous iterations apply also to the following iterations and are omitted.

Iteration 3

Input step

No Input and thus the same matrix

$$U'^{(3)} := \begin{bmatrix} (1,1) & (2,0) \\ (1,2) & (2,1) \end{bmatrix} = U^{(2)}$$

Output step

Pull out $\begin{pmatrix} 2 & 0 \\ 1 & 2 \end{pmatrix}$ { $z_1 = \frac{1}{2}$ }

{ The matrix corresponding to the z_1 quotient is $\begin{pmatrix} 1 & 0 \\ \frac{1}{2} & 1 \end{pmatrix}$ which is reset to $\begin{pmatrix} 2 & 0 \\ 1 & 2 \end{pmatrix}$ with integer entries and which has the same effect on the ratio between the coordinates of the vectors of the resulting $U^{(3)}$ matrix. }

Updates

$$U^{(3)} := \begin{bmatrix} (1,2) & (4,0) \\ (0,4) & (3,2) \end{bmatrix} = U'^{(3)} \circ \begin{pmatrix} 2 & 0 \\ -1 & 2 \end{pmatrix}$$

Iteration 4

Input step

$x_2 = \frac{1}{4}$

{ The matrix corresponding to the x_2 quotient is $\begin{pmatrix} 1 & \frac{1}{4} \\ 0 & 1 \end{pmatrix}$ which is reset to $\begin{pmatrix} 4 & 1 \\ 0 & 4 \end{pmatrix}$ with integer entries and which has the same effect on the ratio between the coordinates of the vectors of the resulting $U^{(4)}$ matrix. }

Updates

$$U^{(4)} := \begin{bmatrix} (4,12) & (19,2) \\ (0,16) & (12,8) \end{bmatrix} = \begin{pmatrix} 4 & 1 \\ 0 & 4 \end{pmatrix} \circ \circ U^{(3)}$$

Output step

No output

Iteration 5

Input step

$y_2 = \frac{1}{4}$ { see remark in step 4 }

Updates

$$U^{(5)} := \begin{bmatrix} (35,50) & (76,8) \\ (12,72) & (48,32) \end{bmatrix} = U^{(4)} \circ \circ \begin{pmatrix} 4 & 0 \\ 1 & 4 \end{pmatrix}$$

Output step

No output

Iteration 6

Input step

$x_3 = 1$ and $y_3 = 1$

$$U'^{(6)} := \begin{bmatrix} (35,50) & (76,8) \\ (47,122) & (124,40) \end{bmatrix} = \begin{pmatrix} 1 & 0 \\ 1 & 1 \end{pmatrix} \circ \circ U^{(5)}$$

$$U''^{(6)} := \begin{bmatrix} (35,50) & (111,58) \\ (47,122) & (171,162) \end{bmatrix} = U'^{(6)} \circ \circ \begin{pmatrix} 1 & 1 \\ 0 & 1 \end{pmatrix}$$

Output step

Pull out $\begin{pmatrix} 2 & 1 \\ 0 & 2 \end{pmatrix}$ $\{z_2 = \frac{1}{2}\}$

Updates

$$U^{(6)} := \begin{bmatrix} (70,65) & (222,5) \\ (94,197) & (342,153) \end{bmatrix} = U''^{(6)} \circ \begin{pmatrix} 2 & -1 \\ 0 & 2 \end{pmatrix}$$

Combining the pull out matrices during the iterations of the algorithm so far we get

$$\begin{pmatrix} 2 & 1 \\ 0 & 2 \end{pmatrix} \cdot \begin{pmatrix} 2 & 0 \\ 1 & 2 \end{pmatrix} \cdot \begin{pmatrix} 1 & 1 \\ 0 & 1 \end{pmatrix} = \begin{pmatrix} 5 & 7 \\ 2 & 6 \end{pmatrix}.$$

The approximations achieved by this output are $\frac{7}{5}$ and $\frac{6}{2} = 3$. On the other hand the values of x and y accumulated at input so far are $x_3 = \frac{\frac{23}{4}}{\frac{9}{4}} = \frac{23}{9}$, $y_3 = \frac{\frac{7}{2}}{\frac{9}{4}} = \frac{14}{9}$ with $\frac{x_3}{y_3} = \frac{23}{14}$ and $\frac{7}{5} < \frac{23}{14} < 3$. The $\frac{x}{y}$ approximation received at input so far is bounded by the output approximations and is closer to the lower bound $\frac{7}{5}$.

The details of the (quite long) proof of the correctness of the algorithm and it's convergence are omitted here. We provide only some basic ideas on which the proof relies.

Assume that $A^{(0)}$ and $B^{(0)}$ are matrices that correspond to the quotients of the input of x and y respectively at the 0th iteration. Assume $C^{(0)}$ corresponds to the output quotient at the t = 0th iteration. In this case the effect of this iteration on the operational unit can be described by means the following computation:

$$\left(\left(A^{(0)}\right) \circ \circ U^{(0)} \circ \circ \left(B^{(0)}\right)\right) \circ \left(C^{(0)}\right).$$

Assume that at the t = 1th iteration the matrices that correspond to the quotients of the input numbers and to the quotient of output are $A^{(1)}$, $B^{(1)}$, and $C^{(1)}$ respectively. In this case the effect of this iteration on the operational unit can be described by means the following computation:

$$\left(\left(A^{(1)}\right) \circ \circ \left(\left(\left(A^{(0)}\right) \circ \circ U^{(0)} \circ \circ \left(B^{(0)}\right)\right) \circ \left(C^{(0)}\right)\right) \circ \circ \left(B^{(1)}\right)\right) \circ \left(C^{(1)}\right).$$

One can prove based on theorem 1 that the above computation is equivalent to the following:

$$\left(\left(\left(A^{(1)}\right) \cdot \left(A^{(0)}\right)\right) \circ \circ U^{(0)} \circ \circ \left(\left(B^{(0)}\right) \cdot \left(B^{(1)}\right)\right)\right) \circ \left(\left(C^{(0)}\right) \cdot \left(C^{(1)}\right)\right).$$

This means that in analyzing the behaviour of the unit one can assume that the unit has received all the x quotients first, then all the y quotients, then it releases all the z quotients at output. Careful analysis of those three steps, based on recursive arguments, shows that the algorithm indeed performs the operation it is designed to perform and the output z, provided on line by the unit in continued fraction form, converges to the required result of the operation.

As mentioned before, arithmetical units based on the continued fraction representation of numbers may have to pass through several input iterations before they can pull-out an output quotient. Such iterations are defined as "delays" in the computation process. In order to minimize this delay phenomenon we relaxed the definition of continued fractions so as to allow rational quotients. In the full thesis we investigate the relation between the length of the delays in a computation, performed by a net of arithmetical units, and various criteria embedded in the computational unit, designed to reduce the delay time, by pulling out rational quotients when integer quotients are not possible. A simulation of a net of such units is shown in the full thesis together with some graphs and statistics, derived from that simulation, on several examples, and illustrating the interrelation between various parameters connected to delay patterns.

In the last section of the full thesis we generalize some results known in the literature for continued fractions with integer quotients to continued fractions with rational quotients and we propose some open problems for further research.

A copy of the full thesis in the English language is available from

octavian@cs.technion.ac.il

References

1. Zippel R. (1993) Effective Polynomial Computation, Kluwer Academic Publishers, Boston, Chapter 2, pp. 11-39
2. Rockett A. M. and Szüsz P. (1992) Continued Fractions, World Scientific Publishing Co. Pte. Ltd, Singapore
3. Vuillemin J. (1987) Exact Real Computer Arithmetic with Continued Fractions, Rapports de Recherche No. 760, INRIA, November
4. Matula D. W. and Kornerup P. (1980) Foundations of Finite Precision Rational Arithmetic, Computing Suppl. 2, Springer-Verlag, Vienna, pp. 85-111
5. Kornerup P. and Matula D. W. (1983) Finite Precision Rational Arithmetic: An Arithmetic Unit, IEEE Transactions on Computers, Vol. c-32, No. 4, April, pp. 378-388
6. Matula D. W. and Kornerup P. (1985) Finite Precision Rational Arithmetic: Slash Number Systems, IEEE Transactions on Computers, Vol. c-34, No. 1, January, pp. 3-18
7. Kornerup P. and Matula D. W. (1985) Finite Precision Lexicographic Continued Fraction Number Systems, IEEE Proceedings of 7th Symposium on Computer Arithmetic, Urbana, Illinois, 4-6 June, 1985, pp. 207-214
8. Kornerup P. and Matula D. W. (1988) An On-Line Arithmetic Unit for Bit-Pipelined Rational Arithmetic, Journal of Parallel and Distributed Computing, Vol. 5, No. 3, pp. 310-330
9. Kornerup P. and Matula D. W. (1989) A Lexicographic Binary Representation of the Rationals, Institut for Matematik og Datalogi, Odense Universitet, Preprints 1989, No. 5, ISSN No. 0903-3920, April
10. Johansen S. P., Kornerup P., and Matula D. W. (1990) A Universal On-line Cell for Interval Arithmetic, Personal Communication, November
11. Orup H. and Kornerup P. (1991) A High-Radix Hardware Algorithm for Calculating the Exponential M^E Modulo N, IEEE Proceedings of 10th Symposium on Computer Arithmetic, Grenoble, France, 26-28 June, 1991, pp. 51-56
12. Kornerup P. (1994) Digit-Set Conversions: Generalizations and Applications, IEEE Transactions on Computers, Vol. 43, No. 5, May, pp. 622-629
13. Gosper R. W. (1972) Item 101 in Hakmem, AIM239, MIT, February, pp. 37-44

Part VI

Combinatorics of Words

On the Index of Sturmian Words

Jean Berstel

Summary. An infinite word x has finite index if the exponents of the powers of primitive words that are factors of x are bounded. F. Mignosi has proved that a Sturmian word has finite index if and only if the coefficients of the continued fraction development of its slope are bounded. Mignosi's proof relies on a delicate analysis of the approximation of the slope by rational numbers. We give here a proof based on combinatorial properties of words, and give some additional relations between the exponents and the slope.

1 Introduction

Sturmian words are infinite words over a binary alphabet that have exactly $n+1$ factors of length n for each $n \geq 0$. It appears that these words admit several equivalent definitions, and can even be described explicitly in arithmetic form. For instance, every Sturmian word has a *slope* associated with it, which is an irrational number in the interval $[0, 1]$.

Sturmian words have a long history. A clear exposition of early work by J. Bernoulli, Christoffel, and A. A. Markov is given in the book by Venkov [30]. The term "Sturmian" has been used by Hedlund and Morse in their development of symbolic dynamics [15–17]. These words are also known as Beatty sequences, cutting sequences, or characteristic sequences. There is a large literature about properties of these sequences (see for example Coven, Hedlund [8], Series [28], Fraenkel *et al.* [14], Stolarsky [29]). From a combinatorial point of view, they have been considered by S. Dulucq and D. Gouyou-Beauchamps [13], Rauzy [25,26], Brown [6], Ito, Yasutomi [18], Crisp *et al.* [7] in particular in relation with iterated morphisms, and by Séébold [27], Mignosi [21]. Sturmian words appear in ergodic theory [24], in computer graphics [5], in crystallography [4], and in pattern recognition. Standard words, and finite factors of Sturmian words are considered in depth by De Luca [12,9,11,10], see also [3]. A survey is [1]. A more systematic presentation of Sturmian words is in preparation ([2]).

The aim of this paper is to present a new proof, with some improvements, of a theorem by Mignosi [21] cited below. Let x be an infinite word, and let $F(x)$ be the sets of its factors (subwords). For $w \in F(x)$, the *index* of w in x is the greatest integer d such that $w^d \in F(x)$, if such an integer exists. Otherwise, w is said to have infinite index.

An infinite word x has *bounded index* if there exists an integer d such that every nonempty factor of x has an index less than or equal to d.

Theorem 1. *A Sturmian word has bounded index if and only if the continued fraction expansion of its slope has bounded partial quotients.*

An initial contribution to this result was by Karhumäki [20] who proved that the Fibonacci word is fourth power free. Mignosi's proof uses involved arguments from number theory. Our proof is combinatorial, and follows an argument by Mignosi and Pirillo [22] in their proof of the sharp bound for the index in the Fibonacci word.

The next section is devoted to a short introduction to Sturmian words. In particular, the standard sequence of a Sturmian word is introduced. The following section gives the proof.

2 Definitions

In this paper, words will be over a binary alphabet $A = \{0, 1\}$.

The *complexity function* of an infinite word x over some alphabet A is the function that counts, for each integer $n \geq 0$, the number $P(x,n)$ of factors of length n in x. A *Sturmian* word is an infinite word s such that $P(s,n) = n+1$ for any integer $n \geq 0$. Sturmian words are aperiodic infinite words of minimal complexity. Indeed, an infinite word of lower complexity is eventually periodic. Since $P(s,1) = 2$, any Sturmian word is over two letters. A *right special* (*left special*) factor of a word x is a word u such that $u0$ and $u1$ ($0u$ and $1u$) are factors of x. Thus a word x is Sturmian if and only if it has exactly one right special factor of each length.

Given two real numbers α and ρ with α irrational and $0 < \alpha < 1$, we define two infinite words

$$s_{\alpha,\rho} : \mathbf{N} \to A, \quad s'_{\alpha,\rho} : \mathbf{N} \to A$$

by

$$\begin{aligned} s_{\alpha,\rho}(n) &= \lfloor \alpha(n+1) + \rho \rfloor - \lfloor \alpha n + \rho \rfloor \\ s'_{\alpha,\rho}(n) &= \lceil \alpha(n+1) + \rho \rceil - \lceil \alpha n + \rho \rceil \end{aligned} \qquad (n \geq 0)$$

The numbers α and ρ are *slope* and the *intercept*. Words $s_{\alpha,\rho}$ and $s'_{\alpha,\rho}$ are called *mechanical*.

Theorem 2. *[17] Let s be an infinite word. The following are equivalent:*

(i) s is Sturmian;
(ii) s is mechanical.

A special case deserves consideration, namely when $\rho = 0$. In this case, $s_{\alpha,0}(0) = \lfloor\alpha\rfloor = 0$, $s'_{\alpha,0}(0) = \lceil\alpha\rceil = 1$, and

$$s_{\alpha,0} = 0c_\alpha, \quad s'_{\alpha,0} = 1c_\alpha$$

where the infinite word c_α is called the *characteristic* word of α. It can be shown that a Sturmian word is characteristic if and only if every prefix is left special.

There is a close relation between the slope of a characteristic word and the combinatorial structure of this word.

Let $(d_1, d_2, \ldots, d_n, \ldots)$ be a sequence of integers, with $d_1 \geq 0$ and $d_n > 0$ for $n > 1$. To such a sequence, we associate a sequence $(s_n)_{n\geq -1}$ of words by

$$s_{-1} = 1, \quad s_0 = 0, \qquad s_n = s_{n-1}^{d_n} s_{n-2} \quad (n \geq 1) \tag{1}$$

The sequence $(s_n)_{n\geq -1}$ is a *standard sequence*, and the sequence $(d_1, d_2, \ldots)$ is its *directive sequence*. Observe that if $d_1 > 0$, then any s_n $(n \geq 0)$ starts with 0; on the contrary, if $d_1 = 0$, then $s_1 = s_{-1} = 1$, and s_n starts with 1 for $n \neq 0$. Every s_{2n} ends with 0, every s_{2n+1} ends with 1.

Example 1. The directive sequence $(1, 1, \ldots)$ gives the standard sequence defined by $s_n = s_{n-1}s_{n-2}$, that is the sequence of finite Fibonacci words. Observe that the directive sequence $(0, 1, 1, \ldots)$ results in the sequence of words obtained from Fibonacci words by exchanging 0 and 1.

Proposition 1. *[14] Let $\alpha = [0, 1 + d_1, d_2, \ldots]$ be the continued fraction expansion of some irrational α with $0 < \alpha < 1$, and let (s_n) be the standard sequence associated to $(d_1, d_2, \ldots)$. Then every s_n is a prefix of c_α and*

$$c_\alpha = \lim_{n\to\infty} s_n\,.$$

Example 2. Consider $\alpha = (\sqrt{3} - 1)/2 = [0, 2, 1, 2, 1, \ldots]$. The directive sequence is $(1, 1, 2, 1, 2, 1, \ldots)$, and the standard sequence starts with $s_1 = 01$, $s_2 = 010$, $s_3 = 01001001, \ldots,$ whence

$$c_{(\sqrt{3}-1)/2} = 01001001010010010010100100 1001\cdots$$

Due to the periodicity of the development, we get for $n \geq 2$ that $s_{n+2} = s_{n+1}^2 s_n$ if n is odd, and $s_{n+2} = s_{n+1}s_n$ if n is even.

3 Index

As usual, a word of the form $w = (xy)^n x$ is written as $w = u^r$, with $u = xy$ and $r = n + |x|/|u|$. The rational number r is the *exponent* of u, and if u is primitive, it is the root of the fractional power w.

Let x be an infinite word. For $w \in F(x)$, the *index* of w in x is the number

$$\mathrm{ind}(w) = \sup\{r \in \mathbf{Q} \mid w^r \in F(x)\}$$

if such an integer exists. Otherwise, w is said to have infinite index. We also define the *prefix index* $\mathrm{pind}(w)$ to be the greatest number r such that w^r is a prefix of x. The prefix index is always finite, provided x is noi periodic, and it is zero when the first letter of w differs from the first letter of x.

Proposition 2. *Every nonempty factor of a Sturmian word s has finite index in s.*

Proof. Assume the contrary. There exist a Sturmian word s and a nonempty factor u of s such that u^n is a factor of s for every $n \geq 1$. Consequently, the periodic word u^ω is in the dynamical system generated by s. Since this system is minimal, $F(s) = F(u^\omega)$, a contradiction. □

An infinite word x has *bounded index* if there exists a number d such that every nonempty factor of x has an index less than or equal to d. If x has bounded index, the upper bound might be irrational. For instance, Mignosi and Pirillo [22] have shown that in the case of the Fibonacci word, this bound is $2+\tau$, where $\tau = (1+\sqrt{5})/2$. We will get this as a consequence of our investigations.

We start with a notation. Let $(s_n)_{n\geq -1}$ be the standard sequence of the characteristic word c_α, with $\alpha = [0, 1+d_1, d_2, \ldots]$. For $n \geq 3$ (and for $n = 2$ if $d_1 \geq 1$), define

$$t_n = s_{n-1}^{d_n - 1} s_{n-2} s_{n-1}$$

and for $n \geq 0$ set

$$p_n = s_{n-1}^{d_n} s_{n-2}^{d_{n-1}} \cdots s_0^{d_1}$$

In particular $p_0 = \varepsilon$. In view of (1), the word t_n is just a conjugate of s_n. More precisely

Lemma 1. *(i) For $n \geq 3$ (and for $n = 2$ if $d_1 \geq 1$), one has*

$$s_n s_{n-1} = s_{n-1} t_n, \qquad s_{n-1} s_n = s_n t_{n-1}\,.$$

(ii) For $n \geq 0$,

$$s_n s_{n-1} = \begin{cases} p_n 10 & \text{if } n \text{ is odd} \\ p_n 01 & \text{if } n \text{ is even.} \end{cases}$$
$$s_{n-1} s_n = \begin{cases} p_n 01 & \text{if } n \text{ is odd} \\ p_n 10 & \text{if } n \text{ is even.} \end{cases}$$

Proof. (i) First,

$$s_n s_{n-1} = s_{n-1}^{d_n} s_{n-2} s_{n-1} = s_{n-1} t_n\,.$$

Next

$$\begin{aligned} s_{n-1} s_n &= s_{n-1}^{d_n} s_{n-1} s_{n-2} \\ &= s_{n-1}^{d_n} s_{n-2} t_{n-1} = s_n t_{n-1} \end{aligned}$$

(*ii*) Since $s_0 = 0$ and $s_{-1} = 1$, the equations hold for $n = 0$. Also,

$$s_1 s_0 = s_0^{d_0} 10 = p_0 10, \quad s_0 s_1 = s_0^{d_0} 01 = p_0 01$$

Next, for $n \geq 2$ and even,

$$s_n s_{n-1} = s_{n-1}^{d_n} s_{n-1} s_{n-2} = s_{n-1}^{d_n} p_{n-1} 01$$

and since $p_n = s_{n-1}^{d_n} p_{n-1}$, one gets the frist formula. The other equations are verified in the same manner. □

Corollary 1. *The words s_n and t_n differ only by their last two letters.*

Proof. In view of the previous lemma

$$s_{n+1} = s_n^{d_{n+1}-1} s_n s_{n-1} = s_n^{d_{n+1}-1} p_n ab$$
$$t_{n+1} = s_n^{d_{n+1}-1} s_{n-1} s_n = s_n^{d_{n+1}-1} p_n ba$$

where $ab = 01$ or $ab = 10$. This proves the claim. □

Example 3. Consider again $\alpha = (\sqrt{3}-1)/2 = [0, 2, 1, 2, 1, \ldots]$ and its directive sequence $(1, 1, 2, 1, 2, 1, \ldots)$. The sequences start with

	$s_0 = 0$	$p_0 = \varepsilon$
$d_1 = 1$	$s_1 = 01$	$p_1 = 0$
$d_2 = 1$	$s_2 = 010$	$p_2 = 010$
$d_3 = 2$	$s_3 = 01001001$	$p_3 = 010010010$
$d_4 = 1$	$s_4 = 01001001010$	$p_4 = 01001001010010010$
$d_5 = 2$	$s_5 = 010010010100100100101001001001$	

The importance of the sequence p_n comes from the following observation. Consider the sequence of integers (q_n) defined by

$$q_{-1} = q_0 = 1 \quad q_{n+1} = d_{n+1} q_n + q_{n-1}$$

so that q_n is precisely the length of s_n.

Proposition 3. *The word p_{n+1} is the heighest rational power of s_n that is a prefix of the standard word c_α. The prefix index of s_n is $1 + d_{n+1} + (q_{n-1} - 2)/q_n$.*

Proof. Clearly, $s_{n+1} s_n$ is a prefix of the characteristic word c_α. Since

$$s_{n+1} s_n = s_n^{d_{n+1}} s_{n-1} s_n = s_n^{1+d_{n+1}} t_{n-1}$$

the word $s_n^{1+d_{n+1}}$ is a prefix of c_α. Observe that t_{n-1} is not a prefix of s_n. Indeed, the word s_{n-1} is prefix of s_n and has the same length as t_{n-1}. Thus the longest common prefix h_n of t_{n-1} and s_{n-1} has length $q_{n-1} - 2$, and since

$$s_{n+1} s_n = s_n^{1+d_{n+1}} t_{n-1} = p_{n+1} ab$$

the longest power of s_n that is prefix of c_α is p_{n+1}. Since $|p_{n+1}| = q_{n+1} + q_n - 2 = (d_{n+1} + 1) q_n + q_{n-1} - 2$, the exponent of s_n is $1 + d_{n+1} + (q_{n-1} - 2)/q_n$. □

Example 4. Consider first the Fibonacci sequence, where all d_n are 1. The q_n are the Fibonacci numbers. The formula shows that the prefix index of s_n is $q_{n+2}/q_n - 2/q_n$, and since $q_{n+2}/q_n \to \tau^2 = 1+\tau$, where τ is the golden ratio, the prefix index is bounded by the sequence $1+\tau$.

Example 5. Consider again $\alpha = (\sqrt{3}-1)/2 = [0,2,1,2,1,\ldots]$. Since its directive sequence has bounded values, the exponent of s_n as a prefix is bounded. It is not difficult to see that the sequence $1 + d_{n+1} + (q_{n-1} - 2)/q_n$ has two accumulation points, namely $2+\sqrt{3}$ and $1+(\sqrt{3}-1)/2$, and thus the prefix index is bounded by $3+2\alpha$.

Proposition 4. *Any Sturmian word contains cubes. More precisely, the index of s_n as a factor in the caracteristic word c_α is at least* $2+d_{n+1}+(q_{n-1}-2)/q_n$.

Proof. Set $\delta_n = 1 + d_{n+1} + (q_{n-1} - 2)/q_n$. We show that s_{n+4} contains a power of s_n of exponent $1+\delta_n$. Indeed,

$$s_{n+4} = s_{n+3}^{d_{n+4}-1} s_{n+3}s_{n+2} = s_{n+3}^{d_{n+4}-1} s_{n+2}t_{n+3}$$

The suffix $s_{n+2}t_{n+3}$ of s_{n+4} contains the desired power. Indeed, s_{n+2} ends with s_n, and t_{n+3} shares a prefix of length $q_{n+3} - 2$ with s_{n+3}. Now p_{n+1} is the prefix of c_α of length $q_{n+1} + q_n - 2$, and since $q_{n+1} + q_n < q_{n+3}$, s_np_{n+1} is a factor of $s_{n+2}t_{n+3}$, and also of s_{n+4}. □

A weak converse also holds.

Proposition 5. *Let w be a primitve factor in $c = c\alpha$, and assume that* ind(w) ≥ 4. *Assume further that w has the maximal index among its conjugates. Then w is one of the s_n.*

Proof. Set $1+d = \text{ind(w)}$, set $w = za$ with a a letter, and let b be the letter preceding the occurrence of w^{d+1}. Then $a \neq b$ since otherwise the conjugate az would have greater index. Thus aw^d and bw^d are factors of c. This means that w^d is a left special factor, and therefore it is a prefix of c.

Let n be the greatest integer such that s_n is a prefix of w^2 (recall that $2 \leq d$). Since s_n is primitive, $s_n \neq w^2$. If w is a prefix of s_n, then either $w = s_n$ and the proposition is proved, or there is a factorization $w = uv$, for non empty u, v, such that $s_n = wu$. Next, s_nw is a prefix of w^3. Thus $s_nwz = w^3$ for some word z. It follows from $s_nwz = wuuvz$ and $w^3 = wuvuv$ that $uv = vu$, which is impossible because w is primitive.

Thus s_n is a proper prefix of w. Now, w^2 is a proper prefix of s_{n+1}, thus also of p_{n+1}; thus $w^2z = s^k$ for some $k > 2$. Thus w and s_n are powers of the same word, and since they are primitive, they are equal. □

Proof of Theorem 1. Since a Sturmian word has the same factors as the characteristic word of same slope, it suffices to prove the result for characteristic words. Let c be the characteristic word of slope $\alpha = [0, 1 + d_1, d_2, \ldots]$. Let $(s_n)_{n \geq -1}$ be the associated standard sequence.

To prove that the condition is sufficient, observe that $s_n^{d_{n+1}}$ is a prefix of c for each $n \geq 1$. Consequently, if the sequence (d_n) of partial quotients is unbounded, the infinite word c has factors of arbitrarily great index.

Conversely, assume that c has unbounded index. Then there are words w of arbitrarily high index. By the preceding proposition, there are, in the standard sequence, words s_n of arbitrarily high prefix index. Since the prefix index of s_n is $1 + d_{n+1} + (q_{n-1} - 2)/q_n$, this means that the partial quotients d_{n+1} are unbounded. This completes the proof. □

4 Concluding remark

There might be a more precise correspondence between the factors and the prefixes of the characteristic word. Also, the precise value of the index of a Sturmian word seems to be more complicated to compute in the general case as for the Fibonacci word.

References

1. J. Berstel, Recent results on Sturmian words, in: J. Dassow, G. Rozenberg, A. Salomaa (eds.) *Developments in Language Theory II*, World Scientific, 1996.
2. J. Berstel, P. Séébold,Sturmian Words, in: M. Lothaire (ed.) *Algebraic Combinatorics on Words*, in preparation.
3. J. Berstel et P. Séébold, A remark on morphic Sturmian words, *Informatique théorique et applications* **28** (1994), 255–263.
4. E. Bombieri, J. E. Taylor, Which distributions of matter diffract? An initial investigation, *J. Phys.* **47** (1986), Colloque C3, 19–28.
5. J. E. Bresenham, Algorithm for computer control of a digital plotter, *IBM Systems J.* **4** (1965), 25–30.
6. T. C. Brown, A characterization of the quadratic irrationals, *Canad. Math. Bull.* **34** (1991), 36–41.
7. D. Crisp, W. Moran, A. Pollington, P. Shiue, Substitution invariant cutting sequences, *J. Théorie des Nombres de Bordeaux* **5** (1933), 123–137.
8. E. Coven, G. Hedlund, Sequences with minimal block growth, *Math. Systems Theory* **7** (1973), 138–153.
9. A. De Luca, Sturmian words: structure, combinatorics, and their arithmetics, *Theoret. Comput. Sci.* **183** (1997), 45–82.
10. A. De Luca, Combinatorics of standard Sturmian words, in: J. Mycielski, G. Rozenberg, A. Salomaa (eds.) *Structures in Logic and Computer Science, Lect. Notes Comp. Sci.* Vol. 1261, Springer-Verlag, 1997, pp 249–267.
11. A. De Luca, Standard Sturmian morphisms, *Theoret. Comput. Sci.* **178** (1997), 205–224.
12. A. De Luca et F. Mignosi, Some combinatorial properties of Sturmian words, *Theoret. Comput. Sci.* **136** (1994), 361–385.

13. S. Dulucq, D. Gouyou-Beauchamps, Sur les facteurs des suites de Sturm, *Theoret. Comput. Sci.* **71** (1990), 381–400.
14. A. S. Fraenkel, M. Mushkin, U. Tassa, Determination of $\lfloor n\theta \rfloor$ by its sequence of differences, *Canad. Math. Bull.* **21** (1978), 441–446.
15. G.A. Hedlund, Sturmian minimal sets, *Amer. J. Math* **66** (1944), 605–620.
16. M. Morse, G.A. Hedlund, Symbolic dynamics, *Amer. J. Math* **60** (1938), 815–866.
17. M. Morse, G.A. Hedlund, Sturmian sequences, *Amer. J. Math* **61** (1940), 1–42.
18. S. Ito, S. Yasutomi, On continued fractions, substitutions and characteristic sequences, *Japan. J. Math.* **16** (1990), 287–306.
19. J. Karhumäki, On strongly cube-free ω-words generated by binary morphisms, in *FCT '81*, pp. 182–189, *Lect. Notes Comp. Sci.* Vol. 117, Springer-Verlag, 1981.
20. J. Karhumäki, On cube-free ω-words generated by binary morphisms, *Discr. Appl. Math.* **5** (1983), 279–297.
21. F. Mignosi, On the number of factors of Sturmian words, *Theoret. Comput. Sci.* **82** (1991), 71–84.
22. F. Mignosi et G. Pirillo, Repetitions in the Fibonacci infinite word, *Theoret. Inform. Appl.* **26**,3 (1992), 199–204.
23. F. Mignosi, P. Séébold, Morphismes sturmiens et règles de Rauzy, *J. Théorie des Nombres de Bordeaux* **5** (1993), 221–233.
24. M. Queffélec, *Substitution Dynamical Systems – Spectral Analysis*, Lecture Notes Math.,vol. 1294, Springer-Verlag, 1987.
25. G. Rauzy, Suites à termes dans un alphabet fini, *Sémin. Théorie des Nombres* (1982–1983), 25-01,25-16, Bordeaux.
26. G. Rauzy, Mots infinis en arithmétique, in: *Automata on infinite words* (D. Perrin ed.), *Lect. Notes Comp. Sci.* **192** (1985), 165–171.
27. P. Séébold, Fibonacci morphisms and Sturmian words, *Theoret. Comput. Sci.* **88** (1991), 367–384.
28. C. Series, The geometry of Markoff numbers, *The Mathematical Intelligencer* **7** (1985), 20–29.
29. K. B. Stolarsky, Beatty sequences, continued fractions, and certain shift operators, *Cand. Math. Bull.* **19** (1976), 473–482.
30. B. A. Venkov, *Elementary Number Theory*, Wolters-Noordhoff, Groningen, 1970.

Repetitions and Boxes in Words and Pictures

Arturo Carpi and Aldo de Luca

Summary. Boxes of a word w are suitable factors of w which allow one to reconstruct the entire word. The initial (resp. terminal) box h_w (resp. k_w) is the shortest unrepeated prefix (resp. suffix) of w. A proper box is any factor of the kind asb, where a, b are letters and s is a bispecial factor of w, i.e. there exist letters x, x', y, y', $x \neq x'$, $y \neq y'$, for which sx, sx', ys and $y's$ are factors of w. In a previous paper we proved that any word is uniquely determined by h_w, k_w and the set of proper boxes. In this paper we extend this theorem to the case of any language. Moreover, a further extension is made to the case of two-dimensional languages (picture languages).

1 Introduction

The study of the combinatorial and structural properties of words is a topic of great interest for formal Language Theory with a great number of applications in Computer Science, Communication Theory and other fields such as Algebra, Physics and Biology. In Computer Science the combinatorial techniques have been used, for instance, for the design of efficient algorithms in Pattern Matching, Computer Graphics and Data Compression.

In some previous papers [2,5] we have introduced an elementary structural analysis of finite words which seems to be very powerful and can be usefully applied for 'data processing' and 'communication' in the case of very long words such as DNA sequences. In fact, the main result of our study shows that with any finite word one can associate some suitable sets of factors which allow one to reconstruct the entire word.

A basic concept in the combinatorics of a word is the 'repetitiveness' of a factor of the word. The analysis of the repetitions in a finite word can be done by the so-called 'special factors' and the 'shortest repeated' prefix and suffix. A factor u of a word w over the alphabet A is *right* (resp. *left*) *special* if u occurs in w at least two times followed on the right (resp. left) by two distinct letters, i.e. there exist two letters $a, b \in A$ such that $a \neq b$ and ua, ub (resp. au, bu) are factors of w. A factor of w which is right and left special is called *bispecial*. Special and bispecial factors have been considered by several authors mainly in the case of infinite words and languages (cf. for instance [1,4,6,7]).

For any word w one can consider the shortest unrepeated prefix h_w of w and the shortest unrepeated suffix k_w of w. Any proper prefix (resp. suffix) of h_w (resp. k_w) is a repeated factor.

Any repeated factor in a word is either a factor of a special factor or a proper factor of h_w or of k_w. However, the knowledge of the special factors of w and of h_w and k_w is not, in general, sufficient to reconstruct the word. In our combinatorial approach a very important notion is that of *box* of a word. The words h_w and k_w are boxes called the *initial* and *terminal box*, respectively. A *proper* box is any factor of w of the kind asb with $a, b \in A$ and s bispecial factor of w. A box is called *maximal* if it is not a factor of another box. In [2] the following noteworthy theorem, called the *maximal box theorem*, was proved: *Any finite word is uniquely determined by the initial box, the terminal box and the set of maximal boxes.*

Another important combinatorial notion is that of *superbox*. A superbox is any factor of w of the kind asb, with a, b letters and such that s is a repeated factor, whereas as and sb are unrepeated. A theorem for superboxes, called the *superbox theorem*, similar to the maximal box theorem was also proved. Moreover, some algorithms allowing one to construct boxes and superboxes and, conversely, to reconstruct the word were given [2].

The notions of special and bispecial factors as well as those of initial, terminal and proper box, can be naturally extended to the case of any language. In Sec. 3 we prove some box theorems for languages. A general box theorem for languages shows that, with the only exception of a trivial case, the set of the factors of a language is uniquely determined by the sets of the initial, terminal and proper boxes. From this theorem one easily derives (cf. Sec. 4) the maximal box theorem and the superbox theorem for finite words. In Sec. 3 we prove a box theorem for rational languages. This theorem shows that, except for some trivial cases, two rational languages have the same set of factors if they have the same set of boxes up to a length which is upperbounded by $q_1 + q_2 - 1$, where q_1 and q_2 are the number of the states of two deterministic automata recognizing the two languages. A consequence of this result is a certain generalization of the Equality theorem of Automata theory.

In Sec. 5 we consider two-dimensional words, called *pictures*, and *picture languages*. The notion of special factor is extended to pictures. Since there is a horizontal and a vertical concatenation, one has to consider right, left, down and up special factors. Thus there are horizontal and vertical bispecial factors. The notion of box can be suitably extended to the case of a picture language. In Sec. 6 a two-dimensional box theorem for picture languages is proved.

2 Preliminaries

Let A be a non-empty set called *alphabet* and A^* the set of all finite words on A, including the *empty word* denoted by ε. We set $A^+ = A^* \setminus \{\varepsilon\}$.

For any word $w \in A^+$, we denote by $|w|$ its *length*. The length of ε is taken equal to 0. For any $p \geq 0$, A^p is the set of all the words of A^* of length p and $A^{[p]}$ is the set of all the words of A^* of length $\leq p$.

A word u is a *factor* of w if there exist $p, q \in A^*$ such that $w = puq$. If $p = \varepsilon$ (resp. $q = \varepsilon$), then u is called *prefix* (resp. *suffix*) of w. We shall denote respectively by $F(w)$, $\mathrm{Pref}(w)$ and $\mathrm{Suff}(w)$ the sets of all factors, prefixes and suffixes of w.

By *language* on the alphabet A we shall mean any non-empty subset L of A^*. A word u is called *factor* (resp. *prefix, suffix*) of L if u is a factor (resp. prefix, suffix) of a word of L. For any language L, we denote by $F(L)$ the set of all factors of L. In a similar way, $\mathrm{Pref}(L)$ and $\mathrm{Suff}(L)$ will denote, respectively, the sets of all prefixes and suffixes of L. We set $\mathrm{alph}(L) = F(L) \cap A$.

If X is a subset of A^*, then $\mathrm{Max}(X)$ will denote the set of the elements of X which are maximal in X with respect to the factor ordering, i.e. any element of $\mathrm{Max}(X)$ is not a factor of another element of X.

Let L be a language on A. A factor u of L is said to be *right* (resp. *left*) *extendable* in L if there exists a letter $a \in A$ such that $ua \in F(L)$ (resp. $au \in F(L)$). We shall denote by U_L the set of all factors of L which cannot be extended on the left in L, i.e.

$$U_L = \{u \in F(L) \mid Au \cap F(L) = \emptyset\}.$$

In a symmetrical way, V_L will denote the set of all factors of L which cannot be extended on the right in L, i.e.

$$V_L = \{u \in F(L) \mid uA \cap F(L) = \emptyset\}.$$

We shall denote by U_L^0 (resp. V_L^0) the set of the elements of U_L (resp. V_L) which are minimal with respect to the prefix (resp. suffix) order. One has:

$$U_L^0 = U_L \setminus U_L A^+, \quad V_L^0 = V_L \setminus A^+ V_L.$$

A factor u of a language L over A is *repeated* in L if there exist two distinct pairs $(p, q), (p', q')$ of words of A^* such that $puq,\ p'uq' \in L$.

A factor u of L is called *right special* if there exist two letters $x, y \in A$, $x \neq y$ such that $sx,\ sy \in F(L)$. Similarly, a factor u of L is called *left special* if there exist two letters $x, y \in A$, $x \neq y$ such that $xs,\ ys \in F(L)$. Clearly a right or left special factor of L is a repeated factor of L. A factor u of L is called *bispecial* if it is both right and left special.

We introduce the set $\mathcal{B}_L$ of the proper boxes of L. A *proper box* α of L is a factor of L of the kind $\alpha = asb$ with $a, b \in A$ and s a bispecial factor of L. Any element of U_L^0 (resp. V_L^0) is called an *initial* (resp. *terminal*) *box*. By *box*, without specification, we mean indifferently an initial, a terminal or a proper box.

3 Box Theorems for Languages

Let L and M be two languages over the alphabet A such that $F(L) \neq F(M)$. A *separating factor* of L and M is any word in the symmetric difference $(F(L) \setminus F(M)) \cup (F(M) \setminus F(L))$. A *minimal separating factor* of L and M is a separating factor of minimal length.

Lemma 1. *Let $L, M \subseteq A^*$ be two languages such that $F(L) \neq F(M)$. Then any minimal separating factor of L and M belongs to the set*

$$A \cup A(U_L^0 \cup U_M^0) \cup (V_L^0 \cup V_M^0)A \cup \mathcal{B}_L \cup \mathcal{B}_M. \tag{1}$$

Proof. Let u be a minimal separating factor of L and M. If u is a letter, then the result is trivial. Let us then suppose that $|u| > 1$ and assume first that $u \in F(L) \setminus F(M)$. We can write $u = asb$ with $a, b \in A$ and $s \in A^*$. From the minimality of u one derives that as and sb are not separating factors of L and M, so that $as, sb \in F(M)$.

If as is not right extendable in M, then $as \in V_M$. However, since s is right extendable in M it follows that $as \in V_M^0$ and $u = asb \in V_M^0 A$. In a symmetrical way, one derives that if sb is not left extendable in M, then $u = asb \in AU_M^0$.

Let us then suppose that as is right extendable in M and sb is left extendable in M. Thus there exist letters $c, d \in A$ such that

$$asc, \ dsb \in F(M).$$

Since $u \notin F(M)$, one has $c \neq b$ and $d \neq a$. From the minimality of u, one has that $sc, ds \in F(L)$. This implies that s is a bispecial factor of L, so that u is a box of L, i.e. $u \in \mathcal{B}_L$.

In a symmetrical way, if $u \in F(M) \setminus F(L)$ one proves that

$$u \in AU_L^0 \cup V_L^0 A \cup \mathcal{B}_M. \quad \square$$

Lemma 2. *Let $L, M \subseteq A^*$ be two languages such that $F(L) \neq F(M)$ and let u be a minimal separating factor of L and M. Set $p = \max\{2, |u|\}$. If the following conditions are satisfied*

1) $U_L^0 \cap A^{p-1} = U_M^0 \cap A^{p-1}, \quad V_L^0 \cap A^{p-1} = V_M^0 \cap A^{p-1}$
2) $\mathcal{B}_L \cap A^p \subseteq F(M), \quad \mathcal{B}_M \cap A^p \subseteq F(L)$,

then $\mathrm{Card}(\mathrm{alph}(L)), \mathrm{Card}(\mathrm{alph}(M)) \leq 1$.

Proof. Let u be a minimal separating factor of L and M. We first suppose that $|u| > 1$. By Lemma 1, u belongs to the set (1). If $u \in \mathcal{B}_L \cup \mathcal{B}_M$, then by Condition 2)

$$u \in F(L) \cap F(M)$$

which contradicts the fact that u is a separating factor of L and M. If $u \in A(U_L^0 \cup U_M^0)$, then by Condition 1)

$$u \in A((U_L^0 \cap A^{p-1}) \cup (U_M^0 \cap A^{p-1})) = A(U_L^0 \cap A^{p-1}) = A(U_M^0 \cap A^{p-1}).$$

This contradicts the fact that $u \in F(L) \cup F(M)$. A similar contradiction is reached if $u \in (V_L^0 \cup V_M^0)A$. Thus, the only remaining possibility is $|u| = 1$, i.e. $u = a \in A$ and $p = 2$.

Without loss of generality, we can assume

$$a \in F(L) \setminus F(M). \tag{2}$$

The word a is right extendable in L, otherwise one would have $a \in V_L^0 \cap A = V_M^0 \cap A \subseteq F(M)$. Thus there is $x \in A$ such that $ax \in F(L)$. One has that $\text{alph}(L) = \{a\}$, otherwise ε would be a bispecial factor of L and, consequently, $ax \in \mathcal{B}_L \cap A^2 \subseteq F(M)$, contradicting (2).

If there exists $b \in \text{alph}(M)$, then $b \neq a$ and $b \in F(M) \setminus F(L)$. Thus, proceeding as before, one gets $\text{alph}(M) = \{b\}$. □

Theorem 1. *Let L and M be languages on the alphabet A. If the following conditions are satisfied:*

1) $U_L^0 = U_M^0$, $V_L^0 = V_M^0$
2) $\mathcal{B}_L \subseteq F(M)$, $\mathcal{B}_M \subseteq F(L)$,

then either $F(L) = a^*$, $F(M) = b^*$, *with* $a, b \in A$ *and* $a \neq b$, *or* $F(L) = F(M)$.

Proof. We shall prove that if $F(L) \neq F(M)$, then $F(L) = a^*$, $F(M) = b^*$, with $a, b \in A$ and $a \neq b$.

By Lemma 2, one has $\text{Card}(\text{alph}(L)), \text{Card}(\text{alph}(M)) \leq 1$. In such a case, either $V_L^0 \neq \emptyset$ and $F(L) = F(V_L^0)$ or $V_L^0 = \emptyset$ and L is an infinite language; similarly, one has that either $V_M^0 \neq \emptyset$ and $F(M) = F(V_M^0)$ or $V_M^0 = \emptyset$ and M is an infinite language. Thus, if $V_L^0 = V_M^0 \neq \emptyset$, one would have

$$F(L) = F(V_L^0) = F(V_M^0) = F(M),$$

which is a contradiction. The only remaining possibility is that $V_L^0 = V_M^0 = \emptyset$ and $L \subseteq a^*$ and $M \subseteq b^*$ are infinite languages, with $a \neq b$. □

One can easily prove that Conditions 1) and 2) in Theorem 1 imply that $\mathcal{B}_L = \mathcal{B}_M$. Since this latter condition trivially implies 2), then it can replace Condition 2) in the statement of the theorem, without loss of generality.

Example 1. Let us consider the two languages $L = a^*b^*$ and $M = \{a^n b^n \mid n \geq 0\}$. One easily verifies that L and M have the same set of right (resp. left) special factors, given by a^* (resp. b^*). Thus, for both the languages, there is a unique bispecial factor, given by ε. This implies that $\mathcal{B}_L = \mathcal{B}_M = \{aa, ab, bb\}$. Moreover, $U_L = U_M = \emptyset$ and $V_L = V_M = \emptyset$. From the above theorem, it follows $F(L) = F(M)$.

Let us now consider a new alphabet $A_0 = A \cup \{\#\}$, where $\# \notin A$. For any language $L \subseteq A^*$, we introduce the language

$$\hat{L} = \#L\#.$$

Theorem 2. *Let L and M be languages on the alphabet A such that*

$$\mathcal{B}_{\hat{L}} \subseteq F(\hat{M}), \quad \mathcal{B}_{\hat{M}} \subseteq F(\hat{L}).$$

Then $L = M$.

Proof. If $L = M = \{\varepsilon\}$, then the result is trivial. Let us then suppose that $L \neq \{\varepsilon\}$. Then ε is a bispecial factor of $\hat{L}$ and

$$U^0_{\hat{L}} = \#A_0 \cap F(\hat{L}) \subseteq \mathcal{B}_{\hat{L}} \subseteq F(\hat{M}),$$

so that $U^0_{\hat{L}} = \#A_0 \cap F(\hat{L}) \cap F(\hat{M})$. In particular, one has $M \neq \{\varepsilon\}$ and therefore, repeating the above argument, $U^0_{\hat{M}} = \#A_0 \cap F(\hat{L}) \cap F(\hat{M}) = U^0_{\hat{L}}$. Similarly, one derives $V^0_{\hat{L}} = V^0_{\hat{M}}$.

Since $\text{Card}(\text{alph}(\hat{L})), \text{Card}(\text{alph}(\hat{M})) > 1$, from Theorem 1 one has $F(\hat{L}) = F(\hat{M})$. This implies that $\hat{L} = \hat{M}$ and then $L = M$. □

We consider now rational languages over a finite alphabet A. One can easily prove that if L is rational, then U_L, V_L and $\mathcal{B}_L$ are rational subsets of A^*. It follows that also U^0_L and V^0_L are rational. The following theorem holds.

Theorem 3. *Let L and M be two rational languages over the alphabet A and q_1 and q_2 be the number of states of two deterministic automata recognizing $F(L)$ and $F(M)$ respectively. Let $q = \max\{2, q_1 + q_2 - 1\}$. If the following conditions are satisfied*

1) $U^0_L \cap A^{[q-1]} = U^0_M \cap A^{[q-1]}, \quad V^0_L \cap A^{[q-1]} = V^0_M \cap A^{[q-1]}$
2) $\text{Max}(\mathcal{B}_L \cap A^{[q]}) \subseteq F(M), \quad \text{Max}(\mathcal{B}_M \cap A^{[q]}) \subseteq F(L)$,

then either $F(L) = a^*$, $F(M) = b^*$, with $a, b \in A$ and $a \neq b$, or $F(L) = F(M)$.

Proof. Let us suppose that $F(L) \neq F(M)$ and show that $F(L) = a^*$ and $F(M) = b^*$. By a classical result of Automata theory [8] one has that the minimal separating factor u of L and M has a length $|u| = p \leq q$. Condition 1) of Lemma 2 is certainly satisfied, since $p \leq q$. Condition 2) of Lemma 2 is also satisfied, since

$$\mathcal{B}_L \cap A^p \subseteq \mathcal{B}_L \cap A^{[q]} \subseteq F(\text{Max}(\mathcal{B}_L \cap A^{[q]})) \subseteq F(M)$$

and, similarly,

$$\mathcal{B}_M \cap A^p \subseteq \mathcal{B}_M \cap A^{[q]} \subseteq F(\text{Max}(\mathcal{B}_M \cap A^{[q]})) \subseteq F(L).$$

By Lemma 2 one has $L \subseteq a^*$, $M \subseteq b^*$, for suitable $a, b \in A$. Let us prove that $a^q \in F(L)$. Indeed, otherwise there would exist a maximal integer t such that $a^t \in F(L)$ and $t < q$. Since a^t is not right extendable in L one has $a^t \in V_L^0 \cap A^{[q-1]} = V_M^0 \cap A^{[q-1]}$. Thus, a^t is the longest word of both L and M, so that $F(L) = F(M) = F(a^t)$, which is a contradiction. From this one derives that $a^q \in F(L)$. Thus $F(L)$ and a^* contain the same words up to the length q. Since a^* is recognized by a deterministic automaton having one state and $q \geq q_1$, it follows that $F(L) = a^*$. In a similar way, one has $F(M) = b^*$. Moreover $a \neq b$, since $F(L) \neq F(M)$. □

Theorem 4. *Let L and M be two rational languages over the alphabet A and q_1 and q_2 be the number of states of two deterministic automata recognizing L and M, respectively. Let $q = q_1 + q_2 + 1$. If*

$$\mathrm{Max}(\mathcal{B}_{\hat{L}} \cap A_0^{[q]}) \subseteq F(\hat{M}), \quad \mathrm{Max}(\mathcal{B}_{\hat{M}} \cap A_0^{[q]}) \subseteq F(\hat{L}),$$

then $L = M$.

Proof. Let us suppose, by contradiction, that $L \neq M$. By the Equality theorem of Automata theory, there exists an element $u \in (L \backslash M) \cup (M \backslash L)$ of length $|u| \leq q_1 + q_2 - 1$. Let $\hat{u} = \#u\#$. One has $|\hat{u}| \leq q$ and $\hat{u} \in (\hat{L} \setminus \hat{M}) \cup (\hat{M} \setminus \hat{L})$. This implies that $\hat{u}$ is a separating factor of $\hat{L}$ and $\hat{M}$. Thus, if v is a minimal separating factor of $\hat{L}$ and $\hat{M}$, then setting $p = \max\{2, |v|\}$, one has $p \leq q$.

We prove that $\hat{L}$ and $\hat{M}$ satisfy the conditions of Lemma 2. Indeed,

$$\mathcal{B}_{\hat{L}} \cap A^p \subseteq \mathcal{B}_{\hat{L}} \cap A^{[q]} \subseteq F(\mathrm{Max}(\mathcal{B}_{\hat{L}} \cap A^{[q]})) \subseteq F(\hat{M})$$

and, similarly,

$$\mathcal{B}_{\hat{M}} \cap A^p \subseteq \mathcal{B}_{\hat{M}} \cap A^{[q]} \subseteq F(\mathrm{Max}(\mathcal{B}_{\hat{M}} \cap A^{[q]})) \subseteq F(\hat{L}).$$

Moreover, by an argument similar to that used in the proof of Theorem 2, one derives $U_{\hat{L}}^0 = U_{\hat{M}}^0$ and $V_{\hat{L}}^0 = V_{\hat{M}}^0$. By Lemma 2, one has Card(alph($\hat{L}$)), Card(alph($\hat{M}$)) ≤ 1, and therefore, necessarily, $L = M = \{\varepsilon\}$, contradicting our assumption. Hence $L = M$. □

4 Box Theorems for Words

In this section we shall apply the results of the previous section to the case when the languages consist of single words.

Let $L = \{w\}$, with $w \in A^*$. We shall denote the set $\mathcal{B}_{\{w\}}$ of the proper boxes of $\{w\}$ simply by $\mathcal{B}_w$. Similarly, we shall write simply U_w^0, V_w^0 instead of $U_{\{w\}}^0$, $V_{\{w\}}^0$. In such a case, U_w^0 and V_w^0 contain a unique element which will be denoted respectively by h_w and k_w.

The following theorem, called the Maximal Box theorem, was proved in [2]. We shall derive it here as a straightforward consequence of Theorem 1.

Theorem 5. (Maximal Box theorem) *Let $f, g \in A^*$ be two words such that*

1) $h_f = h_g, \quad k_f = k_g$,
2) $\mathrm{Max}(\mathcal{B}_f) \subseteq F(g)$,
3) $\mathrm{Max}(\mathcal{B}_g) \subseteq F(f)$.

Then $f = g$.

Proof. By Condition 1), $U_f^0 = U_g^0$ and $V_f^0 = V_g^0$. Moreover, since $F(f)$ and $F(g)$ are finite languages,

$$\mathcal{B}_f \subseteq F(\mathrm{Max}(\mathcal{B}_f)) \subseteq F(g), \quad \mathcal{B}_g \subseteq F(\mathrm{Max}(\mathcal{B}_g)) \subseteq F(f).$$

Finally, since $\{f\}$ and $\{g\}$ are finite languages, it follows from Theorem 1 that $F(f) = F(g)$ and then $f = g$. □

For any word $w \in A^*$ we define as *superbox* of w any factor of the kind asb, with $a, b \in A$, $s \in A^*$, such that s is a repeated factor of w and as, sb are unrepeated factors. We shall denote by $\mathcal{M}_w$ the set of all superboxes of w.

Theorem 6. (Superbox theorem) *Let $f, g \in A^*$ be two words such that*

1) $h_f \in \mathrm{Pref}(g)$, $k_f \in \mathrm{Suff}(g)$, $h_g \in \mathrm{Pref}(f)$, $k_g \in \mathrm{Suff}(f)$,
2) $\mathcal{M}_f \subseteq F(g)$,
3) $\mathcal{M}_g \subseteq F(f)$.

Then $f = g$.

Proof. We consider the alphabet $A_0 = A \cup \{\#\}$ and the word $\hat{f} = \#f\#$. Let us first prove that

$$\mathcal{M}_{\hat{f}} \subseteq \{\#h_f\} \cup \{k_f\#\} \cup \mathcal{M}_f. \tag{3}$$

A superbox of $\hat{f}$ is either a superbox of f or a word of the kind $\#u$ or $v\#$, with $u \in \mathrm{Pref}(f)$ and $v \in \mathrm{Suff}(f)$. Write $u = sa$ with $a \in A$. Since s is repeated and sa is unrepeated, then the only possibility is that $u = h_f$. In a symmetric way one derives that $v = k_f$.

As proved in [2] (see Proposition 21), any proper box of a word w is a factor of a superbox of w, or of h_w or of k_w. Thus

$$\mathcal{B}_{\hat{f}} \subseteq F(\mathcal{M}_{\hat{f}} \cup \{h_{\hat{f}}, k_{\hat{f}}\}). \tag{4}$$

One easily checks that $h_{\hat{f}}$ and $k_{\hat{f}}$ are respectively a prefix of $\#h_f$ and a suffix of $k_f\#$. Hence, from (4), in view of (3) and of Conditions 1) and 2), one obtains

$$\mathcal{B}_{\hat{f}} \subseteq F(\{\#h_f\} \cup \{k_f\#\} \cup \mathcal{M}_f) \subseteq F(\hat{g}).$$

In a symmetrical way, one has $\mathcal{B}_{\hat{g}} \subseteq F(\hat{f})$. By Theorem 2, one has $f = g$. □

We note that a further application of Theorem 1 to infinite words is a new combinatorial proof [3] of the 'uniqueness theorem' for periodic functions of Fine and Wilf, in the discrete case.

5 Two-dimensional Words

A generalization in two dimensions of a finite word is given by the notion of *two-dimensional word*, briefly *2D-word*, (or *picture*) which can be defined as a rectangular array of elements of a given alphabet A. More precisely a 2D-word over the alphabet A of *size* (m, n), with $m, n > 0$, is a matrix

$$w = (w_{ij})_{1 \le i \le m,\ 1 \le j \le n}$$

with $w_{ij} \in A$, $i = 1, \ldots, m$, $j = 1, \ldots, n$. By (m, n)*-word* one means a 2D-word of size (m, n). A $(1, n)$-word (resp. a $(m, 1)$-word) is called a *horizontal* (resp. *vertical*) *word*. A horizontal word coincides with the usual word over A.

We denote by $A^{(m,n)}$ the set of all (m, n)-words over A. Moreover, we set

$$A^{(m,+)} = \bigcup_{n \ge 1} A^{(m,n)}, \quad A^{(+,n)} = \bigcup_{m \ge 1} A^{(m,n)}$$

and

$$A^{(+,+)} = \bigcup_{m,n \ge 1} A^{(m,n)}.$$

One can introduce two binary operations partially defined in $A^{(+,+)}$ called respectively *horizontal* and *vertical concatenation.*

The horizontal (resp. vertical) concatenation is defined for any 2D-words u and v having the same number of rows (resp. columns). We shall denote respectively by $\oplus$ and $\ominus$ the horizontal and vertical concatenations.

Let $u \in A^{(m,n_1)}$ and $v \in A^{(m,n_2)}$. The horizontal concatenation $w = u \oplus v$ of u and v is the matrix $w \in A^{(m,n_1+n_2)}$ defined by

$$w_{hk} = \begin{cases} u_{hk} & \text{if } 1 \le k \le n_1 \\ v_{h,k-n_1} & \text{if } n_1 < k \le n_1 + n_2, \end{cases} \qquad 1 \le h \le m.$$

Let $u \in A^{(m_1,n)}$ and $v \in A^{(m_2,n)}$. The vertical concatenation $w = u \ominus v$ of u and v is the matrix $w \in A^{(m_1+m_2,n)}$ defined by

$$w_{hk} = \begin{cases} u_{hk} & \text{if } 1 \le h \le m_1 \\ v_{h-m_1,k} & \text{if } m_1 < h \le m_1 + m_2, \end{cases} \qquad 1 \le k \le n.$$

It is convenient to consider the set $A^{(*,*)}$ obtained by adjoining to $A^{(+,+)}$ an extra element that we denote by ε and call *empty 2D-word* which is a neutral element with respect to the horizontal and vertical concatenations, i.e. for all $u \in A^{(*,*)}$

$$u \oplus \varepsilon = \varepsilon \oplus u = u, \quad u \ominus \varepsilon = \varepsilon \ominus u = u.$$

Let w be a (m,n)-word over A. A word $u \in A^{(*,*)}$ is a *factor* of w if $u = \varepsilon$ or u is an (h,k)-word such that $h \leq m$, $k \leq n$ and there exist integers p and q, $0 \leq p \leq m-h$, $0 \leq q \leq n-k$ for which

$$u_{ij} = w_{p+i,\ q+j}, \quad 1 \leq i \leq h,\ 1 \leq j \leq k.$$

The set of the factors of w is denoted by $F(w)$.

One easily verifies that u is a factor of w if and only if there exist $\alpha, \beta, \gamma, \delta \in A^{(*,*)}$ such that

$$w = \gamma \ominus (\alpha \oplus u \oplus \beta) \ominus \delta.$$

Similarly, one has that u is a factor of w if and only if there exist $\lambda, \mu, \varrho, \sigma \in A^{(*,*)}$ such that

$$w = \varrho \oplus (\lambda \ominus u \ominus \mu) \oplus \sigma.$$

A *2D-language* (or *picture language*) L over A is any non-empty subset of $A^{(+,+)}$.

A word $u \in A^{(*,*)}$ is a *factor* of L if it is a factor of a word of L. We denote by $F(L)$ the set of the factors of L.

Let u be a factor of L. We say that u is *right* (resp. *left*) *extendable* in L if there exists a vertical word x such that $u \oplus x \in F(L)$ (resp. $x \oplus u \in F(L)$). We say that u is *down* (resp. *up*) *extendable* in L if there exists a horizontal word x such that $u \ominus x \in F(L)$ (resp. $x \ominus u \in F(L)$).

A factor u of L is called *right* (resp. *left*) *special* if there exist two vertical words x, y such that $x \neq y$ and $u \oplus x$, $u \oplus y \in F(L)$ (resp. $x \oplus u$, $y \oplus u \in F(L)$). A factor u of L which is both right and left special is called *horizontal bispecial.*

A factor u of L is called *down* (resp. *up*) *special* if there exist two horizontal words x, y such that $x \neq y$ and $u \ominus x$, $u \ominus y \in F(L)$ (resp. $x \ominus u$, $y \ominus u \in F(L)$). A factor u of L which is both down and up special is called *vertical bispecial.*

Let us now introduce the important notion of *box* of a 2D-language. Let L be a 2D-language of A. A factor u of L is a *box* of L if u can be factorized as

$$u = a \oplus s \oplus b = c \ominus t \ominus d,$$

with a, b vertical words, c, d horizontal words, s a horizontal bispecial factor of L and t a vertical bispecial factor of L. A box of L is *maximal* if it is not a factor of another box of L.

In the following, it is convenient to introduce a special symbol $\# \notin A$ and consider the larger alphabet $A_0 = A \cup \{\#\}$. The symbol # will be used as a 'board marker' for 2D-words. If $w \in A^{(+,+)}$, then $\hat{w}$ will denote the 2D-word w 'surrounded' by #'s, i.e. if the size of w is (m,n), then $\hat{w}$ is the $(m+2, n+2)$-word defined by

$$\hat{w}_{ij} = \begin{cases} w_{i-1,j-1} & \text{if } 2 \leq i \leq m+1,\ 2 \leq j \leq n+1 \\ \# & \text{otherwise.} \end{cases}$$

For any 2D-language L over A, $\hat{L}$ is the language on the alphabet A_0 defined as

$$\hat{L} = \{\hat{w} \mid w \in L\}.$$

Let us observe that, by construction, any element of $\hat{L}$ has size (m, n) with $m, n \geq 2$. The 2D-language $\hat{L}$ has the property that no element u of it is a factor of an element $v \in \hat{L}$ with $v \neq u$. One easily derives that

$$L = \{u \in A^{(+,+)} \mid \hat{u} \in F(\hat{L})\}. \tag{5}$$

Example 2. Let us consider the picture

$$\hat{w} = \begin{array}{cccccccc} \# & \# & \# & \# & \# & \# & \# & \# \\ \# & b & b & a & b & b & b & \# \\ \# & a & a & b & a & a & b & \# \\ \# & b & b & a & b & a & a & \# \\ \# & \# & \# & \# & \# & \# & \# & \# \end{array}$$

Two maximal boxes of $\hat{w}$ are, for instance,

$$\begin{array}{ccc} b & b & a \\ a & a & b \\ b & b & a \end{array} \quad \text{and} \quad \begin{array}{ccc} b & a & a \\ a & b & a \\ \# & \# & \# \end{array}$$

6 A Two-dimensional Maximal Box Theorem

Theorem 7. *Let L and M be two 2D-languages over A. If*

$$\mathcal{B}_{\hat{L}} \subseteq F(\hat{M}) \quad \text{and} \quad \mathcal{B}_{\hat{M}} \subseteq F(\hat{L}),$$

then $L = M$.

Proof. We shall prove that for all $m, n \geq 2$ one has

$$F(\hat{L}) \cap A^{(m,n)} = F(\hat{M}) \cap A^{(m,n)}. \tag{6}$$

The proof is by induction on the integer $p = m+n$. Since ε is both horizontal and vertical bispecial in $\hat{L}$, any (2,2)-word of $F(\hat{L})$ is a box of $\hat{L}$ and, therefore, a factor of $\hat{M}$. Symmetrically, any (2,2)-word of $F(\hat{M})$ is a factor of $\hat{L}$, so that (6) is satisfied for $m = n = 2$.

Let us now suppose $p > 4$. Let $u \in F(\hat{L}) \cap A^{(m,n)}$ with $m, n \geq 2$ and $m + n = p$. We can uniquely factorize u as

$$u = a \oplus s \oplus b = c \ominus t \ominus d,$$

with a, b vertical words, c, d horizontal words and s, t factors of u. If s is a horizontal bispecial factor of $\hat{L}$ and t a vertical bispecial factor of $\hat{L}$, then u is a box of $\hat{L}$ and, therefore, a factor of $\hat{M}$. Thus we can suppose that either s is not a horizontal bispecial factor of $\hat{L}$ or t is not a vertical bispecial factor of $\hat{L}$. We consider only the first eventuality since the second one is symmetrically dealt with. We can assume that s is not a right special factor of $\hat{L}$. The case that s is not a left special factor of $\hat{L}$ is symmetrically dealt with.

Since s is not right special, then $s \neq \varepsilon$. Thus, $n \geq 3$ and $a \oplus s$ is a $(m, n-1)$-word so that, by the inductive hypothesis, $a \oplus s \in F(\hat{M})$. Since $a \oplus s$ is right extendable in $\hat{L}$, its last column contains at least one symbol of A. This implies that $a \oplus s$ is right extendable in $\hat{M}$, too. Thus there exists a vertical word c such that $a \oplus s \oplus c \in F(\hat{M})$. By using again the inductive hypothesis it follows that $s \oplus c \in F(\hat{L})$. Since s is not right special in $\hat{L}$, one derives $b = c$ and therefore

$$u \in F(\hat{M}).$$

By a similar argument one proves that if $u \in F(\hat{M}) \cap A^{(m,n)}$, then $u \in F(\hat{L})$ so that (6) is proved. From this, it follows that

$$F(\hat{L}) = F(\hat{M})$$

and then, by (5), $L = M$. □

Finally, we observe that if we refer to 2D-languages of horizontal words, then from Theorem 7 one can derive Theorem 2.

References

1. M. Béal, F. Mignosi and A. Restivo, Minimal forbidden words and symbolic dynamics, Proc.s STACS '96, in *Lecture Notes in Computer Science*, vol. 1046, pp. 555–566, (Springer Verlag, New York, 1996).
2. A. Carpi and A. de Luca, Words and Special Factors, Preprint 98/33, Dipartimento di Matematica dell'Università di Roma 'La Sapienza', 1998.
3. A. Carpi and A. de Luca, Words and Repeated Factors, Preprint 98/45, Dipartimento di Matematica dell'Università di Roma 'La Sapienza', 1998.
4. J. Cassaigne, Complexité et facteurs speciaux, *Bull. Belg. Math. Soc.* **4** (1997) 67–88.
5. A. de Luca, On the Combinatorics of Finite Words, *Theoretical Computer Science*, Special Issue for the workshop 'Words' (Rouen, Sept. 1997), to appear.
6. A. de Luca and F. Mignosi, Some Combinatorial properties of Sturmian words, *Theoretical Computer Science* **136** (1994) 361–385.
7. A. de Luca and S. Varricchio, Some combinatorial problems of the Thue-Morse sequence and a problem in semigroups, *Theoretical Computer Science* **63** (1989) 333–348.
8. S. Eilenberg, *Automata, Languages, and Machines*, Vol. A (Academic Press, New York, 1974).
9. D. Giammarresi, A. Restivo, S. Seibert and W. Thomas, Monadic Second-order Logic over Rectangular Pictures and Recognizability by Tiling Systems, *Information and Computation* **125** (1996) 32–45.

Small Aperiodic Sets of Triangular and Hexagonal Tiles*

Karel Culik II

Summary. We show that square Wang tiles can be simulated by triangular or hexagonal Wang tiles or by rotating hexagonal tiles. Hence the aperiodic set of 13 square Wang tiles recently constructed by the author yields the smallest known aperiodic sets of all considered types of tiles.

1 Introduction

Wang tiles are unit square tiles with colored edges. A *tile set* is a finite set of Wang tiles. *Tilings* of the infinite Euclidean plane are considered using arbitrarily many copies of the tiles in the given tile set. The tiles are placed on the integer lattice points of the plane with their edges oriented horizontally and vertically. The tiles may not be rotated. A tiling is *valid* if everywhere the contiguous edges have the same color.

Let T be a finite tile set, and $f : \mathbb{Z}^2 \to T$ a tiling. Tiling f is *periodic* with period $(a, b) \in \mathbb{Z}^2 - \{(0,0)\}$ iff $f(x, y) = f(x + a, y + b)$ for every $(x, y) \in \mathbb{Z}^2$. A tile set T is called *aperiodic* iff (i) there exists a valid tiling, and (ii) there does not exist any periodic valid tiling.

R. Berger in his well known proof of the undecidability of the tiling problem [2] refuted Wang's conjecture that no aperiodic set exists, and constructed the first aperiodic set containing 20426 tiles. He shortly reduced it to 104 tiles. A number of researchers, among them some well known people in discrete mathematics, logic and computer science, tried to find a smaller aperiodic set of Wang tiles. Between 1966 and 1978 progressively smaller aperiodic sets were found by D. E. Knuth (92 tiles, 1966, see [7]), H. Läuchli (40 tiles, 1966, see [5, p. 590]), R. M. Robinson (56 tiles, 1967, 35 tiles, 1971), R. Penrose (34 tiles, 1973), R. M. Robinson (32 tiles, 1973, see [5, p. 593] and 24 tiles, 1977, see [5, p. 593]) and finally R. Ammann (16 tiles, 1978, see [9,5, p. 595]). An excellent discussion of these and related results, and in most cases the first published version of them is included in Chapter 10 and 11 of [5].

Recently, J. Kari developed a new method for constructing aperiodic sets that is not based on geometry, as the earlier ones, but on sequential machines that multiply real numbers in the balanced representation by rational constants. The balanced representation is based on Beatty sequences [1]. This

* Supported by the National Science Foundation under Grant No. CCR-9417384

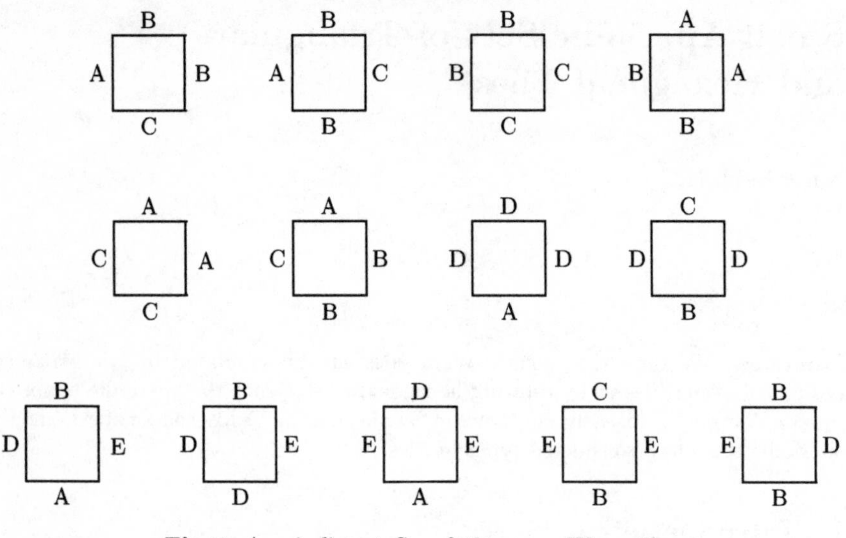

Fig. 1. Aperiodic set S_{13} of 13 square Wang tiles

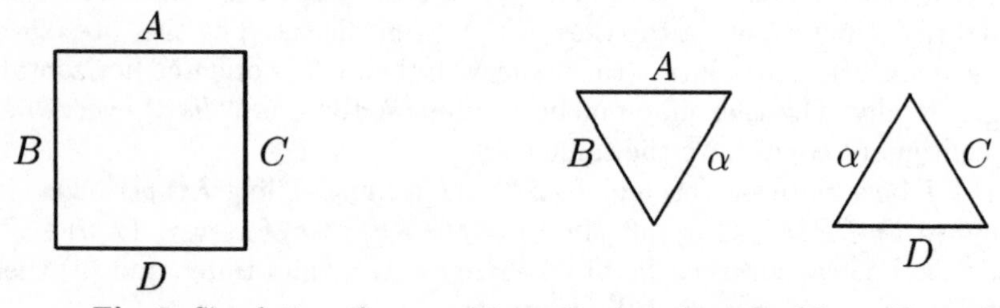

Fig. 2. Simulation of square Wang tiles by triangular Wang tiles

approach makes short and precise correctness arguments possible. He used it to construct a new aperiodic set containing only 14 tiles over 6 colors in [6]. The author added an additional trick in [3] and obtained an aperiodic set T_{13} consisting of 13 tiles over 5 colors shown in Fig. 1.

It is well known that the Euclidean plane allows three different tesselations into regular polygons – namely triangles, squares and hexagons. Hence, besides square Wang tiles, we can consider also triangular and hexagonal Wang tiles, i.e. regular triangles and hexagons with colored edges. The notions of a tile set, valid tiling, periodic tiling, and an aperiodic tile set extends in the obvious way. Again, in the first part of this paper, we assume that the tiles may not be rotated. This makes the triangular Wang tiles somewhat different from the square and hexagonal ones because there are actually two types of the triangular tiles, one pointing up and the other pointing down.

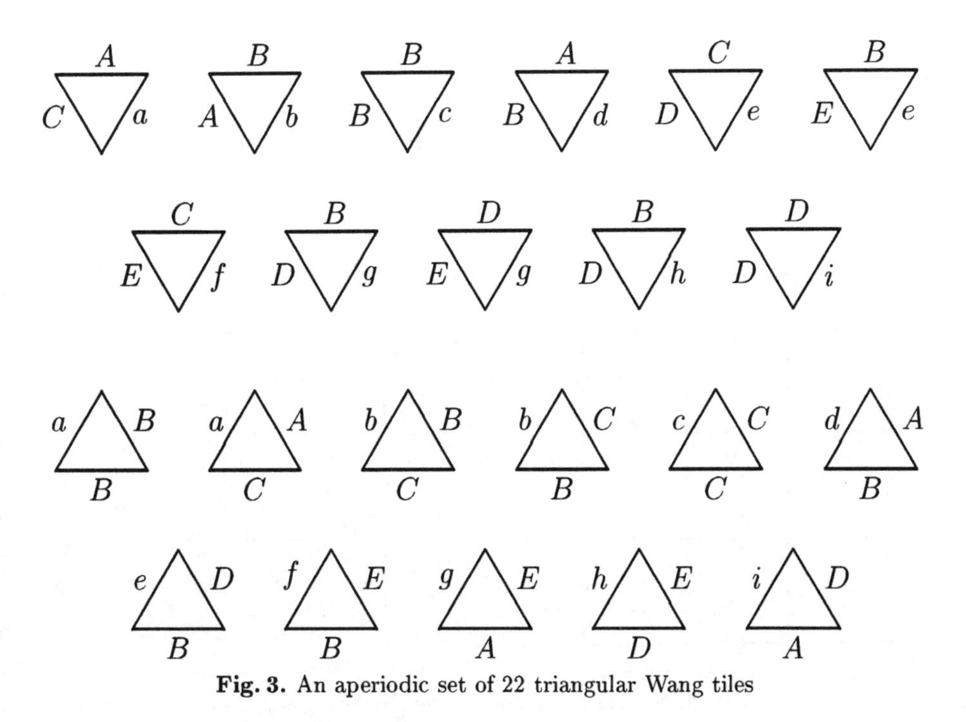

Fig. 3. An aperiodic set of 22 triangular Wang tiles

2 Triangular Wang tiles

It is easy to see that square Wang tiles can be simulated by triangular Wang tiles as indicated in Fig. 2.

Lemma 1 *For each set S of n square Wang tiles there is a set T of at most $2n$ triangular Wang tiles such that there is a bijection between valid tilings by S and valid tilings by T.*

Proof. If we split each square tile in S into two triangular Wang tiles as indicated in Fig. 2 with a unique color α at the new edges we clearly get a set T of $2n$ tiles satisfying the requirement. □

It might be possible to decrease the size of T from Lemma 1 by re-using the color of the new edges if two squares have identical colors at adjacent edges. For example, our aperiodic set S_{13} from Fig. 1 can be simulated by the set T_{22} shown in Fig. 3. Here we "merged" the south-west halves of the first and third tile in the last row and the south-west halves of the last tile in the last two rows.

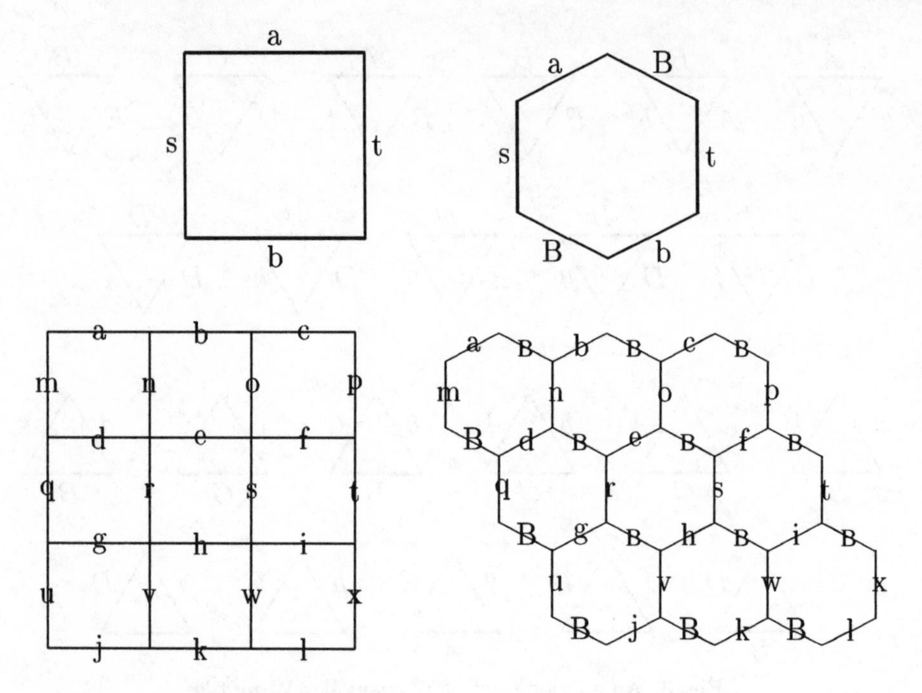

Fig. 4. Simulation of square tiles by hexagonal tiles

3 Hexagonal Wang tiles

It is easy to simulate square Wang tiles by hexagonal Wang tiles as indicated in Fig. 4.

Lemma 2 *For each set S of n square Wang tiles there exists a set of n triangular Wang tiles such that there is a bijection between valid tilings by S and by T.*

Proof. We use just one new color B (blank) on all hexagonal tiles and replace each square tile by a hexagonal one as indicated on the top of Fig. 4. Since the blanks and no other color must be used in the left to right bottom to top diagonal direction there is a one-to-one matching between the tilings by squares and the skewed tilings by hexagons as shown in the example in the bottom half of Fig. 4. □

Hence, for any aperiodic set of square tiles we have an aperiodic set of hexagonal tiles of the same size.

Corollary 1 *There is an aperiodic set of 13 hexagonal tiles.*

From our simulations it follows the following expected result.

Theorem 1 *Given a set of triangular (hexagonal) tiles it is undecidable whether there exists a valid tiling of the whole plane.*

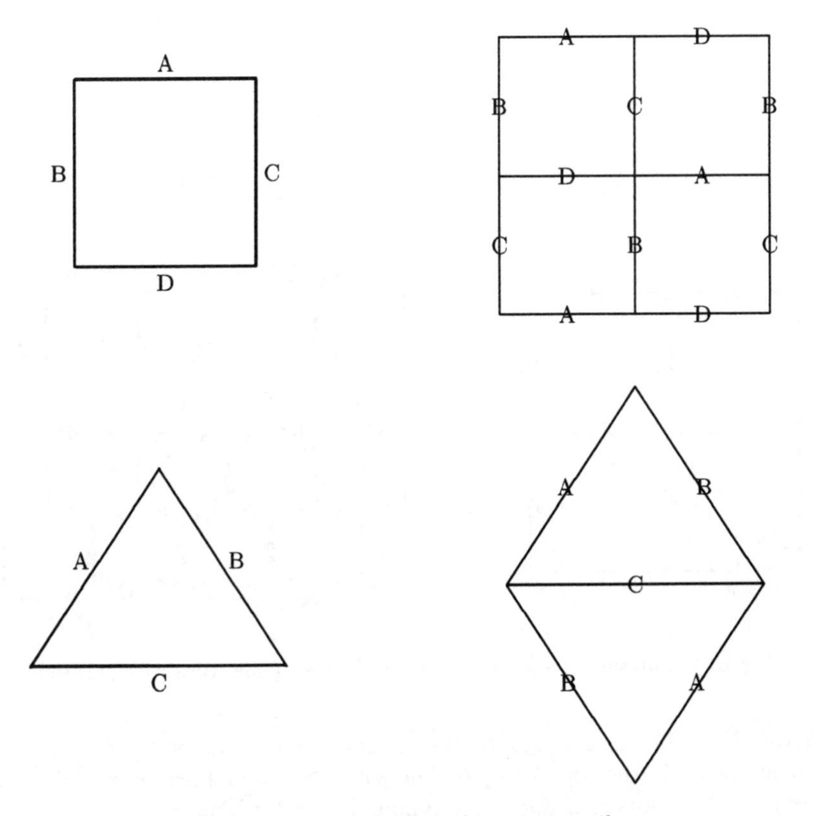

Fig. 5. Square and triangular rotating tiles

4 Rotating tiles

Until now we did not allow tiles to be rotated. Now we relax this restriction. It is well known that in the case of square and triangular rotating tiles the problem of the existence of a valid tiling of the plane and the existence of an aperiodic tile set is trivial.

Using any single square or triangular rotating tile with arbitrary colors we can tile a rectangle with the identical colors on the opposite sides as shown in Fig. 5. Hence, allowing just the rotation by 180°, any single square or triangular tile can tile the whole plane.

However, the situation is different for hexagonal rotating tiles. We can simulate square Wang tiles by hexagonal rotating tiles.

Lemma 3 *For each set S of n square Wang tiles there is a set T of at most $4n$ hexagonal rotating tiles such that there is a bijection between valid tilings by S and valid tilings by T (up to a rotation, i.e. we identify the tilings by T which agree up to a rotation).*

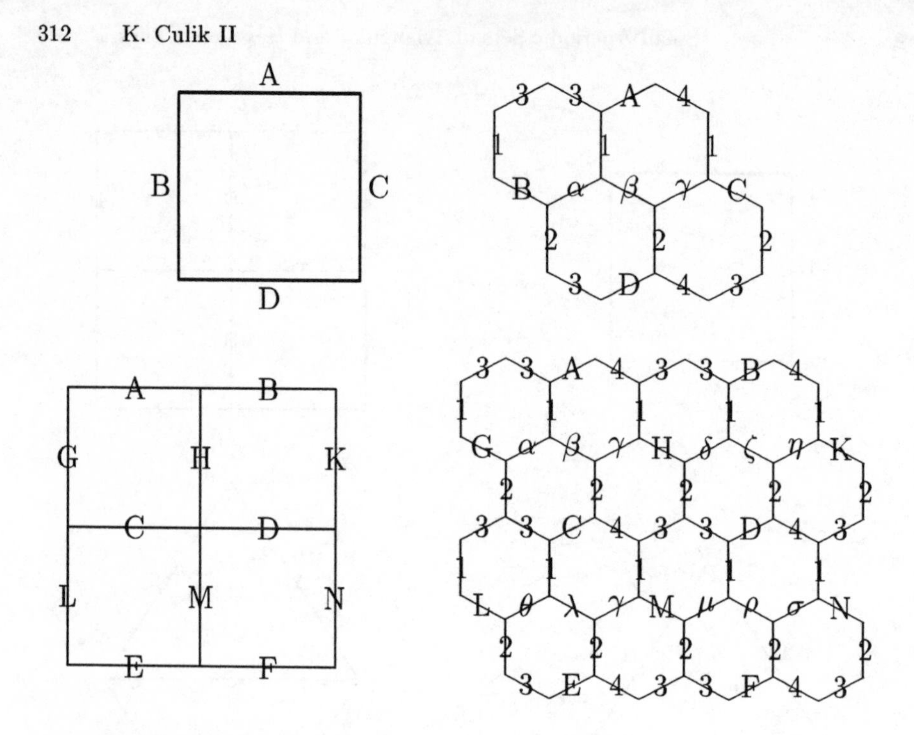

Fig. 6. Simulation of square Wang tiles by hexagonal tiles with rotation

Proof. We replace each square Wang tile in S by four hexagonal rotating tiles as shown in the top of Fig. 6. The colors 2, 3 and 4 are re-used for each tile in S, the colors α, β and γ are unique for each tile in S.

The uniqueness of α, β, γ assures that the four hexagonal tiles simulating one square Wang tile have to always be used together. The colors 2, 3 and 4 assure that the top of the 4-tile block simulating one square tile can be matched only with the bottom of such a block, similarly for the left side and the right side. Therefore in any valid tiling by T all the tiles have to be rotated identically or not at all. The placing of the original colors A, B, C, D assures that the 4-tile blocks from T match iff the simulated square tiles from S match. □

If for two or more simulated square Wang tiles the north-west or the south-east halves agree we can reuse the colors α, β and γ similarly like we reused the color of the cutting edge in section 2. Thus by simulating the aperiodic set T_{13} we get the following.

Corollary 2 *There is an aperiodic set of 44 hexagonal rotating tiles.*

From the undecidability of the square Wang tiling problem clearly follows the following.

Corollary 3 *Given a set H of hexagonal rotating tiles it is undecidable if there exists a valid tiling of the whole plane by H.*

5 Open problems

We obtain relatively small aperiodic sets of triangular Wang tiles, hexagonal Wang tiles and hexagonal rotating tiles, all by simulating our 13-tile aperiodic set of square Wang tiles. In the same way, an aperiodic set of 23 Wang cubes was obtained in [4]. It would be very surprising if a smaller aperiodic set would not exist that takes advantage of the specific shapes in each class.

References

1. S. Beatty, Problem 3173, *Am. Math. Monthly* 33 (1926) 159; solutions in 34, 159 (1927).
2. R. Berger, The Undecidability of the Domino Problem, *Mem. Amer. Math. Soc.* 66 (1966).
3. K. Culik II, An aperiodic set of 13 Wang tiles, *Discrete Mathematics* 160, 245-251 (1996).
4. K. Culik II and J. Kari, An aperiodic set of Wang cubes, *Journal of Universal Computer Science* 1, 675-686 (1995).
5. B. Grünbaum and G.C. Shephard, *Tilings and Patterns*, W.H.Freeman and Company, New York (1987).
6. J. Kari, A small aperiodic set of Wang tiles, *Discrete Mathematics* 160, 259-264 (1996).
7. D. E. Knuth, *The Art of Computer Programming*, Vol.1, p.384, Addison-Wesley, Reading, MA (1968).
8. R. M. Robinson, Undecidability and Nonperiodicity for Tilings of the Plane. *Inventiones Mathematicae* 12, 177-209 (1971).
9. R. M. Robinson, Undecidable tiling problems in the hyperbolic plane, *Inventiones Mathematicae* 44, 259-264 (1978).

Quadratic Word Equations

Volker Diekert and John Michael Robson*

Summary. We consider word equations where each variable occurs at most twice (quadratic systems). The satisfiability problem is NP-hard (even for a single equation), but once the lengths of a possible solution are fixed, then there is a deterministic linear time algorithm to decide whether there is a corresponding solution. If the lengths of a minimal solution were at most exponential, then the satisfiability problem of quadratic systems would be NP-complete.

In the second part we address the problem with regular constraints: The uniform version is PSPACE-complete. Fixing the lengths of a possible solution doesn't make the problem much easier. The non-uniform version remains NP-hard (in contrast to the linear time result above). The uniform version remains PSPACE-complete.

In the third part we show that for quadratic systems the exponent of periodicity is at most linear in the denotational length.

1 Introduction

The existential theory of equations over free monoids is decidable. This major result was obtained by Makanin [10], who showed that the satisfiability of word equations with constants is decidable. For the background we refer to [14], to the corresponding chapter in the Handbook of Formal Languages (Eds. G. Rozenberg and A. Salomaa), [2], or to the forthcoming [3]. In 1990 Schulz [19] showed an important generalization: Makanin's result remains true when adding regular constraints. Thus, we may specify for each word variable x a regular language L_x and we are only looking for solutions where the value of each variable x is in L_x. This was used e.g. when generalizing Makanin's result to free partially commutative monoids, see [4,12].

A lower bound of NP-hardness on the complexity of this problem is not hard to prove. For an upper bound Gutiérrez [5] showed that the problem is in EXPSPACE, and Plandowski stated a better NEXPTIME result in [15]. Both results imply that the satisfiability problem is in 2-DEXPTIME. Plandowski's method is based on another recent result due to Rytter and Plandowski [18] showing that the minimal solution of a word equation is highly compressible in terms of Lempel-Ziv encodings. It is conjectured that

* This work was partially supported by the French German project PROCOPE

the length of a minimal solution is at most exponential in the denotational length of the equation.

In this contribution we only deal with quadratic systems, i.e., systems of word equations where each variable occurs at most twice. These systems are easier to handle. In combinatorial group theory these systems have been introduced by [8], see also [9]. They play an important rôle in the classification of closed surfaces and basic ideas of how to handle quadratic equations go back to [13]. The explicit statement of an algorithm for the solution of quadratic systems of word equations appears in [11].

The satisfiability problem of quadratic systems is still NP-hard (even for a single equation), but once the lengths of a possible solution are fixed, there is a deterministic linear time algorithm to decide whether there is a corresponding solution. A corollary is that if the lengths of a minimal solution of solvable quadratic systems were at most exponential, then the satisfiability problem would be NP-complete. In fact this is strongly conjectured. The conclusion of containment in NP follows also from [18], but the direct method yields a much simpler approach to the special situation of quadratic systems.

In the second part we address the problem with regular constraints. The uniform version is PSPACE-complete. We also show that fixing the lengths of a possible solution doesn't make the problem much easier. The non-uniform version remains NP-hard (in contrast to the linear time result above). The uniform version remains PSPACE-complete.

In the third part we have a closer look at exponent of periodicity for quadratic systems. This is the maximum power of a primitive word which can appear in a minimal solution. We show the exponent of periodicity is at most linear in the denotational length of quadratic systems whereas it can be exponential in general.

The first two parts of the paper are based on the extended abstract [17]. We have however decided to be complementary in the sense that omitted proofs there are given here and full proofs given there are omitted here. The third part on the exponent of periodicity is new. It uses ideas due to Anca Muscholl.

2 Quadratic equations

Let A be an alphabet of constants and let Ω be a set of variables. As usual, $(A \cup \Omega)^*$ means the free monoid over the set $A \cup \Omega$. A *word equation* $L = R$ is a pair $(L, R) \in (A \cup \Omega)^* \times (A \cup \Omega)^*$, and a *system* of word equations is a set of equations $\{L_1 = R_1, \dots, L_k = R_k\}$; its denotational length is defined as $|L_1 R_1 \cdots L_k R_k|$. A *solution* is a homomorphism $\sigma\colon (A \cup \Omega)^* \to A^*$ leaving the letters of A invariant such that $\sigma(L_i) = \sigma(R_i)$ for all $1 \leq i \leq k$. A solution $\sigma\colon \Omega \to A^*$ is called *minimal*, if the sum $\sum_{x \in \Omega} |\sigma(x)|$ is minimal. A system of word equations is called *quadratic*, if each variable occurs at most twice. In the present paper we consider only quadratic systems. The satisfiability problem for word equations is to decide whether a given system of word equations has a solution. Often, this problem is stated only for a single equation, since a system of word equations can be transformed into a single word equation. However this transformation multiplies the number of occurrences of variables. So, since we are dealing with quadratic systems, we cannot use this trick here.

The best known bound on the length of a minimal solution of a quadratic system is doubly exponential [11]. This seems to be quite an overestimation. We have the following conjecture.

Conjecture The length of a minimal solution of a solvable quadratic system of word equations is at most polynomial in the input size.

The value of Theorem 2 would already increase, if only the following much weaker conjecture were true.

Conjecture (weak form) The length of a minimal solution of a solvable quadratic system of word equations is at most exponential in the input size.

The conjectures above are supported by the fact that the exponent of periodicity is at most linear in the denotational length of the system. This result is stated in Theorem 5 below.

3 Complexity results

The first result states that the satisfiability problem of word equations remains NP-hard, even in the restricted case of a single quadratic equation.

Theorem 1. *Let $|A| \geq 2$. The following problem is NP-hard.*
INSTANCE: A quadratic word equation.
QUESTION: Is there a solution $\sigma\colon \Omega \longrightarrow A^*$?

Proof. We give a reduction from 3-SAT. Let $F = C_0 \wedge \cdots \wedge C_{m-1}$ be a propositional formula in 3-CNF over a set of variables Ξ. Each clause has the form

$$C_i = (\tilde{X}_{3i} \vee \tilde{X}_{3i+1} \vee \tilde{X}_{3i+2})$$

where the $\tilde{X}_j$ are literals. We can assume that every variable has both positive and negative occurrences.

First we construct a quadratic system of word equations using word variables

$$\begin{aligned} &c_i,\ d_i,\ 0 \leq i \leq m-1, \\ &x_j,\ 0 \leq j \leq 3m-1, \\ &y_X,\ z_X,\ \text{for each } X \in \Xi. \end{aligned}$$

We use the constants a, b, $a \neq b$. For each clause C_i we have two equations:

$$c_i x_{3i} x_{3i+1} x_{3i+2} = a^{3m} \quad \text{and} \quad c_i d_i = a^{3m-1}.$$

Now let $X \in \Xi$. Consider the set of positions $\{i_1, \ldots, i_k\}$ where $X = \tilde{X}_{i_1} = \cdots = \tilde{X}_{i_k}$ and the set of positions $\{j_1, \ldots, j_n\}$ where $\bar{X} = \tilde{X}_{j_1} = \cdots = \tilde{X}_{j_n}$. We deal with the case $k \leq n$; the case $n \leq k$ is symmetric. With each X we define two more equations:

$$y_X z_X = b \text{ and } x_{i_1} \cdots x_{i_k} y_X a^n b x_{j_1} \cdots x_{j_n} z_X = a^n b a^n b.$$

It is easy to see that the formula is satisfiable if and only if the quadratic system has a solution.

Next, a system of k word equations $L_1 = R_1, \ldots, L_k = R_k$, $k \geq 1$ with $R_1 \cdots R_k \in \{a, b\}^*$ is equivalent to a single equation temporarily using a third constant c: $L_1 c \cdots L_{k-1} c L_k = R_1 c \cdots R_{k-1} c R_k$. Finally, we can eliminate the use of the third letter c without increasing the number of occurrences of any variable by the well known technique of coding the three letters as aba, $abba$ and $abbba$.

The following theorem is the main result of [17]. In a slightly different form it appeared first in an unpublished manuscript of the second author [16].

Theorem 2. *There is a linear time algorithm to solve the following problem (on a unit cost RAM).*

INSTANCE: A quadratic system of word equations with a list of natural numbers $b_x \in \mathbb{N}$, $x \in \Omega$, written in binary.

QUESTION: Is there a solution $\sigma\colon \Omega \longrightarrow A^*$ such that $|\sigma(x)| = b_x$ for all $x \in \Omega$?

Proof. In a linear time preprocessing we can split the system into equations each containing a maximum of three variable occurrences: to see this let $x_1 \cdots x_g = x_{g+1} \cdots x_d$ be a word equation of the system with $1 \leq g < d$, $x_i \in A \cup \Omega$ for $1 \leq i \leq d$. Then the equation is equivalent to:

$$\begin{aligned} x_1 &= y_1, & x_{g+1} &= y_{g+1}, \\ y_1 x_2 &= y_2, & y_{g+1} x_{g+2} &= y_{g+2}, \\ &\vdots & &\vdots \\ y_{g-1} x_g &= y_g, & y_{d-1} x_d &= y_d, \end{aligned}$$
$$y_g = y_d.$$

Here $y_1, \ldots, y_d$ denote new variables, each of them occurring exactly twice. After the obvious simplification of equations with only one variable or constant on each side, we obtain a system where each equation has the form $z = xy$, $x, y, z \in A \cup \Omega$. In fact, using (for the first time) that the lengths b_x

are given, we may assume that $b_x \neq 0$ for all variables $x \in \Omega$ and that each equation has the form $z = xy$, where z is a variable. If m denotes the number of equations, then we can define the input size of a problem instance E as:

$$d(E) = m + \sum_{x \in \Omega} \log_2(b_x).$$

For $x \in A \cup \Omega$ let $|x| = b_x$ if $x \in \Omega$, and $|x| = 1$ if $x \in A$. We are looking for a solution σ such that $|\sigma(x)| = |x|$ for all $x \in A \cup \Omega$. A variable $z \in \Omega$ is called *doubly defined*, if E contains two equations $z = xy$ and $z = uv$. Let $dd(E)$ be the number of doubly defined variables. Define $c = 0.55$ and $k = 3/\ln(1/c)$ (≈ 5.01). Finally, we define the weight of the instance E as follows

$$W(E) = |\Omega| + dd(E) + k \sum_{x \in \Omega} \ln |x|.$$

We show in [17] a process which, in $\mathcal{O}(1)$ operations finds a simplification of the system reducing its weight by at least 1.

Remark 1. The method above yields a most general solution in the following sense. Let E be an instance to the problem of Theorem 2 and assume that E is solvable. Then we produce in linear time a quadratic system over a set of variables Γ (but without doubly defined variables) such that the set of solutions satisfying the length constraints is in a canonical one-to-one correspondence with the set of mappings $\psi : \Gamma \longrightarrow A^*$ where $|\psi(x)| = |b_x|$ for $x \in \Gamma$.

Corollary 1. *If the conjecture (weak form) above is true, then the satisfiability problem for quadratic systems of word equations is NP-complete.*

Remark 2. Given Theorem 1, Corollary 1 follows also from the work of Rytter and Plandowski [18]. They have shown that if the lengths $b_x, x \in \Omega$, are given in binary as part of the input together with a word equation (not necessarily quadratic), then there is a deterministic polynomial time algorithm for the satisfiability problem. Their method is based on Lempel-Ziv encodings and technically involved. Our contribution shows that the situation becomes much simpler for quadratic systems. In particular, we can reduce polynomial time to linear time; and our method is fairly straightforward using variable splitting. In view of the conjectures above it is not clear that the use of Lempel-Ziv encodings can improve the running time for deciding the satisfiability of quadratic systems. The most difficult part is apparently to get an idea of the lengths b_x for $x \in \Omega$. Once these lengths are known (or fixed), the corresponding satisfiability problem for quadratic systems of word equation becomes extremely simple.

4 Regular constraints

There is an interesting generalization of Makanin's result due to Schulz [19]. It says that if a word equation is given with a list of regular languages $L_x \subseteq A^*, x \in \Omega$, then one can decide whether there is a solution $\sigma : \Omega \longrightarrow A^*$ such that $\sigma(x) \in L_x$ for all $x \in \Omega$. In the following we shall assume that regular languages are specified by non-deterministic finite automata (NFA). In the uniform version the NFA are part of the input. In the non-uniform version the NFA are restricted so that each is allowed to have at most k states, where k is a fixed constant not part of the input. Using a recent result of Gutiérrez [5] one can show that the uniform satisfiability problem of word equations with regular constraints can be solved in EXPSPACE (more precisely in DSPACE($2^{\mathcal{O}(d^3)}$), if d denotes the input size), see [3]. So, from the general case it is not really clear whether adding regular constraints makes the satisfiability problem of word equations harder. We give here however some evidence that, indeed, it does. Restricted to quadratic systems the uniform satisfiability problem with regular constraints becomes PSPACE complete.

The non-uniform version is NP-hard and it remains NP-hard, even if the lengths $b_x, x \in \Omega$, are given in unary as part of the input. This is in sharp contrast to Theorem 2. Having regular constraints it is also easy to find examples where the length of a minimal solution increases exponentially.

Theorem 3. *The following problem is PSPACE-complete.*

INSTANCE: A quadratic system of word equations with a list of regular constraints $L_x \subseteq A^*$, $x \in \Omega$.

QUESTION: Is there a solution $\sigma\colon \Omega \longrightarrow A^*$ *such that* $\sigma(x) \in L_x$ *for all* $x \in \Omega$*?*

Moreover, the problem remains PSPACE-complete, if the input is given together with a list of numbers b_x, $x \in \Omega$ *(a number* $b \in \mathbb{N}$ *resp.), written in binary, and if we ask for a solution satisfying in addition the requirement* $|\sigma(x)| = b_x$ *(*$|\sigma(x)| = b$ *resp.) for all* $x \in \Omega$*?*

Proof. The PSPACE-hardness follows directly from a well-known result on regular sets. Let $L_1, \ldots, L_n$ be regular languages specified by NFA. Then the emptiness problem $L_1 \cap \cdots \cap L_n = \emptyset$ is PSPACE-complete, [7]. If the intersection is not empty, then there is a witness of at most exponential length. Let b be this upper bound on the length of a witness. Using a new letter c such that $c \notin A$, we can ask whether the intersection

$$L_1 c^* \cap \cdots \cap L_n c^*$$

contains a word of length b. (Instead of using a new letter we may also use some coding provided $|A| \geq 2$.) The quadratic system is given by n variables $x_1, \ldots, x_n$ and regular constraints $L_{x_i} = L_i c^*$ for $1 \leq i \leq n$. The equations are trivial:

$$x_1 = x_2,\ x_2 = x_3,\ \ldots\ ,\ x_{n-1} = x_n.$$

The PSPACE algorithm for the uniform satisfiability problem is a modification of the proof of Theorem 2. Let r_x be the number of states of the NFA specifying the language L_x. Let $r = \sum_{x \in \Omega} r_x$. Then there are a homomorphism $\varphi : A^* \longrightarrow \mathbb{B}^{r \times r}$, and vectors $I_x, F_x \in \mathbb{B}^r$ such that

$$w \in L_x \iff I_x^T \cdot \varphi(w) \cdot F_x = 1$$

We run the non-deterministic decision algorithm for satisfiability as usual. A variable $x \in \Omega$ is however represented as a pair (x, B_x) where $B_x \in \mathbb{B}^{r \times r}$. When we use a variable for the first time, we check in PSPACE that $B_x \in \varphi(A^*)$. A constant $a \in A$ is represented by the pair $(a, \varphi(a))$. Whenever we meet an equation $z = xy$, we check (in polynomial time) that $B_z = B_x \cdot B_y$.

Theorem 4. *Let $r \geq 4$ be a fixed constant, which is not part of the input. The following problem is NP-complete.*

INSTANCE: A quadratic system of word equations with a list of natural numbers $b_x \in \mathbb{N}$ written in binary, a list of regular constraints $L_x \subseteq A^*$, $x \in \Omega$, such that each language can be specified by some NFA of at most r states, and $|A| \geq 2$.

QUESTION: Is there a solution $\sigma : \Omega \longrightarrow A^*$ such that $|\sigma(x)| = b_x$ and $\sigma(x) \in L_x$ for all $x \in \Omega$?

Moreover, the problem remains NP-hard, if the numbers b_x, $x \in \Omega$, are written in unary, $|A| = 2$, and the system is a single equation.

Proof. The non-uniform problem is in NP, since a test whether a Boolean matrix B belongs to the image $\varphi(A^*)$ can now be performed in polynomial time (in fact, in constant time). Therefore the algorithm mentioned in the proof of Theorem 3 runs in non-deterministic polynomial time.

We have to show that the problem is NP-hard, even if the numbers $b_x, x \in \Omega$, are written in unary. First we use a reduction to a system of equations similar to that from 3-SAT in the proof of Theorem 1. Hence, we start with $F = C_0 \wedge \cdots \wedge C_{m-1}$ and each clause C_i has the form

$$C_i = (\tilde{X}_{3i} \vee \tilde{X}_{3i+1} \vee \tilde{X}_{3i+2})$$

We use word variables:

$$\begin{aligned} c_i &, 0 \leq i \leq m-1, \\ x_j &, 0 \leq j \leq 3m-1, \\ y_X, z_X &, \text{ for each } X \in \Xi. \end{aligned}$$

As above for $X \in \Xi$ let $\{i_1, \ldots, i_k\}$ be the set of positions where $X = \tilde{X}_{i_1} = \cdots = \tilde{X}_{i_k}$ and let $\{j_1, \ldots, j_n\}$ be the set of positions where $\bar{X} = \tilde{X}_{j_1} = \cdots = \tilde{X}_{j_n}$. We define the following equations:

$$\begin{aligned} c_i &= x_{3i} x_{3i+1} x_{3i+2}\,, \quad 0 \leq i \leq m-1 \\ y_X &= x_{i_1} \cdots x_{i_k} z_X x_{j_1} \cdots x_{j_n}\,, \quad X \in \Xi. \end{aligned}$$

The lengths are given by $|c_i| = 3$, $0 \le i \le m-1$, $|x_j| = 1$, $0 \le j \le 3m-1$, $|z_X| = 2$ and $|y_X| = 2+k+n$ for every $X \in \Xi$. The regular constraints are from the following list: $A^*, A^*aA^*, ab \cup ba$, and $a^*b^* \cup b^*a^*$. Allowing several initial states we need at most 4 states for each language. That is why we have chosen $r \ge 4$. Finally, the regular constraints are:

$$\begin{aligned}
&c_i \in A^*aA^* \,, \quad 0 \le i \le m-1, \\
&x_j \in A^* \,, \quad 0 \le j \le 3m-1, \\
&z_X \in ab \cup ba \,, \\
&y_X \in a^*b^* \cup b^*a^* \,.
\end{aligned}$$

Again it is straightforward to see that the word equations have a solution satisfying the constraints if and only if the expression is satisfiable.

The final reduction from a system of equations to a single equation over a binary alphabet is simpler using regular constraints than the one sketched in the proof of Theorem 1. We omit the details.

5 On the exponent of periodicity

Let $E = \{L_1 = R_1, \cdots, L_k = R_k\}$ be a system of word equations. The *exponent of periodicity* of E is defined by

$$\sup\{\alpha \in \mathbb{N} \mid \sigma(x) = up^\alpha v,\ \sigma \text{ is a minimal solution},\ x \in \Omega,\ p \ne \epsilon\}.$$

Example 1. Consider the following quadratic system

$$(x_i = ax_{i-1})_{2 \le i \le n}.$$

Its exponent of periodicity is $n-1$.

It is well-known that the exponent of periodicity is at most singly exponential in the denotational length of the system, see [6]. For quadratic systems we will show a linear bound. A polynomial bound has been proposed by Anca Muscholl and is stated in [1]. Our approach is slightly different and yields the following result.

Theorem 5. *Let E be a quadratic system of word equations. Then the exponent of periodicity is at most linear in the denotational length.*

Proof. The first part is a transformation of a system of word equations to a linear Diophantine system. It relies on the notion of a p-stable normal form and was already used in the original paper by Makanin [10]. This is standard and also explained in detail in [3]. Therefore we only sketch this part:

We may assume that all equations are of the form $z = xy$ with $x,\ y,\ z \in (A \cup \Omega)$. Consider a minimal solution σ and a primitive word $p \in A^+$ such that $\sigma(x) = up^\alpha v$ for some $x \in \Omega$, $u, v \in A^*$ and where α is the exponent of periodicity of the system E. Every word $u \in A^*$ can be written in its p-stable normal form. This is a factorization

$$u = u_0 p^{\alpha_1} u_1 \cdots p^{\alpha_k} u_k$$

such that k is minimal and, if p^2 is a factor of u, then $k > 0$, $u_0 \in A^*p \setminus A^*p^2A^*$, $u_1, \ldots, u_{k-1} \in (A^*p \cap pA^*) \setminus A^*p^2A^*$ and $u_k \in pA^* \setminus A^*p^2A^*$. As the name suggests, the p-stable normal form is unique. Now, an equation $z = xy$ yields an equation $\sigma(z) = \sigma(x)\sigma(y)$. We obtain an equation on the p-stable normal forms:

$$w_0 p^{\gamma_1} w_1 \cdots p^{\gamma_k} w_k p^{\gamma_{k+1}} \cdots p^{\gamma_m} w_m = u_0 p^{\alpha_1} u_1 \cdots p^{\alpha_k} u_k v_0 p^{\beta_1} v_1 \cdots p^{\beta_l} v_l.$$

Reading the exponents α_i, β_i, γ_i as integer variables, we obtain a number of trivial equations $\alpha_1 = \gamma_1$, $\alpha_2 = \gamma_2$, $\cdots$ and at most 2 non-trivial equations. For instance we might have $\gamma_k = \alpha_k + 1$, $\gamma_{k+1} = \beta_1 + 1$. Another possibility is $\gamma_k = \alpha_k + \beta_1 + 3$. In fact, as the reader can easily verify there are nine cases and which one appears depends only on the p-stable normal form of $u_k v_0$. The two cases just mentioned are so to speak *worst cases.*

Putting together the equations derived from each word equation, we obtain a (perhaps huge) linear Diophantine system where every non-zero coefficient is 1 and every variable occurs at most twice. There are $\mathcal{O}(d)$ non-trivial equations and the constant in each equation has absolute value at most 3. Moreover σ provides us with a positive solution and this solution is minimal among all positive solutions (because σ is a minimal solution of E). The trivial equations such as $\alpha_i = \gamma_i$ can be eliminated leaving us with a linear Diophantine system with a rather special structure: there are at most m equations and n variables $X = \{x_1, \cdots, x_n\}$ where m, $n \in \mathcal{O}(d)$. We can write each equation as $\sum a_{i,j} x_j = b_i$ where $|b_i| \leq 3$ and $\sum_i |a_{i,j}| \leq 2$ for each j. This is just a system of 2-type as introduced in the next section. The exponent of periodicity is bounded by the maximum value occurring in all minimal solutions of this system. Therefore Theorem 5 is a direct consequence of Theorem 6 which will be stated and shown in the next section.

Remark 3. The statement of Theorem 5 can be generalized in order to include regular constraints by the same techniques as in [3]. This is left to the interested reader.

6 Linear Diophantine Equations of 2-type

Let $X = \{x_1, \cdots, x_n\}$ be a set of integer variables. A system of linear Diophantine equations is specified by an integer matrix $A = (a_{i,j}) \in \mathbb{Z}^{m \times n}$ and a vector $b \in \mathbb{Z}^m$. Without restriction we have $m = n$. A solution (resp. positive solution) is a vector $x \in \mathbb{Z}^n$ (resp. $x \in \mathbb{N}^n \setminus \{0\}$) such that $Ax = b$.

A positive solution $x \in \mathbb{N}^n \setminus \{0\}$ is minimal if, for all other solutions $x' \in \mathbb{N}^n \setminus \{0\}$, there is at least one component j such that $x_j < x'_j$. The system is called here of *2-type* if for all $1 \leq j \leq k$ we have $\sum_i |a_{i,j}| \leq 2$. We consider only systems of 2-type. These systems correspond to quadratic systems of word equations. The aim of this section is to give an upper bound on the size of all minimal positive solutions. The size of a vector $z \in \mathbb{Z}^n$ is its *norm* defined by

$$||z|| = \sum_{1 \leq i \leq n} |z_i|.$$

Lemma 1. *Every system of 2-type $Ax = b$ with a non-zero integer solution has a non-zero integer solution where $|x_\ell| \leq \max\{2,\ ||b||\}$ for all $1 \leq \ell \leq n$.*

Proof. We give an algorithm to find such a solution. The algorithm works in two phases, *elimination* and *solution*. The elimination phase produces a *core* system of equations in the uneliminated variables such that no variable occurs in two equations and a *defining* equation for each eliminated variable.

Then the solution phase finds a solution to the core system and extends it to the eliminated variables by iterated substitution in the defining equations. As a preliminary step, we replace each $b_i \neq 0$ by a new variable x_{n+i}.

The elimination phase proceeds in stages, each of which eliminates one variable. Initially all equations are in the core, no variables are eliminated, and all variables occur in some equation. If no variable occurs in two equations of the core, then the phase is complete; otherwise suppose that x_i occurs in equations j and j'. It follows that $i \leq n$, and we must have $|\alpha_{i,j}| = |\alpha_{i,j'}| = 1$ so we can rewrite these two equations in the form $x_i = \sum_{k \neq i} \alpha_{j,k} x_k$ and $x_i = \sum_{k \neq i} \alpha_{j',k} x_k$. We replace these two equations in the core by $\sum_{k \neq i}(\alpha_{j,k} - \alpha_{j',k}) x_k = 0$, eliminate x_i and give x_i the defining equation $x_i = \sum_{k \neq i} \alpha_{j,k} x_k$.

The elimination phase produces a set of equations (core and defining) equivalent to the initial system. Now the solution phase starts by replacing the new variables x_{n+i} by the original constants b_i in the core and in the defining equations. The final core equations now have the form

$$\sum_{k \in I} \alpha_k x_k + \sum_{\ell \in J} \beta_\ell x_\ell = b'$$

where $|\alpha_k| = 1$, $|\beta_\ell| \in \{0,\ 2\}$, $|b'| \leq ||b||$, and I and J are subsets of uneliminated variables of $\{1, \ldots, n\}$. (Note that we might have $b' = 0$ although $||b|| \neq 0$, but not vice versa.) The final core equations can be solved independently. Since the system has a solution, each equation can be solved in such a way that $\sum_{k \in I} |x_k| + \sum_{\ell \in J} 2|x_\ell| \leq |b'| \leq ||b||$. Moreover, if $||b|| = 0$, then, since the system has a non-zero solution, at least one core equation has a non-zero solution; for any such equation we have a solution with $\sum_{k \in I} |x_k| + \sum_{\ell \in J} 2|x_\ell| \leq 4$. Hence in all cases, there is a non-zero solution satisfying

$$\sum_{k \in I} |x_k| + \sum_{\ell \in J} 2|x_\ell| + ||b|| \leq \max\{4,\ 2||b||\}.$$

We define for each uneliminated variable x_k and each core equation i a weight $w_{i,k}$. In the initial system $w_{i,k} = |\alpha_{i,k}|$. For the replacement equation obtained in an elimination stage (to have a name let's call the replacement equation r) we define $w_{r,k}$ as $w_{j,k} + w_{j',k}$. Thus we always have $|\alpha_{i,k}| \leq w_{i,k} \leq 2$ and $w_{i,k} = 1$, if $|\alpha_{i,k}| = 1$, for every equation and uneliminated variable.

Claim. For every solution $x \in \mathbb{Z}^n$ and at every stage of the elimination phase, for every ℓ, there is a core equation j such that $2|x_\ell| \leq \sum_k w_{j,k}|x_k|$.

The claim is clearly true for all uneliminated variables since a positive $w_{j,\ell}$ occurs in at least one core equation and either $w_{j,\ell} = 2$ (and the claim becomes trivial) or $w_{j,\ell} = 1$ and then we have

$$|x_\ell| = |\sum_{k \neq \ell} \alpha_{j,k} x_k| \leq \sum_{k \neq \ell} |\alpha_{j,k}||x_k| \leq \sum_{k \neq \ell} w_{j,k}|x_k|.$$

In particular the claim is true for the initial system. Now consider a stage eliminating x_i and replacing equations j and j' by the new core equation r. As can be seen by the defining equation for x_i we have $2|x_i| \leq \sum_{k \neq i} w_{r,k}|x_k|$. So the claim is trivial for i. Thus, the elimination stage can affect the truth of the claim, only if it removes the core equation j satisfying the claim for x_ℓ and $\ell \neq i$. We can make the following calculation:

$$\begin{aligned}
2|x_\ell| &\leq \textstyle\sum_k w_{j,k}|x_k| && \text{before the elimination}\\
&\leq (\textstyle\sum_{k \neq i} w_{j,k}|x_k|) + |x_i| && \text{because } w_{j,i} = 1\\
&\leq (\textstyle\sum_{k \neq i} w_{j,k}|x_k|) + (\textstyle\sum_{k \neq i} \alpha_{j',k}|x_k|) &&\\
&= \textstyle\sum_{k \neq i} (w_{j,k} + \alpha_{j',k})|x_k| &&\\
&\leq \textstyle\sum_{k \neq i} (w_{j,k} + w_{j',k})|x_k| &&\\
&= \textstyle\sum_k w_{r,k}|x_k| && \text{after the elimination.}
\end{aligned}$$

This completes the proof of the claim. Finally the assertion of the lemma follows since for the final core equations we have constructed a solution (yielding a non-zero solution of the initial system) such that

$$2|x_\ell| \leq \sum_k w_{r,k}|x_k| \leq \sum_{k \in I} |x_k| + \sum_{\ell \in J} 2|x_\ell| + ||b|| \leq \max\{4, 2||b||\}.$$

Lemma 2. *Let* $\det(A) \neq 0$. *If* $x \in \mathbb{Z}^n$ *solves* $Ax = b$, *then for all* $1 \leq \ell \leq n$ *we have* $|x_\ell| \leq \max\{2, \ ||b||\}$. *In particular, if* $||b|| \in \mathcal{O}(n)$, *then* $|x_\ell| \in \mathcal{O}(n)$.

Proof. Immediate by Lemma 1 since the solution is unique.

Lemma 3. *Let* $\det(A) = 0$. *Then the homogeneous system* $Ax = 0$ *has a non-zero solution* $x \in \mathbb{Z}^n$ *with* $|x_\ell| \leq 2$ *for all* $1 \leq \ell \leq n$.

Proof. Since the determinant is zero, there is a non-zero integer solution and we use Lemma 1.

Theorem 6. *Let $x \in \mathbb{N}^n \setminus \{0\}$ be a minimal solution of $Ax = b$, $1 \leq \ell \leq n$ and $\|b\| \in \mathcal{O}(n)$. Then $|x_\ell| \in \mathcal{O}(n)$.*

Proof. For $\det(A) \neq 0$ this is clear by Lemma 2. Hence we assume that $\det(A) = 0$. Consider now first the case that $Ax = 0$ has a positive solution with all components at most 2. Since the given solution is minimal, we cannot subtract this positive solution from it and remain in $\mathbb{N}^n \setminus \{0\}$; hence it has some component x_j with value v_j less than or equal to 2. We replace x_j by v_j to obtain a new system $A'x = b'$ for which $(x_1, \cdots, x_{j-1}, x_{j+1}, \cdots x_n)$ is still minimal and $\|b'\| \leq \|b\| + 4$. Repeating as long as $A'x = 0$ has a positive solution with all components at most 2, we eventually are done or we reach a system $A'x = b'$ with $\|b'\| \in \mathcal{O}(n)$ for which all positive solutions of $A'x = 0$ have a component at least 3. Again, if $\det(A') \neq 0$, Lemma 2 gives the result. So we assume $\det(A') = 0$. By Lemma 3 there is still a non-zero solution z of $A'z = 0$ where the absolute value of all components is at most 2. If necessary replacing z by $-z$, we have $z_\ell \geq 0$. There is a maximum non-negative integer λ such that $x' = x + \lambda z$ is a positive solution of $A'x = b'$. The solution x' has some component (say x'_j) with value v_j at most 2 and $x'_\ell \geq x_\ell$. (It might happen that x' is not minimal anymore.) Again we replace x_j by v_j to get a new system where no positive solution of $A'x = 0$ has all components at most 2. (Because otherwise the old system had such a solution with $v_j = 0$.) Repeating this process, we eventually reach a system $A''x = b''$ with $\|b''\| \in \mathcal{O}(n)$ and $\det(A'') \neq 0$ and having a solution x'' with $x''_\ell \geq x_\ell$. Thus using Lemma 2 we have a $\mathcal{O}(n)$ bound on x''_ℓ and so on x_ℓ completing the proof.

The following example is the Diophantine version of Example 1 showing that our bounds are essentially optimal.

Example 2. Consider the following linear Diophantine system of 2-type

$$(x_i - x_{i-1} = 1)_{2 \leq i \leq n}.$$

Its minimal positive solution is $x = (0, 1, \ldots, n-1)$.

References

1. Thierry Arnoux. Untersuchungen zum Makaninschen Algorithmus. Diplomarbeit 1613, Universität Stuttgart, 1998.
2. Christian Choffrut and Juhani Karhumäki. Combinatorics of words. In G. Rozenberg and A. Salomaa, editors, *Handbook of Formal Languages*, volume 1, pages 329–438. Springer-Verlag, Berlin, Heidelberg, New York, 1997.
3. Volker Diekert. Makanin's Algorithm. In M. Lothaire: *Algebraic Combinatorics on Words*. Cambridge University Press. A preliminary version is on the web: http://www-igm.univ-mlv.fr/ berstel/Lothaire/index.html.

4. Volker Diekert, Yuri Matiyasevich, and Anca Muscholl. Solving trace equations using lexicographical normal forms. In P. Degano et al., editors, *Proc. 24th ICALP, Bologna (Italy) 1997*, number 1256 in Lect. Notes Comp. Sci., pages 336–347. Springer-Verlag, Berlin, Heidelberg, New York, 1997.
5. Claudio Gutiérrez. Satisfiability of word equations with constants is in exponential space. In *Proc. 39th FOCS*, pages 112–119, Los Alamitos, California (USA), 1998. IEEE Computer Society Press.
6. Antoni Kościelski and Leszek Pacholski. Complexity of Makanin's algorithm. *J. Assoc. Comput. Mach.*, 43(4):670–684, 1996. Preliminary version in *Proc. 31st FOCS*, Los Alamitos, California (USA), 1990.
7. Dexter Kozen. Lower bounds for natural proof systems. In *Proc. 18th FOCS, Providence, Rhode Island (USA)*, pages 254–266. IEEE Computer Society Press, 1977.
8. Roger C. Lyndon. Equations in free groups. *Transactions of the American Mathematical Society*, 96, 1960.
9. Roger C. Lyndon and Paul E. Schupp. *Combinatorial Group Theory*. Springer-Verlag, Berlin, Heidelberg, New York, 1977.
10. Gennadií S. Makanin. The problem of solvability of equations in a free semigroup. *Mat. Sb.*, 103(2):147–236, 1977. In Russian; English translation in: *Math. USSR Sbornik, 32*, 129–198, 1977.
11. Yuri Matiyasevich. A connection between systems of word and length equations and Hilbert's Tenth Problem. *Sem. Mat. V. A. Steklov Math. Inst. Leningrad*, 8:132–144, 1968. In Russian; English translation in: *Seminars in Mathematics, V. A. Steklov Mathematical Institute, 8*, 61–67, 1970.
12. Yuri Matiyasevich. Some decision problems for traces. In S. Adian et al., editors, *Proc. 4th LFCS, Yaroslavl (Russia) 1997*, number 1234 in Lect. Notes Comp. Sci., pages 248–257. Springer-Verlag, Berlin, Heidelberg, New York, 1997. Invited lecture.
13. Jakob Nielsen. Die Isomorphismen der allgemeinen, unendlichen Gruppe mit zwei Erzeugenden. *Mathematische Annalen*, 78, 1918.
14. Dominique Perrin. Equations in words. In H. Ait-Kaci et al., editors, *Resolution of equations in algebraic structures*, volume 2, pages 275–298. Academic Press, 1989.
15. Wojciech Plandowski. Satisfiability of word equations with constants is in NEXPTIME. In *Proc. 31st STOC, Atlanta, Georgia (USA)*, 1999. To appear.
16. John Michael Robson. Word equations with at most 2 occurrences of each variable. Preprint, LaBRI, Université de Bordeaux I, 1998.
17. John Michael Robson and Volker Diekert. On quadratic word equations. In Ch. Meinel et al., editors, *Proc. 16th STACS, Trier (Germany)*, Lect. Notes Comp. Sci. Springer-Verlag, Berlin, Heidelberg, New York, 1999. To appear.
18. Wojciech Rytter and Wojciech Plandowski. Application of Lempel-Ziv encodings to the solution of word equations. In K. G. Larsen et al., editors, *Proc. 25th ICALP, Aalborg (Denmark)*, number 1443 in Lect. Notes Comp. Sci., pages 731–742. Springer-Verlag, Berlin, Heidelberg, New York, 1998.
19. Klaus U. Schulz. Makanin's algorithm for word equations: Two improvements and a generalization. In K.-U. Schulz, editor, *Proc. IWWERT'90, Tübingen (Germany)*, number 572 in Lect. Notes Comp. Sci., pages 85–150. Springer-Verlag, Berlin, Heidelberg, New York, 1992.

Fair and Associative Infinite Trajectories

Alexandru Mateescu and George Daniel Mateescu

Summary. We study several sets of ω-trajectories that have the following properties: each of them defines an associative and commutative operation of ω-words and, moreover, each of them satisfies a certain condition of fairness.

1 Preliminaries

This paper continues our investigations on the operation of shuffle on trajectories of ω-words and ω-languages, see [3] and [2]. We introduce and investigate several sets of ω-trajectories such that each of these sets satisfies two important features: the shuffle operation associated to the set is associative and it fulfills a certain condition of fairness. Both conditions are important: associativity ensures that the set of ω-words has a structure of semiring where the product is defined by the shuffle on this set of ω-trajectories, whereas the fairness condition ensures good properties for practical use of this parallel composition operation in parallel computations.

The shuffle-like operations considered below are defined using the notion of the ω-trajectory. A first approach of these shuffle-like operations was considered in [4]. An ω-trajectory defines the general strategy to switch from one ω-word to another ω-word. Each set T of ω-trajectories defines in a natural way a shuffle operation over T. Given a set T of ω-trajectories the operation of shuffle over T is not necessarily an associative operation. However, for each set T there exists a smallest set of trajectories $\overline{T}$ such that $\overline{T}$ contains T and, moreover, shuffle over $\overline{T}$ is an associative operations.

The set of nonnegative integers is denoted by ω. If A is a set, then the set of all subsets of A is denoted by $\mathcal{P}(A)$. Let Σ be an alphabet, i.e., a finite nonempty set of elements called *letters*. The free monoid generated by Σ is denoted by Σ^*. Elements in Σ^* are referred to as *words*. The *empty word* is denoted by λ. If $w \in \Sigma^*$, then $|w|$ denotes the length of w. Note that $|\lambda| = 0$. If $a \in \Sigma$ and $w \in \Sigma^*$, then $|w|_a$ denotes the number of occurrences of the symbol a in the word w. The *mirror* of a word $w = a_1a_2 \dots a_n$, where a_i are letters, $1 \leq i \leq n$, is $mi(w) = a_n \dots a_2a_1$ and $mi(\lambda) = \lambda$. A word w is a *palindrome* iff $mi(w) = w$.

An *ω-word* over Σ is a function $f : \omega \longrightarrow \Sigma$. Usually, the ω-word defined by f is denoted as the infinite sequence $f(0)f(1)f(2)\dots$. An ω-word w is *ultimately periodic* iff $w = \alpha vvvvv\ldots$, where α is a (finite) word, possibly empty, and v is a nonempty word. In this case w is denoted as αv^ω. The set of all ultimately periodic ω-words over Σ is denoted by $UltPer(\Sigma)$ or $UltPer$. An ω-word w is referred to as *periodic* iff $w = vvv\ldots$ for some nonempty word $v \in \Sigma^*$. In this case w is denoted as v^ω. The set of all periodic ω-words over Σ is denoted by $Per(\Sigma)$ or Per. Moreover, the set of all periodic ω-words over Σ that have a palindrome as their period is denoted by $PalPer(\Sigma)$ or $PalPer$.

The set of all ω-words over Σ is denoted by Σ^ω. An ω-*language* is a subset L of Σ^ω, i.e., $L \subseteq \Sigma^\omega$. Let w be an ω-word. The set of all (finite) prefixes of w is denoted by $Pref(w)$. The reader is referred to [5] for general results on ω-words.

The *shuffle* operation, denoted by $\sqcup\!\sqcup$, is defined recursively by:

$$(ax \sqcup\!\sqcup by) = a(x \sqcup\!\sqcup by) \cup b(ax \sqcup\!\sqcup y) \qquad \text{and} \qquad x \sqcup\!\sqcup \lambda = \lambda \sqcup\!\sqcup x = \{x\},$$

where $x,\ y \in \Sigma^*$ and $a,\ b \in \Sigma$.

The above operation is extended in a natural way to languages.

2 Shuffle on ω-trajectories

In this section we introduce the notions of the ω-trajectory and shuffle on ω-trajectories. Firstly, we define the shuffle of (finite) words on (finite) trajectories. Let $V = \{r, u\}$ be the set of *versors* in the plane: r stands for the *right* direction, whereas, u stands for the *up* direction.

Definition 1. A *trajectory* is an element t, $t \in V^*$.

Let Σ be an alphabet and let t be a (finite) trajectory, let d be a versor, $d \in V$, let α, β be two (finite) words over Σ.

Definition 2. The shuffle of α with β on the trajectory dt, denoted $\alpha \sqcup\!\sqcup_{dt} \beta$, is recursively defined as follows:

if $\alpha = ax$ and $\beta = by$, where $a, b \in \Sigma$ and $x, y \in \Sigma^*$, then:

$$ax \sqcup\!\sqcup_{dt} by = \begin{cases} a(x \sqcup\!\sqcup_t by), & \text{if } d = r, \\ b(ax \sqcup\!\sqcup_t y), & \text{if } d = u. \end{cases}$$

if $\alpha = ax$ and $\beta = \lambda$, where $a \in \Sigma$ and $x \in \Sigma^*$, then:

$$ax \sqcup\!\sqcup_{dt} \lambda = \begin{cases} a(x \sqcup\!\sqcup_t \lambda), & \text{if } d = r, \\ \emptyset, & \text{if } d = u. \end{cases}$$

if $\alpha = \lambda$ and $\beta = by$, where $b \in \Sigma$ and $y \in \Sigma^*$, then:

$$\lambda \sqcup\!\sqcup_{dt} by = \begin{cases} \emptyset, & \text{if } d = r, \\ b(\lambda \sqcup\!\sqcup_t y), & \text{if } d = u. \end{cases}$$

Finally,

$$\lambda \sqcup\!\sqcup_t \lambda = \begin{cases} \lambda, & \text{if } t = \lambda, \\ \emptyset, & \text{otherwise.} \end{cases}$$

Comment. Note that if $|\alpha| \neq |t|_r$ or $|\beta| \neq |t|_u$, then $\alpha \sqcup\!\sqcup_t \beta = \emptyset$.

Now we define the operation of shuffle of ω-words on ω-trajectories.

Definition 3. An ω-trajectory is an ω-word t over V, i.e., $t \in V^\omega$.

Let α, β be ω-words over Σ. Let t be an ω-trajectory, $t \in V^\omega$.

Definition 4. The shuffle of α with β on the ω-trajectory t is defined as the limit of the sequence $(\alpha' \sqcup\!\sqcup_{t'} \beta')_{t' \in Pref(t)}$, where $\alpha' \in Pref(\alpha)$, $\beta' \in Pref(\beta)$ such that $|\alpha'| = |t'|_r$ and $|\beta'| = |t'|_u$.

One can easily verify that the sequence $(\alpha' \sqcup\!\sqcup_{t'} \beta')_{t' \in Pref(t)}$ has always a limit.

Let Σ be an alphabet and let L_1, L_2 be ω-languages over Σ, i.e., $L_1, L_2 \subseteq \Sigma^\omega$. If T is a set of ω-trajectories, the *shuffle of* L_1 *with* L_2 *on the set* T *of* ω*-trajectories*, denoted $L_1 \sqcup\!\sqcup_T L_2$, is:

$$L_1 \sqcup\!\sqcup_T L_2 = \bigcup_{\alpha \in L_1, \beta \in L_2, t \in T} \alpha \sqcup\!\sqcup_t \beta..$$

3 Associativity and commutativity

The results in this section deal with associativity and commutativity. After a few general remarks, we restrict the attention to the set V_+^ω of ω-trajectories t such that both r and u occur infinitely often in t. (It will become apparent below why this restriction is important.) It turns out that associativity can be viewed as stability under four particular operations, referred to as $\diamond$-operations.

Definition 5. A set T of ω-trajectories is *associative* iff the operation $\sqcup\!\sqcup_T$ is associative. A set T of ω-trajectories is *commutative* iff the operation $\sqcup\!\sqcup_T$ is commutative.

Remark 1. Let $sym : V \longrightarrow V^*$ be the morphism defined as: $sym(r) = u$ and $sym(u) = r$.

It is easy to see that a set T of ω-trajectories is commutative iff $T = sym(T)$.

The following sets of ω-trajectories are associative:

$T_1 = \{r, u\}^\omega$, $T_2 = \{t \in V^\omega \mid |t|_r < \infty\}$, $T_3 = \{\alpha_0\beta_0\alpha_1\beta_1 \ldots \mid \alpha_i, \beta_i$ are of even length, and $\alpha_i \in r^*, \beta_i \in u^*, i \geq 0\}$.

Nonassociative sets of ω-trajectories are for instance:

$T_4 = \{(ru)^\omega\}$, $T_5 = \{t \in V^\omega \mid t$ is a Sturmian ω-word $\}$, $T_6 = \{w_0w_1w_2 \ldots \mid w_i \in L\}$, where $L = \{r^n u^n \mid n \geq 0\}$.

Notation. Let $\mathcal{A}$ be the family of all associative sets of ω-trajectories.

Proposition 1. *If $(T_i)_{i \in I}$ is a family of associative sets of ω-trajectories, then*

$$T' = \bigcap_{i \in I} T_i ,$$

is an associative set of ω-trajectories.

Definition 6. Let T be an arbitrary set of ω-trajectories. The *associative closure* of T, denoted $\overline{T}$, is

$$\overline{T} = \bigcap_{T \subseteq T', T' \in \mathcal{A}} T'.$$

Observe that for all $T, T \subseteq \{r, u\}^\omega$, $\overline{T}$ is an associative set of ω-trajectories and $\overline{T}$ is the smallest associative set of ω-trajectories that contains T.

Notation. Let V_+^ω be the set of all ω-trajectories $t \in V^\omega$ such that t contains infinitely many occurrences both of r and of u.

We present now another characterization of an associative set of ω-trajectories from V_+^ω. However, this characterization is valid only for sets of ω-trajectories from V_+^ω and not for the general case, i.e., not for sets of ω-trajectories from V^ω.

Definition 7. Let W be the alphabet $W = \{x, y, z\}$ and consider the following four morphisms, ρ_i, $1 \leq i \leq 4$, where $\rho_i : W \longrightarrow V_+^\omega$, $1 \leq i \leq 4$, and

$$\rho_1(x) = \lambda \ , \quad \rho_1(y) = r \ , \quad \rho_1(z) = u, \quad \rho_2(x) = r \ , \quad \rho_2(y) = u \ , \quad \rho_2(z) = u,$$

$$\rho_3(x) = r \ , \quad \rho_3(y) = u \ , \quad \rho_3(z) = \lambda, \quad \rho_4(x) = r \ , \quad \rho_4(y) = r \ , \quad \rho_4(z) = u.$$

Next, we consider four operations on the set of ω-trajectories, V_+^ω.

Definition 8. Let $\diamond_i$, $1 \leq i \leq 4$, be the following operations on V_+^ω.

$$\diamond_i : V_+^\omega \times V_+^\omega \longrightarrow V_+^\omega \ , \qquad 1 \leq i \leq 4,$$

Let t, t' be in V_+^ω. By definition:

$$\diamond_1(t, t') = \rho_1((x^\omega \sqcup\!\sqcup_t y^\omega) \sqcup\!\sqcup_{t'} z^\omega), \qquad \diamond_2(t, t') = \rho_2((x^\omega \sqcup\!\sqcup_t y^\omega) \sqcup\!\sqcup_{t'} z^\omega),$$

$$\diamond_3(t', t) = \rho_3(x^\omega \sqcup\!\sqcup_{t'} (y^\omega \sqcup\!\sqcup_t z^\omega)), \qquad \diamond_4(t', t) = \rho_4(x^\omega \sqcup\!\sqcup_{t'} (y^\omega \sqcup\!\sqcup_t z^\omega)).$$

The following theorem was firstly proved in [2].

Theorem 1. *Let T be a set of ω-trajectories, $T \subseteq V_+^\omega$.*

T is an associative set of ω-trajectories. iff T is closed at $\diamond$-operations, i.e., if $t_1, t_2 \in T$, then $\diamond_i(t_1, t_2) \in T$, $1 \leq i \leq 4$.

Remark 2. We restricted our attention only to the set V_+^ω and not to the general case V^ω. However, if T contains a trajectory t that is not in V_+^ω, then $\diamond_1(t,t)$ is not necessarily in V^ω.

Using the above theorem one can easily prove:

Proposition 2. *The following sets of ω-trajectories are associative:*
(i) the set of all periodic ω-trajectories from V_+^ω.
(ii) the set of all periodic ω-trajectories from V_+^ω that have as their period a palindrome.
(iii) the set of all ultimately periodic ω-trajectories from V_+^ω.

4 ω-Trajectories with bounded increase

Definition 9. The trajectory $t = r^{i_1}u^{j_1}r^{i_2}u^{j_2}...r^{i_n}u^{j_n}... \in V_+^\omega$ has a bounded increase iff the sequences $\{i_n\}_{n\geq 1}$ and $\{j_n\}_{n\geq 1}$ are bounded, i.e., there are k_1 and k_2 such that $i_n \leq k_1$ and $j_n \leq k_2$, for all $n \geq 1$.

Notation. $BInc$ denotes the set of all ω-trajectories with bounded increase.

Proposition 3. *An ω-trajectory t has a bounded increase iff there exist two constants c_1 and c_2 such that:*
B1 $|pref_p(t)|_u - |pref_q(t)|_u \leq c_1(|pref_p(t)|_r - |pref_q(t)|_r)$, *whenever* $|pref_p(t)|_r - |pref_q(t)|_r > 0$, *and*
B2 $|pref_p(t)|_r - |pref_q(t)|_r \leq c_2(|pref_p(t)|_u - |pref_q(t)|_u)$, *whenever* $|pref_p(t)|_u - |pref_q(t)|_u > 0$.

Comment. Note that an ω-trajectory t is in $BInc$ iff there are two constants $k_1 > 0$ and $k_2 > 0$ such that, during the parallel composition of two processes on t, after at most k_1 occurrences of actions from the first process, it occurs at least one occurrence of an action from the second process and, after at most k_2 occurrences of actions from the second process, it occurs at least one occurrence of an action from the first process.

Theorem 2. *The set of all ω-trajectories with bounded increase, i.e., $BInc$, is an associative and commutative set of ω-trajectories.*

Proof. First we prove that $BInc$ is a commutative set of ω-trajectories. To see this let t be an ω-trajectory and $\tau = sym(t)$. It is easy to observe that t satisfies **B1** iff τ satisfies **B2** and t satisfies **B2** iff τ satisfies **B1** which proves that $BInc$ is a commutative set of ω-trajectories.

Let t_1, t_2 be two ω-trajectories that satisfy the conditions **B1** and **B2**. First we prove that $\diamond_1(t_1,t_2)$ satisfies also conditions **B1** and **B2**. In the sequel we will use the notation of constants as c_1^1, c_2^1 for t_1 and c_1^2, c_2^2 for t_2.

Let τ be the value of $\diamond_1(t_1,t_2)$. From the definition of the operation $\diamond_1$, it follows that $|pref(\tau)|_r = |pref_{|pref_n(t_2)|_r}(t_1)|_u$ and $|pref(\tau)|_u = |pref_n(t_2)|_u$ for some n.

In order to prove that the condition **B1** is fulfilled, assume that

$$|pref_{|pref_n(t_2)|_r}(t_1)|_u > |pref_{|pref_m(t_2)|_r}(t_1)|_u \tag{1}$$

It follows that $|pref_n(t_2)|_r > |pref_m(t_2)|_r$. Since t_2 also satisfies condition **B1**, we deduce that:

$$|pref_n(t_2)|_u - |pref_m(t_2)|_u \leq c_1^2(|pref_n(t_2)|_r - |pref_m(t_2)|_r) \tag{2}$$

Using again relation (1) and the fact that t_1 also satisfies **B2**, it follows that:

$$|pref_{|pref_n(t_2)|_r}(t_1)|_r - |pref_{|pref_m(t_2)|_r}(t_1)|_r \leq$$
$$\leq c_2^1(|pref_{|pref_n(t_2)|_r}(t_1)|_u - |pref_{|pref_m(t_2)|_r}(t_1)|_u).$$

Now by adding $|pref_{|pref_n(t_2)|_r}(t_1)|_u - |pref_{|pref_m(t_2)|_r}(t_1)|_u$ and using (2) it follows that:

$$|pref_n(t_2)|_u - |pref_m(t_2)|_u \leq$$
$$\leq c_1^2(1 + c_2^1)(|pref_{|pref_n(t_2)|_r}(t_1)|_u - |pref_{|pref_m(t_2)|_r}(t_1)|_u)..$$

Now consider condition **B2**. Assume that $|pref_n(t_2)|_u > |pref_m(t_2)|_u$. Since t_2 satisfies condition **B2** it follows that

$$|pref_n(t_2)|_r - |pref_m(t_2)|_r \leq c_2^2(|pref_n(t_2)|_u - |pref_m(t_2)|_u)$$

Note that the following relations hold:

$$|pref_{|pref_n(t_2)|_r}(t_1)|_u - |pref_{|pref_m(t_2)|_r}(t_1)|_u \leq |pref_n(t_2)|_r - |pref_m(t_2)|_r$$

From the above two inequalities we obtain that:

$$|pref_{|pref_n(t_2)|_r}(t_1)|_u - |pref_{|pref_m(t_2)|_r}(t_1)|_u \leq c_2^2(|pref_n(t_2)|_u - |pref_m(t_2)|_u)$$

Therefore τ satisfies the condition **B2**.

Analogously, one can show that $BInc$ is stable under $\diamond_2$ operation. Therefore, since $BInc$ is commutative, it follows that $BInc$ is associative. □

5 Asymptotic linear ω-trajectories

Definition 10. An ω-trajectory t is linear asymptotic iff the following limit does exist and belongs to $(0, +\infty)$

$$l = lim \frac{|pref_n(t)|_u}{|pref_n(t)|_r}.$$

Notation. The set of all asymptotic linear ω-trajectories is denoted by $ALin$.
Comment. Note that an ω-trajectory t is in $ALin$ iff performing the parallel composition of two processes on t, in the prefix of length n of the resulting sequence, the value of the number of occurrences of actions from the first process divided by the number of occurrences of actions from the second process, is a sequence of real numbers, say $(x_n)_{n>0}$, such that $(x_n)_{n>0}$ has a finite limit l, $l > 0$, when $n \longrightarrow +\infty$. Intuitively, this means that the ω-trajectory is at infinity stable, in such a way that it keeps some "balance" in performing actions from the first/second process. This "balance" is defined as being the limit l.

Theorem 3. *The set of all asymptotic linear ω-trajectories, i.e., $ALin$, is an associative and commutative set of ω-trajectories.*

Proof. Let t be an ω-trajectory and $\tau = sym(t)$. The limit $lim\frac{|pref_n(t)|_u}{|pref_n(t)|_r}$ exists and belongs to $(0, +\infty)$ iff the limit $lim\frac{|pref_n(t)|_r}{|pref_n(t)|_u}$ exists and belongs to $(0, +\infty)$. Therefore $t \in ALin$ iff $\tau \in ALin$ which means that $ALin$ is a commutative set of ω-trajectories.

Let t_1, t_2 be two trajectories in V_+^ω such that

$$l_1 = lim\frac{|pref_n(t_1)|_u}{|pref_n(t_1)|_r}, \text{ and } l_2 = lim\frac{|pref_n(t_2)|_u}{|pref_n(t_2|_r}, l_1, l_2 \in (0, +\infty)$$

We compute the limits corresponding to the operations $\diamond_i$, $1 \leq i \leq 4$. Consider the operation $\diamond_1(t_1, t_2)$. From the definition of $\diamond_1(t_1, t_2)$ it follows that $|pref_m(\diamond_1(t_1, t_2)|_r = |pref_{|pref_n(t_2)|_r}(t_1)|_u$, and $|pref_m(\diamond_1(t_1, t_2)|_u = |pref_n(t_2)|_u$, where n is the number of symbols used from t_2 to obtain a result of length m in $\diamond_1(t_1, t_2)$. Therefore,

$$\frac{|pref_n(t_2)|_u}{|pref_{|pref_n(t_2)|_r}(t_1)|_u} = \frac{|pref_n(t_2)|_u}{|pref_n(t_2)|_r}\frac{|pref_n(t_2)|_r}{|pref_{|pref_n(t_2)|_r}(t_1)|_u} =$$

$$= \frac{|pref_n(t_2)|_u}{|pref_n(t_2)|_r}\frac{|pref_{|pref_n(t_2)|_r}(t_1)|_r + |pref_{|pref_n(t_2)|_r}(t_1)|_u}{|pref_{|pref_n(t_2)|_r}(t_1)|_u} =$$

$$\frac{|pref_n(t_2)|_u}{|pref_n(t_2)|_r}(\frac{|pref_{|pref_n(t_2)|_r}(t_1)|_r}{|pref_{|pref_n(t_2)|_r}(t_1)|_u} + 1)$$

If $n \longrightarrow +\infty$, then it follows that the value of the limit corresponding to $\diamond_1(t_1, t_2)$ is: $l_2(1/l_1 + 1)$.

Similarly, we obtain the limit corresponding to $\diamond_2(t_1, t_2)$, and hence $ALin$ is associative. □

Next proposition shows that the sets $BInc$ and $ALin$ are incomparable with respect to the inclusion. The proof is omitted.

Proposition 4. $BInc - ALin \neq \emptyset$ *and* $ALin - BInc \neq \emptyset$.

6 Quasi linear ω-trajectories

Definition 11. We say that an ω-trajectory t is quasi linear iff there exist two constants, $c_1, c_2 > 0$ such that

$$||pref_n(t)|_u - c_1|pref_n(t)|_r| \leq c_2.$$

Notation. The set of all quasi linear ω-trajectories is denoted by $QLin$.
Comment. An ω-trajectory t is in $QLin$ iff the graph of t is bounded by two parallel lines.

Proposition 5. *The set of all quasi linear ω-trajectories, i.e., $QLin$, is included in both classes $BInc$ and $ALin$.*

Remark 3. From the above proposition, it follows that the class $QLin$ is included in the intersection of classes $BInc$ and $ALin$, i.e., $QLin \subseteq BInc \cap ALin$.

Proposition 6. $(BInc \cap ALin) - QLin \neq \emptyset$.

Theorem 4. *The set of all quasi linear ω-trajectories, $QLin$, is an associative and commutative set of ω-trajectories.*

Proof. Let t be an ω-trajectory and $\tau = sym(t)$. Note that $|pref_n(t)|_u = |pref_n(\tau)|_r$ and $|pref_n(t)|_r = |pref_n(\tau)|_u$. Hence it follows that there are $c_1, c_2 > 0$ such that

$$||pref_n(t)|_u - c_1|pref_n(t)|_r| \leq c_2$$

iff

$$||pref_n(\tau)|_r - c_1|pref_n(\tau)|_u| \leq c_2$$

iff

$$||pref_n(\tau)|_u - \frac{1}{c_1}|pref_n(\tau)|_r| \leq \frac{c_2}{c_1}.$$

Thus $t \in QLin$ iff $\tau \in QLin$. Consequently $QLin$ is a commutative set of ω-trajectories.

Let t_1 and t_2 be two ω-trajectories which satisfy:

$$||pref_n(t_1)|_u - c_1^1|pref_n(t_1)|_r| \leq c_2^1$$

$$||pref_n(t_2)|_u - c_1^2|pref_n(t_2)|_r| \leq c_2^2$$

Thus using the above proposition, $c_1^1 = lim\frac{|pref_n(t_1)|_u}{|pref_n(t_1)|_r}$ and $c_1^2 = lim\frac{|pref_n(t_2)|_u}{|pref_n(t_2)|_r}$

Firstly, assume that: $\tau = \diamond_1(t_1, t_2)$. Consider the notation: $k = |pref_n(t_2)|_r$

$$||pref_n(t_2)|_u - c_1^2(1 + 1/c_1^1)|pref_k(t_1)|_u| =$$

$$||pref_n(t_2)|_u - c_1^2|pref_k(t_1)|_u - \frac{c_1^2}{c_1^1}|pref_k(t_1)|_u| =$$

$$= ||pref_n(t_2)|_u - c_1^2 k + c_1^2|pref_k(t_1)|_r - \frac{c_1^2}{c_1^1}|pref_k(t_1)|_u| \leq$$

$$||pref_n(t_2)|_u - c_1^2 k + \frac{c_1^2}{c_1^1}|c_1^1|pref_k(t_1)|_r - |pref_k(t_1)|_u| \leq c_2^2 + c_1^2 c_2^1/c_1^1$$

Thus τ is in $QLin$. Similarly, $QLin$ is stable under $\diamond_2$. □

Proposition 7. *UltPer is included in QLin and the inclusion is strict.*

Next theorem gives a characterization of those ω-trajectories that are in $QLin$. The theorem is based on an extension of the notion of a Sturmian word, see [1].

Definition 12. The ω-trajectory t is quasi-Sturmian iff there is $c > 0$ such that:

$$||v|_u - |w|_u| \leq c, \forall v, w \in Sub(t), |v| = |w|$$

The set of all quasi-Sturmian ω-trajectories is denoted by QSturm.

Let $t \in V_+^\omega$. Consider the following function:

$$\pi_t : \omega \longrightarrow \omega,\ \pi_t(n) = |pref_n(t)|_u$$

Remark 4. The ω-trajectory t is quasi-Sturmian iff:

$$|\pi_t(n+p) - \pi_t(n) - (\pi_t(m+p) - \pi_t(m))| \leq c, \forall n, m, p \in \omega$$

Proposition 8. *If $t \in V_+^\omega$, $t = r^{i_1}u^{j_1}r^{i_2}u^{j_2}...r^{i_n}u^{j_n}...$ is a quasi-Sturmian ω-trajectory, then the sequence $\{j_n\}_n$ is bounded.*

Next theorem gives an alternative definition of quasi-Sturmian words.

Theorem 5. *QSturm = QLin.*

Proof. Firstly, we prove that $QLin$ is a subset of $QSturm$.
Let $t \in QLin$, i.e., there are $d, c > 0$ such that:

$$|\pi_t(n) - dn| \leq c$$

By using the above relation for $n + p$ and for n we have:

$$d(n+p) - c \leq \pi_t(n+p) \leq d(n+p) + c$$

$$-dn - c \leq -\pi_t(n) \leq -dn + c.$$

By adding these relations it follows:

$$dp - 2c \leq \pi_t(n+p) - \pi_t(n) \leq dp + 2c$$

$$-dp - 2c \leq -(\pi_t(m+p) - \pi_t(m)) \leq -dp + 2c$$

and by adding these inequalities we obtain:

$$-4c \leq \pi_t(n+p) - \pi_t(n) - (\pi_t(m+p) - \pi_t(m)) \leq 4c$$

$$|\pi_t(n+p) - \pi_t(n) - (\pi_t(m+p) - \pi_t(m))| \leq 4c$$

This means that $t \in QSturm$. Now we prove the reverse inclusion.

Claim I. There exists $lim\frac{\pi_t(n)}{n}$ *and, moreover, its value is in* $(0, 1)$.

Proof of Claim I.
Let t be in $QSturm$. Note that $0 < \frac{\pi_t(n)}{n} < 1$ and therefore there exist $lim\ inf\frac{\pi_t(n)}{n}$ and $lim\ sup\frac{\pi_t(n)}{n}$ and,

$$0 \leq lim\ inf\frac{\pi_t(n)}{n} \leq lim\ sup\frac{\pi_t(n)}{n} \leq 1$$

Let x_n and y_n be two sequences such that:

$$lim\frac{\pi_t(x_n)}{x_n} = lim\ inf\frac{\pi_t(n)}{n}$$

$$lim\frac{\pi_t(y_n)}{y_n} = lim\ sup\frac{\pi_t(n)}{n}$$

Since $t \in QSturm$, it follows that:

$$|\pi_t(n+p) - \pi_t(n) - \pi_t(p)| \leq c, \forall n, p \in \omega$$

or

$$\pi_t(n) + \pi_t(p) - c \leq \pi_t(n+p) \leq \pi_t(n) + \pi_t(p) + c, \forall n, p \in \omega$$

By iterating the right part we obtain:

$$\pi_t(nm) \leq n\pi_t(m) + (n-1)c$$

By iterating the left part it follows that:

$$m\pi_t(n) - (m-1)c \leq \pi_t(nm)$$

Therefore:

$$m\pi_t(n) - (m-1)c \leq n\pi_t(m) + (n-1)c$$

Using the above relation for y_n and x_n instead of n and m we obtain:

$$x_n\pi_t(y_n) - (x_n - 1)c \leq y_n\pi_t(x_n) + (y_n - 1)c$$

and dividing by x_ny_n it follows that:

$$\frac{\pi_t(y_n)}{y_n} - \frac{(x_n-1)c}{x_ny_n} \leq \frac{\pi_t(x_n)}{x_n} + \frac{(y_n-1)c}{x_ny_n}$$

Computing the limit we obtain:

$$lim\ sup\frac{\pi_t(n)}{n} \leq lim\ inf\frac{\pi_t(n)}{n}.$$

This means that the limit of $\frac{\pi_t(n)}{n}$ does exist.

In order to prove that $lim\frac{\pi_t(n)}{n} \in (0,1)$ let consider $r^{i_1}u^{j_1}r^{i_2}u^{j_2}...r^{i_q}u^{j_q}$ be a prefix of t, having the length n. Since $t \in QSturm$, i.e., there are $c_1, c_2 > 0$ such that $i_n \leq c_1$ and $j_n \leq c_2$, $\forall n \in \omega$ and the minimum value of $\frac{\pi_t(n)}{n}$ is $\frac{q-1}{qc_1+q-1}$. Therefore $lim\frac{\pi_t(n)}{n} \geq \frac{1}{1+c_1}$. Similarly, it follows that that $lim\frac{\pi_t(n)}{n} \leq \frac{c_2}{1+c_2}$ which complete the proof of the Claim I.

We continue our proof by considering $t \in QSturm$ and $l = lim\frac{\pi_t(n)}{n}$.
From the above proof it follows that:

$$m\pi_t(n) - (m-1)c \leq n\pi_t(m) + (n-1)c$$

and hence:

$$m\pi_t(n) - n\pi_t(m) \leq (n-1)c + (m-1)c$$

By dividing with m we obtain:

$$\pi_t(n) - n\frac{\pi_t(m)}{m} \leq \frac{(n-1)c + (m-1)c}{m}$$

and, if $m \longrightarrow +\infty$, then:

$$\pi_t(n) - ln \leq c$$

Dividing the same relation by n it follows that:

$$m\frac{\pi_t(n)}{n} - \pi_t(m) \leq \frac{(n-1)c + (m-1)c}{n}$$

and if $n \longrightarrow +\infty$, then:

$$lm - \pi_t(m) \leq c$$

Using the last two relations we obtain: $|\pi_t(n) - ln| \leq c$ and thus $t \in QLin$.□

Corollary 1. *The set of all Sturmian ω-words is included in $QLin$ and the inclusion is strict.*

Remark 5. Note that, as a consequence that there are noncountable many Sturmian ω-words, it follows that $QLin$ is a noncountable set, too.

7 ω-Trajectories with bounded prefixes

Definition 13. An ω-trajectory t is referred as prefix bounded iff there exist $a_1, a_2 > 0$ such that:

P1 $|pref_n(t)|_u \leq a_1|pref_n(t)|_r$, whenever $|pref_n(t)|_r > 0$ and
P2 $|pref_n(t)|_r \leq a_2|pref_n(t)|_u$, whenever $|pref_n(t)|_u > 0$.

Notation. The set of all prefix bounded ω-trajectories is denoted by $PrefB$.
Comment. An ω-trajectory is in $PrefB$ iff performing the parallel composition of two processes on t there are two constants $a_1 > 0$ and $a_2 > 0$ such that, in the prefix of length n of the resulting sequence, the value of the number of occurrences of actions from the first process is at most a_2-times the number of occurrences of actions from the second process and symmetrically.

Theorem 6. *The set of prefix bounded ω-trajectories, $PrefB$, is an associative and commutative set of ω-trajectories. Moreover, $PrefB$ strictly includes the sets $BInc$ and $ALin$.*

Combining our results we obtain the main result of this paper:

Theorem 7. *The following sets of ω-trajectories: $PalPer$, Per, $UltPer$, $QLin(=QSturm)$, $BInc \cap ALin$, $ALin$, $BInc$, $PrefB$, V_+^ω are associative, commutative and fair (with respect to a certain type of fairness).*

8 Conclusion

Several sets of ω-trajectories, all of them associative, commutative and fulfilling a certain fairness condition were introduced and their interrelations were established. All these sets define a natural structure of a commutative semiring on the set $\mathcal{P}(\Sigma^\omega)$, the set of all ω-languages over an alphabet Σ. Each multiplicative law of these semirings satisfies a certain fairness condition.

References

[1] J. Berstel, *Recent Results on Sturmian Words*, in Developments in Language Theory, eds. J. Dassow, G. Rozenberg and A. Salomaa, World Scientific, 1996, pp. 13-24.
[2] A. Mateescu and G.D. Mateescu, "Associative shuffle of infinite words", TUCS Technical Report, 104, 1997.
[3] A. Mateescu, G. Rozenberg and A. Salomaa, "Shuffle on Trajectories: Syntactic Constraints", *Theoretical Computer Science* (TCS), 197, 1-2, (1998) 1-56.
[4] D. Park, "Concurrency and automata on infinite sequences", in *Theoretical Computer Science*, ed. P. Deussen, LNCS 104, Springer-Verlag, 1981, pp. 167-183.
[5] D. Perrin and J. E. Pin, *Mots Infinis*, Report LITP 93.40, 1993.

Forbidden Factors in Finite and Infinite Words

Filippo Mignosi, Antonio Restivo, and Marinella Sciortino*

Summary. We study minimal forbidden factors in finite and infinite words. In the case of a finite word w we consider two parameters: the first counts the minimal forbidden factors of w and the second gives the length of the longest minimal forbidden factor of w. We prove sharp upper and lower bounds for both parameters. We prove also that the second parameter is related to the minimal period of w. In the case of an infinite word w we consider the following two functions: $g_w(n)$ that counts the allowed factors of w of length n and $f_w(n)$ that counts the minimal forbidden factors of w of length n. We address the following general problem: which informations about the structure of w can be derived from the pair (g_w, f_w)? We prove that these two functions characterize, up to the automorphism exchanging the two letters, the language of factors of any infinite Sturmian word.

1 Introduction

In many problems concerning the combinatorial analysis of a (finite or infinite) word w and its applications, it is of great interest to consider the set $L(w)$ of factors of w. In the case of a finite word w, it is well known that the language $L(w)$ and the minimal automaton $\mathcal{A}(w)$ recognizing it are used in the design of string-matching algorithms (cf. [7]). In the study and in the classification of infinite words, one defines the *complexity* of an infinite word w as the function that counts, for any natural number n, the words of length n in $L(w)$ (cf. [6]).

In this paper we also consider the words that do not occur as factors of a given word w and that we call the *forbidden* factors of w. We show that the forbidden factors are of fundamental importance in determining the structure of the word itself.

In a more general context, given a factorial language L, a word v is *forbidden* for L, if $v \notin L$. An important point about this notion is that we can introduce a condition of minimality: a word v is a *minimal forbidden word* for L if v is forbidden and all proper factors of v belong to L. We denote by $MF(L)$ the language of minimal forbidden words for L.

* Università degli studi di Palermo, Dipartimento di Matematica ed Applicazioni, Via Archirafi 34, 90123 Palermo. E-mails: `mignosi, restivo, mari @altair.math.unipa.it`

From an algebraic point of view the complement of L in the monoid A^* is an ideal of A^* and the set $MF(L)$ of the minimal forbidden words is its (unique) base. Such concept of minimal forbidden word synthetizes effectively some negative information about a language and plays an important role in several applications. The relevance of this notion in problems in automata theory, text compression and symbolic dynamics is shown in [8], [9], [1] respectively.

In this paper we focus on the study of the set of minimal forbidden factors of a single (finite or infinite) word w. We denote by $MF(w)$ this set. Our results show that the combinatorial properties of $MF(w)$ are an useful tool to investigate the structure of the word w. In particular, in Section 3, we introduce, for any finite word w, the parameters $c(w)$ and $m(w)$, representing the cardinality of $MF(w)$ and the maximal length of words in $MF(w)$ respectively. We give non trivial lower and upper bounds on these parameters. Furthermore we show that the parameter $m(w)$ is related to the minimal period $p(w)$ of the word w.

In Section 4 we introduce, for any infinite word w, besides the complexity function g_w, the function f_w that counts, for any natural number n, the minimal forbidden factors of w of length n. We address the following general problem: which informations does the pair (g_w, f_w) give about the structure of the infinite word w ? The main result of this section states that the set of factors of a Sturmian word $\mathbf{x}$ over the alphabet $\{a, b\}$ is uniquely determined (up to the automorphism exchanging a and b) by the pair of functions $(g_{\mathbf{x}}, f_{\mathbf{x}})$.

2 Minimal forbidden words

For any notation not explicitely defined in this paper we refer to [15] and [13].

Let A be a finite alphabet and let A^* be the set of finite words drawn from the alphabet A, the empty word ϵ included. Let $L \subseteq A^*$ be a *factorial language*, *i.e.* a language satisfying: $\forall u, v \in A^*\ uv \in L \Longrightarrow u, v \in L$. The complement language $L^c = A^* \setminus L$ is a (two-sided) ideal of A^*. Denote by $MF(L)$ the base of this ideal, we have $L^c = A^*MF(L)A^*$.

The set $MF(L)$ is called the set of *minimal forbidden words* for L. A word $v \in A^*$ is forbidden for the factorial language L if $v \notin L$, which is equivalent to say that v occurs in no word of L. In addition, v is minimal if it has no proper factor that is forbidden.

One can note that the set $MF(L)$ uniquely characterizes L, just because

$$L = A^* \setminus A^*MF(L)A^*. \tag{1}$$

The following simple observation provides a basic characterization of minimal forbidden words.

Remark 1. A word $v = a_1a_2 \cdots a_n$ belongs to $MF(L)$ iff the two conditions hold:

- v is forbidden, (*i.e.*, $v \notin L$),
- both $a_1 a_2 \cdots a_{n-1} \in L$ and $a_2 a_3 \cdots a_n \in L$ (the prefix and the suffix of v of length $n-1$ belong to L).

The remark translates into the equality:

$$MF(L) = AL \cap LA \cap (A^* \setminus L). \tag{2}$$

As a consequence of both equalities (1) and (2) we get the following proposition.

Proposition 1. *For a factorial language L, languages L and $MF(L)$ are simultaneously rational, that is, $L \in Rat(A^*)$ iff $MF(L) \in Rat(A^*)$.*

The set $MF(L)$ is an *anti-factorial language* or a *factor code*, which means that it satisfies: $\forall u, v \in MF(L)$ $u \neq v \Longrightarrow u$ is not a factor of v, property that comes from the minimality of words of $MF(L)$.

We introduce a few more definitions. A word $v \in A^*$ *avoids the set* M, $M \subseteq A^*$, if no word of M is a factor of v, (*i.e.*, if $v \notin A^*MA^*$). A language L *avoids* M if every words of L avoid M.

From the definition of $MF(L)$, it readily comes that L is the largest (according to the subset relation) factorial language that avoids $MF(L)$. This shows that for any anti-factorial language M there exists a unique factorial language $L(M)$ for which $M = MF(L)$. The next remark summarizes the relation between factorial and anti-factorial languages.

Remark 2. There is a one-to-one correspondence between factorial and anti-factorial languages. If L and M are factorial and anti-factorial languages respectively, both equalities hold: $MF(L(M)) = M$ and $L(MF(L)) = L$.

3 Forbidden factors of a finite word

Let w be a finite word over the alphabet A. Denote by $L(w)$ the set of factors of w and by $MF(w)$ the corresponding antifactorial language, i.e. $MF(w)$ is the set of minimal forbidden factors of w. It is obvious that $MF(w)$ is a finite set and that it uniquely characterizes the word w. In this section we are interested to relate the combinatorial properties of the set $MF(w)$ with the structure of the word w. We consider in particular the following parameters:

$$c(w) = Card(MF(w))$$

$$m(w) = max\{|v|, v \in MF(w)\}.$$

Example 1. Consider the word $w = acbcabcbc$. One has that

$$MF(w) = \{aa, ba, bb, cc, aca, cba, cac, cbcb, abca, bcbca\}$$

Hence $c(w) = 10$, $m(w) = 5$.

Let $\mathcal{A}(w)$ denote the minimal deterministic automaton recognizing $L(w)$. Recall that the states of $\mathcal{A}(w)$ correspond to the classes of the Nerode equivalence $\approx_w$ induced by the language $L(w)$ and defined as follows:

$$x \approx_w y \text{ if and only if } \forall z \in A^*, xz \in L(w) \Longleftrightarrow yz \in L(w).$$

The zero-class of $\approx_w$ is the set of words $x \in A^*$ such that $xA^* \cap L(w) = \emptyset$ and it corresponds to the sink state of $\mathcal{A}(w)$.

Denote by $P(w)$ the set of proper prefixes of elements of $MF(w)$. The following theorem synthesizes some results obtained in [8].

Theorem 1. *Let w be a finite word over the alphabet A. There is a one-to-one correspondence between the elements of $P(w)$ and the states (different from the sink) of $\mathcal{A}(w)$. Furthermore the elements of $P(w)$ coincide with the words of minimal length in the non-zero classes of the Nerode equivalence $\approx_w$.*

From the relationship between $MF(w)$ and the automaton $\mathcal{A}(w)$ we derive non trivial bounds on the parameter $c(w)$ (cf. [8]). Denote by d the cardinality of the alphabet A and by $d(w)$ the number of the letters of A occurring in w.

Theorem 2. *Let w be a finite word over the alphabet A with $|w| > 2$. The following inequalities hold:*

$$d \le c(w) \le 2(|w| - 2))(d(w) - 1) + d.$$

Proof. The inequality $d \le c(w)$ follows from the remark that for any letter $a \in A$, there exists an integer $n(a)$, such that $a^{n(a)} \in MF(w)$.
In order to state the upper bound, we remark that, by the relationship between $MF(w)$ and $\mathcal{A}(w)$, the number of elements of $MF(w)$ is at most equal to the number of arcs of $\mathcal{A}(w)$ originated at a non sink state and ingoing to the sink state. So we have to count these arcs. From the initial state of $\mathcal{A}(w)$ there are exactly $d - d(w)$ such arcs. From the (unique) state of $\mathcal{A}(w)$ whose outgoing arcs go to the sink state, there are exactly $d(w)$ such arcs. From other states there are at most $d(w) - 1$ such arcs. For $|w| > 2$, it is known (cf. [7]) that $\mathcal{A}(w)$ has at most $2|w| - 2$ states. Therefore:

$$c(w) \le (d - d(w)) + d(w) + (2|w| - 4)(d(w) - 1) = 2(|w| - 2)(d(w) - 1) + d.$$

Remark 3. The inequalities in Theorem 2 are sharp. For any letter a, any word $w \in \{a\}^*$ is such that $c(w) = d$. Indeed, if $w = a^n$, it is easy to prove that $MF(w) = \{a^{n+1}\} \cup A \setminus \{a\}$. Since $d(w) = 1$, one has that $d = c(w) = 2(|w|-2))(d(w)-1)+d$. Notice also that the only words having length smaller than or equal to 2 that are not of the form $w = a^n$, are of the form $w = ab$ with $a, b \in A$, $a \ne b$. In this case $MF(ab) = \{a^2, b^2, ba\} \cup A \setminus \{a, b\}$ and $c(ab) = d + 1$.

The next theorem provides lower and upper bounds on $m(w)$. Recall that, for any real number α, $\lceil \alpha \rceil$ denotes the smallest integer greater than or equal to α.

Theorem 3. *Let w be a finite word over the alphabet A. The following inequalities hold:*

$$\lceil \log_d(|w|) \rceil + 1 \leq m(w) \leq |w| + 1.$$

Furthermore the bounds are actually attained.

Proof. By Theorem 1, we have

$$|w| \leq Card(P(w)) \leq d^{m(w)-1}.$$

It follows that

$$\lceil \log_d(|w|) \rceil + 1 \leq m(w).$$

The inequality $m(w) \leq |w|+1$ follows from the remark that all words of length greater than or equal to $|w| + 1$ are forbidden for $L(w)$. We are proving now that this bounds are sharp. We first show that there exists a word w such that $m(w) = |w| + 1$. Let us consider $w = a^n$. We know that $MF(w) = \{a^{n+1}\} \cup A \setminus \{a\}$. We now show that there exist words that attain the lower bound. Let w be the *De Brujin word of order n* over the alphabet A=$\{a, b\}$ (cf. [10]), i.e. the word w of length $2^n + n - 1$ characterized by the property that $A^n \subseteq L(w)$ and any element of A^n occurs exactly once as factor of w. By the definition we have that, for any $v \in A^n$ such that v is not a suffix of w, there exists only one letter $x \in A$ such that $vx \in L(w)$. We claim that

$$MF(w) = \{v \in A^{n+1} \mid v \notin L(w)\}.$$

By definition of w, we have that

$$\{v \in A^{n+1} \mid v \notin L(w)\} \subseteq MF(w).$$

It remains to prove that $MF(w) \subseteq A^{n+1}$. Let us suppose, by contradiction, that there exists a word $u \in MF(w)$ such that $|u| > n+1$. Let z be the prefix of u of length $|u| - 1 > n$ such that $u = zx$, $x \in A$. By definition $z \in L(w)$ and $zx \notin L(w)$. Let t be the suffix of z of length n. Since there is only one occurrence of t in w, then $tx \notin L(w)$, i.e. tx is forbidden. The fact that tx is a proper suffix of u contradicts the hypothesis that u is a minimal forbidden factor.

In the previous theorem we show that the parameter $m(w)$ reaches its minimum for the *De Brujin words*, i.e. for words that are known to present an "high complexity". On the contrary, the maximum of the same parameter is obtained for very "simple" words, i.e. for words of the form a^n, where a is a letter. This suggests that $m(w)$ can be related, in some sense, to the "complexity" of the word w, and, in particular, that the parameter $\mu(w) = |w| - m(w) + 2$ could be interpreted as a measure of complexity of the word w.

The previous example also suggests that there is a relationship between $m(w)$ and the periodicity of w. In fact the maximum of $m(w)$ is attained for words that present a very strong periodicity, i.e. for words having period 1. The following theorem relates $m(w)$ and the minimal period $p(w)$ of a word w.

Theorem 4. *Let w be a finite word over the alphabet A. We have that*

$$m(w) \geq |w| - p(w) + 2.$$

Proof. Let w be a word having period p. Hence w can be written

$$w = uv = vu',$$

with $|u| = |u'| = p$.
Let y be the first letter of u'. Trivially $uvy \notin L(w)$, i.e. it is forbidden. However $vy \in L(w)$, i.e. it is not forbidden. Then there exists a minimal forbidden word $z \in MF(w)$, such that z is a suffix of uvy and vy is a proper suffix of z. It follows that

$$m(w) \geq |z| \geq |vy| + 1 = |w| - p(w) + 2.$$

Remark 4. The previous inequality is sharp. Indeed, for $w = a^n$, p(w)=1 and m(w)=n+1.

4 Forbidden factors of an infinite word

Let A be a finite alphabet and let w be an infinite word over A. Let us denote by $L(w)$ the language of all factors of w. Recall that an infinite word w is *recurrent* if any factor occurring in w has an infinite number of occurrences. It is easy to verify that if w is recurrent, then $L(w)$ is an *extensible* language, i.e. for any $v \in L(w)$, there exist $x, y \in A$ such that $xv \in L(w)$ and $vy \in L(w)$.

In the study of infinite words an important role is played by *special* and *bispecial* words, defined as follows. Let L be a factorial language. A word $v \in L$ is *special on the left* with respect to B, where $B \subset A$ and $Card(B) \geq 2$, if for any $b \in B$, bv belongs to L. Analogously we define words *special on the right.* Given $B, C \subset A$ such that $Card(B) \geq 2$ and $Card(C) \geq 2$, we say that a word $v \in L$ is *bispecial* with respect to (B, C) if it is special on the left with respect to B and special on the right with respect to C.

In the case of a two letters alphabet A, special and bispecial words have been extensively studied (cf. [3], [11], [12], [5]). Remark that, since $Card(A) = 2$, there is no need to specify the sets B and C (both must be equal to A). In this case let us denote by $BS(L)$ the set of bispecial elements of L.

The following proposition relates bispecial and minimal forbidden words.

Proposition 2. *Let L be a factorial extensible language. For any word $u \in A^*$ the following conditions are equivalent:*

(i) u is bispecial with respect to (B, C) and $buc \notin L$ for some $b \in B$ and $c \in C$,
(ii) $buc \in MF(L)$.

Proof. If u is bispecial with respect to (B, C) then $bu \in L$ and $uc \in L$. Since $buc \notin L$, by Remark 1, $buc \in MF(L)$. We prove that u is left special with respect to B. A symmetric argument proves that u is right special with respect to C. Since $buc \in MF(L)$, $uc \in L$. Since L is extensible there exists a letter x such that $xuc \in L$. This letter x is different from b because $buc \notin L$. If we set $B = \{x, b\}$ then it is easy to verify that u is left special with respect to B.

Example 2. Let us consider the Fibonacci infinite word

$$\mathbf{s} = abaababaabaababaaba\ldots$$

Let L(**s**) be the set of factors of **s**. Then

$BS(L(\mathbf{s})) = \{v \mid v \text{ is a palindrome prefix of } \mathbf{s}\} =$
$= \{\epsilon, a, aba, abaaba, abaababaaba, abaababaabaababaaba, \cdots\}.$

$MF(L(\mathbf{s})) = \{w \mid w = bvb,\ v \text{ is the } n\text{-th palindrome prefix of } \mathbf{s},\ n \text{ is even}\} \cup$
$\cup \{w \mid w = ava,\ v \text{ is the } n\text{-th palindrome prefix of } \mathbf{s},\ n \text{ is odd}\} =$
$= \{bb, aaa, babab, aabaabaa, babaababaabab, aabaababaabaababaabaa, \cdots\}.$

Given an infinite word w over the alphabet A, we can introduce the *complexity function* g_w of w defined as

$$g_w(n) = Card\{v \in L(w) \text{ such that } |v| = n\}$$

and the function f_w defined as

$$f_w(n) = Card\{v \in MF(w) \text{ such that } |v| = n\},$$

where $MF(w) = MF(L(w))$ denotes the set of minimal forbidden factors of w.

In this section we consider the following problem: which informations does the pair (g_w, f_w) give about the structure of the infinite word w ?

Example 2 (continued). In the case of the Fibonacci word, it is known that, for any natural number n,

$g_{\mathbf{s}}(n) = n + 1$, (cf. [6]).

By the explicit form of the set MF(L(s)) given above, we derive that

$$f_{\mathbf{s}}(n) = \begin{cases} 1 \text{ if } n \text{ is a Fibonacci number} \\ 0 \text{ otherwise} \end{cases}$$

From the result of this section we derive that if $\mathbf{s}$ is the Fibonacci word, then $L(\mathbf{s})$ is characterized by the two functions $g_{\mathbf{s}}$ and $f_{\mathbf{s}}$. Actually we will prove that this holds true for all Sturmian words.
Recall that a *Sturmian word* $\mathbf{x}$ is an infinite word over a binary alphabet such that, for any natural number n, $g_{\mathbf{x}}(n) = n + 1$ (cf. [4]).

The main result of this section states that, if $\mathbf{x}$ is a Sturmian word, the language $L(\mathbf{x})$ is uniquely specified (up to the automorphism exchanging the two letters of the alphabet) by the two functions $g_{\mathbf{x}}$ and $f_{\mathbf{x}}$.

Theorem 5. *Let* $\mathbf{x}$ *be a Sturmian word and let* $\mathbf{y}$ *be an infinite word over the alphabet* $\{a, b\}$ *such that* $g_{\mathbf{x}} = g_{\mathbf{y}}$ *and* $f_{\mathbf{x}} = f_{\mathbf{y}}$. *Then* $L(\mathbf{x}) = L(\mathbf{y})$, *up to the automorphism exchanging the two letters* a *and* b.

Remark that, by definition, the fact that, for any natural number n, $g_{\mathbf{x}}(n) = g_{\mathbf{y}}(n)$ implies that $\mathbf{y}$ is also a Sturmian word. It remains to prove that, if two Sturmian words $\mathbf{x}$ and $\mathbf{y}$ are such that $f_{\mathbf{x}} = f_{\mathbf{y}}$, then $L(\mathbf{x}) = L(\mathbf{y})$, up to the automorphism exchanging the two letters a and b.
Recall that a Sturmian word can be also defined by considering the intersections with a squared-lattice of a semi-line having a slope which is an irrational number $\alpha > 0$ (cf. [11]). A vertical intersection is denoted by the letter a, a horizontal intersection by b and the intersection with a corner by ab or ba. It is possible to prove that the language $L(\mathbf{x}_\alpha)$ of factors of a Sturmian word defined in this way depends only on the slope of the line. By this definition it follows that $L(\mathbf{x}_{\frac{1}{\alpha}})$ can be obtained from $L(\mathbf{x}_\alpha)$ by exchanging the two letters a and b. Hence we only need to prove that if $\mathbf{x}_\alpha$ and $\mathbf{x}_\beta$ are two Sturmian words such that $f_{\mathbf{x}_\alpha} = f_{\mathbf{x}_\beta}$ then $\alpha = \beta$ or $\alpha = \frac{1}{\beta}$.
In order to give a detailed proof of this theorem we need some more preliminaries.

We construct the infinite sequence of pairs of words (A_n, B_n), $n \geq 0$, as follows. We set $(A_0, B_0) = (a, b)$. For any $n \geq 0$ the pair (A_{n+1}, B_{n+1}) is obtained from (A_n, B_n) by using one of the following two rules:
First rule: $(A_{n+1}, B_{n+1}) = (A_n, A_n B_n)$.
Second rule: $(A_{n+1}, B_{n+1}) = (B_n A_n, B_n)$.

Let us remind that an irrational number $\alpha > 0$ is characterized by its development in continued fraction $[q_0, q_1, q_2, \cdots]$ with $q_0 \geq 0$ and $q_i > 0$ if $i > 0$ (cf. [14]).

From the theory of continued fractions it follows that given an irrational number $\alpha > 1$ with $[q_0, q_1, q_2, \cdots]$ as development in continued fraction, $q_0 \geq 1$, the irrational number $\frac{1}{\alpha}$ has $[0, q_0, q_1, q_2, \cdots]$ as development in continued fraction.

Let α be an irrational number and let $[q_0, q_1, q_2, \cdots]$ be its development in continued fraction. It has been proved (cf. [16]) that if one applies q_0 times first rule, q_1 times second rule, q_2 times first rule, and so on, then the sequences $\{A_n\}_{n\geq 0}$, $\{B_n\}_{n\geq 0}$ converge to the same infinite Sturmian word $\mathbf{x}_\alpha$ that is associated to the semi-line that have slope α and that starts from the origin. Moreover, if $|A_n| \geq 2$ (resp. $|B_n| \geq 2$) then A_n (resp. B_n) is a prefix of $\mathbf{x}_\alpha$.

Let $\mathbf{x}_\alpha$ be an infinite Sturmian word constructed by the standard method and let $[q_0, q_1, q_2, \cdots]$ be the development in continued fraction of the irrational number α. Let us define the sequences $\{s_n\}_{n\geq 0}$, where $s_0 = B_0$, $s_1 = A_0$ and for all $n \geq 1$, $s_{n+1} = s_n^{q_{n-1}} s_{n-1}$. It's easy to verify that for all $n \geq 0$, $s_{2n} = B_{q_0+q_1+\cdots+q_{2n-2}}$ and $s_{2n+1} = A_{q_0+q_1+\cdots+q_{2n-1}}$. For each $n \geq 0$ let us set $\chi_\alpha(n) = |s_n|$. Hence, for any $n > 0$, $\chi_\alpha(n+1) = q_{n-1}\chi_\alpha(n) + \chi_\alpha(n-1)$. Notice that, since $\chi_\alpha(n) > 0$ for any n and since $q_n > 0$ whenever $n > 0$, above equation implies that the sequence of the $\chi_\alpha(n)$ is a non-decreasing sequence that is strictly increasing starting from $n \geq 2$.

Remark 5. It is easy to verify by induction on n that if $\alpha > 1$ then, for all $n \geq 1$, $\chi_{\frac{1}{\alpha}}(n) = \chi_\alpha(n-1)$ and $\chi_{\frac{1}{\alpha}}(0) = 1$.

Lemma 1. *Let $n \geq 1$ be. We have that*

(1) if $|A_n| < |B_n|$ then $A_n = A_{n-1}$ and $B_n = A_{n-1}B_{n-1}$,
(2) if $|A_n| > |B_n|$ then $B_n = B_{n-1}$ and $A_n = B_{n-1}A_{n-1}$.

Proof. This result can be easily proved by induction on n. Indeed it is clear that, if $|A_n| < |B_n|$ then the $n-th$ approximating pair (A_n, B_n) has been obtained applying the first rule, and so, $A_n = A_{n-1}$ and $B_n = A_{n-1}B_{n-1}$. On the contrary, if $|A_n| > |B_n|$ then the second rule has been applied, and so, $B_n = B_{n-1}$ and $A_n = B_{n-1}A_{n-1}$.

Remark 6. We can observe that for each $n \geq 0$ there always exists an integer $k \geq 0$ such that $min(|A_n|, |B_n|) = \chi_\alpha(k)$ and if $min(|A_{n+1}|, |B_{n+1}|) \neq min(|A_n|, |B_n|)$ then $min(|A_{n+1}|, |B_{n+1}|) = \chi_\alpha(k+1)$. In fact at every step, if we apply the same rule, the minimal element is the same concerning the previous step. Instead, if we change the rule, then the longest element become the smallest element concerning previous step and the length of this element is $\chi_\alpha(k+1)$.

Let us define $A(\mathbf{x}_\alpha) = \bigcup_{n\geq 0}\{A_n\}$ and $B(\mathbf{x}_\alpha) = \bigcup_{n\geq 0}\{B_n\}$ and set $I_\alpha = \{p \in Z, p \geq 2|$ there exists $u \in A(\mathbf{x}_\alpha) \cup B(\mathbf{x}_\alpha)$ such that $p = |u|\}$. Let $(r_i)_{i\geq 1}$ be the sequence of the elements of I_α written in the natural increasing order.

Remark 7. By the definitions it follows that $r_1 = max(|A_1|, |B_1|), \ldots, r_n = max(|A_n|, |B_n|)$.

Proposition 3. *If r_i and r_{i+1} are consecutive integers in I_α then there exists a positive integer k, depending on i, such that $r_{i+1} - r_i = \chi_\alpha(k)$. If k is the greatest such integer, we can also determine two consecutive numbers r_j and r_{j+1}, with $j > i$, such that $r_{j+1} - r_j = \chi_\alpha(k+1)$. Besides, if $k \geq 2$ and i is the smallest index such that $r_{i+1} - r_i = \chi_\alpha(k)$ and j is the smallest index such that $r_{j+1} - r_j = \chi_\alpha(k+1)$ then*

(1) $j - i = q_{k-1}$, where q_{k-1} is the $(k-1)-th$ element of the development in continued fraction of α,
(2) for all integer j' such that $i \leq j' < j$, $r_{j'+1} - r_{j'} = \chi_\alpha(k)$.

Proof. From Remark 7, $r_i = max(|A_i|, |B_i|)$ and $r_{i+1} = max(|A_{i+1}|, |B_{i+1}|)$. Let us assume that $r_{i+1} = |A_{i+1}|$ (the case $r_{i+1} = |B_{i+1}|$ can be proved analogously). According to Lemma 1, as $|A_{i+1}| > |B_{i+1}|$, $B_{i+1} = B_i$ and $A_{i+1} = B_i A_i$. Therefore $|A_{i+1}| - r_i = min(|A_i|, |B_i|) = \chi_\alpha(k)$ for some k. The existence of k comes from Remark 6 and the fact that there exists a greatest one that satisfies the equality derives from the strictly monotonicity of the sequence $\chi_\alpha(n)$ with $n \geq 2$. Therefore $r_{i+1} - r_i = \chi_\alpha(k)$. In the same way $r_{i+2} - r_{i+1} = \chi_\alpha(d)$, for some positive integer d. As a conseguence of the previous argument $r_{i+2} = max(|A_{i+2}|, |B_{i+2}|)$ and so $\chi_\alpha(d) = min(|A_{i+1}|, |B_{i+1}|)$. According to Remark 6, if the $(i+1)-th$ pair has been obtained applying the previous rule, then $min(|A_{i+1}|, |B_{i+1}|) = min(|A_i|, |B_i|)$ and so $\chi_\alpha(k) = \chi_\alpha(d)$. From the choice of k and d it follows that $k = d$. On the contrary if we change rule then $min(|A_{i+1}|, |B_{i+1}|) \neq min(|A_i|, |B_i|)$. By Remark 6 we have that $\chi_\alpha(d) = \chi_\alpha(k+1)$ and so $d = k+1$. If we want to find r_j, we only need to take the smallest number in I_α greater than r_i such that $r_{j+1} - r_j \neq min(|A_i|, |B_i|) = \chi_\alpha(k)$, with r_{j+1} consecutive of r_j in I_α. This argument proves the point (2). The proof of the point (1) follows from the fact that the number of indices j' with $i \leq j' \leq j$ such that $r_{j'+1} - r_{j'} = \chi_\alpha(k)$ is exactly the number of times that the same rule has been applied to obtain s_{k+1}, number that is q_{k-1}.

Theorem 6. *If $f_{\mathbf{x}_\alpha} = f_{\mathbf{x}_\beta}$ then $\alpha = \beta$ or $\alpha = \frac{1}{\beta}$.*

Proof. By [11, Proposition 9] it is easy to prove that s is bispecial in $L(\mathbf{x}_\alpha)$ if and only if s is a palindrome prefix of $\mathbf{x}_\alpha$. By [11, Proposition 7], s is a palindrome prefix of $\mathbf{x}_\alpha$ if and only if $s \in A(\mathbf{x}_\alpha)(ba)^{-1} \cup B(\mathbf{x}_\alpha)(ab)^{-1}$. As $L(\mathbf{x}_\alpha)$ is an extensible factorial language, by Proposition 2, if $w \in MF(L(\mathbf{x}_\alpha))$ then there exists $u \in A(\mathbf{x}_\alpha) \cup B(\mathbf{x}_\alpha)$ such that $|w| = |u|$ and conversely if $u \in A(\mathbf{x}_\alpha) \cup B(\mathbf{x}_\alpha)$, $|u| \geq 2$, then there exists $w \in MF(L(\mathbf{x}_\alpha))$ such that $|w| = |u|$ (more precisely the bispecial prefix s of u of length $|u| - 2$ is the "central" factor of $|w|$). It follows that $f_{\mathbf{x}_\alpha}$ is a zero-one function that coincides with the characteristic function of the set I_α. In the same way $f_{\mathbf{x}_\beta}$ is a zero-one function that coincides with the characteristic function of the set I_β. By the hypothesis we have that $I_\alpha = I_\beta$. Let us denote with $(r_i(\alpha))_{i \geq 1}$ the sequence of elements of I_α and with $(r_i(\beta))_{i \geq 1}$ the sequence of elements of I_β. We have that for all $i \geq 1$, $r_i(\alpha) = r_i(\beta)$, and, consequently, $r_{i+1}(\alpha) - r_i(\alpha) = r_{i+1}(\beta) - r_i(\beta)$. Let us distinguish four cases:

1. $\alpha > 1$ and $\beta > 1$. We have that $r_2(\alpha) - r_1(\alpha) = \chi_\alpha(1) = 1 = \chi_\beta(1) = r_2(\beta) - r_1(\beta)$. According to Proposition 3, for all $k \geq 0$, $\chi_\alpha(k) = \chi_\beta(k)$.
2. $\alpha < 1$ and $\beta < 1$. We have that $r_2(\alpha) - r_1(\alpha) = \chi_\alpha(2) = 1 = \chi_\beta(2) = r_2(\beta) - r_1(\beta)$. Hence, for all $k \geq 0$, $\chi_\alpha(k) = \chi_\beta(k)$.
3. $\alpha > 1$ and $\beta < 1$. In this case $r_2(\alpha) - r_1(\alpha) = \chi_\alpha(1) = 1 = \chi_\beta(2) = r_2(\beta) - r_1(\beta)$. Hence for all $k \geq 0$, $\chi_\alpha(k) = \chi_\beta(k+1)$.
4. $\alpha < 1$ and $\beta > 1$. Analogously to the previous case $r_2(\alpha) - r_1(\alpha) = \chi_\alpha(2) = 1 = \chi_\beta(1) = r_2(\beta) - r_1(\beta)$. It follows that, for all $k \geq 0$, $\chi_\alpha(k+1) = \chi_\beta(k)$.

Therefore in the cases 1 and 2 we can easily prove that $\alpha = \beta$. Let us denote with q_n^α and q_n^β the $n-th$ element of the development in continued fraction respectively of α and β. For all $n \geq 0$ we have that
$\chi_\alpha(n+2) = q_n^\alpha \chi_\alpha(n+1) + \chi_\alpha(n)$ and
$\chi_\beta(n+2) = q_n^\beta \chi_\beta(n+1) + \chi_\beta(n)$.
By making the difference between the left-hand sides and the right-hand sides respectively of the two equations, we have that $(q_n^\alpha - q_n^\beta)\chi_\alpha(n+1) = 0$. Hence, for all $n \geq 0$, $q_n^\alpha = q_n^\beta$ and then $\alpha = \beta$.

Let us consider now the case 3 (the case 4 is analogous). By Remark 5, we have that $\chi_\alpha(k) = \chi_\beta(k+1) = \chi_{\frac{1}{\beta}}(k)$ with $\alpha > 1$ and $\frac{1}{\beta} > 1$. According to what previously proved it follows that $\alpha = \frac{1}{\beta}$.

We conclude the paper by remarking that the previous theorem is related to another result showing that the dynamical systems associated to two Sturmian words $\mathbf{x}_\alpha$ and $\mathbf{x}_\beta$ such that $\alpha \neq \beta$ and $\alpha \neq \frac{1}{\beta}$ are not isomorphic (cf. [2]).

References

1. M.P. Béal, F. Mignosi, A. Restivo. *Minimal Forbidden words and Symbolic Dynamics*, in STACS'96, C. Puech and R. Reischuk (Eds.), Lecture Notes in Comp. Science, **1046**, Springer, Berlin 1996, 555-566.
2. M.P. Béal, F. Mignosi, A. Restivo, M. Sciortino. *Forbidden words in Symbolic Dynamics* (in preparation).
3. J. Berstel. *Fibonacci Words - a Survey*, in "The Book of L", G. Rozenberg, A. Salomaa eds., Springer-Verlag 1986.
4. J. Berstel, P. Séebold. *Sturmian Words*, Chapter 2 of " Algebraic Combinatorics on Words", J. Berstel, D. Perrin, eds., Cambridge University Press, to appear. Available at `http://www-igm.univ-mlv.fr/ berstel`.
5. J. Cassaigne. *Complexité et Facteurs Spéciaux*, Actes des Journées Montoises, 1994.

6. C. Choffrut, J. Karhumäki. *Combinatorics on words*, Chapter 6 of vol. 1 of "Handbook of Formal Languages", G. Rozenberg, A. Salomaa (Eds.), Springer-Verlag 1997.
7. M. Crochemore, C. Hancart. *Automata for matching patterns*, Chapter 9 of vol. 2 of "Handbook of Formal Languages", G. Rozenberg, A. Salomaa (Eds.), Springer-Verlag 1997, 399-462.
8. M. Crochemore, F. Mignosi, A. Restivo. *Automata and forbidden words*, Inf. Proc. Lett., **67**(1998), 111-117
9. M. Crochemore, F. Mignosi, A. Restivo, S. Salemi. *Data Compression using antidictionaries*, Technical Report, IGM-98-9, Institut Gaspard-Monge, 1998.
10. N. G. De Brujin. *A combinatorial problem*, Nederl. Akad. Wetensch. Proc., **49** (1946), 758-764.
11. A. de Luca, F. Mignosi. *Some Combinatorial Properties of Sturmian Words*, Theor. Comp. Science, **136** (1994), 361-385.
12. A. de Luca, L. Mione. *On Bispecial Factors of the Thue-Morse Word*, Inf. Proc. Lett., **49** (1994), 179-183.
13. S. Eilenberg. *Automata, Languages, Machines*, Vol. A, Academic Press, 1974.
14. G. H. Hardy, E. M. Wright. *An Introduction to the Theory of Numbers*, Oxford University Press.
15. M. Lothaire. *Combinatorics on Words*, Addison-Wesley, Reading, MA 1983.
16. G. Rauzy, *Mots infinis en arithmétique* in "Automata on Infinite words", M. Nivat and D. Perrin (Eds.), Lecture Notes in Comp. Science, **192**, Springer-Verlag, 1984.

Part VII

Novel Directions

Reversible Molecular Computation in Ciliates*

Lila Kari, Jarkko Kari, and Laura F. Landweber

Summary. In [10] we proved that a model for the guided homologous recombinations that take place during gene rearrangement in ciliates has the computational power of a Turing machine, the accepted formal model of computation. In this paper we change some of the assumptions and propose a new model that *(i)* allows recombinations between and within circular strands in addition to recombinations between linear and circular strands and within linear strands, *(ii)* relies on a *commutative* splicing scheme, and *(iii)* where all recombinations are reversible. We prove that the modified model, the *reversible guided recombination system*, has the computational power of a Turing machine. This indicates that, in principle, some unicellular organisms may have the capacity to perform any computation carried out by an electronic computer.

1 Basic notions and notations

The process we model is gene rearrangement in ciliates, a diverse group of 8000 or more unicellular eukaryotes (nucleated cells) named for their wisp-like covering of cilia. They possess two types of nuclei: an active *macronucleus* (soma) and a functionally inert *micronucleus* (germline) which contributes only to sexual reproduction. The somatically active macronucleus forms from the germline micronucleus after sexual reproduction, during the course of development. The genomic copies of some protein-coding genes in the micronucleus of hypotrichous ciliates are obscured by the presence of intervening non-protein-coding DNA sequence elements (internally eliminated sequences, or *IES*s). These must be removed before the assembly of a functional copy of the gene in the somatic macronucleus. Furthermore, the protein-coding DNA segments (macronuclear destined sequences, or *MDS*s) in species of *Oxytricha* and *Stylonychia* are sometimes present in a permuted order relative to their final position in the macronuclear copy. For example, in *O. nova*, the micronuclear copy of three genes (Actin I, α-telomere binding protein, and DNA polymerase α) must be reordered and intervening DNA sequences

* Research partially supported by Grant R2824A01 of the Natural Sciences and Engineering Research Council of Canada to L.K., N.S.F. Grant CCR 97-33101 to J.K., and a Burroughs-Wellcome Fund New Investigator Award in Molecular Parasitology to L.F.L.

removed in order to construct functional macronuclear genes. Most impressively, the gene encoding DNA polymerase α (DNA pol α) in *O. trifallax* is apparently scrambled in 50 or more pieces in its germline nucleus [9]. Destined to unscramble its micronuclear genes by putting the pieces together again, *O. trifallax* routinely solves a potentially complicated computational problem when rewriting its genomic sequences to form the macronuclear copies [10].

This process of unscrambling bears a remarkable resemblance to the DNA algorithm Adleman [1] used to solve a seven-city instance of the Directed Hamiltonian Path Problem. The developing ciliate macronuclear "computer" (Figure 1) apparently relies on the information contained in short direct repeat sequences to act as minimal guides in a series of homologous recombination events. These guide-sequences act as splints, and the process of recombination results in linking the protein-encoding segments (MDSs) that belong next to each other in the final protein coding sequence. As such, the unscrambling of sequences that encode DNA polymerase α accomplishes an astounding feat of cellular computation. Other structural components of the ciliate chromatin presumably play a significant role, but the exact details of the mechanism are still unknown [10].

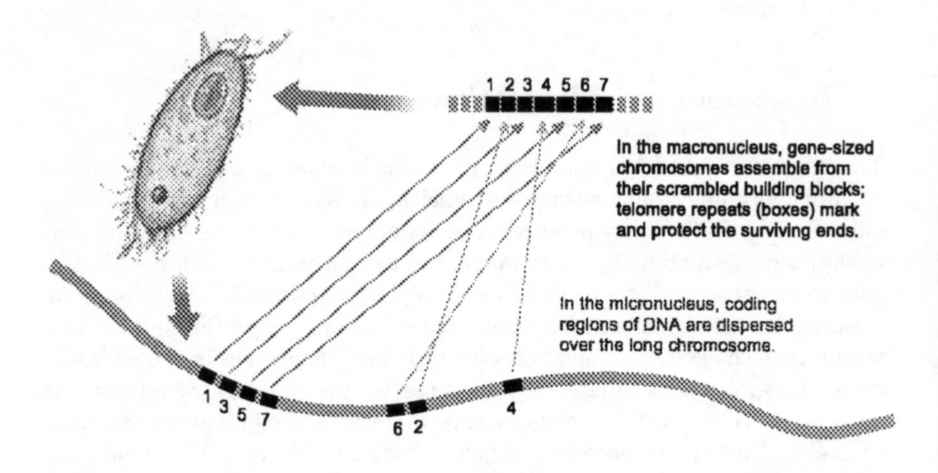

Fig. 1. Overview of gene unscrambling. Dispersed coding MDSs 1-7 reassemble during macronuclear development to form the functional gene copy (top), complete with telomere addition to mark and protect both ends of the gene.

Before introducing the formal model, we summarize our notation. An alphabet Σ is a finite, nonempty set. A sequence of letters from Σ is called a string (word) over Σ and in our interpretation corresponds to a linear strand. The words are denoted by lowercase letters such as u, v, α_i, x_{ij}. The length of a word w is denoted by $|w|$ and represents the total number of occurrences of letters in the word. A word with 0 letters in it is called an empty word and is denoted by λ. The set of all possible words consisting of letters from Σ is denoted by Σ^*, and the set of all nonempty words by Σ^+. We also define circular words over Σ by declaring two words to be equivalent if and only if (iff) one is a cyclic permutation of the other. In other words, w is equivalent to w' iff they can be decomposed as $w = uv$ and $w' = vu$, respectively. Such a circular word $\bullet w$ refers to any of the circular permutations of the letters in w. Denote by $\Sigma^\bullet$ the set of all circular words over Σ.

A rewriting system $TM = (S, \Sigma \cup \{\#\}, P)$ is called a *Turing machine*, [14], iff:

(i) S and $\Sigma \cup \{\#\}$ (with $\# \notin \Sigma$ and $\Sigma \neq \emptyset$) are two disjoint alphabets referred to as the *state* and the *tape* alphabets.

(ii) Elements s_0 and s_f of S, and B of Σ are the *initial* and *final* state, and the *blank symbol*, respectively. Also a subset T of Σ is specified and referred to as the *terminal* alphabet. It is assumed that T is not empty.

(iii) The productions (rewriting rules) of P are of the forms

(1) $s_i a \longrightarrow s_j b$ (overprint)
(2) $s_i ac \longrightarrow as_j c$ (move right)
(3) $s_i a\# \longrightarrow as_j B\#$ (move right and extend workspace)
(4) $cs_i a \longrightarrow s_j ca$ (move left)
(5) $\#s_i a \longrightarrow \#s_j Ba$ (move left and extend workspace)
(6) $s_f\, a \longrightarrow s_f$
(7) $a\, s_f \longrightarrow s_f$

where s_i and s_j are states in S, $s_i \neq q_f$, $s_j \neq q_f$, and a, b, c are in Σ. For each pair (s_i, a), where s_i and a are in the appropriate ranges, P either contains no productions (2) and (3) (resp.(4) and (5)) or else contains both (3) and (2) for every c (resp.contains both (5) and (4) for every c). There is no pair (s_i, a) such that the word $s_i a$ is a subword of the left side in two productions of the forms (1), (3), (5).

A configuration of the TM is of the form $\#w_1 s_i w_2\#$, where $w_1 w_2$ represents the contents of the tape, #s are the boundary markers, and the position of the state symbol s_i indicates the position of the read/write head on the tape: if s_i is positioned at the left of a letter a, this indicates that the read/write head is placed over the cell containing a. The TM changes from one configuration to another according to its rules. For example, if the current configuration is $\#ws_i aw'\#$ and the TM has the rule $s_i a \longrightarrow s_j b$, this means that the read/write head positioned over the letter a will write b over it, and change its state from s_i to s_j. The next configuration in the derivation will be thus $\#ws_j bw'\#$.

The Turing machine TM *halts* with a word w iff there exists a derivation that, when started with the read/write head positioned at the beginning of w eventually reaches the final state, i.e. if $\#s_0 w\#$ derives $\#s_f\#$ by succesive applications of the rewriting rules (1) - (7) The language $L(TM)$ *accepted* by TM consists of all words over the terminal alphabet T for which the TM halts. Note that TM is *deterministic*: at each step of the rewriting process, the application of at most one production is possible.

In the proof of our main result we will implicitely and repeatedly rely on the following lemma.

Lemma. *([5], Theorem 2.1, p.444) Let* $TM = (S, \Sigma \cup \{\#\}, P)$ *be a Turing machine. If* $\#s_0w\#$ *derives* $\#s_f\#$ *using forward and backward applications of rewriting rules of the TM, then w is accepted by the TM.*

2 Computational power of reversible guided recombination systems

In [10] we have introduced the notion of a *guided recombination system* that models the process taking place during gene rearrangement, and proved that such systems have the computational power of a Turing machine, the most widely used theoretical model of electronic computers.

In this section we change some of the assumptions of [10] and propose a new model that: (1) allows inter- and intra-molecular recombinations between and within circular strands in addition to linear-circular intermolecular recombinations and intramolecular recombinations within linear strands, (2) allows the splicing scheme to be commutative, and (3) allows each recombination operation to be reversible by default. We prove that the modified model, the *reversible guided recombination system*, has the computational power of a Turing machine.

The following string operations model intra- and intermolecular recombinations by assuming that homologous recombination is influenced by the presence of certain contexts, i.e., either the presence of an IES or an MDS flanking a junction sequence x_{ij}. The observed dependence on the old macronuclear sequence for correct IES removal in *Paramecium* suggests that this is the case ([11]). This restriction captures the fact that the guide sequences do not contain all the information for accurate splicing during gene unscrambling.

We define the contexts that restrict the use of recombinations by a *splicing scheme*, [6], [7], a pair $(\Sigma, \sim)$ where Σ is the alphabet and $\sim$, the pairing relation of the scheme, is a binary relation between triplets of nonempty words satisfying the following condition: If $(p, x, q) \sim (p', y, q')$ then $x = y$.

In the splicing scheme $(\Sigma, \sim)$ pairs $(p, x, q) \sim (p', x, q')$ now define the contexts necessary for a recombination between the repeats x.

Let us consider a splicing scheme where the binary relation is *commutative*. Then we define *reversible guided recombination* as

$$\{uxwxv\} \Longleftrightarrow \{uxv, \bullet wx\}, \text{ where } u = u'p, w = qw' = w''p', v = q'v',$$

and $p, q, p', q', u, x, w, v, u', w', w'' \in \Sigma^*$, $x \neq \lambda$. This constrains recombination to occur only if the restrictions of the splicing scheme concerning x are fulfilled, i.e., the first occurrence of x is preceded by p and followed by q and its second occurrence is preceded by p' and followed by q'.

Intramolecular recombinations can also take place within a circular strand, and intermolecular recombinations can occur between two circular strands. This leads to the definition of *circular guided recombination* as

$$\{\bullet uxwxv\} \Longleftrightarrow \{\bullet uxv, \bullet wx\}, \text{ where } u = u'p, w = qw' = w''p', v = q'v',$$

and $p, q, p', q', u, x, w, v, u', w', w'' \in \Sigma^*$, $x \neq \lambda$.

The above operations resemble the "splicing operation" introduced by Head in [6] and "circular splicing" ([7], [15], [13]). [12], [2] and subsequently [16] showed that these models have the computational power of a universal Turing machine. (See [8] for a review.)

The particular case where all the contexts are empty, i.e, $(\lambda, x, \lambda) \sim (\lambda, x, \lambda)$ for all $x \in \Sigma^+$ corresponds to the case where recombination may occur between every repeat sequence, regardless of the contexts. These unguided (context-free) recombinations are computationally not very powerful: we have proved that they can only generate regular languages.

If we use the classical notion of a set, we can assume that the strings entering a recombination are available for multiple operations. Similarly, there would be no restriction on the number of copies of each strand produced by recombination. However, we can also assume some strings are only available in a limited number of copies. Mathematically this translates into using *multisets*, where one keeps track of the number of copies of a string at each moment. In the style of [4], if $\mathbf{N}$ is the set of natural numbers, a multiset of Σ^* is a mapping $M : \Sigma^* \longrightarrow \mathbf{N} \cup \{\infty\}$, where, for a word $w \in \Sigma^*$, $M(w)$ represents the number of occurrences of w. Here, $M(w) = \infty$ means that there are unboundedly many copies of the string w. The set $\text{supp}(M) = \{w \in \Sigma^* | \; M(w) \neq 0\}$, the *support of* M, consists of the strings that are present at least once in the multiset M. The definition can be naturally extended to a multiset consisting of linear as well as circular strands.

We now define a *reversible guided recombination system* that captures the series of dispersed homologous recombination events that take place during these gene rearrangements in ciliates.

Definition *A reversible guided recombination system is a triple* $R = (\Sigma, \sim, A)$ *where* $(\Sigma, \sim)$ *is a splicing scheme, and* $A \in \Sigma^+$ *is a linear string called the axiom.*

A reversible guided recombination system R defines a *derivation relation* that produces a new multiset from a given multiset of linear and circular strands, as follows. Starting from a "collection" (multiset) of strings with a certain number of available copies of each string, the next multiset is *derived from* the first one by an intra- or inter-molecular recombination between existing strings. The strands participating in the recombination are "consumed" (their multiplicity decreases by 1) whereas the products of the recombination are added to the multiset (their multiplicity increases by 1).

For two multisets S and S' in $\Sigma^* \cup \Sigma^\bullet$, we say that S derives S' and we write $S \Longrightarrow_R S'$, iff one of the following two cases hold:

(1) there exist $\alpha \in \text{supp}(S)$, $\beta, \bullet\gamma \in \text{supp}(S')$ such that

– $\{\alpha\} \Longrightarrow \{\beta, \bullet\gamma\}$ according to an intramolecular recombination step in R,

– $S'(\alpha) = S(\alpha) - 1$, $S'(\beta) = S(\beta) + 1$, $S'(\bullet\gamma) = S(\bullet\gamma) + 1$;

(2) there exist $\alpha', \bullet\beta' \in \text{supp}(S)$, $\gamma' \in \text{supp}(S')$ such that

– $\{\alpha', \bullet\beta'\} \Longrightarrow \{\gamma'\}$ according to an intermolecular recombination step in R,

– $S'(\alpha') = S(\alpha') - 1$, $S'(\bullet\beta') = S(\bullet\beta') - 1$, $S'(\gamma') = S(\gamma') + 1$.

In the above definition of the derivation relation α and β are either both linear or both circular strands in $\Sigma^* \cup \Sigma^\bullet$.

Those linear strands which, by repeated recombinations with initial and intermediate strands eventually produce the axiom, form the language of the reversible guided recombination system. Formally,

$$L_k(R) = \{w \in \Sigma^* |\ \{w\} \Longrightarrow_R^* S \text{ and } A \in \text{supp}(S)\},$$

where the multiplicity of w equals k. Note that $L_k(R) \subseteq L_{k+1}(R)$ for any $k \geq 1$.

Theorem. *Let L be a language over T^* accepted by a Turing machine $TM = (S, \Sigma \cup \{\#\}, P)$e. Then there exist an alphabet Σ', a sequence $\pi \in \Sigma'^*$, depending on L, and a reversible guided recombination system R such that a word w over T^* is in L if and only if $\#^6 q_0 w \#^6 \pi$ belongs to $L_k(R)$ for some $k \geq 1$.*

Proof. Consider that the rules of P are ordered in an arbitrary fashion and numbered. Thus, if TM has m rules, a rule is of the form $i : u_i \longrightarrow v_i$ where $1 \leq i \leq m$.

We construct a reversible guided recombination system $R = (\Sigma', \sim, A)$ and a sequence $\pi \in \Sigma'^*$ with the required properties. The alphabet is $\Sigma' = S \cup \Sigma \cup \{\#\} \cup \{\$_i |\ 0 \leq i \leq m+1\}$. The axiom, i.e., the target string to be achieved at the end of the computation, consists of the final state of the TM bounded by markers:

$$A = \#^{n+2} q_f\, \#^{n+2} \$_0 \$_1 \ldots \$_m \$_{m+1},$$

where n is the maximum length of the left-hand side or right-hand side words of any of the rules of the Turing machine, i.e., $n = 4$.

The sequence π consists of the catenation of the right-hand sides of the TM rules bounded by markers, as follows:

$$\pi = \$_0\ \$_1 e_1 v_1 f_1 \$_1\ \$_2 e_2 v_2 f_2 \$_2 \ldots \$_m e_m v_m f_m \$_m\ \$_{m+1},$$

where $i: \; u_i \longrightarrow v_i$, $1 \le i \le m+1$ are the rules of TM and $e_i, v_i \in \Sigma \cup \{\#\}$.

If a word $w \in T^*$ is accepted by the TM, a computation starts then from a strand of the form $\#^{n+2} q_0 w \#^{n+2} \pi$, where we will refer to the subsequence starting with $\$_0$ as the "program", and to the subsequence at the left of $\$_0$ as the "data".

We construct the relation $\sim$ such that:

– The right-hand sides of rules of TM can be excised from the program as circular strands which then interact with the data.

– When the left-hand side of a TM rule appears in the data, the application of the rule can be simulated by the insertion of the circular strand encoding the right-hand side, followed by the deletion of the left-hand side.

To this aim, for each rule $i : u \longrightarrow v$ of the TM, we introduce in $\sim$ the pairs

$$(A) \qquad (ceu, f, \$_i ev) \sim (\$_i ev, f, d),$$

$$(B) \qquad (c, e, uf\$_i) \sim (uf\$_i, e, vfd),$$

$$(C) \qquad (\$_{i-1}, \$_i, evf) \sim (evf, \$_i, \$_{i+1}),$$

for all $c \in \{\#\}^* \Sigma^*$, $d \in \Sigma^* \{\#\}^*$, $|c| = |d| = n$, $e, f \in \Sigma \cup \{\#\}$.

Pairs (C) allow recombinations of the type

$$\{x\$_{i-1}\$_i evf\$_i\$_{i+1}y\} \Longleftrightarrow \{x\$_{i-1}\$_i\$_{i+1}y, \;\; \bullet\$_i evf\}$$

i.e., allow excision of right-hand sides or TM rules from the program.

Pairs (A) allow recombinations of the type

$$\{xceufdy, \;\; \bullet\$_i evf\} \Longleftrightarrow \{xceuf\$_i evfdy\}$$

that insert the right-hand side of a TM rule next to its left-hand side occurring in the data.

Finally, pairs (B) allow recombinations of the type

$$\{xceuf\$_i evfdy\} \Longleftrightarrow \{xcevfdy, \;\; \bullet\$_i euf\}$$

that excise the left-hand side of the TM rules in the data.

Note that (C) serves the role of excising the necessary right-hand side of a TM rule from the program while the tandem (A)-(B) accomplishes the replacement of a left-hand side of a TM rule in the data by its right-hand side, simulating thus a TM rewriting step. Indeed, for any $x, y \in \Sigma'$ we can simulate a derivation step of the TM as follows:

$$\{xceufdy\$_0 \ldots \$_{i-1}\$_i evf\$_i\$_{i+1} \ldots \$_{m+1}\} \Longrightarrow_R$$

$$\{xceufdy\$_0 \ldots \$_{i-1}\$_i\$_{i+1} \ldots \$_{m+1}, \bullet\$_i evf\} \Longrightarrow_R$$

$$\{xceuf\$_i evfdy\$_0 \ldots \$_{i-1}\$_i\$_{i+1} \ldots \$_{m+1}\} \Longrightarrow_R$$

$$\{xcevfdy\$_0 \ldots \$_{i-1}\$_i\$_{i+1} \ldots \$_{m+1}, \; \bullet\$_i euf\}.$$

The first step is an intramolecular recombination using contexts (C) around the repeat $\$_i$ to excise $\bullet\$_i evf$. Note that if the current strand does not contain a subword $\$_i evf\$_i$, this can be obtained from another copy of the original linear strand, which is initially present in k copies. The second step is an intermolecular recombination using contexts (A) around the repeat f, to insert $\$_i evf$ after $ceuf$. The third step is an intramolecular recombination using contexts (B) around the direct repeat e to delete $\$_i euf$ from the linear strand. Thus, the "legal" insertion/deletion succession that simulates one TM derivation step claims that any u in the data, that is surrounded by at least $n+1$ letters on both sides may be replaced by v. This explains why in our choice of axiom we needed $n+1$ extra symbols $\#$ to provide the contexts allowing recombinations to simulate all TM rules, including (3) and (5).

From the fact that a TM derivation step can be simulated by recombination steps we deduce that, if the TM accepts a word w, then we can start a derivation in R from

$$\#^{n+2} q_0 w \#^{n+2} \pi = \#^{n+2} q_0 w \#^{n+2} \$_0 \$_1 \ldots \$_i e_i v_i f_i \$_i \ldots \$_m \$_{m+1}$$

and reach the axiom by only using recombinations according to R. This means that our word is accepted by R, that is, it belongs to $L_k(R)$ for some k. Note that if some rules of the TM have not been used in the derivation then they can be excised in the end, and that k should be large enough so that we do not exhaust the set of rewriting rules.

For the converse implication, it suffices to prove that starting from the strand $\#^{n+2} q_0 w \#^{n+2} \pi$, no other recombinations except those that excise rules of TM from the program and those that simulate steps of the TM (forwards or backwards) in the data are possible in R.

In the beginning of the derivation we start with no circular strands and k copies of the linear strand

$$\#^{n+2} q_0 w \#^{n+2} \$_0 \$_1 e_1 v_1 f_1 \$_1 \ldots \$_{i-1} \$_i e_i v_i f_i \$_i \$_{i+1} \ldots \$_{m+1}, w \in T^*,$$

where $i : u_i \longrightarrow v_i$ are TM rules, $e_i, f_i \in \Sigma \cup \{\#\}$, $1 \leq i \leq m$.

Assume now that the current multiset contains linear strands of the form $\delta_0 \pi$, where $\delta_0 \in \Sigma'^*$ contains only one state symbol and no $\$_i$ symbols and

$$\pi = \$_0 r_1 r_2 \ldots r_m \$_{m+1},$$

with r_i either encoding the right-hand side of a rule or being the remnant of a rule, i.e., $r_i \in \{\$_i e_i v_i f_i \$_i\} \cup \{\$_i\}$, $1 \leq i \leq m$. Moreover, assume that the circular strands present in the multiset are of the form $\bullet\$_i e_i v_i f_i$, or $\bullet\$_i e_i u_i f_i$ with $i : u_i \longrightarrow v_i$ a rule of the TM and $e_i, f_i \in \Sigma \cup \{\#\}$.

Then,

(i) Due to the way we constructed the recombination system, circular strands cannot interact with each other.

Indeed, an insertion or deletion using (C) would require the presence of more than one $\$_i$-marker in a circular strand, which contradicts our assumptions. An insertion/deletion using (A) or (B) would require the presence in some circular strand of a subword of the form $ceuf$ (respectively $evfd$) the length of which is at least $n+3$, greater than the length of any $\$_i$-free subword of existent circular strands.

(ii) We cannot use (A) or (B) to insert or delete in the program.

Indeed, an insertion in π using (A) (respectively (B)) could happen in one of the following cases. In the first case, π contains a word $ceufd$ (respectively $cevfd$) of minimum length of $2n+3$. This, however, is more than the length of the longest subword over $\Sigma \cup \{\#\}$ in π. The other possibility would be that π contains the subsequence $\$_i evf\$_i ev$ (respectively $uf\$_i euf\$_i$), which contradicts the fact that no marker $\$_i$ appears alone in π, as π always contains at least two consecutive markers. Consequently, insertions in π using (A) or (B) cannot occur.

A deletion from π using (A) or (B) would imply the presence in π of $ceuf\$_i ev$ (respectively $ceuf\$_i$) of minimum length of $n+3$, which is greater than $n+2$, the maximum length of a $\$_i$-free subword in the program.

(iii) We cannot use (C) to insert or delete in the data. Indeed such recombinations would require the presence in either δ_0 or in a circular strand with more than one $\$_i$-marker, which is impossible.

Arguments *(i)* - *(iii)* show that the only possible recombinations are either insertions/deletions in the program using (C) or insertions/deletions in the data using (A) and (B).

Deletions in π according to (C) result in the release of circular strands $\bullet\$_i evf$ that encode right-hand sides of TM rules. Insertions according to (C) only mean that circular strands encoding right sides of a TM rule can be inserted back after their excision, pointing out the reversibility of the constructed system. Indeed the only other possibility of inserting into π using (C) would require the presence in π of $evf\$_i evf$ - contradiction with the form of π.

The next step is to show that the only possible insertions/deletions in the data are those simulating a rewriting step of TM in forwards or backwards direction using a TM rewriting rule or its reverse.

Indeed,

(1) It is not possible to delete in δ_0 using (A) or (B), as these operations would require the presence of a $\$_i$ in δ_0. Therefore only insertions in δ_0 using (A) or (B) are possible.

An insertion is possible only if δ_0 contains the left- or right-hand side of some rewrite rule $u \longrightarrow v$. In the first case, an insertion using (A) is possible, in the second case, an insertion using (B) is possible.

In any case, after the insertion in the data, $xceufdy\pi$ (respectively $xcevfdy\pi$) has become $xceuf\$_i evfdy\pi$.

Let δ_1 be the new updated data, i.e., the current linear strand is of the form

$$\delta_1 \, \pi = xceuf\$_i evfdy \, \pi.$$

Note that, as δ_0 contains only one state symbol and no marker $\$_i$, the newly formed word δ_1 contains only two state symbols (read/write heads), one in u and one in v, and only one marker $\$_i$. (Here we use the fact that every rule $u \longrightarrow v$ of the TM has exactly one state symbol on each side.)

(2) Starting now from $\delta_1 \pi$,

(2a) No insertion in δ_1 using (A) or (B) may take place.

Indeed, another insertion in δ_1 using (A) or (B) can happen in one of two cases. The case requiring that $\$_i ev f \$_i ev$ (respectively $uf\$_i euf\$_i$) be present in δ_1 immediately leads to a contradiction as δ_1 does not contain two $\$_i$-markers.

The other case is also impossible as the marker $\$_i$ "breaks" the contexts necessary for further insertions. Indeed, this second case requires the presence in δ_1 of a word of the form $ceufd$ (respectively $cevfd$). This implies that the read/write head symbol should be both followed and preceded by at least $(n+1)$ letters different from $\$_i$. In δ_1, the first read/write head is in u and the number of letters in $\Sigma \cup \{\#\}$ following it is at most $|u| - 1 + |f| \leq n - 1 + 1 = n$, which is not enough. The second read/write head is in v and the number of letters preceding it is at most $|e| + |v| - 1 \leq 1 + n - 1 = n$, which is not enough.

(2b) We can delete in δ_1 using (A) or (B).

There is only one possible location for either deletion, and they result in replacing $xceuf\$_i evfdy\pi$ by $xceufdy\pi$ or $xcevfdy\pi$ respectively.

All in all, in two steps (1)-(2), the only possible recombinations on the data either replace u with v, v with u, or keep the strand unchanged, where $u \longrightarrow v$ is a rewrite rule of the TM.

The arguments above imply that the only possible operations on the data simulate legal rewritings of the TM by tandem recombination steps that necessarily follow each other. Together with the arguments that the only operations affecting the program are excisions of circular strands encoding TM rules, and that the circular TM rules do not interact with each other, this proves the converse implication.

From the definition of the Turing machine we see that n, the maximum length of a word occurring in a TM rule, equals 4, which completes the proof of the theorem.

Q.E.D.

The conclusion that this formal model of the process of gene unscrambling in ciliates can simulate a Turing machine suggests that such processes may have many potential uses in solving computational problems that occur in complex evolving systems.

Acknowledgements. Rani Siromoney and Gilles Brassard for suggestions, Gheorghe Păun, Erik Winfree for comments. Grzegorz Rozenberg for discussion.

References

1. Adleman, L. M. 1994. Molecular computation of solutions to combinatorial problems. *Science* 266: 1021-1024.
2. Csuhaj-Varju, E., Freund, R., Kari, L., and G. Paun. 1996. DNA computing based on splicing: universality results. In Hunter, L. and T. Klein (editors). *Proceedings of 1st Pacific Symposium on Biocomputing.* World Scientific Publ., Singapore, Pages 179-190.
3. Denninghoff, R.W and Gatterdam, R.W. 1989. On the undecidability of splicing systems. *International Journal of Computer Mathematics*, 27, 133-145.
4. Eilenberg, S. 1984. *Automata, Languages and Machines.* Academic Press, New York.
5. Harju, T. and Karhumaki, J. 1997. Morphisms. In *Handbook of Formal Languages*, vol.1, (Rozenberg, G. and Salomaa, A. Eds.), Springer Verlag, Berlin, Pages 439-510.
6. Head, T. 1987. Formal language theory and DNA: an analysis of the generative capacity of specific recombinant behaviors. *Bull. Math. Biology* 49: 737-759.
7. Head, T, 1991. Splicing schemes and DNA. In *Lindenmayer systems* (Rozenberg, G. and Salomaa, A., Eds.). Springer Verlag, Berlin. Pages 371-383.
8. Head, T., Paun, G. and Pixton, D. 1997. Language theory and molecular genetics. In *Handbook of Formal Languages* (Rozenberg, G. and Salomaa, A. Eds.), vol. 2, Springer Verlag, Berlin. Pages 295-358.
9. Hoffman, D.C., and Prescott, D.M. 1997. Evolution of internal eliminated segments and scrambling in the micronuclear gene encoding DNA polymerase α in two *Oxytricha* species. *Nucl. Acids Res.* 25: 1883-1889.
10. Landweber, L.F., and Kari, L. 1999. Universal molecular computation in ciliates. To appear in *Evolution as Computation* (Landweber, L.F. and Winfree, E., Eds.), Springer-Verlag.
11. Meyer, E. and Duharcourt, S. 1996. Epigenetic Programming of Developmental Genome Rearrangements in Ciliates. *Cell* vol. 87, 9-12.
12. Paun, G. 1995. On the power of the splicing operation. *Int. J. Comp. Math* 59, 27-35.
13. Pixton, D., 1995. Linear and circular splicing systems. Proceedings of the *First International Symposium on Intelligence in Neural and Biological Systems*, IEEE Computer Society Press, Los Alamos. Pages 181-188.
14. Salomaa, A. 1973. *Formal Languages.* Academic Press, New York.
15. Siromoney, R., Subramanian, K.G. and Rajkumar Dare, V. 1992. Circular DNA and splicing systems. In Parallel Image Analysis. *Lecture Notes in Computer Science* 654, Springer Verlag, Berlin. Pages 260-273.
16. Yokomori, T., Kobayashi, S., and Ferretti, C. 1997. Circular Splicing Systems and DNA Computability. In *Proc. of IEEE International Conference on Evolutionary Computation'97.* Pages 219-224.

Logic, Probability, and Rough Sets

Zdzisław Pawlak

Summary. The paper analyzes some properties of decision rules in the framework of rough set theory.and knowledge discovery systems. With every decision rule two conditional probabilities are associated called the certatinty and coverage factors, respectively. It is shown that these coefficients satisfy the Bayes' theorem. This relationship can be used as a new approach to Bayesian reasoning, without referring to prior and posterior probabilities, inherently associated with classical Bayesian inference. Decision rules are implications and the relationship between implications and Bayes' theorem first was revealed by Lukasiewicz in connection with his multivaled logic.

1 Introduction

This paper is concerned with some considerations concerning inference rules and decision rules, from the rough set perspective.

Decision rules play an important rule in various branches of AI, e.g., data mining, machine learning, decision support and others. Interference rules play a fundamental role in logic, and attracted attention of logicians and philosophers for many years. From logical point of view both, decision rules and interference rules are implications.

However, there are essential differences between using decision rules in AI and inference rules in logic. Inference rules (*modus ponens*) must be true, in order to guarantee to draw true conclusions from true premises. In contrast, in AI decision rules are meant as prescription of decisions that must be taken when some conditions are satisfied. In this case, in order to express to what degree the decision can be trusted, instead of truth, a *credibility factor* of the decision rule is associated. The relationship between truth and probability first was investigated by Łukasiewicz [3, 16], who showed that probabilistic interpretation of implication leads to Bayes' theorem.

In rough set theory decision rules are of special interest, since they are inherently connected with the basic concepts of the theory – approximations and partial dependencies. The rough set approach bridges to some extend the logical and AI views on decision rules, and can be seen as generalization of Łukasiewicz's probabilistic logic associated with multivalued logic [3, 16]. Some considerations on this subject can be also found in [22, 23, 24, 25, 30].

For more information about rough set theory and its generalizations the reader is advised to consult the enclosed references. In particular an overview of current state of the theory and its applications can be found in [29].

2 The Łukasiewicz's Approach

In this section we present briefly basic ideas of Łukasiewicz's approach to multivalued logics as probabilistic logic.

Łukasiewicz associates with every so called *indefinite proposition* of one variable x, $\Phi(x)$ a true value $\pi(\Phi(x))$, which is the ratio of the number of all values of x which satify $\Phi(x)$, to the number of all possible values of x. For example, the true value of the proposition "x is greater than 3" for $x = 1, 2, \ldots, 5$ is $2/5$. It turns out that assuming the following three axioms

1) Φ is false if and only if $\pi(\Phi) = 0$;
2) Φ is true if and only if $\pi(\Phi) = 1$;
3) if $\pi(\Phi \rightarrow \Psi) = 1$ then $\pi(\Phi) + \pi(\sim \Phi \wedge \Psi) = \pi(\Psi)$;

one can show that

4) if $\pi(\Phi \equiv \Psi) = 1$ then $\pi(\Phi) = \pi(\Psi)$;
5) $\pi(\Phi) + \pi(\sim \Phi) = 1$;
6) $\pi(\Phi \vee \Psi) = \pi(\Phi) + \pi(\Psi) - \pi(\Phi \wedge \Psi)$;
7) $\pi(\Phi \wedge \Psi) = 0$ iff $\pi(\Phi \vee \Psi) = \pi(\Phi) + \pi(\Psi)$.

Obviously, the above properties have probabilistic flavour.

The idea that implication should be associated with conditional probability is attributed to Ramsey (cf. [1]), but as mentioned in the introduction, is can be traced back to Łukasiewicz [3, 16], who first formulated this idea in connection with his multivalued logic and probability. More extensive study of connection of implication and conditional probability can be found in [1].

3 Rough Sets – the Intuitive Background

Rough set theory is based on the indiscernibility relation. The indiscernibility relation identifies objects displaying the same properties, i.e., groups together elements of interest into *granules* of indiscernible (similar) objects. These granules, called *elementary sets* (*concepts*), are basic building blocks (concepts) of knowledge about the universe. For example, if our *universe of discourse* were patients suffering from a certain disease, then patients displaying the same symptoms would be indiscernible in view of the available information and form clusters of similar patients.

Union of elementary concepts is referred to as a *crisp* or *precise* concept (set); otherwise a concept (set) is called *rough, vague* or *imprecise*. Thus rough concepts cannot be expressed in terms of elementary concepts. However, they can be expressed approximately by means of elementary concepts using the idea of the *lower* and the *upper approximation* of a concept. The lower approximation of the concept is the union of all elementary concepts which are included in the concept, whereas the upper approximation is the union of all elementary concepts which have nonempty intersection with the concept, i.e., the lower and the upper approximations of a concept are the union of all elementary concepts which are surely and possibly included in the concept, respectively. The difference between the lower and the upper approximation of the concept is its *boundary region*. Hence a concept is rough if it has nonempty boundary region.

Approximations are basic operations in rough set theory. They are used to deal with rough (vague) concepts, since in the rough set approach we replace rough concepts by pairs of precise concepts – the lower and the upper approximations of the rough concept. Thus approximations are used to express precisely our knowledge about imprecise concepts.

The problem of expressing vague concepts in terms of precise concepts in rough set theory can be also formulated differently, by employing the idea of dependency (partial) between concepts. We say that a concept (set) depends totally on a set of concepts if it is the union of all those concepts; if it is the union of some concepts it depends partialy on these concepts. Thus partaial dependency can be also used to express vague concepts in terms of precise concepts.

Both, approximations and dependencies are defined using decision rules, which are implications in the form *"if...then..."*

Approximations, dependencies and decision rules are basic tools of rough set theory and will be discussed in details in the next sections.

4 Database

Rough set theory is mainly ment to be used to data analysis, therfore in what follows the theory will be formulated not in generale terms but with refererence to data. Hence we will start our consideration from a database. Intuitevily by the database we will understand a data table whose columns are are labelled by attributes (e.g., color, temerature, etc.), rows are lablled by objects of interst (e.g., patienents, states, processe etc.) and entries of the table are attrinute values (e.g., red, high, etc.). A very simple example of a database is shown below: The table contains data about six cars where F, C, P and M denote fuel consumption, color, selling price and marketabiality, respectively.

Formally the database is defined as follows.

By a *database* we will understand a pair $S = (U, A)$, where U and A, are finite, nonempty sets called the *universe*, and a set of *attributes* respectively. With every attribute $a \in A$ we associate a set V_a, of its *values*, called the

Table 1. An example of a database

Car	*F*	*C*	*P*	*M*
1	*med.*	*black*	*med.*	*poor*
2	*high*	*white*	*med.*	*poor*
3	*med.*	*white*	*low*	*poor*
4	*low*	*black*	*med.*	*good*
5	*high*	*red*	*low*	*poor*
6	*med.*	*white*	*low*	*good*

domain of a. Any subset B of A determines a binary relation $I(B)$ on U, which will be called an *indiscernibility relation*, and is defined as follows:

$(x, y) \in I(B)$ if and only if $a(x) = a(y)$ for every $a \in A$, where $a(x)$ denotes the value of attribute a for element x.

It can easily be seen that $I(B)$ is an equivalence relation. The family of all equivalence classes of $I(B)$, i.e., partition determined by B, will be denoted by $U/I(B)$, or simple U/B; an equivalence class of $I(B)$, i.e., block of the partition U/B, containing x will be denoted by $B(x)$.

If (x, y) belongs to $I(B)$ we will say that x and y are *B-indiscernible* or indiscernible with respect to B. Equivalence classes of the relation $I(B)$ (or blocks of the partition U/B) are refereed to as *B-elementary sets* or *B-granules.*

For example, cars 1, 3 and 6 are pairwise indiscernible with respect to the attribute F. If $B = \{C, P\}$ and $x = 3$, then $B(x) = \{3, 6\}$.

Instead of an equivalence relation as a basis for rough set theory many authors proposed another relations, e.g., a tolerance relation, an ordering relations and others. However in this paper we will stay by the equivalence relation.

5 Approximations of Sets

Having defined the notion of a database we are now in the position to put forth our basic notions of approximation of a set by other sets, which is defined next.

Let us define two following operations on sets:

$$B_*(X) = \bigcup_{x \in U} \{B(x) : B(x) \subseteq X\},$$

$$B^*(X) = \bigcup_{x \in U} \{B(x) : B(x) \cap X \neq \emptyset\},$$

assigning to every $X \subseteq U$ two sets $B_*(X)$ and $B^*(X)$ called the *B-lower* and the *B-upper approximation* of X, respectively.

Hence, the B-lower approximation of a concept is the union of all B-granules that are included in the concept, whereas the *B-upper* approximation of a concept is the union of all B-granules that have a nonempty intersection with the concept. The set

$$BN_B(X) = B^*(X) - B_*(X)$$

will be referred to as the *B-boundary region* of X.

If the boundary region of X is the empty set, i.e., $BN_B(X) = \emptyset$, then X is *crisp* (*exact*) with respect to B; in the opposite case, i.e., if $BN_B(X) \neq \emptyset$, X is referred to as *rough* (*inexact*) with respect to B.

For example, for the set of cars $X = \{4, 6\}$ selling well the B-lower and the B-upper approximations of X are $\{4\}$ and $\{3, 4, 6\}$, respectively, where $B = \{F, C, P\}$.

6 Dependency of Attributes

Another important issue in data analysis is discovering dependencies between attributes. Intuitively, a set of attributes D *depends totally* on a set of attributes C, denoted $C \Rightarrow D$, if all values of attributes from D are uniquely determined by values of attributes from C. In other words, D depends totally on C, if there exists a functional dependency between values of D and C.

We would need also a more general concept of dependency, called a *partial dependency* of attributes. Intuitively, the partial dependency means that only some values of D are determined by values of C.

Formally dependency can be defined in the following way. Let D and C be subsets of A.

We will say that D *depends on* C in a *degree* k ($0 \leq k \leq 1$), denoted $C \Rightarrow_k D$, if

$$k = \gamma(C, D) = \frac{\sum_{X \in U/D} card(C_*(X))}{card\, U}.$$

If $C \Rightarrow_k D$, we will call $C-$ condition and $D-$ decision attributes, respectively. Any database with distinguished condition and decision attributes is usually called a *decision table.*

If $k = 1$ we say that D *depends totally* on C, and if $k < 1$, we say that D *depends partially* (in a *degree* k) on C, and if $k = 0$, *does not depend on* C

The coefficient k expresses the ratio of all elements of the universe, which can be properly classified to blocks of the partition U/D, employing attributes C and will be called the *degree of the dependency.*

For example, the degree of dependency between the set of condition attributes $\{F, C, P\}$ and the decision attribute $\{M\}$ in Table 1 is 2/3.

The notion of dependency of attributes is used to express relationships hidden in the database. It expresses the global properties of the database in contrast to approximations which express local properties of the database.

7 Decision Rules

Let S be a database and let C and D be condition and decision attributes, respectively.

By Φ, Ψ etc. we will denote logicals formulas built up from attributes, attribute-values and logical connectives (*and, or, not*) in a standard way. We will denote by $|\Phi|_S$ the set of all object $x \in U$ satisfying Φ in S and refer to as the *meaning* of Φ in S.

The expression $\pi_S(\Phi) = \frac{card(|\Phi|_S)}{card(U)}$ can be interpreted the probability that the formula Φ is true in S.

A *decision rule* is an expression in the form *"if...then..."*, written $\Phi \to \Psi$; Φ and Ψ are refered to as *conditions* and *decisions* of the rule respectively.

A decision rule $\Phi \to \Psi$ is *admissible* in S if $|\Phi|_S$ is the union of some C-elementary sets, $|\Psi|_S$ is the union of some D-elementary sets and $|\Phi \wedge \Psi|_S \neq \emptyset$. In what follows we will consider admissible decision rules only.

With every decision rule $\Phi \to \Psi$ we associate the conditional probability that Ψ is true in S given Φ is true in S with the probability $\pi_S(\Phi)$, called the *certainty factor* and defined as follows:

$$\pi_S(\Psi|\Phi) = \frac{card(|\Phi \wedge \Psi|_S)}{card(|\Phi|_S)},$$

where $|\Phi|_S \neq 0$ denotes the set of all objects satisfying Φ in S.

Besides, we will also need a *coverage factor* [46]

$$\pi_S(\Phi|\Psi) = \frac{card(|\Phi \wedge \Psi|_S)}{card(|\Phi|_S)},$$

which is the conditional probability that Φ is true in S given Ψ is true in S with the probability $\pi_S(\Psi)$.

For example, *(F,low) and (C,black) and (P,med.)* $\to$ *(M,good)* is an admissible rule in Table 1 and the certainty and coverage factors for this rule are 1/2 and 1/4, respectively.

Let $\{\Phi_i \to \Psi\}_n$ be a set of decision rules such that all conditions Φ_i are pairwise mutally exclusive, i.e., $|\Phi_i \wedge \Phi_j|_S = \emptyset$, for any $1 \leq i, j \leq n$, $i \neq j$, and

$$\sum_{i=1}^{n} \pi_S(\Phi_i|\Psi) = 1. \tag{1}$$

Then the following property holds:

$$\pi_S(\Psi) = \sum_{i=1}^{n} \pi_S(\Psi|\Phi_i) \cdot \pi_S(\Phi_i). \tag{2}$$

For any decision rule $\Phi \to \Psi$ the following formula is valid:

$$\pi_S(\Phi|\Psi) = \frac{\pi_S(\Psi|\Phi) \cdot \pi_S(\Phi)}{\sum_{i=1}^{n} \pi_S(\Psi|\Phi_i) \cdot \pi_S(\Phi_i)}. \tag{3}$$

Formula (2) is well known as the formula for total probability and it can be seen as generalization of axiom 3) in Łukasiewicz's probabilistic logic whereas formula (3) is the Bayes' theorem.

This means that any database, with distinguished condition and decision attributes (a decision table) or any set of implications satisfying condition (1) satisfies the Bayes' theorem. Thus databases or set of decision rules can be perceived as a new model for the Bayes' theorem. Let us note that in both cases we do not refer to prior or posterior probabilities and the Bayes' theorem simple reveals some patterns in data. This property can be used to reason about data, by inverting implications valid in the database.

8 Conclusions

The papers shows that any set of data satisfying some simple conditions satisfies the Bayes' theorem – without referring to prior and posterior probabilities inherently associated with Baysian statistical inference philosophy. The result bridges somehow rough set theory and some ideas provided by Lukasiewicz in context of his multivalued logic and probability.

References

1. Adams E. W. (1975) The Logic of Conditionals, an Application of Probability to Deductive Logic. D. Reidel Publishing Company, Dordrecht, Boston
2. Bandler W., Kohout L. (1980) Fuzzy Power Sets and Fuzzy Implication Operators. Fuzzy Sets and Systems 4:183–190
3. Borkowski L. (Ed.) (1970) Jan Łukasiewicz – Selected Works, North Holland Publishing Company, Amsterdam, London, Polish Scientific Publishers, Warszawa
4. Greco S., Matarazzo B., Słowiński R. (1998) Fuzzy Similarity Relation as a Basis for Rough Approximations. In: Polkowski L., Skowron A. (Eds.) Rough Sets and Current Trends in Computing, Lecture Notes in Artificial Intelligence, 1424 Springer, First International Conference, RSCTC'98, Warsaw, Poland, June, Proceedings, 283–297
5. Grzymała-Busse, J. (1986) On the Reduction of Knowledge Representation Systems. In: Proc. of the 6th International Workshop on Expert Systems and their Applications 1, Avignon, France, April 28–30, 1986 463–478
6. Hu X., Cercone N. (1995) Mining Knowledge Rules from Databases: A Rough Set Approach. In: Proceedings of the 12th International Conference on Data Engineering, New Orleans, 96–105
7. Hu X., Cercone N., Ziarko W. (1997) Generation of Multiple Knowledge from Databases Based on Rough Set Theory. In: Lin T.Y., Cercone N. (Eds.) Rough Sets and Data Mining. Analysis of Imprecise Data. Kluwer Academic Publishers, Boston, Dordrecht, 109–121
8. Hu X., Shan N., Cercone N., Ziarko W. (1994) DBROUGH: A Rough Set Based Knowledge Discovery System. In: Ras Z. W., Zemankova M. (Eds.) Proceedings of the Eighth International Symposium on Methodologies for Intelligent Systems

(ISMIS'94), Charlotte, NC, October 16–19, 1994, Lecture Notes in Artificial Intelligence 869, Springer-Verlag, 386–395
9. Kretowski M., Stepaniuk J.(1996) Selection of Objects and Attributes, a Tolerance Rough Set Approach. In: Preceedings of the Poster Session of Ninth International Symposium on Methodologies for Intelligent Systems, June 10–13, Zakopane, Poland, 169–180
10. Lin T. Y. (Ed.) (1994) Proceedings of the Third International Workshop on Rough Sets and Soft Computing (RSSC'94). San Jose State University, San Jose, California, USA, November 10–12, 1994
11. Lin T. Y. (Ed.) (1995) Proceedings of the Workshop on Rough Sets and Data Mining at 23rd Annual Computer Science Conference. Nashville, Tennessee, March 2, 1995
12. Lin T. Y. (Ed.) (1996) Journal of the Intelligent Automation and Soft Computing 2/2, (special issue)
13. Lin T. Y. (Ed.) (1996) International Journal of Approximate Reasoning 15/4, (special issue)
14. Lin T. Y., Cercone N. (Eds.) (1997) Rough Sets and Data Mining. Analysis of Imprecise Data. Kluwer Academic Publishers, Boston, Dordrecht
15. Lin T. Y., Wildberger A. M. (Eds.) (1995) Soft Computing: Rough Sets, Fuzzy Logic, Neural Networks, Uncertainty Management, Knowledge Discovery. Simulation Councils, Inc., San Diego, CA
16. Łukasiewicz J. (1913) Die logischen Grundlagen der Wahrscheinlichkeitsrechnung. Krakow
17. Handbook of Philosophical Logic, Vol.1, Elements of Classical Logic, Gabbay D. and Guenthner F., Kluwer Academic Publishers, Dordrecht, Boston, London, 1983.
18. Magrez P., Smets P. (1975) Fuzzy Modus Ponens: A new Model Suitable for Applications in Knowledge-Based Systems. Information Journal of Intelligent Systems 4:181-200
19. Orłowska E. (Ed.) (1997) Incomplete Information: Rough Set Analysis. Physica-Verlag, Heidelberg
20. Pal S. K., Skowron A. (Eds.) Fuzzy Sets, Rough Sets and Decision Making Processes. Springer–Verlag, Singapore (in preparation)
21. Pawlak, Z. (1991) Rough Sets – Theoretical Aspects of Reasoning about Data. Kluwer Academic Publishers, Boston, Dordrecht
22. Pawlak Z. (1998) Reasoning about Data – a Rough Set Perspective. In: Polkowski L., Skowron A. (Eds.) Rough Sets and Current Trends in Computing, Lecture Notes in Artificial Intelligence, 1424 Springer, First International Conference, RSCTC'98, Warsaw, Poland, June, Proceedings, 25–34
23. Pawlak Z. (1998) Granurality of Knowledge, Indiscernibility and Rough Sets. In: IEEE Conference on Evolutionary Computation, pages 100–103
24. Pawlak Z. (1998) Rough Modus Ponens. In: Proceedings of Seventh International Conference, International Processing and Management of Uncertainty in Knowledge-Based (IPMU), Paris, France, July 6–10, 1998, 1162–1166
25. Pawlak Z. (1998) Sets, Fuzzy Sets and Rough Sets. In: Proceedings of the 5. International Wokrshop Fuzzy-Neuro Systems'98 (FNS'98), Munich, Germany, March 19–20, 1998, 1–9
26. Pawlak Z. Skowron, A. (1994) Rough Membership Functions. In: Yaeger R. R., Fedrizzi M., and Kacprzyk J. (Eds.) Advances in the Dempster Shafer Theory of Evidence, John Wiley & Sons, Inc., New York, 251–271

27. Polkowski L., Skowron A. (1994) Rough Mereology. In: Proceedings of the Symposium on Methodologies for Intelligent Systems, Charlotte, NC, October 16–19, Lecture Notes in Artificial Intelligence 869, Springer-Verlag, Berlin 1994, 85–94; see also: Institute of Computer Science, Warsaw University of Technology, ICS Research Report 44/94
28. Polkowski L., Skowron A. (Eds.) (1998) Rough Sets and Current Trends in Computing, Lecture Notes in Artificial Intelligence 1424, Springer-Verlag, First International Conference, RSCTC'98, Warsaw, Poland, June, Proceedings,
29. Polkowski L., Skowron A. (Eds.) (1998) Rough Sets in Knowledge Discovery. Physica-Verlag, Vol. 1, 2
30. Skowron A. (1994) Menagement of Uncertainty in AI: A Rough Set Approach. In: Proceedings of the conference SOFTEKS, Springer Verlag and British Computer Society, 69–86
31. Słowiński R. (Ed.) (1992) Intelligent Decision Support – Handbook of Applications and Advances of the Rough Sets Theory. Kluwer Academic Publishers, Boston, Dordrecht
32. Skowron A., Stepaniuk J. (1994) Generalized Approximations Spaces. In: Proceedings of the Third International Workshop on Rough Sets and Soft Computing, San Jose, November 10–12, 1994, 156–163
33. Słowiński R. (1992) A Generalization of the Indiscernibility Relation for Rough Sets Analysis. In: Słowiński R., Stefanowski J. (Eds.) Proceedings of the First International Workshop on Rough Sets: State of the Art and Perspectives. Kiekrz – Poznań, Poland September 2–4, 1992, 68–70
34. Słowiński R. (1992) A Generalization of the Indiscernibility Relation for Rough Set Analysis of Quantitative Information. Rivista di Matematica per le Scienze Economiche e Sociali 15/1:65–78
35. Słowiński R. (1992) Rough Sets with Strict and Weak Indiscernibility Relations. In: Proceedings of IEEE International Conference on Fuzzy Systems, IEEE 92CH3073-4, San Diego, California, 695–702
36. Słowiński R., Stefanowski J. (Eds.) (1993) Foundations of Computing and Decision Sciences 18/3–4:155–396 (special issue)
37. Słowiński R., Stefanowski J. (1994) Rough Classification with Valued Closeness Relation. In: Diday E., Lechevallier Y., Schrader M., Bertrand P., and Burtschy B. (Eds.) New Approaches in Classification and Data Analysis, Springer-Verlag, Berlin, 482–488
38. Słowiński R., Vanderpooten D. (1997) Similarity Relation as a Basis for Rough Approximations. Institute of Computer Science, Warsaw University of Technology, ICS Research Report **53/95** (1995); see also: Wang P. P. (Ed.) Advances in Machine Intelligence & Soft–Computing, Bookwrights, Raleigh, NC, 17–33
39. Słowiński R., Vanderpooten D. A Generalized Definition of Rough Approximations Based on Similarity. IEEE Transactions on Data and Knowledge Engineering (to appear)
40. Stepaniuk J. (1998) Approximations Spaces in Extensions of Rough Set Theory. In: Polkowski L., Skowron A. (Eds.) Rough Sets and Current Trends in Computing, Lecture Notes in Artificial Intelligence, 1424 Springer, First International Conference, RSCTC'98, Warsaw, Poland, June, Proceedings, 290–297
41. Stepaniuk J., Kretowski M. (1995) Decision system based on tolerance rough sets. In: Proceedings of the Fourth International Workshop on Intelligent Information Systems, Augustów, Poland, June 5–9, 1995, 62–73

42. Tsumoto S. (Ed.) (1996) Bulletin of International Rough Set Society 1/1
43. Tsumoto S. (Ed.) (1997) Bulletin of International Rough Set Society 1/2
44. Tsumoto S. (Ed.) (1998) Bulletin of International Rough Set Society 2/1
45. Tsumoto S. Kobayashi S., Yokomori T., Tanaka H., and Nakamura A. (Eds.) (1996) Proceedings of the Fourth International Workshop on Rough Sets, Fuzzy Sets, and Machine Discovery (RSFD'96). The University of Tokyo, November 6–8, 1996
46. Tsumoto S. (1998) Modelling Medical Diagnostic Rules Based on Rough Sets. In: Polkowski L., Skowron A. (Eds.), Rough Sets and Current Trends in Computing, Lecture Notes in Artificial Intelligence 1424, Springer-Verlag, First International Conference, RSCTC'98, Warsaw, Poland, June, Proceedings, 475–482
47. Wang P. P. (Ed.) (1995) Proceedings of the International Workshop on Rough Sets and Soft Computing at Second Annual Joint Conference on Information Sciences (JCIS'95), Wrightsville Beach, North Carolina, 28 September - 1 October 1995
48. Wang P. P. (Ed.) (1997) Proceedings of the Fifth International Workshop on Rough Sets and Soft Computing (RSSC'97) at Third Annual Joint Conference on Information Sciences (JCIS'97). Duke University, Durham, NC, USA, Rough Set & Computer Science 3, March 1–5, 1997
49. Zadeh L. (1977) Fuzzy Sets as a Basis for a Theory of Possibility. Fuzzy Sets and Systems 1:3–28
50. Zadeh L. (1983) The Role of Fuzzy Logic in in the Management of Uncertainty in Expert Systems. Fuzzy Sets and Systems 11:199–277
51. Ziarko W. (1987) On Reduction of Knowledge Representation. In: Proc. 2nd International Symp. on Methodologies of Intelligent Systems, Charlotte, NC, North Holland, 99–113
52. Ziarko W. (1988) Acquisition of Design Knowledge from Examples. Math. Comput. Modeling 10:551–554
53. Ziarko W. (1994) Rough Sets and Knowledge Discovery: An Overview. In: Ziarko W. (Ed.) Rough Sets, Fuzzy Sets and Knowledge Discovery (RSKD'93). Workshops in Computing, Springer-Verlag & British Computer Society, London, Berlin, 11–15
54. Ziarko W. (Ed.) (1993) Proceedings of the Second International Workshop on Rough Sets and Knowledge Discovery (RSKD'93). Banff, Alberta, Canada, October 12–15, 1993
55. Ziarko W. (1993) Variable Precision Rough Set Model. Journal of Computer and System Sciences 46/1:39–59
56. Ziarko W. (Ed.) (1994) Rough Sets, Fuzzy Sets and Knowledge Discovery (RSKD'93). Workshops in Computing, Springer-Verlag & British Computer Society, London, Berlin
57. Ziarko W. (1995) Introduction to the Special Issue on Rough Sets and Knowledge Discovery. In: Ziarko W. (Ed.) Computational Intelligence: An International Journal 11/2:223–226 (special issue)

List of Contributors

Jean Berstel
Institut Gaspard Monge
Université Marne-la-Vallée
F-77454 Marne-la-Vallée, France
jean.berstel@univ-mlv.fr

Janusz A. Brzozowski
Department of Computer Science
University of Waterloo
Waterloo, Ontario N2L 3G1, Canada
brzozo@uwaterloo.ca

Cristian S. Calude
Department of Computer Science
University of Auckland
Private Bag 92019
Auckland, New Zealand
cristian@cs.auckland.ac.nz

Arturo Carpi
Istituto di Cibernetica del CNR
via Toiano, 6
I-80072 Arco Felice (NA), Italy
arturo@cib.na.cnr.it

Christian Choffrut
Université Paris 7
LIAFA
2 Pl. Jussieu
F-75251 Paris Cedex 05, France
cc@liafa.jussieu.fr

Richard J. Coles
Department of Computer Science
University of Auckland
Private Bag 92019
Auckland, New Zealand
coles@cs.auckland.ac.nz

Karel Culik II
Department of Computer Science
University of South Carolina
Columbia, S.C. 29208, USA
culik@cs.sc.edu

Jürgen Dassow
Otto-von-Guericke-Universität
Magdeburg
Fakultät für Informatik
Postfach 4120
D-39016 Magdeburg, Germany
dassow@cs.uni-magdeburg.de

Volker Diekert
Institut für Informatik
Universität Stuttgart
Breitwiesenstr. 20-22
D-70565 Stuttgart, Germany
diekert@informatik.
uni-stuttgart.de

Joost Engelfriet
Institute of Computer Science
Leiden University
P.O.Box 9512
2300 RA Leiden, The Netherlands
engelfri@wi.LeidenUniv.nl

Dora Giammarresi
Dipartimento di Matematica
Applicata e Informatica
Università Ca' Foscari di Venezia
via Torino, 155
I-30173 Venezia Mestre, Italy
dora@dsi.unive.it

Serge Grigorieff
Université de Paris 7
2 Pl. Jussieu
F-72251 Paris Cedex 05, France
also
L.L.A.I.C
Université de Clermont-Ferrand
France
seg@ufr-info-p7.jussieu.fr

Vesa Halava
Department of Mathematics and
Turku Centre for Computer Science
Lemminkäisenkatu 14 A
FIN-20520 Turku, Finland
vehalava@cs.utu.fi

Tero Harju
Department of Mathematics and
Turku Centre for Computer Science
University of Turku
FIN-20014 Turku, Finland
harju@utu.fi

Juha Honkala
Department of Mathematics and
Turku Centre for Computer Science
University of Turku
FIN-20014 Turku, Finland
juha.honkala@utu.fi

Hendrik Jan Hoogeboom
Institute of Computer Science
Leiden University
P.O.Box 9512
2300 RA Leiden, The Netherlands
hoogeboom@wi.LeidenUniv.nl

Juraj Hromkovič
RWTH Aachen
Lehrstuhl Informatik I
Ahornstraße 55
D-52056 Aachen, Germany
jh@i1.informatik.
rwth-aachen.de

Oscar H. Ibarra
Department of Computer Science
University of California
Santa Barbara, CA 93106, USA
ibarra@cs.ucsb.edu

Balázs Imreh
Department of Informatics
József Attila University
Árpád tér 2
H-6720 Szeged, Hungary
imreh@inf.u-szeged.hu

Masami Ito
Faculty of Science
Kyoto Sangyo University
Kyoto 603, Japan
ito@ksuvx0.kyoto-su.ac.jp

Tao Jiang
Department of
Computing and Software
McMaster University
Hamilton, Ontario L8S 4K1, Canada
jiang@cas.mcmaster.ca

Helmut Jürgensen
Department of Computer Science
University of Western Ontario
London, Ontario N6A 5B7, Canada
helmut@uwo.ca
and
Institut für Informatik
Universität Potsdam
Am Neuen Palais 10
D-14469 Potsdam, Germany
helmut@cs.uni-potsdam.de

Juhani Karhumäki
Department of Mathematics and
Turku Centre for Computer Science
University of Turku
FIN-20014 Turku, Finland
karhumak@cs.utu.fi

Lila Kari
Department of Computer Science
University of Western Ontario
London, Ontario N6A 5B7, Canada
lkari@csd.uwo.ca

Jarkko Kari
Department of Computer Science
15 MLH, University of Iowa
Iowa City, IA 52242, USA
jjkari@cs.uiowa.edu

Werner Kuich
Abteilung für
Theoretische Informatik
Institut für
Algebra und Diskrete Mathematik
Technische Universität Wien
Wiedner Hauptstraße 8–10
A–1040 Wien, Austria
kuich@tuwien.ac.at

Laura F. Landweber
Department of Ecology and
Evolutionary Biology
Princeton University
Princeton, NJ 08544 – 1003, USA
LFL@princeton.edu

Michel Latteux
CNRS URA 369
L.I.F.L.
Université des Sciences et
Technologies de Lille
U.F.R. I.E.E.A. Informatique
F-59655 Villeneuve d'Ascq Cedex
France
latteux@lifl.fr

Ming Li
Department of Computer Science
University of Waterloo
Waterloo, Ontario N2L 3G1, Canada
mli@math.uwaterloo.ca

Leonid P. Lisovik
Department of Cybernetics
Kiev State University
Kiev, 252017, Ukraine
Lis@cyber.kiev.ua

Aldo de Luca
Dipartimento di Matematica
Università di Roma "La Sapienza"
Piazzale Aldo Moro, 2
I-00185 Roma, Italy
adl@sole.cib.na.cnr.it

Vincenzo Manca
Università degli studi di Pisa
Dipartimento di Informatica
Corso Italia 40, I-56125 Pisa, Italy
mancav@di.unipi.it

Solomon Marcus
Romanian Academy
Calea Victoriei 125
RO-71102 Bucureşti, Romania
smarcus@stoilow.imar.ro
solomon@imar.ro

Carlos Martín-Vide
Research group in Mathematical Linguistics and Language Engineering
Rovira i Virgili University
Pl. Imperial Tàrraco 1
E-43005 Tarragona, Spain
cmv@astor.urv.es

Alexandru Mateescu
Department of Mathematics and
Turku Centre for Computer Science
FIN-20014 Turku, Finland
and

Department of Mathematics
University of Bucharest
Romania
mateescu@utu.fi

George Daniel Mateescu
Faculty of Mathematics
University of Bucharest
Academiei 14 sector 1
RO-70109 Bucharest, Romania
gmateesc@pcnet.pcnet.ro

Robert McNaughton
Department of Computer Science
Rensselaer Polytechnic Institute
Troy, NY 12180-3590, USA
mcnaught@cs.rpi.edu

Giovanna Melideo
Dipartimento di
Informatica e Sistemistica
Universitá di Roma "La Sapienza"
Italy
melideo@dis.uniroma1.it

Filippo Mignosi
Università degli studi di Palermo
Dipartimento di
Matematica ed Applicazioni
Via Archirafi 34
I-90123 Palermo, Italy
mignosi@altair.math.unipa.it

Taishin Yasunobu Nishida
Faculty of Engineering
Toyama Prefectural University
Kosugi-machi
939-0398 Toyama, Japan
nishida@pu-toyama.ac.jp

Cesidia Pasquarelli
Dipartimento di Matematica
Universitá di L'Aquila
Italy
pasquare@univaq.it

Gheorghe Páun
Institute of Mathematics of
the Romanian Academy
PO Box 1-764
RO-70700 Bucureşti, Romania
gpaun@imar.ro

Zdzisław Pawlak
Institute for
Theoretical and Applied Informatics
Polish Academy of Sciences
ul. Bałtycka 5
44-000 Gliwice, Poland
zpw@ii.pw.edu.pl

Azaria Paz
Technion IIT
Faculty of Computer Science
Haifa, Israel
paz@cs.nyu.edu

Wojciech Plandowski
Instytut Informatyki
Uniwersytet Warszawski
Banacha 2
02–097 Warszawa, Poland
wojtekpl@mimuw.edu.pl

Jean-Luc Ponty
L.I.F.A.R.
Université de Rouen
Faculté des Sciences et
des Techniques
Place Émile Blondel
F-76821 Mont-Saint-Aignan Cedex
France
ponty@dir.univ-rouen.fr

Antonio Restivo
Università degli studi di Palermo
Dipartimento di
Matematica ed Applicazioni
Via Archirafi 34
I-90123 Palermo, Italy
restivo@altair.math.unipa.it

John Michael Robson
LaBRI
Université Bordeaux I
351, cours de la Libération
F-33405 Talence Cédex, France
robson@labri.u-bordeaux.fr

Yves Roos
CNRS URA 369
L.I.F.L.
Université des Sciences et
Technologies de Lille
U.F.R. I.E.E.A. Informatique
F-59655 Villeneuve d'Ascq Cedex
France
yroos@lifl.fr

Wojciech Rytter
Instytut Informatyki
Uniwersytet Warszawski
Banacha 2
02–097 Warszawa, Poland
and
Department of Computer Science
University of Liverpool
England
rytter@mimuw.edu.pl

Kai Salomaa
Department of Computer Science
University of Western Ontario
London, Ontario N6A 5B7, Canada
ksalomaa@csd.uwo.ca

Marinella Sciortino
Università degli studi di Palermo
Dipartimento di
Matematica ed Applicazioni
Via Archirafi 34
I-90123 Palermo, Italy
mari@altair.math.unipa.it

Octavian Soldea
Technion IIT
Faculty of Computer Science
Haifa, Israel
octavian@cs.Technion.AC.IL

Jianwen Su
Department of Computer Science
University of California
Santa Barbara, CA 93106, USA
su@cs.ucsb.edu

Wolfgang Thomas
RWTH Aachen
Lehrstuhl Informatik VII
D-52056 Aachen, Germany
thomas@informatik.
rwth-aachen.de

Stefano Varricchio
Dipartimento di Matematica
Universitá di Roma "Torvergata"
Viale della Riverca Scientifica
I-00133 Roma, Italy
varricch@mat.uniroma2.it

Paul Vitányi
CWI, Kruislaan 413
1098 Amsterdam, The Netherlands
Paul.Vitanyi@cwi.nl

Derick Wood
Department of Computer Science
Hong Kong University of Science &
Technology
Clear Water Bay, Kowloon
Hong Kong SAR
dwood@cs.ust.hk

Sheng Yu
Department of Computer Science
University of Western Ontario
London, Ontario N6A 5B7, Canada
syu@csd.uwo.ca

Springer and the environment

At Springer we firmly believe that an international science publisher has a special obligation to the environment, and our corporate policies consistently reflect this conviction.
We also expect our business partners – paper mills, printers, packaging manufacturers, etc. – to commit themselves to using materials and production processes that do not harm the environment. The paper in this book is made from low- or no-chlorine pulp and is acid free, in conformance with international standards for paper permanency.

Springer

Printing: Saladruck, Berlin
Binding: Buchbinderei Lüderitz & Bauer, Berlin